Novel Research about Biomechanics and Biomaterials Used in Hip, Knee and Related Joints

Novel Research about Biomechanics and Biomaterials Used in Hip, Knee and Related Joints

Editors

J. Philippe Kretzer
Catherine Van Der Straeten

MDPI • Basel • Beijing • Wuhan • Barcelona • Belgrade • Manchester • Tokyo • Cluj • Tianjin

Editors
J. Philippe Kretzer
Laboratory of Biomechanics and Implant Research, Clinic for Orthopedics and Trauma Surgery
Heidelberg University Hospital
Heidelberg
Germany

Catherine Van Der Straeten
Health, innovation and research institute
Universitair Ziekenhuis Gent
Gent
Belgium

Editorial Office
MDPI
St. Alban-Anlage 66
4052 Basel, Switzerland

This is a reprint of articles from the Special Issue published online in the open access journal *Materials* (ISSN 1996-1944) (available at: www.mdpi.com/journal/materials/special_issues/biomech_biomater_hip).

For citation purposes, cite each article independently as indicated on the article page online and as indicated below:

LastName, A.A.; LastName, B.B.; LastName, C.C. Article Title. *Journal Name* **Year**, *Volume Number*, Page Range.

ISBN 978-3-0365-1281-5 (Hbk)
ISBN 978-3-0365-1280-8 (PDF)

Contents

materials

MDPI

Editorial

Special Issue "Novel Research about Biomechanics and Biomaterials Used in Hip, Knee and Related Joints"

Jan Philippe Kretzer [1,*] and Catherine Van Der Straeten [2]

1 Laboratory of Biomechanics and Implant Research, Clinic for Orthopedics and Trauma Surgery, Heidelberg University Hospital, 69118 Heidelberg, Germany
2 Health Innovation and Research Institute, Ghent University Hospital, C. Heymanslaan 10, 9000 Gent, Belgium; catherine.vanderstraeten@ugent.be
* Correspondence: philippe.kretzer@med.uni-heidelberg.de; Tel.: +49-(0)-6221/5629209

check for updates

Citation: Kretzer, J.P.; Van Der Straeten, C. Special Issue "Novel Research about Biomechanics and Biomaterials Used in Hip, Knee and Related Joints". *Materials* **2021**, *14*, 2222. https://doi.org/10.3390/ma14092222

Received: 22 April 2021
Accepted: 25 April 2021
Published: 26 April 2021

Joint replacement is a very successful medical treatment. However, the survivorship of hip, knee, shoulder, and other implants is variable depending on the material and the loading conditions. The degradation of materials, but also the immune response against degradation products or an altered tissue loading condition as well as infections remain key factors of their failure.

Current research in biomechanics and biomaterials is trying to overcome the existing limitations. This includes new implant designs and materials, bearings concepts and tribology, kinematical concepts, surgical techniques, and anti-inflammatory and infection prevention strategies. A careful evaluation of new materials and concepts is required in order to fully assess the strengths and weaknesses and improve the quality and outcomes of joint replacements. Therefore, extensive research and clinical trials are essential.

The main aspects that are addressed in this Special Issue are related to new materials, the design and manufacturing considerations of implants [1–10], implant wear and its potential clinical consequences [11–16], implant fixation [17–19], infection-related material aspects [20,21] and taper-related research topics [22–24].

This Special Issue gives an overview of the ongoing research in these fields. Contributions are solicited from researchers working in the fields of biomechanics, biomaterials, and bio- and tissue-engineering.

We hope you enjoy reading this book.

Funding: This research received no external funding.

Conflicts of Interest: The authors declare no conflict of interest.

References

1. de Ruiter, L.; Cowie, R.M.; Jennings, L.M.; Briscoe, A.; Janssen, D.; Verdonschot, N. The effects of cyclic loading and motion on the implant-cement interface and cement mantle of peek and cobalt-chromium femoral total knee arthroplasty implants: A preliminary study. *Materials* **2020**, *13*, 3323. [CrossRef] [PubMed]
2. Bormann, T.; Mai, P.T.; Gibmeier, J.; Sonntag, R.; Muller, U.; Kretzer, J.P. Corrosion behavior of surface-treated metallic implant materials. *Materials* **2020**, *13*, 2011. [CrossRef]
3. Kebbach, M.; Darowski, M.; Krueger, S.; Schilling, C.; Grupp, T.M.; Bader, R.; Geier, A. Musculoskeletal multibody simulation analysis on the impact of patellar component design and positioning on joint dynamics after unconstrained total knee arthroplasty. *Materials* **2020**, *13*, 2365. [CrossRef] [PubMed]
4. Correa Filho, L.; Schmidt, S.; Goyenola, C.; Skjoldebrand, C.; Engqvist, H.; Hogberg, H.; Tobler, M.; Persson, C. The effect of n, c, cr, and nb content on silicon nitride coatings for joint applications. *Materials* **2020**, *13*, 1896. [CrossRef] [PubMed]
5. Wölfle-Roos, J.; Amet, B.K.; Fiedler, J.; Michels, H.; Kappelt, G.; Ignatius, A.; Dürselen, L.; Reichel, H.; Brenner, R. Optimizing manufacturing and osseointegration of ti6al4v implants through precision casting and calcium and phosphorus ion implantation? In vivo results of a large-scale animal trial. *Materials* **2020**, *13*, 1670.
6. Vogel, D.; Klimek, M.; Saemann, M.; Bader, R. Influence of the acetabular cup material on the shell deformation and strain distribution in the adjacent bone-a finite element analysis. *Materials* **2020**, *13*, 1372. [CrossRef]

7. Hettich, G.; Schierjott, R.A.; Epple, M.; Gbureck, U.; Heinemann, S.; Mozaffari-Jovein, H.; Grupp, T.M. Calcium phosphate bone graft substitutes with high mechanical load capacity and high degree of interconnecting porosity. *Materials* **2019**, *12*, 3471. [CrossRef]
8. Adler, D.; Akbar, M.; Spicher, A.; Goerke, S.A.; Schmoelz, W. Biomechanical study of a novel, expandable, non-metallic and radiolucent cf/peek vertebral body replacement (vbr). *Materials* **2019**, *12*, 2732. [CrossRef]
9. Almansour, H.; Sonntag, R.; Pepke, W.; Bruckner, T.; Kretzer, J.P.; Akbar, M. Impact of electrocautery on fatigue life of spinal fusion constructs-an in vitro biomechanical study. *Materials* **2019**, *12*, 2471. [CrossRef]
10. Inyang, A.O.; Vaughan, C.L. Functional characteristics and mechanical performance of pcu composites for knee meniscus replacement. *Materials* **2020**, *13*, 1886. [CrossRef]
11. Eckert, J.A.; Mueller, U.; Walker, T.; Schwarze, M.; Jaeger, S.; Kretzer, J.P. Bearing thickness is not a predictive factor for damage and penetration in oxford unicompartmental knee arthroplasty-a retrieval analysis. *Materials* **2020**, *13*, 4589. [CrossRef]
12. Braun, S.; Jaeger, S.; Sonntag, R.; Schroeder, S.; Kretzer, J.P. Quantitative measurements of backside wear in acetabular hip joint replacement: Conventional polyethylene versus cross-linked polyethylene. *Materials* **2020**, *13*, 1854. [CrossRef]
13. Bormann, T.; Jager, S.; Kretzer, J.P.; Nebel, L.; Clarius, L.; Omlor, G.; Bitsch, R.; Lehner, B. Retrieval analysis of modern knee tumor megaendoprosthesis shows considerable volumetric metal wear generated at the rotating hinge. *Materials* **2020**, *13*, 1519. [CrossRef]
14. Cowie, R.M.; Pallem, N.M.; Briscoe, A.; Jennings, L.M. Third body wear of uhmwpe-on-peek-optima. *Materials* **2020**, *13*, 1264. [CrossRef]
15. Cheng, X.; Dirmeier, S.C.; Hasselt, S.; Baur-Melnyk, A.; Kretzer, J.P.; Bader, R.; Utzschneider, S.; Paulus, A.C. Biological reactions to metal particles and ions in the synovial layer of mice. *Materials* **2020**, *13*, 1044. [CrossRef]
16. Reiner, T.; Klotz, M.C.; Seelmann, K.; Hertzsch, F.; Innmann, M.M.; Streit, M.R.; Nees, T.A.; Moradi, B.; Merle, C.; Kretzer, J.P.; et al. Long-term results of a second-generation, small-diameter, metal-on-metal bearing in primary total hip arthroplasty at 14-year follow-up. *Materials* **2020**, *13*, 557. [CrossRef] [PubMed]
17. Beckmann, N.A.; Bitsch, R.G.; Schonhoff, M.; Siebenrock, K.A.; Schwarze, M.; Jaeger, S. Comparison of the primary stability of porous tantalum and titanium acetabular revision constructs. *Materials* **2020**, *13*, 1783. [CrossRef] [PubMed]
18. Jaeger, S.; Uhler, M.; Schroeder, S.; Beckmann, N.A.; Braun, S. Comparison of different locking mechanisms in total hip arthroplasty: Relative motion between cup and inlay. *Materials* **2020**, *13*, 1392. [CrossRef]
19. Beckmann, N.A.; Bitsch, R.G.; Bormann, T.; Braun, S.; Jaeger, S. Titanium acetabular component deformation under cyclic loading. *Materials* **2019**, *13*, 52. [CrossRef] [PubMed]
20. Lunz, A.; Sonntag, R.; Kretzer, J.P.; Jaeger, S.; Bormann, T.; Streit, M.R.; Beckmann, N.A.; Lehner, B.; Omlor, G.W. Hip spacers with a metal-on-cement articulation did not show significant surface alterations of the metal femoral head in two-stage revision for periprosthetic joint infection. *Materials* **2020**, *13*, 3882. [CrossRef]
21. Fabritius, M.; Al-Munajjed, A.A.; Freytag, C.; Julke, H.; Zehe, M.; Lemarchand, T.; Arts, J.J.; Schumann, D.; Alt, V.; Sternberg, K. Antimicrobial silver multilayer coating for prevention of bacterial colonization of orthopedic implants. *Materials* **2020**, *13*, 1415. [CrossRef]
22. Xu, J.; Sonntag, R.; Kretzer, J.P.; Taylor, D.; Forst, R.; Seehaus, F. Model-based roentgen stereophotogrammetric analysis to monitor the head-taper junction in total hip arthroplasty in vivo-and they do move. *Materials* **2020**, *13*, 1543. [CrossRef] [PubMed]
23. Grupp, T.M.; Baxmann, M.; Jansson, V.; Windhagen, H.; Heller, K.D.; Morlock, M.M.; Knaebel, H.P. How to proceed with asymptomatic modular dual taper hip stems in the case of acetabular revision. *Materials* **2020**, *13*, 1098. [CrossRef] [PubMed]
24. Rieker, C.B.; Wahl, P. What the surgeon can do to reduce the risk of trunnionosis in hip arthroplasty: Recommendations from the literature. *Materials* **2020**, *13*, 1950. [CrossRef] [PubMed]

materials

MDPI

Article

Bearing Thickness Is Not a Predictive Factor for Damage and Penetration in Oxford Unicompartmental Knee Arthroplasty—A Retrieval Analysis

Johannes Adrian Eckert †, Ulrike Mueller †, Tilman Walker, Martin Schwarze, Sebastian Jaeger and Jan Philippe Kretzer *

Laboratory of Biomechanics and Implant Research, Clinic for Orthopedics and Trauma Surgery, Heidelberg University Hospital, Schlierbacher Landstraße 200a, 69118 Heidelberg, Germany; Johannes.Eckert@med.uni-heidelberg.de (J.A.E.); ulmuel@hotmail.de (U.M.); Tilman.Walker@med.uni-heidelberg.de (T.W.); Martin.Schwarze@med.uni-heidelberg.de (M.S.); Sebastian.Jaeger@med.uni-heidelberg.de (S.J.)

* Correspondence: philippe.kretzer@med.uni-heidelberg.de

† These authors contributed equally.

Received: 19 August 2020; Accepted: 13 October 2020; Published: 15 October 2020

check for updates

Abstract: The medial Oxford unicompartmental knee arthroplasty (OUKA) shows good survivorship, as well as clinical results. Aseptic loosening, however, remains one of the main reasons for revision and polyethylene debris is known to cause aseptic loosening. The role of bearing thickness in total as well as unicondylar knee arthroplasty has been the subject of controversial discussions, especially the longevity of lower thickness bearings in total knee arthroplasty was questioned. The purpose of this study was to assess the influence of bearing thickness on time to revision, damage pattern, penetration, and volumetric material loss. A cohort of 47 consecutively retrieved medial OUKA bearings was analyzed with conventional direct light microscopy applying the Hood damage analysis, as well as measuring the penetration depth. In this retrieval cohort, a difference on survival time, damage, penetration, as well as volumetric material loss could not be seen. We conclude that low as well as high thickness bearings can safely be used in OUKA without any relevant differences in terms of wear and damage.

Keywords: oxford unicompartmental knee arthroplasty; bearing thickness; retrieval analysis

---

1. Introduction

The Oxford unicompartmental knee arthroplasty (OUKA) has been shown to generate good clinical results. Due to the increased use of highly crosslinked polyethylene (HXLPE), the number of revisions due to wear has decreased [1–3] (the Annual Report of the Australian National Joint Registry notes only 1.4% of direct tibial insert wear [4]). However, aseptic loosening—which can be caused by wear—is still the major cause for revisions of UKA [4,5], the 15-year revision rate is around 11% [6]. The use of a spherical femoral component and a fully congruous meniscal bearing to increase the contact area theoretically reduces the potential for polyethylene wear [7]. Mobile bearing unicondylar designs, however, can exhibit wear at the femoral, as well as the tibial aspect of the inlay [8,9]. Especially in cases of suboptimal bearing positioning, the wear can increase drastically [10]. To date, there is no scientific consent as to which inlay thicknesses generate better outcomes. Longer times to revision in Oxford UKA have been described for 3 and 4 mm bearings, respectively, compared to thicker bearings [10]. In addition, the use of thinner inlays allows for a reduced loss of bonestock and usage of thicker bearings has a higher risk for overcorrection of the mechanical axis, resulting in a higher post-operative valgus [11]. In incongruent articulations, however, the wear rate of polyethylene is increased when it is used in a thin layer [12,13], and bearing failures have been described for inlays <6 mm, caused by wear [14,15]. Accordingly, the FDA has issued a Class II Special Controls Guidance Document in 2003, recommending a minimum thickness of 6 mm at the lowest point below the condyle (or alternatively the demonstration of the ability of the said bearing to survive 10 million cycles in a physiological knee simulator). Unfortunately, the information provided by manufacturers with regards to the actual inlay thickness often appears to be inaccurate or misleading [16].

In this retrieval analysis of a series of 47 consecutively explanted medial Oxford unicondylar knee replacements, we assessed the types of damage using the modified Hood score and analyzed penetration depth and volumetric changes during the lifetime of the retrieved bearings [17]. We wanted to assess whether the use of higher-thickness inlays (5–6 mm in this cohort) leads to higher wear than the use of low-thickness inlays (3–4 mm) and if different types of damage are seen in the two groups.

2. Materials and Methods

From October 2013 to August 2018, 47 ultra-high molecular weight polyethylene (UHMWPE) meniscal bearings of the medial Oxford unicompartmental knee arthroplasty (OUKA) from 46 patients were consecutively explanted and included in this retrieval analysis. All retrievals were conducted at our hospital. Of the 47 inlays, the majority (n = 41) was of the newer 'anatomic' generation, only six were old type inlays. Eight of the inlays were small (all anatomic), 19 were medium-sized (16 anatomic, three non-anatomic), and 20 were large (17 anatomic, three non-anatomic). No XS inlays were explanted in the period examined. Of the revised knees, 21 were left knees, 26 were right knees. In 38 cases, the primary procedure was performed in our hospital, the remaining eight procedures were performed elsewhere. Primary implantations were conducted from October 2002 to September 2017. The manufacturer changed the type of UHMWPE in 2005. In this cohort, five of the retrieved inlays were implanted before, the rest after the change in material. The UHMWPE used before 2005 was Hi-Fax 1900, the PE currently in use is GUR 1050. A publication by Mohammad et al. showed no significant differences in wear between the two types of UHMWPE using a RSA analysis [18]. Microscopical images of the types of damage were made using a Keyence VHX-5000 reflected-light microscope.

The reasons for revision in descending order were progression of disease in other compartments (n = 19; 30%), followed by aseptic loosening (n = 9; 14%) and bearing dislocation (n = 5; 8%). Other reasons for revision were periprosthetic fractures, persistent postoperative pain, instability, infection, etc.

Following an institutional review board (IRB)-approved protocol, all orthopedic devices were collected, cleaned, and cataloged in an electronic database for long-term storage in an inert environment.

Surface damage of the retrieved meniscal bearings was assessed using light stereomicroscope analysis at magnifications from ×10 to ×20. The tibial bearings were divided into three regions (front-anterior, front-posterior, and back; see Figure 1). For these regions, the wear was quantified using the modified Hood-method [17], a method which has since been used for most macroscopic retrieval analyses [19,20]. Eight types of damage were assessed: Burnishing, scratching, pitting, delamination, surface deformation, abrasion, third-body embedded debris, and edge loading (Figure 2). Grading was performed by two graders (JAE, JPK) blinded to the clinical and radiographic data. Scores of 1, 2, and 3 indicated damage areas of <10%, 10% to 50%, and >50% of the surface area, respectively. The total score for each bearing was the sum of the eight individual damage scores over the three zones. The maximum score for each surface was 72.

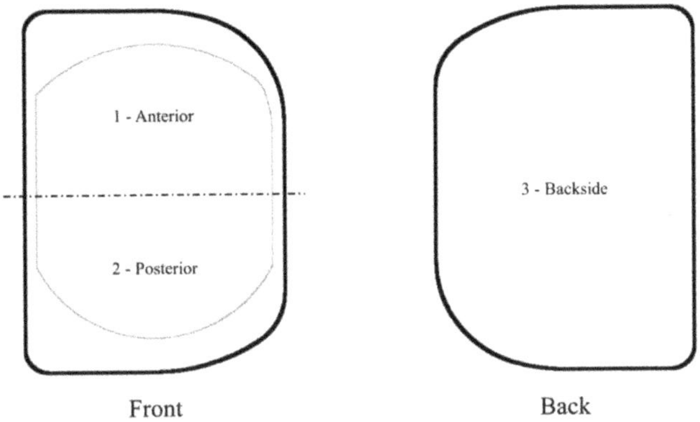

Figure 1. The front and back of the meniscal bearings were divided into three sections. Surface damage grading scores were assigned for each section using light stereoscopic examination and a subjective grading score.

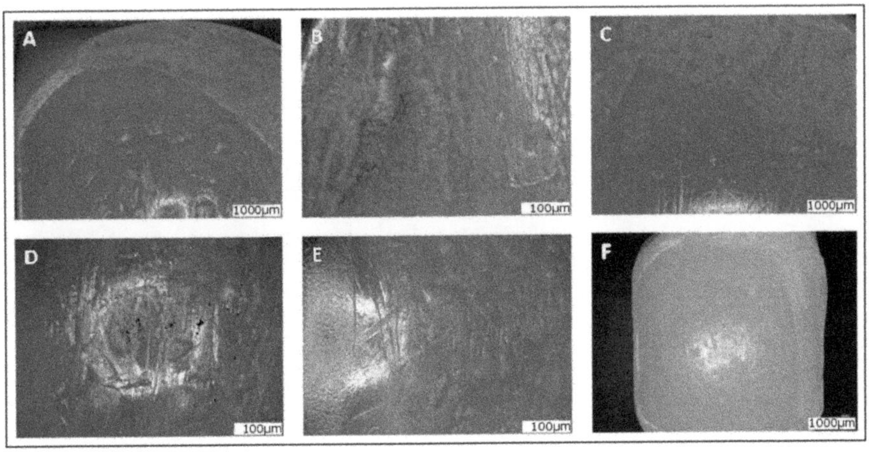

Figure 2. (**A**) Edge loading and abrasion in the anterior medial border of the inlay, pitting throughout the inlay, small scratches; (**B**) delamination at the anterior medial border of the inlay (noticeable at the brown change in color little pitting and scratching; (**C**) abrasion, scratching, and little pitting; (**D**) embedded particles (small black dots, possibly metal debris), a lot of scratching and pitting; (**E**) burnishing (see changes of regular manufacturing marks on the left side to a polished surface); (**F**) surface deformation, edge loading, abrasion, and a little pitting and scratching.

The thicknesses of the tibial bearings as specified by the manufacturer for the current implant design of the OUKA range from 3 to 9 mm. The thicknesses of the retrieved bearings in this study ranged from 3 to 6 mm, 3 and 4 mm inserts were classified as low thickness inserts, 5 and 6 mm were classified as high thickness inserts. The manufacturer does not further specify the thicknesses, however, the nominal thickness of each bearing is always bearing thickness +0.5 mm. The mean variance of +0.130 mm was added, as the nominal thickness does not resemble the true thickness of a new inlay. We have found a mean variance of +0.130 mm in 10 brand new inlays, measured for this purpose.

To establish the linear penetration, the minimum thickness of each retrieved bearing was subtracted from the initial bearing thickness. The thickness was measured using a digital dial gauge (MarCator 1075 R, Mahr, Göttingen, Germany). A similar dial-gauge technique was reported by Psychoyios et al. to determine the minimum thickness of similar bearings [21]. Each bearing was mounted on a precision comparator stand with a universal measuring arm and a probe with a spherical tip of 4 mm in diameter was used. The dial gauge was determined to have a mean error of 0.0005 mm. Each bearing was measured once by two independent observers (JAE and JPK) and the mean of these values was taken as the minimum thickness of the bearing surface. Linear penetration was then calculated using the following formula:

Linear penetration (LP) (mm) = TN + TC − TR

TN = Nominal thickness = Bearing size + 0.5 mm

TC = Tolerance correction thickness = 0.13 mm

TR = Thickness at retrieval = Measured thickness (mm)

As the inlay is a fully congruent bearing, covered in the backside base area by the tibial plate and in the top area by the femoral component, the approximated volumetric wear rate was additionally assessed as follows:

Approximated volumetric wear rate (AWR) = ((base area + top area)/2) × LP.

The base areas were established dependent on the size and the design (anatomic/non-anatomic) of the bearing. Due to the necessity of a correction factor, especially in short time periods to revision, a negative approximated volumetric change—meaning an increase in thickness of the retrieved bearing—can occur. This was the case for $n = 4$ bearings (IDs 126, 199, 268, 1168). Four other inlays (IDs 92, 100, 675, 1092) had a time to revision <1 month. These eight bearings were excluded from the analysis of the penetration rate, as well as the approximated volumetric change.

3. Results

The data collected showed a comparable result between the two independent examiners (R = 0.82). The predominant type of damage was burnishing, followed by scratching and pitting. Edge loading, abrasion, and surface deformation were less common, while delamination and embedded particles were seen very rarely (Figure 3). Burnishing showed the highest number for all portions of the bearing, scratching—while seen in every specimen—was equally distributed over the Hood score. Pitting was also evenly distributed, yet some inlays did not exhibit any signs of pitting.

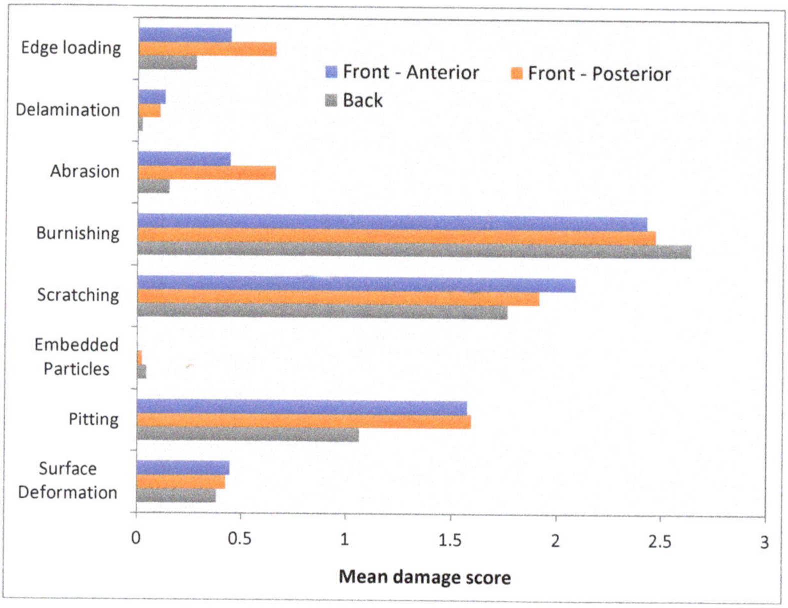

Figure 3. Hood damage score for all parts of the inlay (front-anterior and -posterior; back).

An ANOVA was conducted to compare damages for the four different bearing sizes. The mean total damage scores did not differ significantly in the four groups in this cohort (p = 0.86; Figure 4).

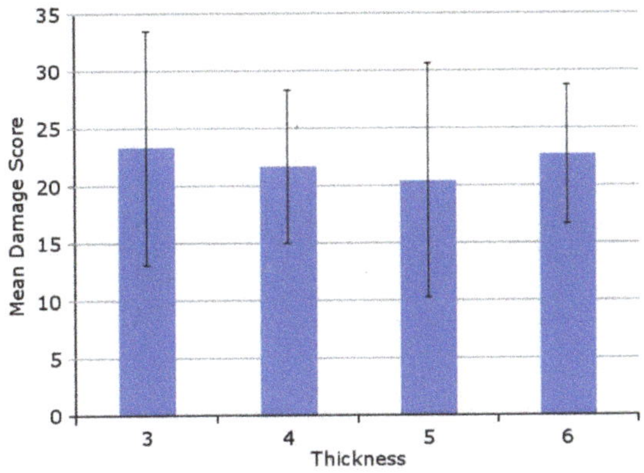

Figure 4. The distribution of the mean total damage score for the four inlay sizes (n = 47).

The lowest damage score could be seen at the backside of the implant, while front-anterior and front-posterior showed comparable results. As expected, the total damage score increased with the ongoing time to revision (Figure 5), although there was only a weak correlation (R = 0.3; R^2 = 0.09, $p < 0.05$).

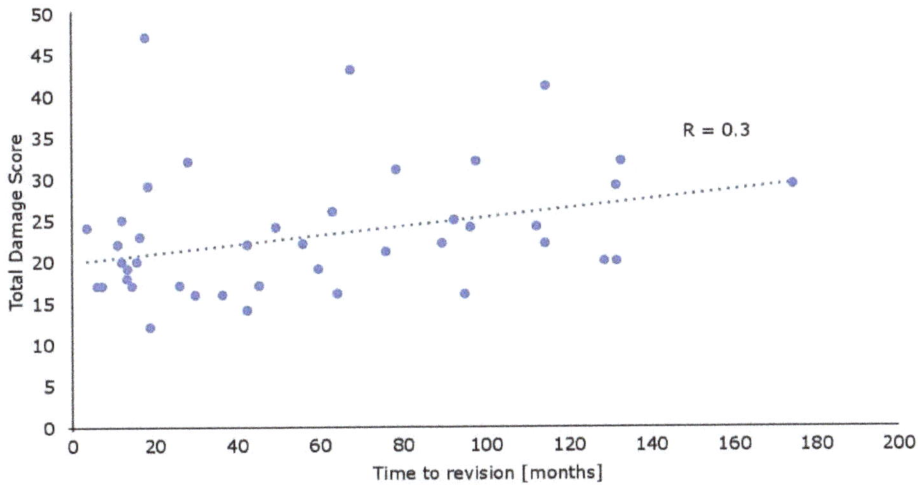

Figure 5. With the increasing time to revision, the total damage score increased ($p < 0.05$), but this correlation was weak (R = 0.3).

The linear wear of the inlay increased with time (Figure 6).

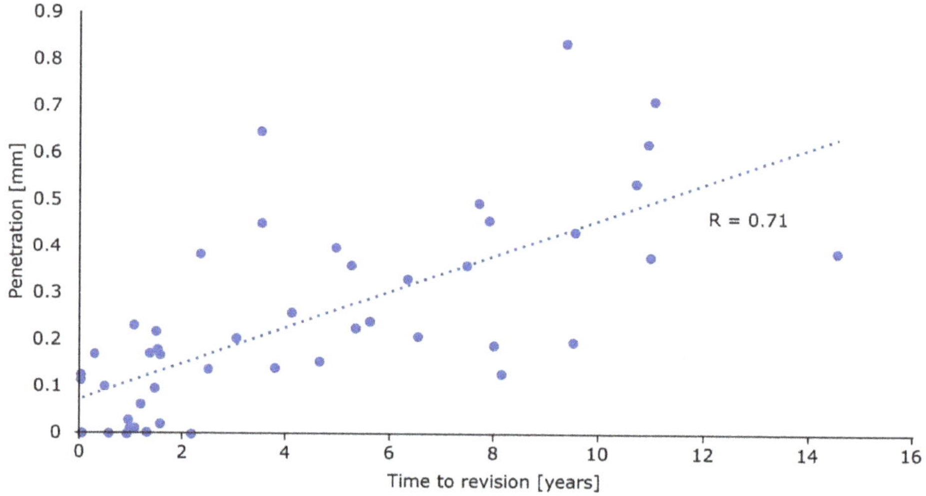

Figure 6. There is a strong correlation between the time to revision and penetration (R = 0.71; $p < 0.001$; $n = 47$).

A Shapiro-Wilk test was performed to check for normal distribution with a p > 0.05. The Shapiro-Wilk test revealed non-normal distribution and a Wilcoxon signed-rank test was performed for the subsequent analyses. The low-thickness inlays showed a mean penetration of 37.5 μm per year, whereas the high-thickness inlays exhibited lower penetration of 27.6 μm per year, the difference was not significant (Figure 7, p = 0.173).

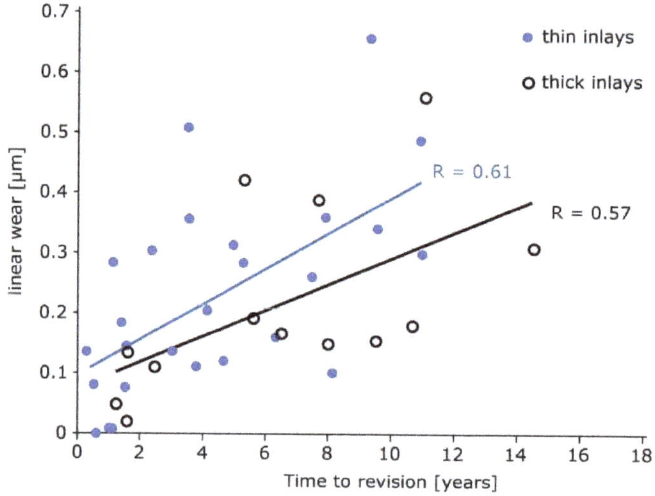

Figure 7. Linear wear shows comparable results for thin, as well as thick inlays (p = 0.173).

The total damage score and the linear penetration did not show any correlation (R = 0.137; p = 0.383). The approximated wear volume for all inlays increased over time (R = 0.53, p = 0.001). The mean approximated wear volume for low-thickness inlays was 26.3 mm^3 per year (R = 0.59, p = 0.001) and for high-thickness inlays 18.0 mm^3 per year (0.53, p = 0.001) without significant differences (Figure 8, p = 0.196).

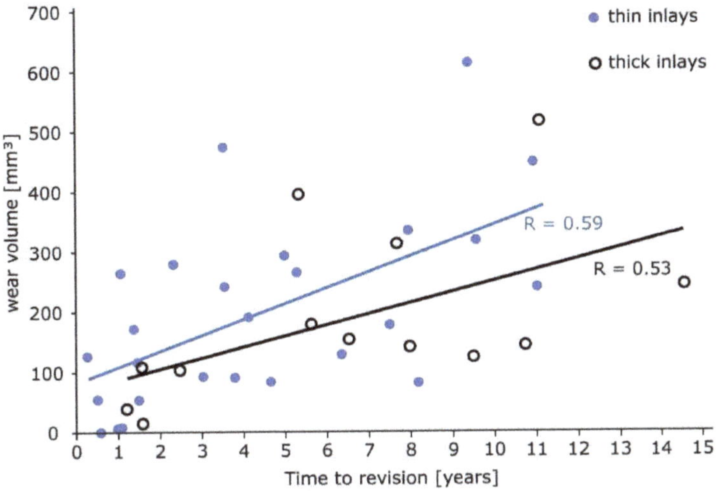

Figure 8. Thin and thick inlays show comparable wear volumes over time (p = 0.196).

Finally, there was no significant difference in survival time for the two different groups (Figure 9, p = 0.264).

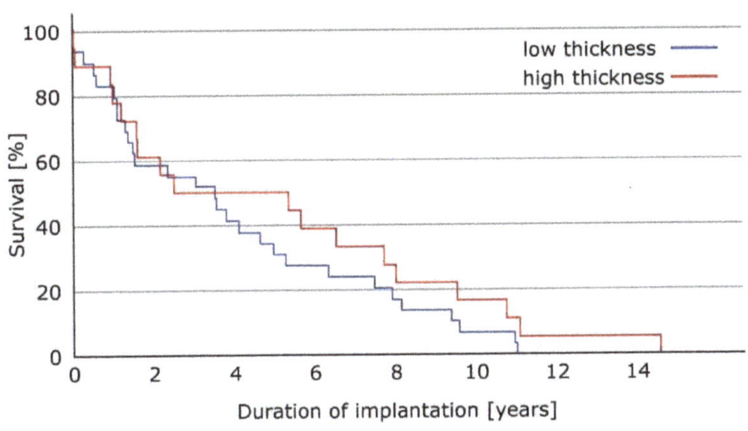

Figure 9. Kaplan-Meier blot for the two groups, showing no significant differences in survival (p = 0.264).

4. Discussion

In this retrieval analysis, the most common damage forms in descending order were burnishing, scratching, and pitting. This is consistent with earlier studies showing the same pattern of damage in total knee replacement [22,23]. However, also different damage patterns have been described and the pattern seems to be strongly influenced by the type of bearing [24,25]. In fully congruent meniscal bearing, extreme forms of damage such as edge loading are only expected to occur in small subsets of specimen, as they are rarely seen in cases of dislocation of subluxation of the inlays [26], while others such as embedded particles are also associated with the implantation technique. In this cohort, most types of damage distributed equally between the different surfaces (anterior, posterior, and back—i.e., tibial articulating surface). Abrasion was rarely seen at the backside which is to be expected, given the fully congruent nature of the tibial articulating portion of the bearing. In UKA as well as TKA, it has been advised against the use of thin inlays, due to the higher risk of thinning, increased wear, and risk for revision favoring the use of thicker inlays [27,28]. However, in our study we could find no such connections. While the linear penetration and the approximated volumetric wear rates were higher for the low thickness bearings, these findings were insignificant. Likewise, the survival rates for both groups showed comparable results. The total damage assessed by the Hood score and the penetration per month did not show any correlation. While counter-intuitive, this has been described by other groups before [29,30].

The penetration of 37.5 μm per year for low thickness and 27.6 μm per year for high thickness bearings is in range with studies published previously. Engh et al. described rates of approximately 50–60 μm per year for the medial and 30 μm for the lateral compartment in a type of total knee replacement [22], which is in line with results reported by Kop and Swarts who reported 54 μm for the superior surface of LCS knees [31]. For the Oxford unicompartmental knee arthroplasty, Horsager et al. described wear rates of 40 and 50 μm per year for the cemented and cementless versions, respectively, which is slightly lower than what we have seen in our cohort [32]. The aforementioned study was conducted using the radiostereometic analysis (RSA), which has been shown to produce accurate results in total knee replacement wear analysis [33,34]. However, no study comparing wear assessed by RSA and subsequently by the retrieval analysis has been conducted so far, to the best of our knowledge, and a certain error in measurement has to be accepted. In the present study, a direct measurement of the retrieved inlays was possible, which can be assumed as the most valid form to measure inlay penetration.

There are several limitations to our study: Since not all of the prostheses were implanted in our hospital, we could not assess clinical and radiological data. We analyzed both anatomical and non-anatomical bearings and the polyethylene used in the different generations of bearings does differ. Furthermore, as is the case for most retrieval studies, this was conducted as a retrospective analysis. Accordingly, the radiological examinations were not standardized to the need of the analyzer but to the need of the clinician. Hence, we could not properly assess pre- and post-operative valgus/varus, which might have an influence on the types and extent of damage. Furthermore, a correlation between patient satisfaction and experimental findings could not be investigated. Finally, as knee revision surgery is a complex and rarely standardized procedure, damages to the bearing and the rest of the implant can occur during the aforementioned surgery, although the differences between surgically induced damage and damage through wear and tear can be distinguished quite easily as done in this study.

To the best of our knowledge, this is the largest independent series of consecutively retrieved bearings of the medial Oxford unicompartmental knee arthroplasty to date. We could not find any significant differences in the two analyzed groups (high thickness—5 and 6 mm—and low thickness—3 and 4 mm—inlays) with regards to damage score, types of damage, penetration, approximated wear rate, and survival time.

5. Conclusions

In this retrospective retrieval analysis, we wanted to assess whether bearing thickness influences the clinical outcome and biomechanical aspects, with a special focus on time to revision and wear. We could not find significant differences for low- and high-thickness inlays. This study suggests that inlays with the analyzed bearing thicknesses can safely be used in OUKA without an increased risk for a decreased time to revision due to higher wear.

Author Contributions: Conceptualization, J.A.E. and J.P.K.; methodology, U.M.; software, S.J. and T.W.; validation, J.A.E., U.M., and J.P.K.; formal analysis, J.A.E.; investigation, J.A.E. and J.P.K.; resources, J.P.K.; data curation, J.A.E. and U.M.; writing—original draft preparation, J.A.E.; writing—review and editing, U.M. and J.P.K.; visualization, M.S.; supervision, J.P.K.; project administration, J.P.K. All authors have read and agreed to the published version of the manuscript.

Funding: This research received no external funding.

Acknowledgments: We would like to thank Linda Stange for language editing the manuscript.

Conflicts of Interest: The authors declare no conflict of interest.

References

1. Dyrhovden, G.S.; Lygre, S.H.L.; Badawy, M.; Gøthesen, Ø.; Furnes, O. Have the causes of revision for total and unicompartmental knee arthroplasties changed during the past two decades? *Clin. Orthop. Relat. Res.* **2017**, *475*, 1874–1886. [CrossRef] [PubMed]
2. Brown, T.S.; Van Citters, D.W.; Berry, D.J.; Abdel, M.P. The use of highly crosslinked polyethylene in total knee arthroplasty. *Bone Jt. J.* **2017**, *99*, 996–1002. [CrossRef] [PubMed]
3. Chakravarty, R.; Elmallah, R.D.K.; Cherian, J.J.; Kurtz, S.M.; Mont, M.A. Polyethylene Wear in Knee Arthroplasty. *J. Knee Surg.* **2015**, *28*, 370–375. [CrossRef] [PubMed]
4. *Annual Report Australian Orthopaedic Association*. 2017. Available online: https://aoanjrr.sahmri.com/de/annual-reports-2017 (accessed on 14 October 2020).
5. Jennings, J.M.; Kleeman-Forsthuber, L.T.; Bolognesi, M.P. Medial unicompartmental arthroplasty of the knee. *J. Am. Acad. Orthop. Surg.* **2019**, *27*, 166–176. [CrossRef] [PubMed]
6. Mohammad, H.R.; Strickland, L.; Hamilton, T.W.; Murray, D.W. Long-term outcomes of over 8,000 medial Oxford Phase 3 Unicompartmental Knees—A systematic review. *Acta Orthop.* **2018**, *89*, 101–107. [CrossRef] [PubMed]
7. Simpson, D.J.; Gray, H.; D'Lima, D.; Murray, D.W.; Gill, H.S. The effect of bearing congruency, thickness and alignment on the stresses in unicompartmental knee replacements. *Clin. Biomech.* **2008**, *23*, 1148–1157. [CrossRef]
8. Kretzer, J.P.; Jakubowitz, E.; Reinders, J.; Lietz, E.; Moradi, B.; Hofmann, K.; Sonntag, R. Wear analysis of unicondylar mobile bearing and fixed bearing knee systems: A knee simulator study. *Acta Biomater.* **2011**, *7*, 710–715. [CrossRef]
9. Manson, T.T.; Kelly, N.H.; Lipman, J.D.; Wright, T.M.; Westrich, G.H. Unicondylar knee retrieval analysis. *J. Arthroplast.* **2010**, *25*, 108–111. [CrossRef]
10. Pandit, H.; Hamilton, T.W.; Jenkins, C.; Mellon, S.J.; Dodd, C.A.F.; Murray, D.W. The clinical outcome of minimally invasive Phase 3 Oxford unicompartmental knee arthroplasty: A 15-year follow-up of 1000 UKAs. *Bone Jt. J.* **2015**, *97*, 1493–1500. [CrossRef]
11. Kim, S.-J.; Bae, J.-H.; Lim, H.C. Factors Affecting the postoperative limb alignment and clinical outcome after oxford unicompartmental knee arthroplasty. *J. Arthroplast.* **2012**, *27*, 1210–1215. [CrossRef]
12. Bartel, D.L.; Bicknell, V.L.; Wright, T.M. The effect of conformity, thickness, and material on stresses in ultra-high molecular weight components for total joint replacement. *J. Bone Jt. Surg. Am.* **1986**, *68*, 1041–1051. [CrossRef]
13. Bartel, D.L.; Rawlinson, J.J.; Burstein, A.H.; Ranawat, C.S.; Flynn, W.F. Stresses in polyethylene components of contemporary total knee replacements. *Clin. Orthop. Relat. Res.* **1995**, *317*, 76–82.
14. Engh, G.A.; Dwyer, K.A.; Hanes, C.K. Polyethylene wear of metal-backed tibial components in total and unicompartmental knee prostheses. *J. Bone Jt. Surg. Br.* **1992**, *74*, 9–17. [CrossRef] [PubMed]
15. McAuley, J.P.; Engh, G.A.; Ammeen, D.J. Revision of failed unicompartmental knee arthroplasty. *Clin. Orthop. Relat. Res.* **2001**, 279–282. [CrossRef]
16. Edwards, S.A.; Pandit, H.G.; Ramos, J.L.; Grover, M.L. Analysis of polyethylene thickness of tibial components in total knee replacement. *J. Bone Jt. Surg. Am.* **2002**, *84*, 369–371. [CrossRef]
17. Hood, R.W.; Wright, T.M.; Burstein, A.H. Retrieval analysis of total knee prostheses: A method and its application to 48 total condylar prostheses. *J. Biomed. Mater. Res.* **1983**, *17*, 829–842. [CrossRef]
18. Mohammad, H.R.; Campi, S.; Kennedy, J.A.; Judge, A.; Murray, D.W.; Mellon, S.J. Long-term in vivo wear of different bearing types used for the oxford unicompartmental knee replacement. *Bone Jt. Res.* **2019**, *8*, 535–543. [CrossRef]
19. Cerquiglini, A.; Henckel, J.; Hothi, H.; Rotigliano, N.; Hirschmann, M.T.; Hart, A.J. 3D patient imaging and retrieval analysis help understand the clinical importance of rotation in knee replacements. *Knee Surg. Sports Traumatol. Arthrosc.* **2018**, *26*, 3351–3361. [CrossRef]

20. Tone, S.; Hasegawa, M.; Pezzotti, G.; Puppulin, L.; Sudo, A. Effect of e-beam sterilization on the in vivo performance of conventional UHMWPE tibial plates for total knee arthroplasty. *Acta Biomater.* **2017**, *55*, 455–465. [CrossRef]
21. Psychoyios, V.; Crawford, R.W.; O'Connor, J.J.; Murray, D.W. Wear of congruent meniscal bearings in unicompartmental knee arthroplasty: A retrieval study of 16 specimens. *J. Bone Jt. Surg. Br.* **1998**, *80*, 976–982. [CrossRef]
22. Engh, G.A.; Zimmerman, R.L.; Parks, N.L.; Engh, C.A. Analysis of wear in retrieved mobile and fixed bearing knee inserts. *J. Arthroplast.* **2009**, *24*, 28–32. [CrossRef] [PubMed]
23. Harman, M.; Affatato, S.; Spinelli, M.; Zavalloni, M.; Stea, S.; Toni, A. Polyethylene insert damage in unicondylar knee replacement: A comparison of in vivo function and in vitro simulation. *Proc. Inst. Mech. Eng. H* **2010**, *224*, 823–830. [CrossRef] [PubMed]
24. Lu, Y.C.; Huang, C.H.; Chang, T.K.; Ho, F.Y.; Cheng, C.K.; Huang, C.H. Wear-pattern analysis in retrieved tibial inserts of mobile-bearing and fixed-bearing total knee prostheses. *J. Bone Jt. Surg. Br.* **2010**, *92*, 500–507. [CrossRef]
25. Ho, F.-Y.; Ma, H.-M.; Liau, J.-J.; Yeh, C.-R.; Huang, C.-H. Mobile-bearing knees reduce rotational asymmetric wear. *Clin. Orthop. Relat. Res.* **2007**, *462*, 143–149. [CrossRef]
26. Kendrick, B.J.L.; Longino, D.; Pandit, H.; Svard, U.; Gill, H.S.; Dodd, C.A.F.; Murray, D.W.; Price, A.J. Polyethylene wear in Oxford unicompartmental knee replacement. *J. Bone Jt. Surg. Br. Vol.* **2010**, *92*, 367–373. [CrossRef] [PubMed]
27. Lingaraj, K.; Morris, H.; Bartlett, J. Polyethylene thickness in unicompartmental knee arthroplasty. *Knee* **2011**, *18*, 165–167. [CrossRef] [PubMed]
28. Pijls, B.G.; Van der Linden-Van der Zwaag, H.M.J.; Nelissen, R.G.H.H. Polyethylene thickness is a risk factor for wear necessitating insert exchange. *Int. Orthop.* **2012**, *36*, 1175–1180. [CrossRef]
29. Knowlton, C.B.; Bhutani, P.; Wimmer, M.A. Relationship of surface damage appearance and volumetric wear in retrieved TKR polyethylene liners: Retrieved tkr polyethylene damage patterns and volumetric wear. *J. Biomed. Mater. Res. Part B Appl. Biomater.* **2017**, *105*, 2053–2059. [CrossRef]
30. Cerquiglini, A.; Henckel, J.; Hothi, H.; Moser, L.B.; Eskelinen, A.; Hirschmann, M.T.; Hart, A.J. Retrieval analysis of contemporary antioxidant polyethylene: Multiple material and design changes may decrease implant performance. *Knee Surg. Sports Traumatol. Arthrosc.* **2019**, *27*, 2111–2119. [CrossRef]
31. Kop, A.M.; Swarts, E. Quantification of polyethylene degradation in mobile bearing knees: A retrieval analysis of the Anterior-Posterior-Glide (APG) and Rotating Platform (RP) Low Contact Stress (LCS) knee. *Acta Orthop.* **2007**, *78*, 364–370. [CrossRef]
32. Horsager, K.; Madsen, F.; Odgaard, A.; Fink Jepsen, C.; Rømer, L.; Kristensen, P.W.; Kaptein, B.L.; Søballe, K.; Stilling, M. Similar polyethylene wear between cemented and cementless Oxford medial UKA: A 5-year follow-up randomized controlled trial on 79 patients using radiostereometry. *Acta Orthop.* **2019**, *90*, 67–73. [CrossRef] [PubMed]
33. Van Ijsseldijk, E.A.; Valstar, E.R.; Stoel, B.C.; Nelissen, R.G.H.H.; Reiber, J.H.C.; Kaptein, B.L. The robustness and accuracy of in vivo linear wear measurements for knee prostheses based on model-based RSA. *J. Biomech.* **2011**, *44*, 2724–2727. [CrossRef]
34. Van Ijsseldijk, E.A.; Harman, M.K.; Luetzner, J.; Valstar, E.R.; Stoel, B.C.; Nelissen, R.G.H.H.; Kaptein, B.L. Validation of a model-based measurement of the minimum insert thickness of knee prostheses: A retrieval study. *Bone Jt. Res.* **2014**, *3*, 289–296. [CrossRef]

materials

MDPI

Article

Hip Spacers with a Metal-on-Cement Articulation Did Not Show Significant Surface Alterations of the Metal Femoral Head in Two-Stage Revision for Periprosthetic Joint Infection

Andre Lunz [1,*], Robert Sonntag [2], J. Philippe Kretzer [2], Sebastian Jaeger [2], Therese Bormann [2], Marcus R. Streit [1], Nicholas A. Beckmann [1], Burkhard Lehner [1] and Georg W. Omlor [1]

1 Clinic for Orthopedics and Trauma Surgery, Center for Orthopedics, Trauma Surgery and Spinal Cord Injury, Heidelberg University Hospital, Schlierbacher Landstrasse 200a, 69118 Heidelberg, Germany; streitmarcus@gmail.com (M.R.S.); nicholas.beckmann@med.uni-heidelberg.de (N.A.B.); burkhard.lehner@med.uni-heidelberg.de (B.L.); georg.omlor@med.uni-heidelberg.de (G.W.O.)

2 Laboratory of Biomechanics and Implant Research, Clinic for Orthopedics and Trauma Surgery, Heidelberg University Hospital, Schlierbacher Landstrasse 200a, 69118 Heidelberg, Germany; Robert.Sonntag@med.uni-heidelberg.de (R.S.); philippe.kretzer@med.uni-heidelberg.de (J.P.K.); Sebastian.Jaeger@med.uni-heidelberg.de (S.J.); therese.bormann@med.uni-heidelberg.de (T.B.)

* Correspondence: andre.lunz@med.uni-heidelberg.de

Received: 31 July 2020; Accepted: 31 August 2020; Published: 2 September 2020

check for updates

Abstract: Two-stage revision is considered the gold standard treatment for chronic periprosthetic joint infection (PJI). During the interim period, between explantation of the infected hip endoprosthesis and revision arthroplasty, individually formed articulating hip spacers made of polymethylmethacrylate (PMMA) bone cement can be used to provide better soft tissue preservation, local antibiotic release, and improved postoperative mobilization. If effective prevention from luxation is achieved, hip function and hence overall patient satisfaction is improved. Zirconium oxide particles inside conventional PMMA bone cement, however, are known to enhance third-body wear, which may cause alterations of the metal head in the articulating spacer and hence become a serious risk for the patient. Therefore, the aim of our study was to analyze whether the articular surface of cobalt-chrome (CoCr) femoral heads is significantly altered in the setting of a metal-on-cement articulation during the interim period of two-stage revision for PJI. We analyzed a consecutive series of 23 spacer cases and compared them with femoral heads from two series of conventional hip arthroplasty revisions with metal-on-polyethylene articulations and different time intervals in situ. To investigate metallic wear, the femoral heads were thoroughly examined, and their surface roughness was measured and analyzed. We found no significant differences between the two conventional hip arthroplasty groups, despite their very different times in situ. Furthermore, the individually different times in situ within the spacer group had no significant impact on surface roughness, either. Compared with the spacer group, the surface roughness of the metal femoral heads from both conventional hip arthroplasty groups were even higher. Within the spacer group, roughness parameters did not show significant differences regarding the five predefined locations on the metal head. We conclude that metal-on-cement articulations do not cause enhanced surface alterations of the metal femoral head and hence do not limit the application in articulating hip spacers in the setting of two-stage revision for PJI.

Keywords: periprosthetic joint infection; cement spacer; articulating spacer; hip spacer; two-stage revision; surface alteration; surface roughness; third-body wear; zirconium oxide particles; metal-on-cement articulation

1. Introduction

Periprosthetic joint infection (PJI) is a dreaded complication of total joint replacement that brings numerous challenges to the affected patients and the orthopedic surgeons. One-stage revision has recently gained more support because of its good clinical results and numerous advantages in carefully selected patients. A widely accepted consensus was reached on several strict exclusion criteria such as unknown or multiresistant organism-based infections, relevant comorbidities, sinus tract, and other relevant soft tissue complications [1–3]. Therefore, two-stage revision is still widely considered the gold standard treatment in the setting of chronic and complex periprosthetic joint infections. Insall et al. first described this technique in 1983 [4]. Since then, numerous studies have reported that two-stage revision has been used successfully and can result in infection eradication rates beyond 90% [5–7]. The treatment regimen includes aggressive debridement, taking of numerous samples for histopathological and microbiological analysis, and removal of all implants in the first stage. To avoid a Girdlestone hip, where the implant or joint resection is conducted without replacement, an antibiotic loaded polymethylmethacrylate (PMMA) cement spacer can be implanted for the interim period. The advantage of a spacing device is not only that the soft tissue envelope will be better preserved at the time of reimplantation, but also that articulating spacers help prevent arthrofibrosis, limb shortening, and osteolysis, leading to an overall improved outcome [7–9]. Mobile hip spacers create an articulation between the femoral head and acetabulum, similar to the articulation in conventional total hip arthroplasty, resulting in improved hip function during the interim period and consequently after final reimplantation [10–12]. Several techniques for constructing mobile spacers have been described, [12–14] providing a metal-on-polyethylene articulation with the major disadvantage of using a polyethylene (PE) liner in the setting of infection, leading to an increased risk of biofilm formation on its surface. An articulating spacer technique using an individually formed metal-on-cement articulation can be used as an alternative. Here, a custom-made antibiotic-loaded cemented acetabular socket is formed to provide a congruent articulating surface with a conventional metal femoral head. A deliberately loosely cemented unsophisticated femoral stem serves to increase the strength of the construct and at the same time reduces the risk of periprosthetic fracture. In practice, this spacer technique allows early patient mobilization with a good range of motion and at least partial weight-bearing activity directly from postoperative day one [own unpublished data].

The aim of the current study was to analyze whether the articular surface of the metal (CoCr) femoral head is altered in the setting of a metal-on-cement articulation. This major concern arises because PMMA bone cement contains hard zirconium-oxide-particles, which are known to be responsible for enhanced third-body wear after total joint replacement [15–18]. Since we have had good clinical experience with this described articulating spacer technique, our aim was to evaluate its biomechanical usability. We hypothesized that surface roughness of the femoral heads from spacers would not increase significantly compared to conventional metal-on-polyethylene articulations, allowing the conclusion that no relevant wear has occurred. We therefore analyzed and compared the surface roughness of metal femoral heads from articulating cement spacers with metal femoral heads from conventional hip arthroplasty revisions with metal-on-polyethylene articulations.

2. Materials and Methods

All revision procedures were performed at the same university institution. First, a consecutive series, from November 2017 to August 2019, of 23 articulating hip spacer cases was analyzed. All spacers had a mobile metal-on-cement articulation and were implanted for the interim period in the setting of PJI. Two series of conventional hip arthroplasty revision cases with metal-on-polyethylene articulations with heterogeneous in situ time intervals were analyzed and compared as a reference. All implants were collected on a standardized base. All operations were performed only for medical reasons by specialized surgeons at our certified over-regional joint arthroplasty center. All patients have given written consent for their retrieved implants to be used for different research purposes. This study was approved by the local Medical Ethics Committee (S-091/2018).

2.1. Clinical Data

The study included 23 femoral head components from performed spacer explantations in the setting of two-stage revision for PJI (Table 1, No. 1–23). The interim period between explantation and reimplantation was variable and based on clinical, radiological, and laboratory evidence that infection had been overcome. In 6 cases, the collected intraoperative samples during explantation surgery only showed histopathological evidence of infection without microbiological detection of any microorganisms and in 3 cases more than one bacterium was found. In another case a fungus was established as the cause of infection. During second-stage surgery the microbiologic samples of 4 cases stayed positive, as reported in Table 2.

The mean interim period and hence mean time of spacers in situ was 79 (14–253) days. The spacer components came from 13 men and 10 women. Mean patient age was 70 years at the time of spacer implantation. As this spacer technique allows early patient mobilization, we tried to standardize weight bearing as much as possible by giving the same instructions to all patients and by using the help of a specially trained physical therapist performing standardized mobilization protocols to all patients.

Table 1. Clinical data of the 3 groups: articulating spacer patients (No. 1–23), conventional hip arthroplasty patients with short implant time in situ (No. A–C), and long time in situ (No. D–L).

No.	Patient Age (years)	Patient Gender (m = male, f = female)	Femoral Head Size (mm)	Time In Situ (d = days, y = years)
1	76	m	32	44 d
2	67	f	32	17 d
3	68	m	32	86 d
4	78	m	28	84 d
5	67	m	28	14 d
6	57	m	28	71 d
7	54	f	28	28 d
8	69	f	28	140 d
9	61	m	28	113 d
10	82	m	28	117 d
11	63	f	28	130 d
12	64	f	32	68 d
13	65	m	28	253 d
14	69	f	28	51 d
15	58	f	28	112 d
16	68	m	28	77 d
17	69	f	28	76 d
18	75	m	28	35 d
19	65	f	28	26 d
20	81	f	28	84 d
21	78	m	28	62 d
22	78	m	28	57 d
23	75	m	28	70 d
Mean (Range)	**69 (54–82)**	**m = 13, f = 10**	**28 mm = 19, 32 mm = 4**	**79 d (14–253)**
A	71	m	32	350 d
B	67	f	32	459 d
C	77	m	32	89 d
Mean (Range)	**72 (67–77)**	**m = 2, f = 1**	**32 mm = 3**	**299 d (89–459)**
D	77	m	32	9 y
E	84	m	32	15 y
F	78	m	28	10 y
G	74	f	28	7 y
H	74 76	f	28	15 y
I	76	f	28	17 y
J	85	m	32	5 y
K	77	m	28	13 y
L	76	m	32	8 y
Mean (Range)	**78 (74–85)**	**m = 6, f = 3**	**28 mm = 5, 32 mm = 4**	**11 years (5–17)**

Table 2. Microbiological results from collected intraoperative samples during a) first- and b) second-stage surgery in the articulating spacer group.

	(a) First-Stage Surgery (Explantation)		(b) Second-Stage Surgery (Reimplantation)	
Microorganism	Number	Percentage (%)	Number	Percentage (%)
None	6	**26.1**	19	82.6
Coagulase-positive staphylococci	2	**8.7**		
Staph. aureus	2	8.7		
Coagulase-negative staphylococci	11	**47.8**	3	13
Staph. lugdunensis	2	8.7		
Staph. capitis	2	8.7		
Staph. epidermidis	7	30.4	3	13
Other	4	**17.4**		
Bacillus spec.	1	4.3		
Cutibacterium acnes	3	13		
Fungal infection	1	**4.3**	1	4.3
Candida albicans	1	4.3	1	4.3
Polymicrobial infections	3	13		

In all 23 cases, the metal-on-cement articulation was constructed using a 28 or 32 mm CoCr femoral head (S-2XL, DePuy Synthes, West Chester, PA, USA) with a corresponding femoral stem (Weber Stem CS/CM/SM, Zimmer, Warsaw, IN, USA) and a custom-made acetabular socket formed out of 40–80 g of antibiotic-loaded PMMA cement (Palacos cement, Heraeus, Hanau, Germany) with 3 g Vancomycin powder per 40 g of cement, as shown in Figure 1 (own unpublished data).

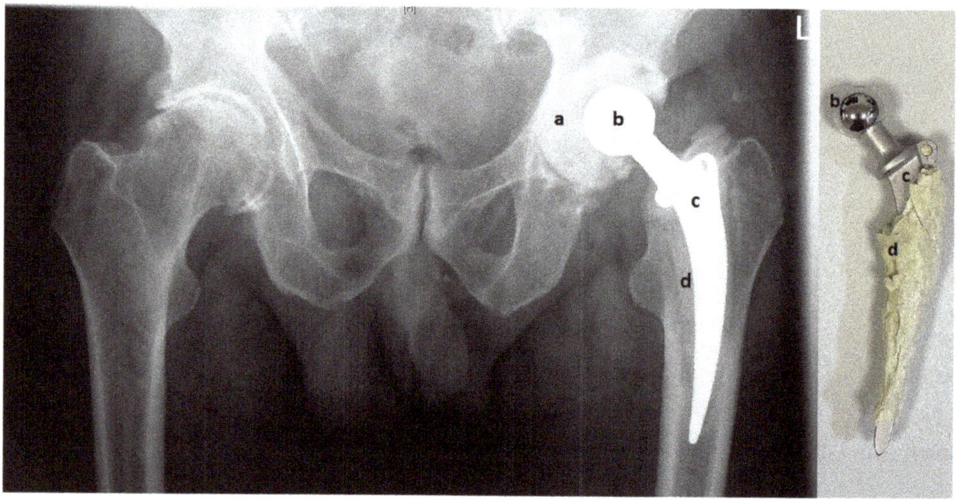

Figure 1. Pelvic X-ray with articulating spacer in situ (left) and the same spacer after explantation (right). The custom-made acetabular socket (**a**) is made of antibiotic-loaded PMMA (polymethylmethacrylate) bone cement and creates the metal-on-cement articulation with the CoCr femoral head (**b**). PMMA bone cement (**d**) is also used to fix the femoral stem into the femur (**c**).

After a sufficient number of samples are taken and radical debridement is completed, the spacer is constructed as follows: First, the acetabular shell is loosely formed by hand and put as dough into the acetabular groove. Before PMMA polymerization has occurred, surplus cement is removed at the acetabular edges and an articulation groove is formed in the middle of the acetabular spacer using a plunger with a slightly larger diameter of its head than the original metal head of the later articulation. To form a smooth groove, the plunger is continuously rotated and moved until PMMA polymerization has finished and the acetabular component is fixed. Later, this allows free movement of the metal head inside the cement groove. Then, as the second step, the femoral component is built by applying cement on the metal stem. To facilitate later removal, no cement is applied around the tip of the stem. After a couple of minutes, the cement becomes more solid and the stem is carefully pushed into the femur. To avoid cement penetration into the bone, the stem is continuously moved back and forwards until PMMA polymerization has completed. From our experience, this "deliberately loose cementing technique" provides sufficient fixation in the bone and allows easy removal in the second stage. After both spacer components (acetabular and femoral) are in place, a metal head is put on the stem and the reduction maneuver is carefully performed. Finally, it is extremely important to check for sufficient hip stability in flexion, extension, and rotation to minimize the risk of spacer luxation. If the articulation of the metal head inside the cement groove is not stable enough (e.g., it can be easily luxated), the spacer cannot be retained and must be replaced by a new one.

To allow comparisons with metal-on-polyethylene articulations in short- and long-term settings, the study also included two series from conventional hip arthroplasty revisions with a CoCr femoral head and a metal-on-polyethylene articulation (Table 1, No. A–C and No. D–L), with short and long implant periods in situ. All revision surgeries were carried out due to confirmed or highly suspected PJI. The mean period in situ was 299 (89–459) days and 11 (5–17) years in the short and in the long implant period group, respectively. Mean age of the patients at explantation was 72 and 78 years, respectively, as reported in Table 1.

2.2. Femoral Head Analysis

To determine alterations of the metal head as a sign for possible metallic wear, the femoral heads were analyzed for surface roughness and deviations in shape caused by in situ wear. A tactile roughness measurement instrument (Perthomether M2 profilometer; Mahr, Göttingen, Germany) was used to characterize the surface of the femoral heads with 12 nm of accuracy at 5 predefined locations on the implants, as shown in Figure 2. Care was taken to avoid measuring areas that most probably had been damaged during revision surgery and not by in situ wear. All measurements were performed according to ISO 4287 [19] with a scanning length of 1.25 mm. The parameters, which were used to describe and interpret the surface alterations of the femoral heads, are described in Table 3. The common parameter average roughness (Ra) and average maximum profile height (Rz) provide a general basis to evaluate and interpret surface topographies. Since the definition of Ra is based on arithmetic averaging and Rz on an average of five maximum heights, these values are by definition unable to distinguish between peaks and valleys. Thus, the core roughness depth (Rk) that has been established to characterize load-carrying surfaces has been used [20,21].

Table 3. This table summarizes the used surface characteristics parameters [19–21].

Parameter	Definition	Description	Standard
Ra	Arithmetic average profile roughness	Arithmetic average of the absolute values of the roughness profile ordinates	EN ISO 4287
Rz	Average maximum height of the profile	Average of the 5 highest peaks and 5 deepest valleys in the profile	EN ISO 4287
RSm	Mean line peak spacing	The mean spacing between the profile peaks over the sampling length	EN ISO 4287
Rp	Maximum peak height of the profile	The height of the highest peak above the mean line within the sampling length	EN ISO 4287
Rk	Core roughness depth	The depth of the roughness core profile within the evaluation length, excluding the height of protruding peaks and deep valleys. Rk is obtained from the material ratio curve (Abbott curve)	EN ISO 13565

The metallic surface of the CoCr femoral heads was further examined, and light optical images were taken with a digital microscope (VHX-5000, Keyence, Osaka, Japan) at a magnification ×200.

2.3. Statistical Analysis

All descriptive data is presented as the arithmetic mean and standard deviation. The Kruskal–Wallis test was used to compare the means between the groups. Pearson correlations were used to determine statistical correlations between clinical parameters (time in situ, patient age, femoral head size) and roughness parameters. The level of significance was set at $p < 0.05$ for all statistical tests. The statistical analyses were performed using SPSS software (version 25.0; SPSS Inc, Chicago, IL, USA).

3. Results

3.1. Roughness Measurements

First, differences in the spacer group between roughness measurements of the five investigated locations on the femoral head (Figure 2) were analyzed. No significant statistical differences were found, as shown in Table 4.

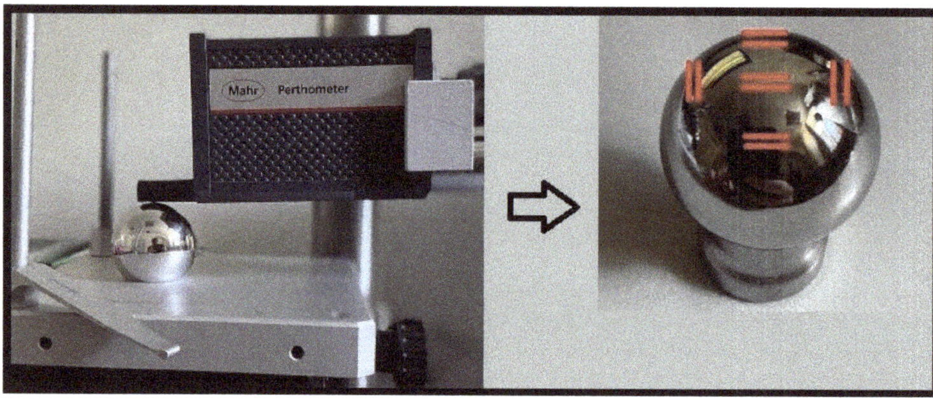

Figure 2. Shows the 5 predefined locations where the roughness measurements were performed.

Table 4. Analysis of roughness parameters in the spacer group at the 5 predefined locations. Statistical significance is assumed for $p < 0.05$ and was not found.

	Roughness Parameters in μm (Mean ± Standard Deviation)				
Predefined Measurement Places on the Femoral Heads	**Ra**	**Rz**	**RSm**	**Rp**	**Rk**
No. 1–2	0.054 ± 0.01	0.230 ± 0.07	255.5 ± 157.3	0.200 ± 0.08	0.064 ± 0.02
No. 3–4	0.052 ± 0.01	0.222 ± 0.11	372.8 ± 301.8	0.195 ± 0.09	0.063 ± 0.02
No. 5–6	0.046 ± 0.01	0.208 ± 0.07	280.0 ± 258.8	0.175 ± 0.06	0.058 ± 0.01
No. 7–8	0.050 ± 0.02	0.221 ± 0.15	177.8 ± 118.5	0.175 ± 0.09	0.066 ± 0.03
No. 9–10	0.048 ± 0.01	0.212 ± 0.10	286.6 ± 203.5	0.176 ± 0.07	0.062 ± 0.02
***p* value**	**0.153**	**0.541**	**0.072**	**0.342**	**0.654**

The surface roughness parameters between the spacer group and the two conventional hip arthroplasty revision groups were compared. Since significant differences were found, a further pairwise comparison was conducted. No significant differences were shown regarding the two conventional hip arthroplasty groups and their different implant times in situ. Compared with the spacer group, the surface roughness of the femoral heads in the metal-on-polyethylene bearings was significantly increased with regard to all measured roughness parameters, except for Ra, which was also higher, but without reaching significance. The results are summarized in Tables 5 and 6.

Table 5. Comparison of roughness parameters between the spacer group and two conventional hip arthroplasty revision groups (PE-Group Short = short implant time in situ, PE-Group Long = long implant time in situ).

	Roughness Parameters in μm (Mean ± Standard Deviation)				
	Ra	**Rz**	**RSm**	**Rp**	**Rk**
Spacer Group	0.050 ± 0.01	0.221 ± 0.12	273.2 ± 222.3	0.183 ± 0.08	0.063 ± 0.02
PE-Short Group	0.061 ± 0.03	0.312 ± 0.14	189.5 ± 95.7	0.233 ± 0.06	0.084 ± 0.04
PE-Long Group	0.056 ± 0.03	0.355 ± 0.19	128.9 ± 106.1	0.244 ± 0.13	0.089 ± 0.09
***p*-value**	**0.062**	**<0.01 ***	**<0.01 ***	**<0.01 ***	**<0.01 ***

*: The level of significance was set at $p < 0.05$

Table 6. Pairwise Comparison between all 3 Groups (Spacer, PE-Short, PE-Long).

	Rz	**RSm**	**Rp**	**Rk**
Pairwise Comparison	*p* value	*p* value	*p* value	*p* value
Spacer vs. PE-Short	<0.01 *	<0.01 *	<0.01 *	<0.01 *
Spacer vs. PE-Long	<0.01 *	<0.01 *	<0.01 *	<0.01 *
PE-Short vs. PE-Long	0.532	0.234	0.756	0.811

*: The level of significance was set at $p < 0.05$

Furthermore, light optical images were taken of the metallic articular surface using a magnification × 200. An example of this is shown in Figure 3, in which surface alterations in the spacer group are markedly less prominent in comparison to both conventional hip arthroplasty groups. A conventional hip endoprosthesis with a metal-on-metal articulation was used as a reference and showed even more surface roughness evident as more scratches and grooves than all three groups.

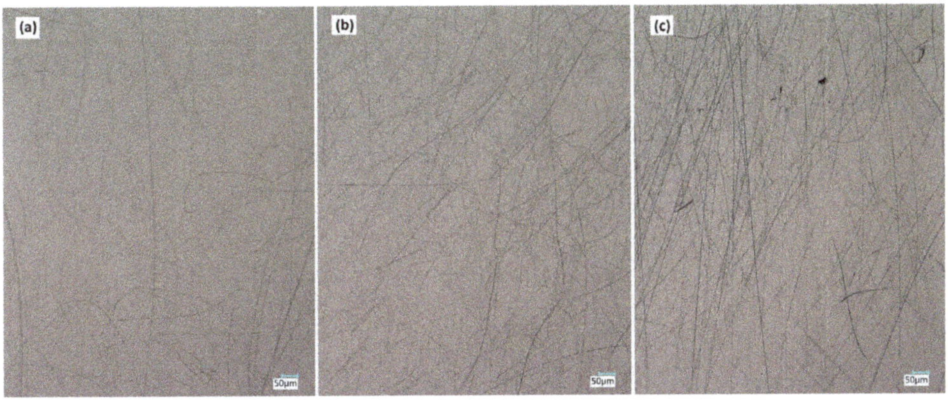

Figure 3. Representative examples of light optical images of the metal femoral heads at a magnification ×200. (**a**) Spacer with metal-on-cement articulation, (**b**) conventional hip arthroplasty with metal-on-polyethylene articulation, and (**c**) conventional hip arthroplasty with metal-on-metal articulation.

3.2. Correlations to Clinical Data

We further looked for correlations between roughness parameters within the spacer group and available clinical parameters, such as patient age, time of the implant in situ, and femoral head size. Again, no significant statistical correlations were found (Table 7).

Table 7. Calculated p values for statistical correlations between roughness parameters and clinical parameters in the spacer group. Statistical significance is assumed for $p < 0.05$ and was not found.

	Roughness Parameters				
Clinical Parameters	Ra	Rz	RSm	Rp	Rk
Time In Situ	0.507	0.767	0.658	0.932	0.816
Patient Age	0.864	0.267	0.748	0.863	0.254
Femoral Head Size	0.088	0.087	0.811	0.469	0.052

4. Discussion

This study analyzed and compared the surface roughness of metal femoral heads from explanted articulating spacers and conventional hip arthroplasty revisions. All surgery was performed because of confirmed or highly suspected PJI. We assumed that a higher surface roughness would most likely lead to more metallic and cement wear. All analyzed articulating spacers had metal-on-cement articulations, which could be prone to enhanced third-body wear because of hard zirconium oxide particles in the PMMA bone cement. The goal of our study was to evaluate biomechanical usability against the background of this assumption.

Interestingly, no significant differences between the two conventional hip arthroplasty groups were found, despite their very different times in situ. Compared with the spacer group, the surface roughness of both conventional hip arthroplasty groups was significantly higher. This allows the conclusion that in the setting of two-stage revision for PJI, metal-on-cement articulations in articulating spacers show less surface modification on the metallic head than metal-on-polyethylene articulations in conventional hip arthroplasties. The amplified light optical images of the metal femoral heads confirm these results. Femoral heads from the spacer group show unambiguously fewer surface alterations than both conventional hip arthroplasty groups. The problem of enhanced third-body wear in joint replacement, known to be caused by zirconium oxide particles in PMMA bone cement, seems to be neglectable in the setting of two-stage revision for PJI. This is probably because of the rather short interim period and thus time in situ.

Wear is also thought to be influenced by the patient's activity level and thus implant exposure in situ. The articulating spacer system allows hip joint movement (usual range of motion: extension/flexion: 0/0/90°) and partial weight bearing (usually 20 kg body weight). This is an improvement when compared to a Girdlestone hip, but remains more restricted than in patients who received conventional total hip replacement. Limited activity and partial weight bearing in the spacer group could also be an important reason for the differences in surface roughness compared with the conventional hip arthroplasty groups. Nevertheless, these findings justify the described spacer technique using a metal-on-cement articulation despite the enhanced risk for early (low level) third-body wear.

The roughness parameters within the spacer group showed no significant differences regarding the five analyzed locations on the femoral head. In contrast, patients suffering from osteoarthritis have very well described locations on the femoral head, where increased cartilage wear is found. This implies that the measured surface roughness, and therefore wear, was most probably too minute to permit significant differences in the setting of articulating spacers.

No significant correlations between clinical data (implant time in situ, patient age, or femoral head size) and roughness parameters within the spacer group were found. As discussed, one would assume that longer times in situ would enhance implant exposure and thus create increased surface roughness and wear. Once again, the uniformly low patient activity level during the interim period was probably responsible for this lack of significance. Furthermore, the period of time of all spacers in situ was well under a year and thus possibly not long enough to create significant surface alterations.

To summarize, CoCr femoral heads from articulating spacers showed decreased surface roughness parameters in comparison to retrieved metal femoral heads from conventional hip arthroplasties. The small amount of surface alterations on the CoCr femoral heads allows the assumption that metallic wear is considerably low. These findings implicate that metal-on-cement articulations in interim spacers are biomechanically usable in the setting of two-stage revision for PJI.

Author Contributions: Conceptualization, J.P.K., R.S., B.L., N.A.B., S.J., T.B., M.R.S., A.L. and G.W.O.; methodology, R.S., J.P.K., A.L. and G.W.O.; investigation, A.L. and G.W.O.; resources, J.P.K. and G.W.O.; writing—original draft preparation, A.L. and G.W.O.; writing—review and editing, G.W.O., R.S., J.P.K., S.J., T.B., N.A.B., M.R.S., A.L. and B.L.; supervision, J.P.K., G.W.O. and R.S. All authors have read and agreed to the published version of the manuscript.

Funding: This research received no external funding.

Conflicts of Interest: The authors declare no conflict of interest.

References

1. Svensson, K.; Rolfson, O.; Kärrholm, J.; Mohaddes, M. Similar risk of re-revision in patients after one- or two-stage surgical revision of infected total hip arthroplasty: An analysis of revisions in the Swedish hip arthroplasty register 1979–2015. *J. Clin. Med.* **2019**, *8*, 485. [CrossRef] [PubMed]
2. Pangaud, C.; Ollivier, M.; Argenson, J.N. Outcome of single-stage versus two-stage exchange for revision knee arthroplasty for chronic periprosthetic infection. *EFORT Open Rev.* **2019**, *4*, 495–502. [CrossRef] [PubMed]
3. Nguyen, M.; Sukeik, M.; Zahar, A.; Nizam, I.; Haddad, F.S. One-stage exchange arthroplasty for periprosthetic hip and knee joint infections. *Open Orthop. J.* **2016**, *10*, 646–653. [CrossRef] [PubMed]
4. Insall, J.N.; Thompson, F.M.; Brause, B.D. Two-stage reimplantation for the salvage of infected total knee arthroplasty. 1983. *J. Bone Jt. Surg. Am.* **2002**, *84*, 490. [CrossRef] [PubMed]
5. Lu, J.; Han, J.; Zhang, C.; Yang, Y.; Yao, Z. Infection after total knee arthroplasty and its gold standard surgical treatment: Spacers used in two-stage revision arthroplasty. *Intractable Rare Dis. Res.* **2017**, *6*, 256–261. [CrossRef]
6. Kalore, N.V.; Gioe, T.J.; Singh, J.A. Diagnosis and management of infected total knee arthroplasty. *Open Orthop. J.* **2011**, *5*, 86–91. [CrossRef] [PubMed]
7. Charette, R.S.; Melnic, C.M. Two-stage revision arthroplasty for the treatment of prosthetic joint infection. *Curr. Rev. Musculoskelet. Med.* **2018**, *11*, 332–340. [CrossRef] [PubMed]
8. Fink, B.; Vogt, S.; Reinsch, M.; Buchner, H. Sufficient release of antibiotic by a spacer 6 weeks after implantation in two-stage revision of infected hip prostheses. *Clin. Orthop. Relat. Res.* **2011**, *469*, 3141–3147. [CrossRef] [PubMed]
9. Masri, B.A.; Duncan, C.P.; Beauchamp, C.P. Long-term elution of antibiotics from bone-cement: An in vivo study using the prosthesis of antibiotic-loaded acrylic cement (PROSTALAC) system. *J. Arthroplast.* **1998**, *13*, 331–338. [CrossRef]
10. Durbhakula, S.M.; Czajka, J.; Fuchs, M.D.; Uhl, R.L. Spacer endoprosthesis for the treatment of infected total hip arthroplasty. *J. Arthroplast.* **2004**, *19*, 760–767. [CrossRef] [PubMed]
11. Younger, A.S.; Duncan, C.P.; Masri, B.A. Treatment of infection associated with segmental bone loss in the proximal part of the femur in two stages with use of an antibiotic-loaded interval prosthesis. *J. Bone Jt. Surg. Am.* **1998**, *80*, 60–69. [CrossRef] [PubMed]
12. Wentworth, S.J.; Masri, B.A.; Duncan, C.P.; Southworth, C.B. Hip prosthesis of antibiotic-loaded acrylic cement for the treatment of infections following total hip arthroplasty. *J. Bone Jt. Surg. Am.* **2002**, *84* (A Suppl. 2), 123–128. [CrossRef]
13. Haddad, F.S.; Masri, B.A.; Garbuz, D.S.; Duncan, C.P. The treatment of the infected hip replacement. The complex case. *Clin. Orthop. Relat. Res.* **1999**, 144–156. [CrossRef] [PubMed]
14. Lausmann, C.; Citak, M.; Hessling, U.; Wolff, M.; Gehrke, T.; Suero, E.M.; Zahar, A. Preliminary results of a novel spacer technique in the management of septic revision hip arthroplasty. *Arch. Orthop. Trauma Surg.* **2018**, *138*, 1617–1622. [CrossRef] [PubMed]
15. Davidson, J.A.; Poggie, R.A.; Mishra, A.K. Abrasive wear of ceramic, metal, and UHMWPE bearing surfaces from third-body bone, PMMA bone cement, and titanium debris. *Biomed. Mater. Eng.* **1994**, *4*, 213–229. [CrossRef] [PubMed]

16. Paulus, A.C.; Franke, M.; Kraxenberger, M.; Schröder, C.; Jansson, V.; Utzschneider, S. PMMA third-body wear after unicondylar knee arthroplasty decuples the UHMWPE wear particle generation in vitro. *Biomed. Res. Int.* **2015**, *2015*, 575849. [CrossRef] [PubMed]
17. Que, L.; Topoleski, L.D. Third-body wear of cobalt-chromium-molybdenum implant alloys initiated by bone and poly(methyl methacrylate) particles. *J. Biomed. Mater. Res.* **2000**, *50*, 322–330. [CrossRef]
18. Schroeder, C.; Grupp, T.M.; Fritz, B.; Schilling, C.; Chevalier, Y.; Utzschneider, S.; Jansson, V. The influence of third-body particles on wear rate in unicondylar knee arthroplasty: A wear simulator study with bone and cement debris. *J. Mater. Sci. Mater. Med.* **2013**, *24*, 1319–1325. [CrossRef] [PubMed]
19. Volk, R. *Rauheitsmessung: Theorie und Praxis*; Beuth Verlag: Berlin, Germany, 2018.
20. Mueller, U.; Braun, S.; Schroeder, S.; Sonntag, R.; Kretzer, J.P. Same same but different? 12/14 stem and head tapers in total hip arthroplasty. *J. Arthroplast.* **2017**, *32*, 3191–3199. [CrossRef] [PubMed]
21. Bulhoff, M.; Reinders, J.; Zeifang, F.; Raiss, P.; Muller, U.; Kretzer, J.P. Surface and form alterations in retrieved shoulder hemiarthroplasty. *J. Shoulder Elb. Surg.* **2017**, *26*, 521–528. [CrossRef] [PubMed]

materials

MDPI

Article

The Effects of Cyclic Loading and Motion on the Implant–Cement Interface and Cement Mantle of PEEK and Cobalt–Chromium Femoral Total Knee Arthroplasty Implants: A Preliminary Study

Lennert de Ruiter [1], Raelene M. Cowie [2], Louise M. Jennings [2], Adam Briscoe [3], Dennis Janssen [1,*] and Nico Verdonschot [1,4]

1 Orthopaedic Research Laboratory, Radboud university medical center, Radboud Institute for Health Sciences, P.O. Box 9101, 6500 HB Nijmegen, The Netherlands; lennert.deruiter@radboudumc.nl (L.d.R.); Nico.Verdonschot@radboudumc.nl (N.V.)

2 Institute of Medical and Biological Engineering, School of Mechanical Engineering, University of Leeds, Leeds LS2 9JT, UK; r.cowie@leeds.ac.uk (R.M.C.); l.m.jennings@leeds.ac.uk (L.M.J.)

3 Invibio Ltd., Thornton Cleveleys FY5 4QD, UK; a.briscoe@invibio.com

4 Laboratory of Biomechanical Engineering, University of Twente, P.O. Box 217, 7500 AE Enschede, The Netherlands

* Correspondence: dennis.janssen@radboudumc.nl

Received: 21 June 2020; Accepted: 22 July 2020; Published: 26 July 2020

check for updates

Abstract: This study investigated the fixation of a cemented PEEK femoral TKA component. PEEK and CoCr implants were subjected to a walking gait cycle for 10 million cycles (MC), 100,000 cycles or 0 cycles (unloaded control). A method was developed to assess the fixation at the cement–implant interface, which exposed the implants to a fluorescent penetrant dye solution. The lateral condyles of the implants were then sectioned and viewed under fluorescence to investigate bonding at the cement–implant interface and cracking of the cement mantle. When tested for 100,000 cycles, debonding of the cement–implant interface occurred in both PEEK (61%) and CoCr (13%) implants. When the duration of testing was extended (10 MC), the percentage debonding was further increased for both materials to 88% and 61% for PEEK and CoCr, respectively. The unloaded PEEK specimens were 79% debonded, which suggests that, when PEEK femoral components are cemented, complete bonding may never occur. Analysis of cracks in the cement mantle showed an absence of full-thickness cracks in the unloaded control group. For the 100,000-cycle samples, on average, 1.3 and 0.7 cracks were observed for PEEK and CoCr specimens, respectively. After 10 MC, these increased to 24 for PEEK and 19 for CoCr. This was a preliminary study with a limited number of samples investigated, but shows that, after 10 MC under a walking gait, substantial debonding was visible for both PEEK and CoCr implants at the cement–implant interface and no significant difference in the number of cement cracks was found between the two materials.

Keywords: total knee arthroplasty; polyetheretherketone; fixation; debonding; implant–cement interface; PMMA

1. Introduction

In cemented total knee arthroplasty (TKA), fixation is achieved by mechanical interlock of the implant with the bone via a layer of polymethylmethacrylate (PMMA) bone cement. During surgery, the doughy cement is typically applied to bone and/or implant surfaces after which the implant is pushed into place. The result is an implant that is well-fixed to the underlying structure, which clinically has demonstrated good long-term survival [1–4]. To ensure adequate long-term fixation, the implant–cement interface can be strengthened by a number of options. Femoral components are usually designed with cement pockets for macro-interlock, and a surface texture is added to enhance the fixation strength through micro-interlock. Standard application of these features has amounted to decades of evidence of firm and reliable fixation [1–4]. Additional efforts have been made in the past to further enhance implant fixation by making the material adhere to the cement via chemical bonds rather than just shape-match at a macro- and microscale [5–9]. However, technologies such as PMMA or silane pre-coatings in cemented arthroplasty have not been largely adopted in conventional implant designs, at least in part due to the absence of evidence on clinical efficacy [10,11]. Hence, most femoral TKA implants on the market rely on mechanical interlock of the implant to the cement and consequent fixation of the cement to the bone [12–15].

For conventional femoral component materials (cobalt chrome, CoCr), a micro-texture is often applied to the fixation surface (Figure 1), and, with this modification, there have been few clinical reports of debonding at the cement–implant interface [16–21]. An increasing interest in the investigation of different materials and manufacturing techniques for joint replacements however brings about the potential for different failure modes of the implant. PEEK-OPTIMA™, for example, has been considered as an alternative to CoCr in the femoral component of a TKA to give a metal-free implant [22–28]. The lower modulus of a PEEK implant compared to CoCr may help to reduce stress shielding but may also change the distribution of forces at the cement–implant interface, which may influence implant fixation. There are potential advantages of investigating different implant materials. With PEEK, for example, the injection molding process used in manufacturing can apply macro- and micro-textures to the fixation surfaces in a one-stage manufacturing technique (Figure 1). A previous study into fixation strength of a PEEK implant with modifications of the fixation surface including the addition of macro- and micro-textures demonstrated an altered distribution of forces at the cement–implant interface compared to CoCr implants. Despite a decrease in fixation strength of PEEK femoral components, the failure modes of the different implant materials were similar, and it was concluded that the bond between implant and cement may be sufficiently strong for clinical use [26].

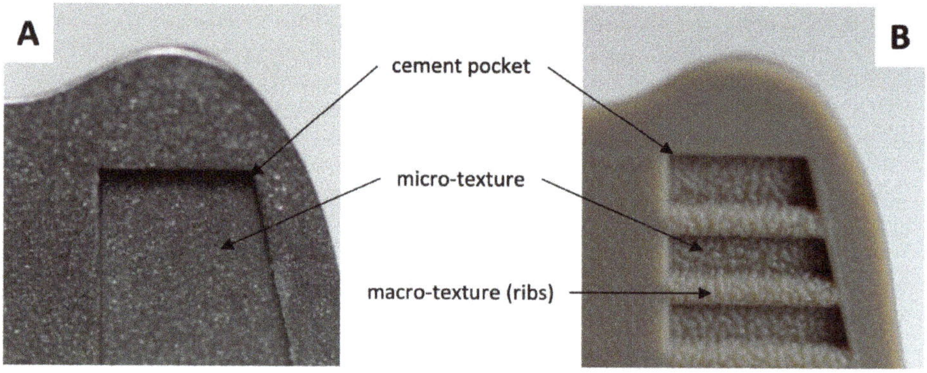

Figure 1. Surface texture of the CoCr (**A**) and PEEK (**B**) implants. The additional cement-bonding features of the PEEK implant are visible, comprising a macroscopic rib features within the cement pockets and a microscopic pattern superimposed over the surface. Adapted from de Ruiter et al. (2017) [26].

The fixation at the cement–implant interface is understudied. However, debonding of the femoral component may lead to gross implant loosening, abrasion at the cement–implant interface and failure of the cement mantle. Previous mechanical testing indicated that gross loosening of a PEEK femoral component is very unlikely if the fixation surface has been optimized to provide sufficient mechanical interlock [26]. However, micromotions of a loosened implant could cause abrasion of the cement–implant interface leading to the subsequent release of PEEK and/or cement particles, which may accelerate wear debris induced osteolysis and/or lead to failure of the cement mantle and could produce cement particles leading to third-body wear [29,30]. These potential complications underline the importance of understanding the implant–cement interface, particularly for a PEEK femoral TKA component.

The aim of this study was to determine the quality of the implant–cement interface of a PEEK femoral TKA component and compare it to a CoCr implant. Implants were subjected to clinically relevant loading and motions for up to 10 million cycles (MC) in a knee simulator and a method was developed to assess the bonding between the implant and cement and the integrity of the cement mantle. It was hypothesized that, due to the difference in thermal conductivity and modulus of the implant materials, the bonding at the cement–implant interface and the cracking of the cement would differ between implant materials and that PEEK would show more debonding and cracks in the cement mantle than CoCr. This was a preliminary study to establish a method to evaluate the implant–cement interface and as such was carried out with a small sample size.

2. Materials and Methods

2.1. Materials

Mid-size (size C) injection molded PEEK-OPTIMA™ femoral components (collaboration partners Maxx Orthopedics Inc., Plymouth Meeting, PA, USA and Invibio Knee Ltd., Thornton-Cleveleys, UK) and MAXX freedom knee (CoCr) femoral components (Maxx Orthopedics Inc., Plymouth Meeting, PA, USA) were used in this study. The implants had a similar geometry, although the macro-features and texture on the fixation surface differed (Figure 1) and was optimized for each femoral component material. The samples were cemented to custom made polyoxymethylene (Delrin®) fixtures using Palacos R&G cement (Heraeus, Hanau, Germany). Delrin was chosen as the substrate due to having mechanical properties suitable for 10 MC wear simulation and a low porosity as the testing was carried out in various liquids. While the machined surface of the Delrin fixture may not be clinically relevant, the fixtures were consistent for all samples and the fixture–implant interface was not of interest in this study. The geometry of the Delrin substrate was designed using CAD and based on the geometry of the implant, allowing for a 1-mm cement mantle. The subsequent CAD model was CNC-machined using a five-axis machine. The cement used was the same as that used clinically. It was mixed manually, and the doughy cement was applied to the implant in excess. The implant was pressurized onto the Delrin fixture and shims were used to create a cement mantle with a consistent 1-mm thickness. The process was carried out at room temperature with the same technique applied irrespective of the implant material.

When tested under physiological loading and motion, the femoral components were coupled with Size C all-polyethylene tibial components (Maxx Orthopedics Inc., Mahwah, NY, USA). To assess the debonding of the femoral component–cement interface, the implants were immersed and/or tested in a fluorescent penetrant dye (WB200, Sherwin Babbco) in saline solution at 1:10 concentration.

2.2. Experimental Design

Three experimental groups with three samples in each were defined, as shown in Table 1. Group 1 components were soaked in the penetrant dye for 27.8 h (equivalent to the duration of 100,000 gait cycles carried out at 1 Hz). This was carried out for PEEK implants only, to assess the initial bonding between the cement and implant ("unloaded control"). Only PEEK implants were investigated in this experimental group because a previous in-house cadaveric experiment demonstrated poor initial fixation for PEEK specimens and good initial fixation between the implant and cement for CoCr components. It was assumed that loading of the CoCr device was required to disrupt the fixation at the cement–implant interface and to encourage the dye into the interface. Group 2 ("Gait control") comprised both PEEK and CoCr femoral components, which, following cementing onto fixtures, underwent physiological gait loading and motion (Figure 2) in the penetrant dye for 100,000 cycles at 1 Hz (27.8 h) using a six-station ProSim electropneumatic knee simulator (Simulation Solutions, Stockport, UK). The simulator has six degrees of freedom, and four axes of motion were controlled during the test: Axial Force (AF), Flexion Extension (FE), Tibial Rotation (TR) and Anterior Posterior displacement (AP). Performing 100,000 cycles facilitated dye uptake in debonded or cracked areas. To determine the number of gait cycles required for the dye to enter the cement–implant interface, an experiment was carried out in which PEEK implants were cemented to a bone-analog foam and loaded uniaxially (260–2600 N at a frequency of 1 Hz) in the fluorescent dye for 100,000 cycles. At the conclusion of this preliminary study (Figure 3), for PEEK implants, a fluorescent line could be seen between the cement and implant where the dye had entered the interface. It was assumed that dye uptake would be higher when the sample was loaded in a simulator and subjected to simultaneous loading and motion rather than the uniaxial loading used during method development. It was not feasible to use a bone-analog foam as the substrate for the 10 MC simulation, thus, to maintain consistency between samples, Delrin was used as the substrate throughout. Group 3 comprised PEEK and CoCr femoral components that had been previously tested for 10 MC in a ProSim knee simulator under physiological loading and motion to represent the kinetics and kinematics at the bearing surface of the tibiofemoral joint during a walking gait cycle. High flexion activities and forces at the patellofemoral joint were not considered in this study. The experimental wear simulation study was carried out under "Leeds high kinematic" conditions (Figure 2) in a lubricant of 25% bovine serum supplemented with 0.03% sodium azide solution against all-polyethylene tibial components [31]. These test conditions were similar to those previously described by Cowie et al. (2016) [23]. Following wear simulation, the samples were cleaned using detergent ensuring the PEEK and CoCr implants were treated the same in side-by-side studies. This test group ("10 MC gait") subsequently underwent a further 100,000 cycles under the same loading and motion while immersed in the penetrant dye. All groups were thus exposed to the dye solution for the same duration.

Table 1. The experimental groups and sample size for each femoral component material.

	Group 1	Group 2	Group 3
	Unloaded control	Gait control	10 MC gait
PEEK	3	3	3
CoCr	-	3	3

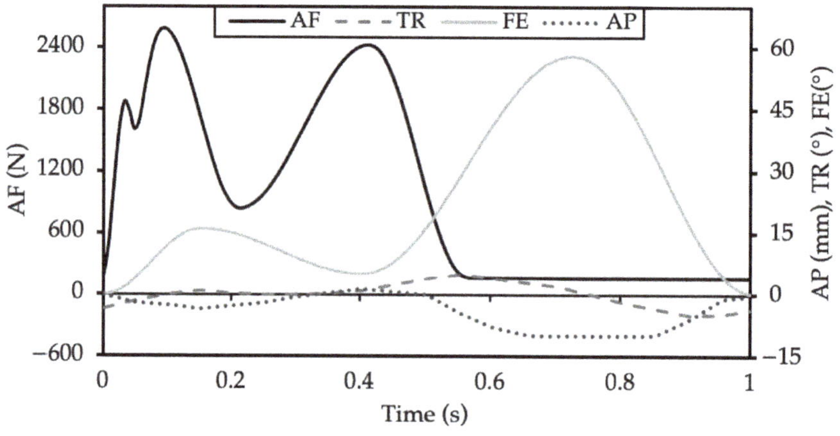

Figure 2. Input profiles for cyclic loading on the knee simulator. The parameters are axial force (AF), tibial rotation (TR), flexion-extension angle (FE) and anterior-posterior displacement (AP).

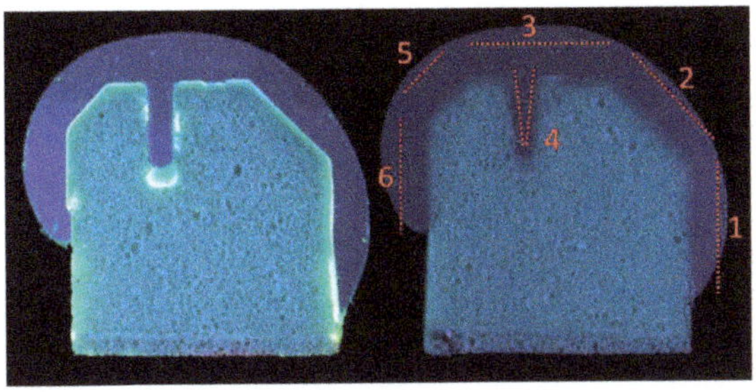

Figure 3. Proof-of-concept lateral condyle PEEK specimens cemented on bone-analog foam after 100,000 cycles in fluorescent dye: (Left) the implant under UV lighting showing complete interface dye fluorescence (score 3); and (Right) the same sample without UV lighting and how the interface was divided into 6 regions for analysis (**1**, anterior flange; **2**, anterior chamfer; **3**, distal area; **4**, peg; **5**, posterior chamfer; **6**, posterior flange).

2.3. Analysis of Dye Penetration (Fluorescence)

Having been immersed in the dye, the specimens were sectioned in the sagittal plane through the center of the lateral condyle with a cutting blade under water cooling. The lateral condyle was chosen over the medial condyle because, when cross-sectioned, all internal implant faces are visible (Figure 3); for the medial condyle, the geometry of the implant means that when cross-sectioned through the center of the condyle, the anterior chamfer cannot be seen. A UV light was used to excite the fluorescent dye (320–420 μm) and imaging performed using a generic microscope at 1.5 × 10 magnification. A scoring system was devised to assess whether fluorescent dye was visible at the implant–cement interface. The femoral components were divided into six distinct regions for analysis (Figure 3): the anterior flange, the anterior chamfer, the distal area, the peg, the posterior chamfer and the posterior flange. No differentiation was made for the intensity of the UV-light since debonding was assumed to be complete in regions where fluorescence was observed regardless of the light intensity. Scoring was carried out manually by two scorers and each area was ranked between 0 and 3 (0: no fluorescence; 1: up to 33% of the interface fluorescent; 2: up to 67% of the interface fluorescent; and 3: complete fluorescence). There was a high level of inter-observer reliability resulting in few discrepancies between the two scorers. When differences were identified, researchers deliberated and agreed on a final score. Separate area scores were summed for the entire interface and averaged to obtain a single score for each specimen.

2.4. Analysis of Cement Damage (Full-Thickness Cracks)

Cement damage was scored for each of the six regions on the femoral component described in Figure 3 by assessing the number and location of cracks that crossed the full thickness of the cement mantle. Again, two scorers independently examined the implants. Few differences between scorers were identified and were reviewed and debated until a consensus was reached. The dataset was checked and corrected for double hits in overlapping images. The number of cracks in each region was averaged for all specimens in the study group. Average scores were compared between PEEK and CoCr, for both the entire cement mantle and each separate region.

2.5. Statistics

The data are presented as the mean (±standard deviation) for both fluorescence and full-thickness cracks. Statistical analysis was carried out in SPSS 24.0 (IBM Corp, Armonk, NY, USA) using a *t*-test to compare PEEK and CoCr for each experimental group, under the hypothesis that PEEK would show more fluorescence and cracks than CoCr. Groups were analyzed with a 0.05 significance level.

3. Results

3.1. Analysis of Dye Penetration (Fluorescence)

Unloaded PEEK control specimens (Group 1) showed high levels of dye penetration at the implant–cement interface without the components undergoing loading and motion (Figure 4). On average, 79% (±11%) of the PEEK–cement interface was fluorescent after being soaked in dye (unloaded) for 28 h (Figure 5). After 100,000 gait cycles, for the PEEK gait controls (Group 2), the average fluorescence area was 61% (±23%). This was lower than the unloaded controls (Group 1), but with a larger variability between samples. The CoCr Group 2 gait control samples showed limited dye penetration at the interface (13% (±6%)) after 100,000 gait cycles (Figure 4). After an extended number (10 MC) of test cycles under physiological loading and motion (Group 3), the implant–cement interfaces were easily distinguishable for both femoral component materials (Figure 4). For Groups 2 and 3, the PEEK components showed significantly ($p < 0.05$) more fluorescence than the CoCr implants. Comparing the gait controls (Group 2) to the implants loaded for an extended number of cycles (Group 3), the PEEK femoral components showed a slight increase in percentage fluorescence, from 61% (±23%) to 88% (±5%), while the CoCr implants displayed a steep increase after 10 MC of simulation, from 13% (±6%) to 62% (±6%) interface fluorescence (Figure 5). The variability between the samples for the Group 3 implants was lower than the other groups for both material types.

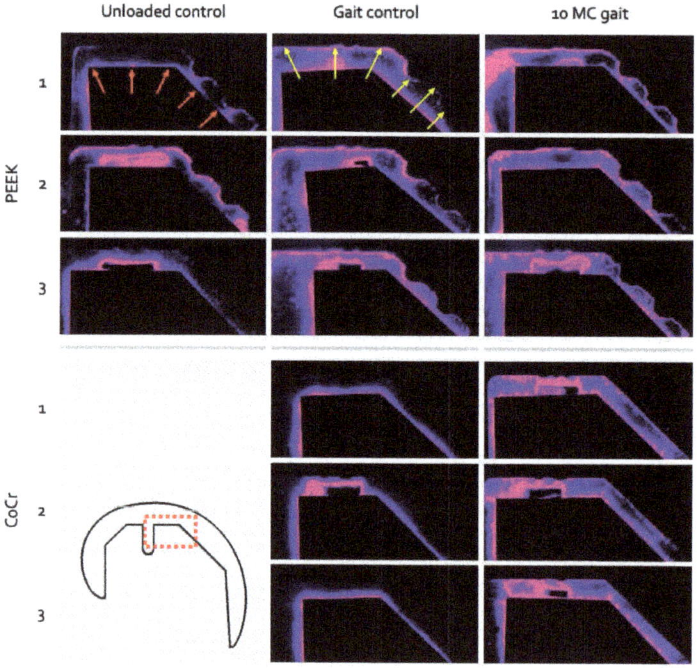

Figure 4. Test specimens under UV lighting show dye fluorescence intensity as blue-to-pink coloration with the pink area highlighting the highest dye uptake. Both cement–Delrin (red arrows) and implant–cement (yellow arrows) interfaces are visible. A section of the anterodistal cement mantle is shown at the three time intervals to demonstrate the appearance of dye penetration in both the PEEK and CoCr group. The variability within the PEEK soaked control group is noticeable, ranging from near-full bonding (Specimen 3) to complete debonding (Specimen 2). A clear evolution of dye penetrance in the CoCr can be seen, where no implant–cement interface is visible in the loaded controls, but full interface fluorescent is visible in the Group 3 samples.

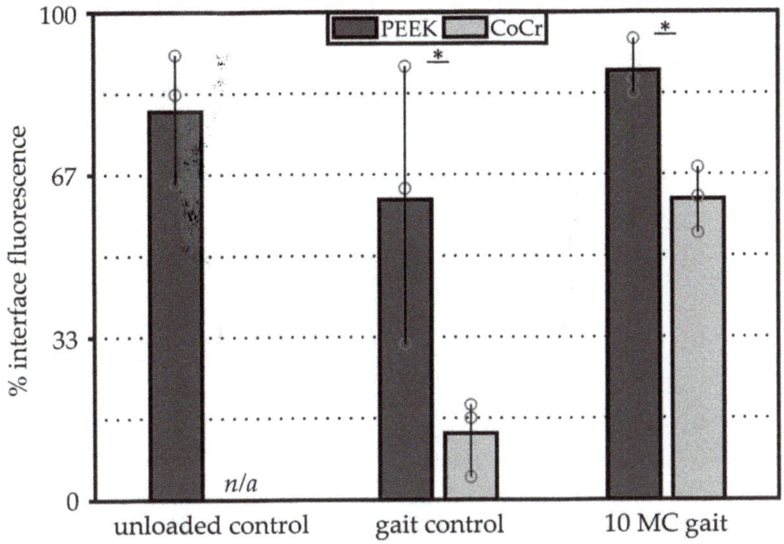

Figure 5. Average interface fluorescence scores of the complete implant–cement interfaces at difference intervals. The bars show the range of observations with the individual specimens shown as circles (statistical analysis compares PEEK to CoCr; * denotes $p < 0.05$).

3.2. *Analysis of Cement Damage*

Full-thickness cracks were observed in the cement mantles against both the PEEK and CoCr femoral components for the gait controls and 10 MC test groups. The locations of these cracks however were markedly different with cracks at the interface chamfers more often observed with a CoCr component than with a PEEK implant. With CoCr, the cracks tended to run the full thickness of the cement mantle, as opposed to with PEEK where the cracks at the chamfers were mostly incomplete (Figure 6A). In the PEEK femoral components, at the apex of the ridges which were incorporated into the cement pockets to enhance fixation, full-thickness cracks were common in the Group 3 femoral components which had been previously tested for 10 MC (Figure 6B). Both CoCr and PEEK reconstructions showed similar crack patterns in the anterior and posterior flange areas. In this region, the cracks underneath the PEEK components generally resulted in full-thickness cracks, while those underneath the CoCr implants showed both full-thickness cracks and numerous small cracks (Figure 6C).

No full-thickness cracks were observed in the unloaded control PEEK femoral components (Figure 7). In the gait control femoral components, some full-thickness cracks appeared: 1.3 (±1.9) and 0.7 (±0.9) cracks on average for PEEK and CoCr, respectively. This difference however was not significant ($p > 0.05$). After 10 MC under gait conditions, the number of cracks in both reconstructions had substantially increased. The average number of cracks in the cement layer below the PEEK femoral components was 24 (±4.5), while the CoCr reconstructions demonstrated a mean of 19 (±3.7) full-thickness cracks. Again, this difference was not significant ($p > 0.05$).

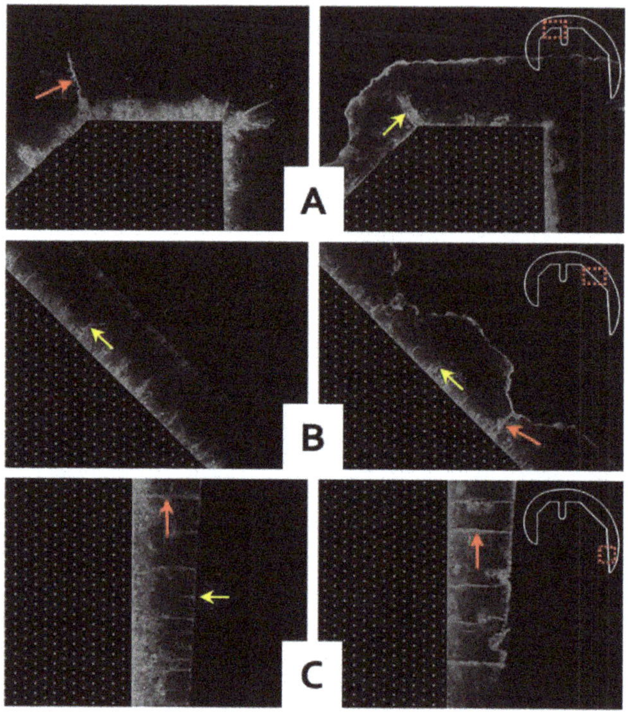

Figure 6. Representative images of CoCr (left) and PEEK (right) reconstructions after 10 MC, showing dye penetration in cement mantle cracks at three locations (**A**–**C**) around the femoral component as indicated by the red squares in the detail-figures. The red arrows indicate full-thickness cracks in the PMMA cement, the yellow arrows incomplete cracks. The reduced cement mantle thickness, caused by the ridges in the PEEK surface, is visible in (**B**) (right).

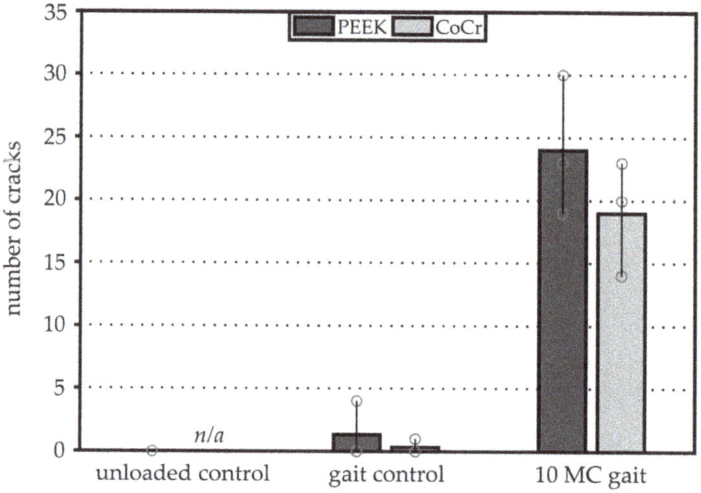

Figure 7. The average number of full-thickness cracks for the three experimental groups. The bars show the range of observations with the individual specimens shown as circles.

Further examination of the crack locations after 10 MC revealed that most cracks occurred in the cement mantle below the anterior flange of the implant (Figure 8). For both femoral component materials, the average number of cracks was 12.7 in this region. For the CoCr components, few cracks were visible around the anterior chamfer, while the PEEK implants showed cracks in the cement in this area at the apex of the fixation ridges, as shown in Figure 6B. The posterior flange cement area also showed more cracking with the PEEK implant than with CoCr. After 10 MC, the femoral component material did not influence the number of cracks in the distal area, peg or posterior chamfer regions ($p > 0.05$).

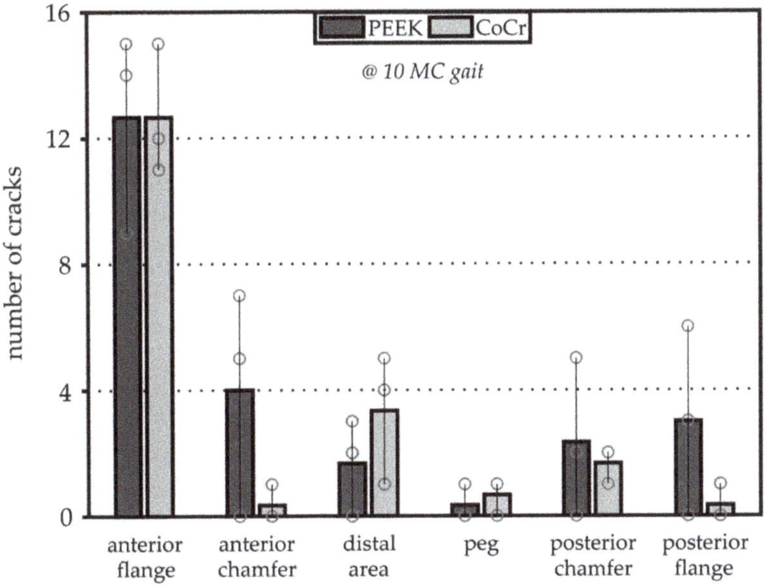

Figure 8. The average number of full-thickness cracks separated for each region after 10 MC. The bars show the range of observations with the individual specimens shown as circles.

4. Discussion

The aim of this study was to assess and compare the femoral component cement–implant interface for TKAs manufactured from PEEK and CoCr. The two femoral component materials were shown to have different effects on the cement mantle due to differences in the material properties of the components and variations in both the geometry and topography of the fixation surface. In summary, the CoCr implants showed less dye penetration at the cement–implant interface than PEEK components, indicating superior adhesion between CoCr and cement. However, debonding of the cement–implant interface was evident for all implants when tested for high numbers of cycles. The integrity of the cement mantle was also analyzed. Cracks were evident in the cement beneath both PEEK and CoCr femoral components. There was no significant difference between the number of cracks in the cement–CoCr or cement–PEEK interface, but the location of the cracks differed depending on the implant material.

4.1. Analysis of Dye Penetration

The PEEK unloaded soaked control specimens showed that PMMA cement did not fully bond to the PEEK implant. Dye penetration did not differ between unloaded (Group 1) PEEK femoral components and those tested for 10 MC, (Group 3) specimens. Therefore, 10 MC experimental simulation had no additional effect on the bonding state of the PEEK implant–cement interface and in no samples was gross implant loosening observed. A previous pull-off fixation study by De Ruiter et al. (2017) concluded that PEEK femoral components with the same profile on the fixation surface as used in this study have adequate fixation strength [26]. The information from these two studies combined suggests that debonding of the PEEK–cement interface does not necessarily limit the long-term mechanical fixation of the construct as a whole, although, in both studies, non-physiological bone surrogates were used. This is further emphasized by the fact that substantial interface debonding was seen with CoCr implants following testing for 10 MC. CoCr femoral components are known to be mechanically stable when implanted for periods in excess of 10 years. Clinical evidence from successful polyethylene (PE) implants show similar debonding scenarios. All-polymer PE tibial TKA components, for example, have been available for several decades and have very positive outcomes, despite PE having no intrinsic bond with the cement [32–34]. Similarly, cemented all-polymer PE acetabular cups for hip arthroplasty, which are dependent on surface textures for fixation, also show excellent survival rates [1,3,35], and a clinical trial of a Delrin femoral component showed a low incidence of loosening after 10 years implantation [36].

The cause of the immediate PEEK–cement interface debonding can be attributed to the lack of adhesion between the cement and implant, which means fixation primarily relies on mechanical macro- and micro-interlock with the surface topographical features on the implant fixation surface. The poorer thermal conductivity of PEEK compared to CoCr may reduce the dissipation of heat produced as the PMMA cement cures which may lead to shrinkage of the cement contributing to debonding of the PEEK–cement interface. In addition, in clinical practice, the bond may be further influenced by contamination of the interface by blood or fat, as well as poor cementing technique or timing. From that perspective, the clean surfaces and absence of time pressure under which the samples were prepared for this study represent the optimal conditions for obtaining a well-fixed implant–cement interface.

4.2. Analysis of Cement Cracks

Macroscopic damage evaluation showed that full thickness cracks were present in both CoCr and PEEK reconstructions tested under a gait cycle. Following 10 MC gait simulation, the mean number of full-thickness cracks in the cement mantle was approximately 20% higher for PEEK implants compared to CoCr; however, this difference was not significant ($p > 0.05$). With the addition of potential stress risers (ridges in the fixation surface), which lead to a thinner cement mantle at the ridge-locations, the difference between the PEEK and CoCr components was expected to be larger. However, perhaps with the CoCr component, a different failure mechanism occurred. It is postulated that the higher stiffness of CoCr compared to PEEK also gives the potential for higher local stresses in the cement–CoCr implant interface as the femoral component is less compliant. Previous studies have shown the stiff CoCr component to generate high stress peaks in the cement underneath the proximal anterior flange [24–26,37]. Many small cracks were visible in this region. Analysis of these small cracks was beyond the scope of this study but there is potential for these small cracks to grow, which may further increase cement damage in the anterior flange [20,24,25]. The numbers of cracks in cement at the anterior flange and posterior chamfer areas of the cement mantle were similar in both implants; however, there was a greater number of cracks in the PEEK–cement interface in the anterior chamfer and the posterior flange compared to the CoCr–cement interface.

4.3. Limitations

There are a number of limitations associated with this study which should be considered in the clinical interpretation of these findings. Firstly, the study was performed on an experimental wear simulator. As such, it is designed to mimic the kinetics and kinematics at the implant bearing surface as opposed to the femoral/fixture (bone) interface. The femoral components were cemented onto custom-shaped Delrin blocks that were mounted into the simulator. This replaced the in-vivo cement–bone interface with a cement–Delrin interface. It is acknowledged that Delrin may not represent bone in terms of its porosity, elastic modulus and surface texture. However, for this study, which involved extensive wear simulation (>6 months) in a biological lubricant applying forces up to 2.8 kN, it was considered appropriate and was easy to section to allow analysis of the cement–implant interface. The intention was not to study the Delrin–cement interface and, therefore, the surface was consistent (as machined) for all experimental groups, and, other than pockets to accommodate the pegs, no additional features were introduced into the surface of the Delrin, which provided a shape-lock fixation. The different mechanical properties of Delrin compared to bone may however have changed the load distribution in the cement mantle and how this influenced interface debonding and cement mantle damage is unknown. The 10 MC of gait simulation, equivalent to the loading the implant undergoes following approximately 10 years use in a moderately active patient [38], assumes that, throughout the duration of implantation, the implant remains fully supported with no resorption of the underlying bone. It is not known whether the relatively sharp corners at the chamfers of the Delrin block may have stimulated crack propagation or whether the smooth Delrin surface may have reduced stress risers created by individual trabeculae, which may play an important role in crack initiation [39]. Hence, the exact effect of the use of Delrin is unclear and future studies could consider a more physiologically relevant substrate. In terms of the cementing of the implants, there were further limitations as it was prepared using a manual mixing technique, which is inferior to the vacuum mixing process routinely used in the clinic. Manual mixing of cement gives rise to the potential for entrapment of air during cement preparation, which may cause pores. During analysis of the cement mantles, however, no pores were found in the cement layer, which may be attributed to the well-controlled laboratory conditions in which the reconstructions were prepared.

In addition, the simulator included only tibiofemoral contact. When both the tibiofemoral and patellofemoral joints are considered, in vivo forces are higher, since patellofemoral contact is a major contributor to the total loads on the femoral component particularly at higher flexion angles [24]. However, the increased forces due to patellofemoral contact do not necessarily lead to an increased risk of debonding. A study by Berahmani et al. (2016) into micromotions behind cementless femoral knee components concluded that the patellofemoral contact decreased the micromotions in the anterior flange up to 22% [40]. The study also investigated a gait cycle only. Although high-flexion activities are carried out less frequently than level gait, more strenuous activities including stair climbing, standing up and squatting may influence the study findings; however, the comparative nature of this study comparing the cement–implant interface of PEEK to that of CoCr is a strength.

Furthermore, only three samples were investigated in each experimental group; this was limited by the extended duration of the studies within excess of six months continual testing required to prepare the 10-MC gait samples. Future work should consider larger implant sizes and perhaps extending the number of timepoints investigated especially for the CoCr implants to gain a better understanding of when debonding of these implants occurs and including a CoCr Group 1 investigation to better understand the initial fixation of CoCr implants. However, increasing the sample size may necessitate automation of the analysis protocols to minimize variability between samples. The protocol used merely considers the loading the implants undergo but not whether degeneration of the cement occurs during ageing. Further method development would be required to understand this and whether the mechanical properties of cement change with time. To minimize errors between experimental groups associated with ageing effects, the PEEK and CoCr implants were tested in parallel.

Finally, the study outcome parameters, dye penetration and cement cracking, are a local 2D representation of the full cement mantle and are thus subject to extrapolation error. The lateral condyle was chosen to represent all areas from anterior to posterior flange which does not cover the entire surface area. However, cracks in this section did propagate further laterally and medially into the cement mantle, and dye visualized at this location must have travelled from external boundaries, which supports the extrapolation applied here. This research did not include a microscopic damage assessment of the implant or cement interface surfaces after 10 MC of experimental simulation. One of the hypothesized results of long-term interface micromotions is the formation of wear particles. Once formed, these could travel into the joint space where they may initiate inflammatory processes which could contribute to wear debris induced osteolysis leading to implant loosening or, if larger particles were to migrate between the articulating surfaces, they could act as a third-body particle and may accelerate bearing surface damage or wear [29,41,42].

5. Conclusions

This study aimed to develop a method to assess the cement–implant interface bonding and the integrity of the cement mantle. This was done using a fluorescent penetrant dye and was then used to assess CoCr and PEEK femoral components which had undergone up to 10 million walking gait cycles. The study showed poor initial bonding of the PEEK–cement interface; however, after 10 MC simulation, the bonding of the implant remained similar to that of the controls. For CoCr implants, good fixation was measured for the gait control samples, but, after 10 MC, substantial implant–cement interface debonding occurred. After 10 MC, there was no significant difference in implant–cement debonding for the femoral component materials investigated, nor were there significant differences in macroscopic damage of the cement mantle. Further investigations either using a more physiologically relevant simulation system or through either animal studies or a clinical trial may be necessary to confirm these findings.

Author Contributions: Conceptualization, all authors; methodology, L.d.R., R.M.C., and D.J.; formal analysis, L.d.R.; investigation, L.d.R. and R.M.C.; resources, N.V., L.M.J., and A.B.; data curation, L.d.R.; writing—original draft preparation, L.d.R. and R.M.C.; writing—review and editing, all authors; visualization, L.d.R.; supervision, N.V. and L.M.J.; and funding acquisition, N.V. All authors have read and agreed to the published version of the manuscript.

Funding: This research was funded by Invibio Ltd., Lancashire, UK.

Acknowledgments: The authors would like to thank Leon Driessen and Richard van Swam for their support in preparation and scoring of the specimens.

Conflicts of Interest: A.B. is a paid employee of Invibio Ltd. and had a role in the study design and review of the manuscript. L.M.J., N.V., and D.J. have received research funding support from Invibio Ltd. N.V. is a consultant to Invibio Ltd. The other authors declare no conflict of interest. The funders had no role in the collection, analyses, or interpretation of data.

References

1. Australian Orthopaedic Association National Joint Replacement Registry (AOANJRR). *Hip, Knee & Shoulder Arthroplasty: 2017 Annual Report*; AOA: Adelaide, Australia, 2017.
2. Gandhi, R.; Tsvetkov, D.; Davey, J.R.; Mahomed, N.N. Survival and clinical function of cemented and uncemented prostheses in total knee replacement: A meta-analysis. *J. Bone Jt. Surg. Br.* **2009**, *91*, 889–895. [CrossRef] [PubMed]
3. National Joint Registry for England, Wales, Northern Ireland and the Isle of Man. *2017 Annual Report*; NJR: Hemel Hempstead, UK, 2017.
4. Hopley, C.D.J.; Dalury, D.F. A systematic review of clinical outcomes and survivorship after total knee arthroplasty with a contemporary modular knee system. *J. Arthroplast.* **2014**, *29*, 1398–1411. [CrossRef] [PubMed]
5. Erli, H.J.; Marx, R.; Paar, O.; Niethard, F.U.; Weber, M.; Wirtz, D.C. Surface pretreatments for medical application of adhesion. *Biomed. Eng. Online* **2003**, *2*, 15. [CrossRef] [PubMed]
6. Marx, R.; Qunaibi, M.; Wirtz, D.C.; Niethard, F.U.; Mumme, T. Surface pretreatment for prolonged survival of cemented tibial prosthesis components: Full- vs. surface-cementation technique. *Biomed. Eng. Online* **2005**, *4*, 61. [CrossRef]
7. Yerby, S.A.; Paal, A.F.; Young, P.M.; Beaupre, G.S.; Ohashi, K.L.; Goodman, S.B. The effect of a silane coupling agent on the bond strength of bone cement and cobalt—Chrome alloy. *J. Biomed. Mater. Res.* **2000**, *49*, 127–133. [CrossRef]
8. Ohashi, K.L.; Yerby, S.A.; Dauskardt, R.H. Effects of an adhesion promoter on the debond resistance of a metal–polymethylmethacrylate interface. *J. Biomed. Mater. Res.* **2001**, *54*, 419–427. [CrossRef]

9. Mann, K.; Bhashyam, S. Mixed-mode fracture toughness of the cobalt-chromium alloy/polymethlymethacrylate cement interface. *J. Orthop. Res.* **1999**, *17*, 321–328. [CrossRef]
10. Ong, A.; Wong, K.L.; Lai, M.; Garino, J.; Steinberg, M.E. Early failure of precoated femoral components in primary total hip arthroplasty. *J. Bone Jt. Surg. Am.* **2002**, *84*, 786–792. [CrossRef]
11. Bini, S.A.; Chen, Y.; Khatod, M.; Paxton, E.W. Does pre-coating total knee tibial implants affect the risk of aseptic revision? *Bone Jt. J.* **2013**, *95*, 367–370. [CrossRef]
12. Ohashi, K.; Romero, A.C.; McGowan, P.D.; Maloney, W.; Dauskardt, R. Adhesion and reliability of interfaces in cemented total hip arthroplasties. *J. Orthop. Res.* **1998**, *16*, 705–714. [CrossRef]
13. Goodheart, J.R.; Miller, M.A.; Mann, K.A. In Vivo Loss of Cement-Bone Interlock Reduces Fixation Strength in Total Knee Arthroplasties. *J. Orthop. Res.* **2014**, *32*, 1052–1060. [CrossRef] [PubMed]
14. Srinivasan, P.; Miller, M.A.; Verdonschot, N.; Mann, K.A.; Janssen, D. Experimental and computational micromechanics at the tibial cement-trabeculae interface. *J. Biomech.* **2016**, *49*, 1641–1648. [CrossRef] [PubMed]
15. Van de Groes, S.; de Waal-Malefijt, M.; Verdonschot, N. Probability of mechanical loosening of the femoral component in high flexion total knee arthroplasty can be reduced by rather simple surgical techniques. *Knee* **2014**, *21*, 209–215. [CrossRef] [PubMed]
16. Bollars, P.; Luyckx, J.; Innocenti, B.; Labey, L.; Victor, J.; Bellemans, J. Femoral component loosening in high-flexion total knee replacement. *J. Bone Jt. Surg. Br.* **2011**, *93*, 1355–1361. [CrossRef]
17. Bergschmidt, P.; Dammer, R.; Zietz, C.; Finze, S.; Mittelmeier, W.; Bader, R. Adhesive strength of total knee endoprostheses to bone cement—Analysis of metallic and ceramic femoral components under worst-case conditions. *Biomed. Tech.* **2016**, *61*, 281–289. [CrossRef]
18. Zelle, J.; Janssen, D.; Peeters, S.; Brouwer, C.; Verdonschot, N. Mixed-mode failure strength of implant-cement interface specimens with varying surface roughness. *J. Biomech.* **2011**, *44*, 780–783. [CrossRef]
19. Arnholt, C.M.; MacDonald, D.W.; Malkani, A.L.; Klein, G.R.; Rimnac, C.M.; Kurtz, S.M.; Kocagoz, S.B.; Gilbert, J.L. Corrosion Damage and Wear Mechanisms in Long-Term Retrieved CoCr Femoral Components for Total Knee Arthroplasty. *J. Arthroplast. Arthroplast.* **2016**, *31*, 2900–2906. [CrossRef]
20. Zelle, J.; van de Groes, S.; de Waal Malefijt, M.; Verdonschot, N. Femoral loosening of high-flexion total knee arthroplasty: The effect of posterior cruciate ligament retention and bone quality reduction. *Med. Eng. Phys.* **2014**, *36*, 318–324. [CrossRef]
21. Han, H.S.; Kang, S.-B.; Yoon, K.S. High incidence of loosening of the femoral component in legacy posterior stabilised-flex total knee replacement. *J. Bone Jt. Surg. Br.* **2007**, *89*, 1457–1461. [CrossRef]
22. Rankin, K.E.; Dickinson, A.S.; Briscoe, A.; Browne, M. Does a PEEK Femoral TKA Implant Preserve Intact Femoral Surface Strains Compared With CoCr? A Preliminary Laboratory Study. *Clin. Orthop. Relat. Res.* **2016**, *474*, 2405–2413. [CrossRef]
23. Cowie, R.M.; Briscoe, A.; Fisher, J.; Jennings, L.M. PEEK-OPTIMA^{TM} as an alternative to cobalt chrome in the femoral component of total knee replacement: A preliminary study. *Proc. Inst. Mech. Eng. Part H J. Eng. Med.* **2016**, *230*, 1008–1015. [CrossRef] [PubMed]
24. De Ruiter, L.; Janssen, D.; Briscoe, A.; Verdonschot, N. The mechanical response of a polyetheretherketone femoral knee implant under a deep squatting loading condition. *Proc. Inst. Mech. Eng. Part H J. Eng. Med.* **2017**, *231*, 1204–1212. [CrossRef] [PubMed]
25. De Ruiter, L.; Janssen, D.; Briscoe, A.; Verdonschot, N. A preclinical numerical assessment of a polyetheretherketone femoral component in total knee arthroplasty during gait. *J. Exp. Orthop.* **2017**, *4*, 3. [CrossRef] [PubMed]
26. De Ruiter, L.; Janssen, D.; Briscoe, A.; Verdonschot, N. Fixation strength of a polyetheretherketone femoral component in total knee arthroplasty. *Med. Eng. Phys.* **2017**, *49*, 157–162. [CrossRef]
27. Du, Z.; Zhu, Z.; Yue, B.; Li, Z.; Wang, Y. Feasibility and Safety of a Cemented PEEK-on-PE Knee Replacement in a Goat Model: A Preliminary Study. *Artif. Organs* **2018**, *42*, E204–E214. [CrossRef]
28. Meng, X.; Du, Z.; Wang, Y. Characteristics of wear particles and wear behavior of retrieved PEEK-on-HXLPE total knee implants: A preliminary study. *RSC Adv.* **2018**, *8*, 30330–30339. [CrossRef]
29. Gallo, J.; Goodman, S.B.; Konttinen, Y.T.; Wimmer, M.A.; Holinka, M. Osteolysis around total knee arthroplasty: A review of pathogenetic mechanisms. *Acta Biomater.* **2013**, *9*, 8046–8058. [CrossRef]
30. Cowie, R.M.; Pallem, N.M.; Briscoe, A.; Jennings, L.M. Third Body Wear of UHMWPE-on-PEEK-OPTIMA^{TM}. *Materials* **2020**, *13*, 1264. [CrossRef]

31. McEwen, H.M.J.; Barnett, P.I.; Bell, C.J.; Farrar, R.; Auger, D.D.; Stone, M.H.; Fisher, J. The influence of design, materials and kinematics on the in vitro wear of total knee replacements. *J. Biomech.* **2005**, *38*, 357–365. [CrossRef]
32. Ranawat, A.S.; Mohanty, S.S.; Goldsmith, S.E.; Rasquinha, V.J.; Rodriguez, J.A.; Ranawat, C.S. Experience with an all-polyethylene total knee arthroplasty in younger, active patients with follow-up from 2 to 11 years. *J. Arthroplast.* **2005**, *20*, 7–11. [CrossRef]
33. Gioe, T.J.; Stroemer, E.S.; Santos, E.R.G. All-polyethylene and metal-backed tibias have similar outcomes at 10 years: A randomized level II evidence study. *Clin. Orthop. Relat. Res.* **2007**, *455*, 212–218. [CrossRef] [PubMed]
34. Gioe, T.J.; Bowman, K.R. A randomized comparison of all-polyethylene and metal-backed tibial components. *Clin. Orthop. Relat. Res.* **2000**, *380*, 108–115. [CrossRef] [PubMed]
35. Haft, G.F.; Heiner, A.D.; Dorr, L.D.; Brown, T.D.; Callaghan, J.J. A biomechanical analysis of polyethylene liner cementation into a fixed metal acetabular shell. *J. Bone Jt. Surg. Am.* **2003**, *85*, 1100–1110. [CrossRef] [PubMed]
36. Moore, D.; Freeman, M.; Revell, P.; Bradley, G.; Tuke, M. Can a total knee replacement prosthesis be made entirely of polymers? *J. Arthroplast.* **1998**, *13*, 388–395. [CrossRef]
37. Zelle, J.; Janssen, D.; Van Eijden, J.; de Waal Malefijt, M.; Verdonschot, N. Does high-flexion total knee arthroplasty promote early loosening of the femoral component? *J. Orthop. Res.* **2011**, *29*, 976–983. [CrossRef]
38. Schmalzried, T.P.; Szuszczewicz, E.S.; Northfield, M.R.; Akizuki, K.H.; Frankel, R.E.; Belcher, G.; Amstutz, H.C. Quantitative assessment of walking activity after total hip or knee replacement. *J. Bone Jt. Surg. Am.* **1998**, *80*, 54–59. [CrossRef]
39. Race, A.; Miller, M.A.; Ayers, D.C.; Mann, K.A. Early cement damage around a femoral stem is concentrated at the cement/bone interface. *J. Biomech.* **2003**, *36*, 489–496. [CrossRef]
40. Berahmani, S.; Janssen, D.; Wolfson, D.; De Waal Malefijt, M.; Fitzpatrick, C.K.; Rullkoetter, P.J.; Verdonschot, N. FE analysis of the effects of simplifications in experimental testing on micromotions of uncemented femoral knee implants. *J. Orthop. Res.* **2016**, *34*, 812–819. [CrossRef]
41. Lorber, V.; Paulus, A.C.; Buschmann, A.; Schmitt, B.; Grupp, T.M.; Jansson, V.; Utzschneider, S. Elevated cytokine expression of different PEEK wear particles compared to UHMWPE in vivo. *J. Mater. Sci. Mater. Med.* **2014**, *25*, 141–149. [CrossRef]
42. Paulus, A.C.; Haßelt, S.; Jansson, V.; Giurea, A.; Neuhaus, H.; Grupp, T.M.; Utzschneider, S. Histopathological Analysis of PEEK Wear Particle Effects on the Synovial Tissue of Patients. *Biomed. Res. Int.* **2016**, *2*, 2198914. [CrossRef]

materials

MDPI

Article

Musculoskeletal Multibody Simulation Analysis on the Impact of Patellar Component Design and Positioning on Joint Dynamics after Unconstrained Total Knee Arthroplasty

Maeruan Kebbach [1,*], Martin Darowski [1], Sven Krueger [2], Christoph Schilling [2], Thomas M. Grupp [2,3], Rainer Bader [1] and Andreas Geier [1,4]

1. Biomechanics and Implant Technology Research Laboratory, Department of Orthopaedics, Rostock University Medical Center, 18057 Rostock, Germany; MartinMarkus.Darowski@med.uni-rostock.de (M.D.); rainer.bader@med.uni-rostock.de (R.B.); andreas.geier@med.uni-rostock.de (A.G.)
2. Aesculap AG Research and Development, 78532 Tuttlingen, Germany; sven.krueger@aesculap.de (S.K.); christoph.schilling@aesculap.de (C.S.); thomas.grupp@aesculap.de (T.M.G.)
3. Department of Orthopaedic Surgery, Ludwig Maximilians University Munich, Physical Medicine and Rehabilitation, Campus Grosshadern, 81377 Munich, Germany
4. Department of Modern Mechanical Engineering, Waseda University, Tokyo 169-0072, Japan

* Correspondence: maeruan.kebbach@med.uni-rostock.de; Tel.: +49-381-498-8985

Received: 28 February 2020; Accepted: 16 May 2020; Published: 21 May 2020

Abstract: Patellofemoral (PF) disorders are considered a major clinical complication after total knee replacement (TKR). Malpositioning and design of the patellar component impacts knee joint dynamics, implant fixation and wear propagation. However, only a limited number of studies have addressed the biomechanical impact of the patellar component on PF dynamics and their results have been discussed controversially. To address these issues, we implemented a musculoskeletal multibody simulation (MMBS) study for the systematical analysis of the patellar component's thickness and positioning on PF contact forces and kinematics during dynamic squat motion with virtually implanted unconstrained cruciate-retaining (CR)-TKR. The patellar button thickness clearly increased the contact forces in the PF joint (up to 27%). Similarly, the PF contact forces were affected by superior–inferior positioning (up to 16%) and mediolateral positioning (up to 8%) of the patellar button. PF kinematics was mostly affected by the mediolateral positioning and the thickness of the patellar component. A medialization of 3 mm caused a lateral patellar shift by up to 2.7 mm and lateral patellar tilt by up to 1.6°. However, deviations in the rotational positioning of the patellar button had minor effects on PF dynamics. Aiming at an optimal intraoperative patellar component alignment, the orthopedic surgeon should pay close attention to the patellar component thickness in combination with its mediolateral and superior–inferior positioning on the retropatellar surface. Our generated MMBS model provides systematic and reproducible insight into the effects of patellar component positioning and design on PF dynamics and has the potential to serve as a preoperative analysis tool.

Keywords: joint replacement; knee joint; total knee arthroplasty; patellar component; musculoskeletal multibody simulation; patellofemoral joint

1. Introduction

Total knee replacement (TKR) is an established and effective surgical procedure for progressive osteoarthritis. TKR is currently performed over 700,000 times a year in the USA and this number is expected to grow exponentially worldwide [1,2]. Nevertheless, the rate of satisfied patients is 80%, which is rather low [2–4]. The patellofemoral (PF) joint represents a crucial part after total knee arthroplasty, and persistent PF pain remains a common postoperative complication with or without patellar resurfacing [5–7]. Complications include anterior knee pain, patellar maltracking, fracture, and patellar component loosening [8–10]. In this context, patella resurfacing is an important intraoperative factor: in the USA, more than 80% of primary TKRs are performed with this technique [11]. During surgery, the accurate positioning of the patellar component remains challenging because intraoperative alignment involves considerable inaccuracies [5,9,12,13]. The incidence of PF disorders ranges from 7% to 30% after a minimum of two postoperative years [14]. Patellar component malalignment is related to increased retropatellar loading and abnormal patellar kinematics [15,16]. Regarding the high number of intraoperative parameters, the dynamic interplay of the articulating joint partners of PF and tibiofemoral joint need to be analyzed in a systematic and reproducible manner [15,17,18].

Clinical observations alone cannot entirely explain unsatisfactory patient outcomes as they are often limited to the retrospective analysis of the influence of intraoperative parameters on pain and functional outcome [5,7,11–13,19,20]. However, clinical observations could be correlated with the mechanical loading of the joint to identify the underlying biomechanical causative chain. Therefore, knowledge of the biomechanical influence of intraoperative positioning of the patellar component during typical movements of daily living is essential to understand the underlying mechanical causes of implant failure and to improve the postoperative outcome. In this regard, the squat motion is known as one of the most dissatisfying motions after TKR [19].

Despite improvements in surgical instruments and techniques, many of the causes of revisions and patient dissatisfaction are directly related to implant component malalignment [6,18,21,22]. The morphology of the knee joint differs between both genders and different ethnicities [23–26]. These differences cannot only affect the implant component size selection but also the trochlear groove geometry, as well as the positioning of the patellar component between patients. Likewise, interindividual morphological joint variation is known to influence knee joint dynamics [27] and functionality of TKRs [26,28]. In this regard, Chen et al. [29] showed that the internal rotation of the femoral component and the varus malpositioning of the tibial component led to unfavorable postoperative loading conditions. Moreover, the malpositioning of the implant components often results in the overloading of the articulating joint compartments [18,30,31]. Different designs (dome-shaped, modified dome or anatomic design) and positions of the patellar button have been shown to affect knee joint dynamics [6,28,32–37]. An increased patellar component size was reported to affect the patellofemoral kinematics and postoperative outcome [38,39]. For instance, variations in the mediolateral and superior–inferior position of the patellar component resulted in different PF contact forces and kinematics [6,28,30,32,33,40,41]. Furthermore, the PF dynamics is influenced by the patellar component design [6,28] and the considerable impact of the patellar component thickness on PF kinematics and knee flexion has been widely reported [6,21,35,42–45]. For example, Abolghasemian et al. [43] reported a flexion loss of 1.28° with every millimeter of increased patellar component thickness. Bracey et al. [42] performed a similar study on 10 cadaveric knees and showed a flexion loss of 1.2° for each 2-mm increase of the patellar component thickness which resulted in a lateral patellar shift of more than 2 mm and a lateral patellar tilt of more than 4° whereas the patellar rotation remained nearly unchanged [43]. Bengs et al. [35] showed for 31 CR-TKRs with four different patellar thicknesses that on average the flexion was decreased by 3° for every 2-mm increase of the patellar component thickness but had no major impact on patella subluxation or tilt. Additionally, PF overstuffing has been reported with increased polyethylene wear, PF maltracking, and aseptic loosening of the patellar component [6,35,42,46,47]. Biomechanical studies have analyzed various mechanical aspects of patellar maltracking [15,32,33], e.g., medialization of the patellar component decreased the PF contact force [32,36,48,49] and altered the PF kinematics [36,41,48,50]. In a biomechanical study, Anglin et al. [48] found for a 2.5 mm component medialization a mean change in lateral patellar shift of 1.9 mm and a mean change in lateral tilt of 3.2°. These studies have contributed to understanding the positioning and design of the patellar component from a biomechanical perspective.

However, most experimental studies have only investigated passive knee flexion without active muscle forces, simulated rather uncommon motion patterns, commonly the knee rig configuration [28,33,42,51], or assumed a static quadriceps force, usually some predefined maximum value [28]. Furthermore, only a few studies addressing the biomechanical impact of different patellar component positions and designs on PF dynamics could be identified [15,18,28,33]. Instead, studies have revealed that the optimum position of the patellar component remains controversial [5,28,33,43,49], although evidently important for the postoperative outcome [28]. Some research groups have recommended medialization [8,41,48,50,52], while others have suggested centralization [32]. Moreover, biomechanical studies analyzing the superior–inferior position of the patellar component have reported contradictory findings [5,32,33,48], indicating a lack of understanding of this issue [5,49]. Hence, the PF joint mechanics have not yet been sufficiently quantified and understood so far [5,33]. The effects of various surgical parameters remain unclear and somewhat controversial. Using computational models, the influence of surgical and implant design parameters on knee joint dynamics can be investigated more comprehensibly [29,33,53–56].

Therefore, the current computational study aimed to systematically analyze and determine the biomechanical impact of patellar component malpositioning and design on PF dynamics during a dynamic squat motion using musculoskeletal multibody simulation (MMBS) in which a detailed knee joint model resembled the loading of a virtually implanted unconstrained cruciate-retaining (CR)-TKR including a dome patellar button. The findings could contribute to improving surgical techniques, preventing postoperative complications, and reducing wear propagation.

2. Materials and Methods

Our MMBS model is based on the experimental dataset of the *4th Grand Challenge Competition to Predict In Vivo Knee Loads* [57], which is a standardized dataset used by the research community to validate musculoskeletal models. This dataset comprises the CT scans (pre- and post-op) of a male subject (age = 88 years, height = 168 cm, and weight = 66.7 kg) who underwent TKR due to primary osteoarthritis and received an instrumented cruciate-retaining (CR) TKR (P.F.C. Sigma, DePuy Synthes, Raynham, MA, USA). This implant design was imported without changing the original standard size. The implanted TKR has a first-generation tray design eKnee and allows measuring the in vivo tibiofemoral contact forces during activities of daily living using an integrated telemetric force sensor. Concerning the knee implant, it represents a fixed-bearing, unconstrained bicondylar design which was implanted into the right knee of the patient. The femoral component had an asymmetrical dual radius design and was composed of cobalt–chromium alloy. The tibial component was composed of titanium alloy. Regarding the patellar component, an all-polyethylene dome-shaped component with three fixation pegs was used. The dataset enables the computational reconstruction of the 3D bone segments of the lower right extremity (pelvis, femur, patella, tibia, fibula, and pes), as well as the implant component geometries (femoral component, patellar button, tibial insert, and tibial tray). Moreover, motion capturing marker trajectories for activities of daily living are included, resulting in a comprehensive database for kinematic and kinetic model validation. A detailed description of this dataset can be found in Fregly et al. [57].

2.1. Overview of the Deployed Workflow of Musculoskeletal Multibody Simulation

The workflow for developing our subject-specific MMBS model for the computational analysis of the PF joint after TKR is depicted in Figure 1. Briefly, the relevant bone geometries were reconstructed from preoperative CT scans [57,58], which allowed the virtual implantation of TKR implant geometries. The implementation of the musculoskeletal geometry was based on the data reported in the TLEM 2.0 anthropometric database [42]. Additionally, the origin and insertion of muscles and ligaments were verified by an experienced orthopedic surgeon. Contacting surfaces were modeled by means of the polygonal contact model [59] to enable physiological-like joint dynamics. In this regard, relevant ligament structures were modeled as sets of nonlinear force elements [60,61] to resemble the respective anatomy of the ligaments. The inverse kinematics analysis was performed on a realistic squatting motion as reported in the SimTK data set [57], in which recorded marker trajectories applied to a patient allowed the calculation of the relative joint coordinates in the MMBS model deploying a global optimization procedure [62,63]. Finally, an inverse dynamics analysis coupled with a static optimization [55,62,64–66] allowed the computation of the individual muscle forces for the forward dynamic prediction of knee joint dynamics, e.g., tibiofemoral and PF contact forces, tibiofemoral and PF kinematics, as well as muscle and ligament forces under the presence of surgical parameter variations related to TKR.

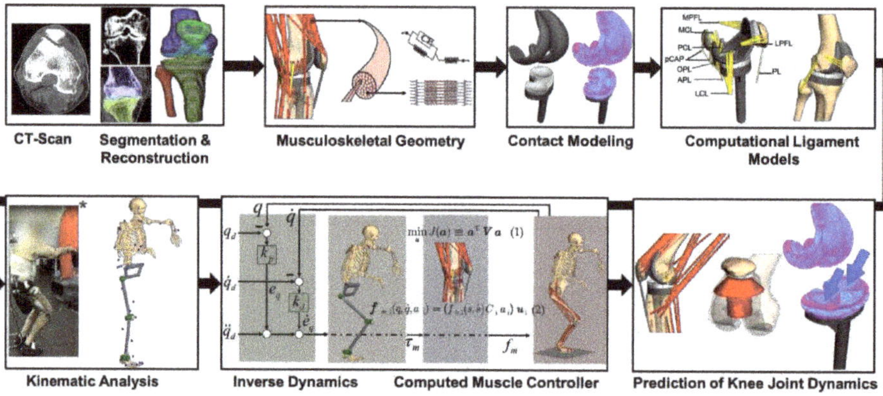

Figure 1. Workflow for generating the musculoskeletal multibody simulation model of the lower extremity with a total knee replacement. The illustration marked with * was taken from [57]. Permission to publish is granted under a CC BY open access license.

2.2. *Musculoskeletal Multibody Simulation Model with a Cruciate-Retaining Total Knee Replacement*

The MMBS model for the detailed analysis of the PF joint was generated based on a previously presented and validated MMBS model of the lower right extremity [65], Figure 2A,B. Implant and bone geometries, as well as relevant soft tissue structures, were modeled in the multibody software SIMPACK (V9.7, Dassault Systèmes Deutschland GmbH, Gilching, Germany). A variant of the computed muscle controller (CMC) with static optimization for individual muscle force prediction was implemented in MATLAB/Simulink® (v8.1, 2013a, The MathWorks Inc., Natick, MA, USA) and interfaced the MMBS model via TCP/IP-communication for forward dynamic co-simulation. All simulations were performed on an off-the-shelf computer (Intel® Xeon E5-1650 v4 CPU @3.60 GHz, 32 GB RAM).

The reconstructed bone segments were mutually connected by ideal joints as described in [62,65]. The respective centers of rotation of the ideal joints were determined by fitting cylinders or spheres into the cartilage surfaces of the articulating joint compartments. To ensure physiological-like roll-glide dynamics in the knee joint, both tibio- and patellofemoral joint compartments were modeled with six degrees of freedom (DoF) by implementing a polygon contact model [59], thereby resembling the complex articulation of the freeform implant surfaces. The lower left extremity was modeled as a symmetry condition in the sagittal plane. More precisely, the movement of the pelvis perpendicular to the sagittal plane was restrained by a spring-damper force element connecting the symmetry plane of the pelvis to the sagittal plane, thereby representing the lower left extremity during a symmetrical squat motion [65]. Regarding the squat motion, the patient started from a standing position and the range of knee flexion was about 0°–90°. The mass properties of the bone segments and soft tissue structures were calculated using regression equations as a function of the patient body weight [67].

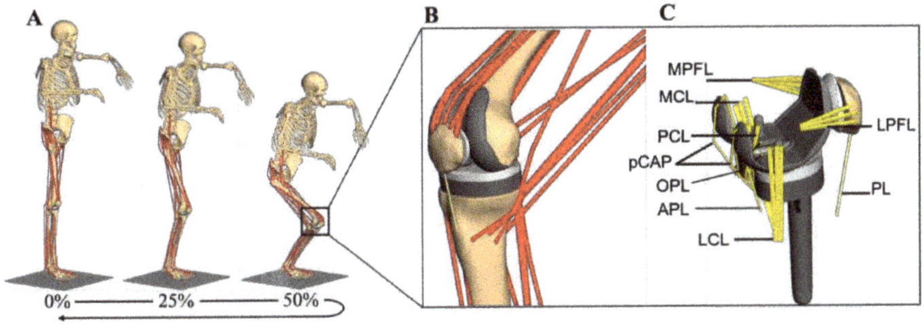

Figure 2. The developed musculoskeletal multibody simulation model with a cruciate-retaining total knee replacement in the lower right extremity during a dynamic squat motion combining musculoskeletal motion dynamics, knee implants with articular contact definitions, muscles, and ligaments (**A**). Detailed representation of the knee joint with implant components and muscle structures, including muscle wrapping. Note that ligaments are not shown for the sake of clarity (**B**). Investigated implant components with ligament structures of the tibio- and patellofemoral joint (**C**).

For the description and comparison of the joint dynamics, standardized coordinate systems were established [68–70] which further allowed the identification of attachment points of relevant muscle and ligament structures as described in *Twente Lower Extremity Model 2.0* anthropometric database [71]. Additionally, the attachment sites of muscles and ligaments were verified by an experienced orthopedic surgeon. Muscles were implemented in the form of unidimensional Hill force elements [55,71,72]. The muscles were further subdivided into several structural bundles based on the attachment area to account for the wide attachment surfaces [71]. Muscle deflection phenomena around the bones were incorporated either as segment-fixed via points or using wrapping surfaces [73] where required.

For a physiological representation of the knee joint dynamics, the MMBS model comprised, next to the explicit contact surface modeling of both the joint compartments, the implementation of all relevant ligamentous soft tissue structures (Figure 2C) with nonlinear force-strain relation (as nonlinear springs) [60,61]. Precisely, we implemented the posterior cruciate ligament (PCL), medial collateral ligament (MCL), lateral collateral ligament (LCL), oblique popliteal ligament (OPL), arcuate popliteal ligament (APL), posterior capsule (pCAP), medial patellofemoral ligament (MPFL), lateral patellofemoral ligament (LPFL), and the patellar ligament (PL) according to their anatomic descriptions, i.e., all ligaments were modeled as bundles of strands extending between the origin and insertion as described in Smith et al. [74]. The force elements for the representation of the ligaments followed a nonlinear elastic characteristic with a slack region as reported by Blankevoort et al. [61]. As it concerns the ligament parameterization, we first generated a MMBS model resembling the passive knee flexion, since the passive knee joint dynamics majorly depend on the ligaments and their parameterization (initial parameter sets were taken from the literature [61,74]). Then, by repeatedly simulating the knee flexion motion, the ligament parameters were adjusted with respect to the joint quantities for validation of the passive knee flexion. Once the appropriate ligament parameters were identified, the very same, parameterized ligament apparatus was transferred to the MMBS model resembling the squat motion and model validation was performed as described below (Section 2.5. *Validation of the musculoskeletal multibody simulation model*). Due to its high stiffness, the patellar ligament (PL) was modeled as a rigid coupling element between *Apex patellae* and *Tuberositas tibiae* with a fixed length [54,75]. The anterior cruciate ligament was virtually resected as it is sacrificed during CR-TKR surgery. A complete summary of the mechanical material properties for the ligaments [60,61,66,74] is provided in the Supplementary Information (Appendix A, Table A1). The PF joint was characterized by a patellar length of 43.6 mm, a patellar tendon length of 59.7 mm, and a tibial tuberosity-trochlear groove distance of 9.61 mm. The anthropomorphic details of the joint were in the normal range for

Caucasians with native patellae [76,77]. Moreover, the retropatellar surface area after resection was in the normal range [78].

Hence, the developed MMBS model of the knee joint, allows for the systematic and reproducible analysis of the TF and PF joint dynamics with its emphasis on the detailed representation of the articulating endoprosthesis components and ligamentous structures during dynamic, muscle-induced full-body motion.

2.3. Kinematic Analysis

The motion capture data comprises the trajectories of reflective skin markers applied to the patient's body [57] as depicted in Figure 1. These marker trajectories were used as input for an optimization algorithm [79] to derive the generalized joint coordinates $q, \dot{q}, \ddot{q}$ that describe the timely trajectories of the MMBS model's DoF. More precisely, the skin markers were modeled as moving reference points and coupled to the respective anatomical landmark, i.e., related segment-fixed points on the bone surface using spring-damper force elements. In this manner, the inverse kinematics problem could be resolved by the minimization of the spring-damper potential, thereby minimizing the error between motion capture data and the MMBS model motion. The desired joint trajectories $q_d, \dot{q}_d, \ddot{q}_d$ were then used as input for the forward dynamic simulation of the muscle-induced squat motion for numerous MMBS model variations as described in the Section 2.6. *In silico study on the effect of the patellar component design and positioning on patellofemoral joint dynamics after TKR*. In this manner, we generated a nominal configuration of the MMBS model based on the optimal surgical technique [57] which was further used to predict different postoperative situations of the patellar component design and position. Finally, the implant configurations were verified by an experienced orthopedic surgeon.

2.4. Forward Dynamic Musculoskeletal Multibody Simulation of a Squat Motion

A variant of the CMC algorithm [66] with static optimization for individual muscle force prediction was implemented to track the desired joint coordinates, as derived from the experimental motion capture data, by generating coordinated muscle forces. This controller has been recently described and verified to enable in vivo knee kinematics and kinetics in patient-specific musculoskeletal models [55,65,66,74,80]. Within CMC, the inverse dynamics model serves for input-output linearization of the neuro-musculoskeletal system, which is further superposed with generic feedback control to accurately track the experimentally obtained motion maneuvers $q_d, \dot{q}_d, \ddot{q}_d$. Subsequently, the computed joint torques τ_m necessary to drive the DoFs $q, \dot{q}, \ddot{q}$ are distributed over the available muscle actuators $\boldsymbol{f}_{m,i}(\cdot)$ by means of static optimization [64]:

$$\min_{a} J(\boldsymbol{a}) \equiv \boldsymbol{a}^T \boldsymbol{V} \boldsymbol{a}, \text{ subject to } \boldsymbol{D}\, \boldsymbol{a} = \tau_m \text{ and } 0 \le a_i \le 1. \tag{1}$$

The aforementioned muscle distribution problem was solved for optimal muscle activation levels a_i^* by minimizing an energy-optimal quadratic cost function $J(\boldsymbol{a})$ in which the diagonal weight matrix $\boldsymbol{V} = diag(V_1, \ldots, V_n)$ includes the muscle volumes V_i of each muscle unit and a_i is bounded by its physiological limits $0 \le a_i \le 1$. Moreover, the optimization problem is constrained by the set of linear equations $\boldsymbol{Da} = \tau_m$ where D represents the contribution of the respective muscle to the respective joint as described in a previous study [65]. We deployed a Hill-type muscle of the form:

$$\boldsymbol{f}_{m,i}(q, \dot{q}, a_i) = (f_{lv,i}(s, \dot{s})\; C_i\, a_i)\; \boldsymbol{u}_i \text{ with } C_i = A_i\, \sigma_i \text{ and } i = 1, \ldots, n, \tag{2}$$

where n is the number of muscle units, $f_{lv,i}(\cdot)$ is the force-length-velocity relation with muscle length s and muscle contraction velocity $\dot{s}$, A_i describes the physiological cross-sectional area, $\boldsymbol{u}_i$ is the muscle's unit direction vector, and σ_i is the maximal isometric muscle stress which was set to $\sigma_i = 1$ MPa [72]. Note that within this work, the force-length-velocity factor was assumed to $f_{lv,i}(\cdot) = 1$ as properties of the activated muscle structures and the activation dynamics have little influence on the prediction of muscle forces [64]. Therefore, the defined muscle force element depends on its theoretical maximum force $A_i\ \sigma_i$, its activation level a_i, and on the resulting moment arm only.

2.5. Validation of the Musculoskeletal Multibody Simulation Model

The predicted TF contact forces were validated using in vivo measured knee forces measured during squat motion exercises [57]. Furthermore, the TF contact force as well as the quadriceps force and PF contact force was compared with literature data [81–84] which is in Supplementary Information (Appendix B, Figure A1).

Since the in vivo measurements provided by the *4th Grand Challenge Competition to Predict In Vivo Knee Loads* [57] are limited to the contact forces acting in the TF joint, TF and PF kinematics were validated based on the in vitro and in silico results reported by Woiczinski et al. [85]. They measured TF and PF dynamics of 15 fresh frozen specimens with implanted CR-TKR during a squat motion using a knee rig [85].

2.6. In Silico Study on the Effect of the Patellar Component Design and Positioning on Patellofemoral Joint Dynamics after TKR

The validated MMBS model, as described previously, has been used as the nominal configuration corresponding to the optimal surgical technique [57] and served as a reference for subsequent model variations. Accordingly, with respect to the nominal model, we analyzed six patellar component configurations: spin of the patellar component ±5°, tilt of the patellar component ±5°, flexion-extension of patellar component ±5°, superior–inferior positioning of patellar component ±3 mm, and mediolateral positioning of patellar component ±3 mm. The position of the patellar component was changed based on a standardized coordinate system [70]. For instance, the shift was varied by medial or lateral movement of the center of the patellar component along a mediolateral axis defined as fixed to the patella. Moreover, we investigated the most important patellar button design parameter by increasing/decreasing the thickness of the patellar component ±2 mm.

The variations for each parameter configuration (Figure 3) were systematically chosen according to the reported findings from clinical [36,48,49], in vitro [15,16,32], and in silico [28,33] studies. The effect of each configuration was statistically evaluated with respect to the nominal simulation model as a function of the knee flexion angle over the full range of motion of the dynamic squat motion. Note that the calculated forces from the TF contact models were expressed with reference to the tibial component system to allow a direct comparison with the in vivo measured knee forces as reported by Fregly et al. [57].

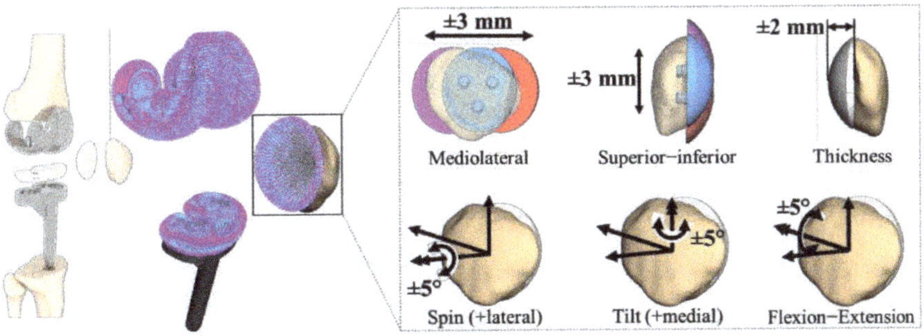

Figure 3. In silico study on the effect of the patellar component designs and positionings. Mediolateral positioning of patellar component ±3 mm, superior–inferior positioning of patellar component ±3 mm, and patellar button design parameter by increasing/decreasing the thickness of the patellar component ±2 mm. The rotational positioning was analyzed by the spin of the patellar component ±5°, tilt of the patellar component ±5°, and flexion-extension of patellar component ±5°.

2.7. Statistical Metrics

The predicted TF contact force of the MMBS model was compared with the in vivo measured knee force [57]. The prediction accuracy was quantified by mean absolute deviation (MAD) as a measure of magnitude differences, root-mean-square error (RMSE) as a measure for the difference between values predicted by the numerical model and the values actually observed from the experimental setup, Pearson correlation coefficient (r^2) as a measure of shape differences, and coefficient of determination (R^2) as a measure of magnitude and shape differences. For clarity, the mathematical definitions of the statistic metrics are provided in the Supplementary Information (Appendix C).

3. Results

3.1. Validation of the Musculoskeletal Multibody Simulation Model

The validation of the TF contact force, as well as the quadriceps force and PF contact force using literature data [81–84], is presented in Supplementary Information (Appendix B, Figure A1).

The predicted and in vivo measured medial, lateral, and total TF contact forces during two-leg squat motion are depicted in Figure 4. Their quantification in terms of statistic metrics (Table 1) for the predicted contact forces of the MMBS model was in good agreement with a satisfactory level of accuracy (RMSE = 0.39 body weight (BW), R^2 = 0.94, r^2 = 0.97) in terms of the magnitude and the general trend. The medial contact force was slightly overestimated (RMSE = 0.35 BW). The predicted TF contact forces in the lateral compartment exhibited an excellent agreement (RMSE = 0.10 BW). Therefore, the MMBS model closely captured the overall pattern and timing (R^2 = 0.94, r^2 = 0.97) of the in vivo measured TF contact force, indicating sufficient model fidelity for contact force prediction.

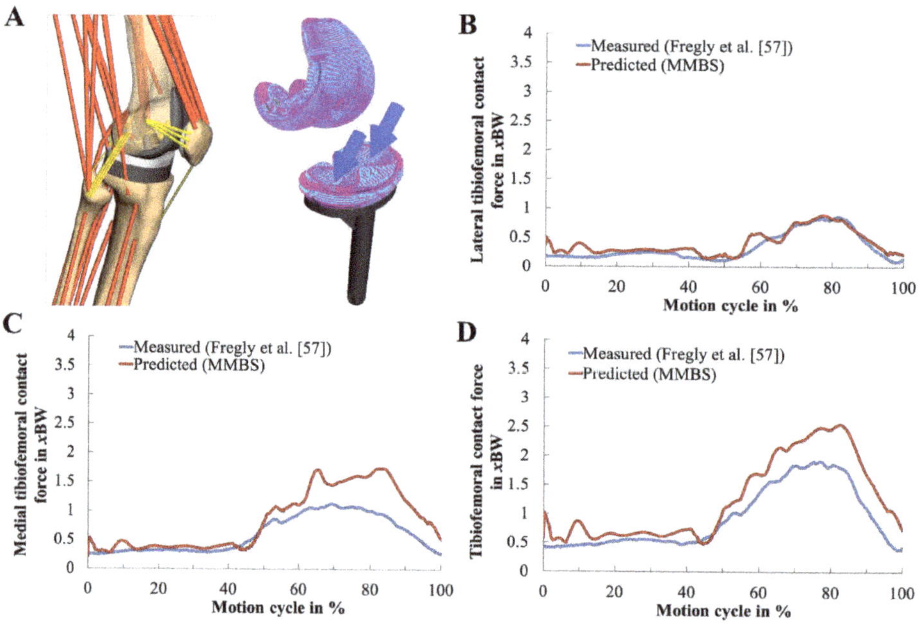

Figure 4. Musculoskeletal multibody simulation of the patellofemoral joint during the dynamic squat motion (**A**). Model validity was confirmed by comparing the reported lateral (**B**), medial (**C**), and total (**D**) tibiofemoral contact forces (in unit of body weight BW) of the in vivo measurements (blue, [57]) to our predictions (red).

Table 1. Kinetic validation of the tibiofemoral joint. Quantitative analysis of the predicted medial, lateral, and total tibiofemoral contact forces with reference to in vivo measurements during the two-leg squat motion cycle as reported by Fregly et al. [57]. Units are expressed in Newton (N) and body weight (xBW). Mean absolute deviation (MAD), root-mean-square error (RMSE), Pearson correlation coefficient (r^2), and coefficient of determination (R^2).

Value	MAD [N]	MAD [xBW]	RMSE [N]	RMSE [xBW]	r^2	R^2
Total knee contact force	209.68	0.32	255.19	0.39	0.97	0.94
Medial contact force	163.02	0.25	226.65	0.35	0.94	0.89
Lateral contact force	46.67	0.07	68.28	0.10	0.95	0.91

Moreover, the TF and PF kinematics were validated based on the in vitro and in silico results reported by Woiczinski et al. [85]. Specifically, anterior-posterior translation, tibial rotation, patellar tilt, patellar rotation, and patellar shift were compared (Figure 5). The PF and TF kinematics were reproduced with good agreement to the experimental and simulation data [85]. In general, the MMBS model predicted TF and PF kinematic patterns with a reasonable level of accuracy [85]. Overall, good agreement was observed between our MMBS model and reported data.

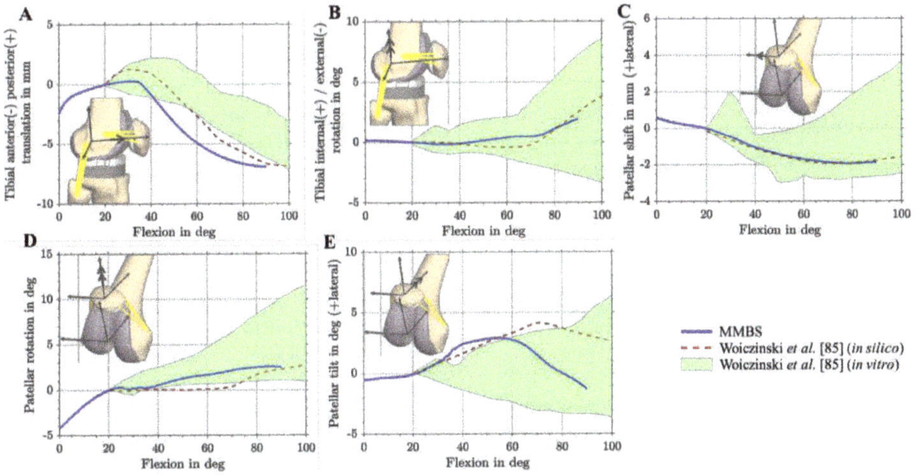

Figure 5. Tibio- and patellofemoral kinematics during dynamic squat motion. In silico (red dotted line) and in vitro (green area) [85] comparison of tibiofemoral and patellofemoral kinematics with kinematics obtained from musculoskeletal multibody simulation (MMBS) model (blue line). Comparison of anterior-posterior tibial translation with reference to the femur (**A**). Tibial internal/external rotation with reference to the femur (**B**). Patellar shift (**C**). Patellar rotation (**D**). Patellar tilt (**E**).

3.2. Effect of Patellar Component Design and Positioning on Patellofemoral Joint Dynamics after Unconstrained Total Knee Arthroplasty

PF dynamics was strongly affected by the position and design of the patellar component (Figures 6–10). Generally, the PF contact forces increased with increasing knee flexion angle due to the progressive involvement of the *M. quadriceps femoris* that enables such deep flexion angles. However, depending on the positioning and design of the patellar component, it is possible to reduce or increase the PF contact force. In this context, the analysis showed that the PF contact force was strongly affected by the patellar component thickness (RMSE = 440 N), considerably affected by superior–inferior positioning (RMSE = 199 N), and only moderately affected by mediolateral (RMSE = 98 N) positioning (Figure 10).

The variation in the superior–inferior positioning of the patellar component is depicted in Figure 6. An inferior position of the patellar component led to a reduction of the PF contact force (RMSE = 199 N) by up to 16% (Figure 6A). The maximum PF contact forces of the inferior, reference, and superior positions were 1626 N, 1947 N, and 2048 N, respectively. Superior positioning tended to cause a more medial tilting of the patella during flexion. An inferior position of the patellar component slightly increased the medial patellar shift, whereas a superior position of the patellar component decreased the lateral patellar tilt and shift (Figure 6B,C).

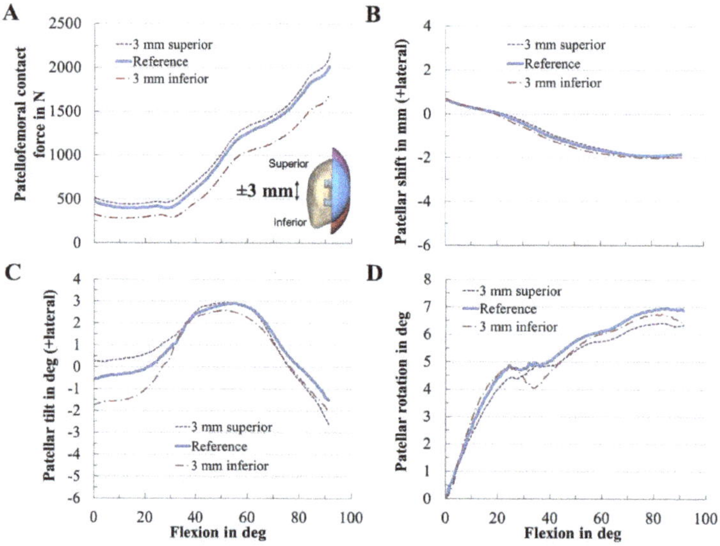

Figure 6. Effect of the superior–inferior positioning of the patellar component on patellofemoral dynamics. Effect of superior–inferior positioning of the patellar component on patellofemoral contact force (**A**), patellar shift (**B**), patellar tilt (**C**), and patellar rotation (**D**).

Similarly, the mediolateral positioning of the patellar component clearly affected the PF dynamics. Concerning the PF contact force, medialization and lateralization of the patellar component affected the PF contact forces (Figure 7A), i.e., a medialized patellar component decreased the maximum PF contact force during knee flexion by up to 8%. For instance, at 90° knee flexion, the PF contact force of the medial, reference, and lateral positions were 1798 N, 1947 N, and 2082 N, respectively. Furthermore, the mediolateral position of the patellar component clearly influenced PF kinematics regarding patellar shift, patellar tilt, and rotation. Mediolateral positioning of the patellar component shifted and tilted the patella path in the opposite direction, e.g., a medialized patella resulted in a more lateral shift and tilt of the patella and vice versa. Similarly, the patellar tilt was most sensitive to the mediolateral position of the patellar component. The more the patellar component was medialized, the more the patella tilted laterally with respect to the femur (Figure 7C).

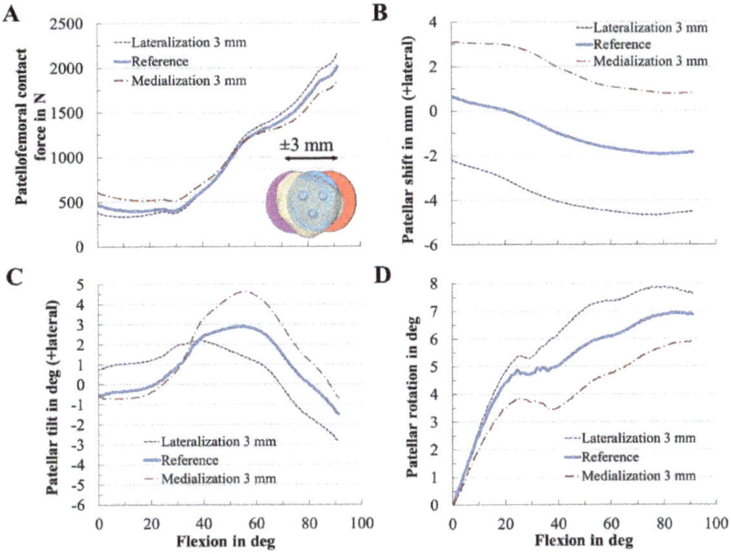

Figure 7. Effect of the mediolateral positioning of the patellar component on patellofemoral dynamics. Effect of mediolateral positioning of the patellar component on patellofemoral contact force (**A**), patellar shift (**B**), patellar tilt (**C**), and patellar rotation (**D**).

The variation in the thickness of the patellar component dramatically changed the PF dynamics (Figure 8). Generally, lower PF contact force values were seen in thinner components, while with an increase in component thickness, the contact force increased by up to 27% (Figure 8A). For the nominal configuration, the peak PF contact force was 1947 N, and the corresponding value obtained from the other variations were 2480 N (+2 mm), and 1344 N (−2 mm). The contact force was 176% of that of the thinner component. There was a direct relationship between the maximum contact force and patellar component thickness (Figure 8E): an increasing component thickness leads to higher maximum contact forces following a linear trend ($R^2 = 0.9976$). Furthermore, we found that for every 1 mm of increased patellar thickness, PF contact force increased by 13.7% (0.41 BW). There was also an effect of the patellar component thickness on patellar tracking, e.g., an increased patellar component thickness led to an increased lateral patellar tilt (Figure 8C). Contrarily, the effect on the patellar shift was smaller (Figure 8B).

The rotational positioning of the patellar component influenced PF dynamics only slightly in our TKR design (Figure 9). Tilting the patellar component, however, corresponded to a nonsymmetrical rotation of the patellar dome and therefore affected patellar tilt and shift. A medial tilt of the patellar component slightly increased the PF contact force, the patella shifted, and tilted more laterally. Flexion-extension of the patellar component slightly affected the PF contact force, while the spin of the patellar component left the PF dynamics nearly unchanged (Figure 10).

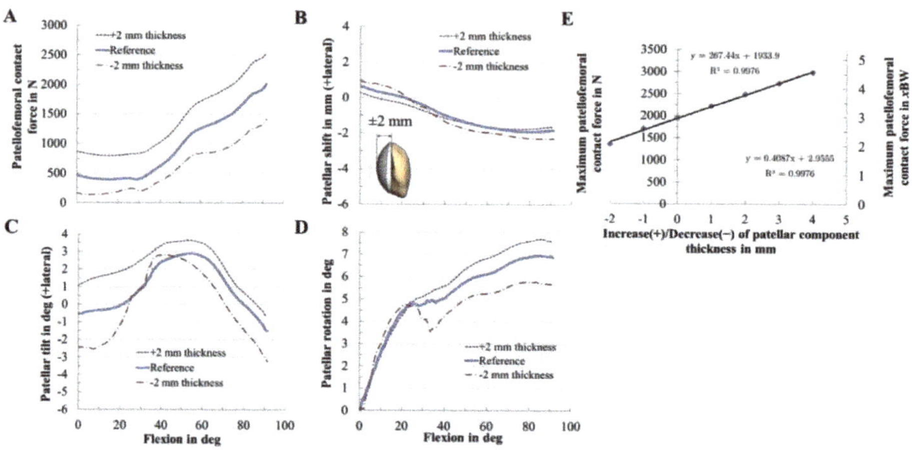

Figure 8. Effect of the patellar component thickness on patellofemoral dynamics. Effect of patellar component thickness on patellofemoral contact force (**A**), patellar shift (**B**), patellar tilt (**C**), and patellar rotation (**D**). Relationship of the maximum patellofemoral contact force for different patellar component thickness (**E**).

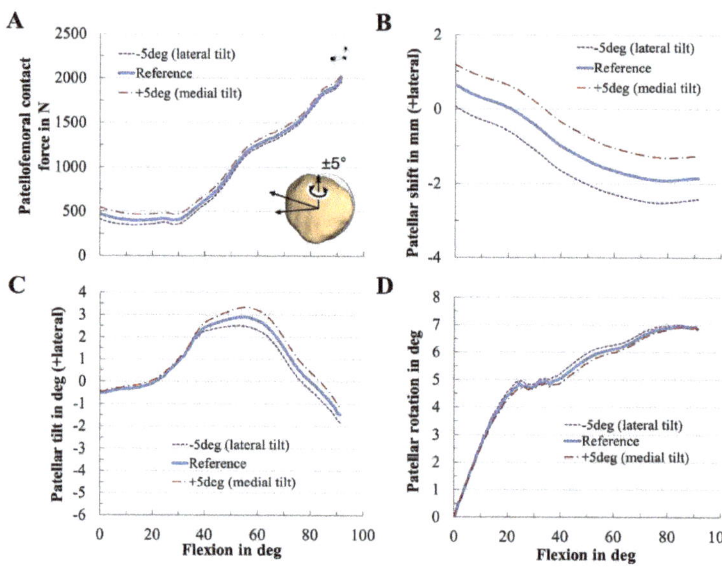

Figure 9. Effect of the tilting of the patellar component on patellofemoral dynamics. Effect of mediolateral positioning of the patellar component on patellofemoral contact force (**A**), patellar shift (**B**), patellar tilt (**C**), and patellar rotation (**D**).

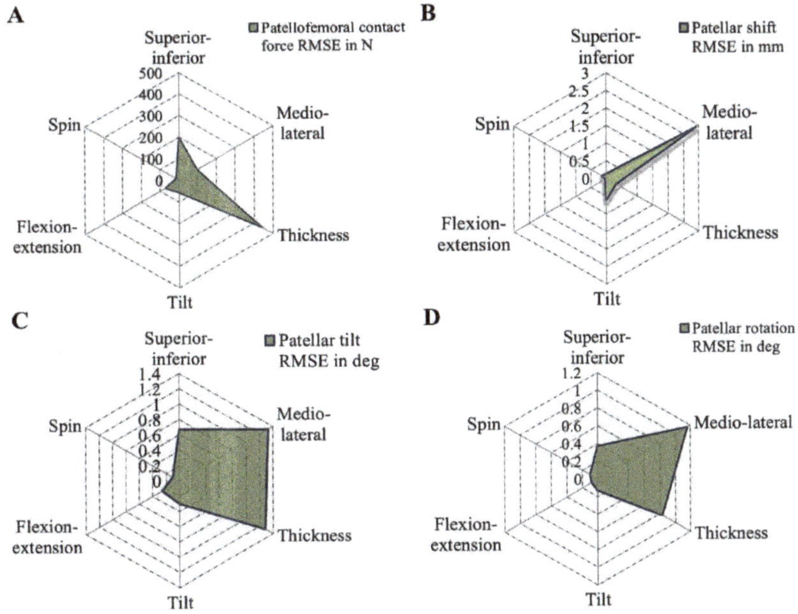

Figure 10. Radar chart representing the root-mean-square errors (RMSE) of patellofemoral contact force (**A**), patellar shift (**B**), patellar tilt (**C**), and patellar rotation (**D**) for different patellar component designs and positions: superior–inferior position, mediolateral position, patellar component thickness, tilt, flexion-extension, and spin.

3.3. *Effect of Coupling Patellar Component Design and Positioning Parameters on Patellofemoral Joint Dynamics after Unconstrained Total Knee Arthroplasty*

Additionally to the presented findings, we analyzed the coupling of the most critical patellar component design parameter with all other patellar component positioning parameters, i.e., we combined the increase of the thickness of 2 mm with all other positioning parameters (Figure 11). Similarly, the increase in the thickness combined with the rotational positioning of the patellar component influenced PF dynamics only slightly in our TKR design. However, an increase in the thickness combined with changing the mediolateral position of the patellar component mainly affected the PF kinematics. For instance, a solely increase in the thickness slightly changed the patellar shift (Figure 8B), whereas combined with a medialized patellar component the patellar shifted more laterally (Figure 11E).

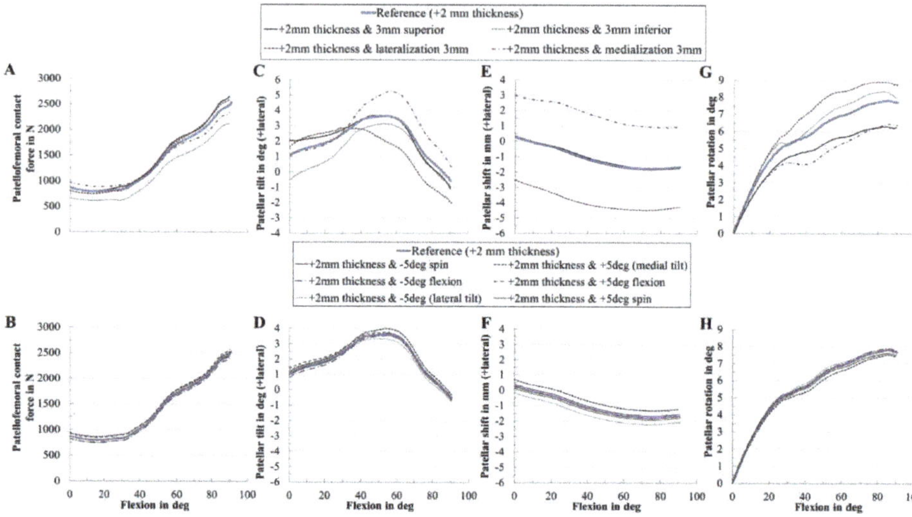

Figure 11. Effect of the increase of the patellar component thickness by +2 mm with all positioning parameters on patellofemoral dynamics. Effect of patellar component thickness combined with the positioning parameters on patellofemoral contact force (**A**,**B**), patellar tilt (**C**,**D**), patellar shift (**E**,**F**), and patellar rotation (**G**,**H**).

4. Discussion

In our computational study, we evaluated the effects of different intraoperative positionings and designs of the patellar component on both PF kinematics and kinetics during a dynamic squat motion of a subject with implanted CR-TKR with a dome patellar button. We developed a musculoskeletal knee joint model within a multibody simulation framework capable of analyzing the PF kinematics and kinetics. The results presented imply that even minor variations in the position and design of the patellar component can have a crucial effect on PF dynamics (Figures 6–10). PF contact forces were mostly affected by patellar component thickness (up to 27%), patellar component superior–inferior positioning (up to 16%), and mediolateral patellar component positioning (up to 8%). Concerning PF kinematics, the patellar component thickness as well as the mediolateral position of the patellar component revealed the greatest impact. Based on our findings, we can conclude that the malalignment in mediolateral as well as superior–inferior direction and the thickness of the resurfaced patella are the most important intraoperative parameters affecting PF dynamics. Furthermore, we could show that the translational positioning is more critical than rotational positioning regarding the resulting PF contact force. Generally, our findings are in good agreement with those of previous experimental and clinical studies [5,6,28,32,35,36,48,49,86].

However, similar to other in vitro and in silico studies, our approach has some limitations. Our MMBS does not reflect a diverse population of different patients. Moreover, only the two-leg squat motion was considered. Additional simulations which cover more activities (e.g., rising from a chair, normal gait) will be helpful in the future for a more robust investigation and to support our presented findings as total knee kinematics are task-dependent [87]. However, the investigations were performed for a two-leg squat motion because this motion covers a wider range of knee flexion compared to gait and it is reported to be one of the most dissatisfying motions after total knee arthroplasty [19]. One limitation which should be considered when transferring our in silico findings to clinical practice is associated with ligament modeling as accurate representation of ligament parameters is still challenging, and ligaments' mechanical properties might change during surgery. In the present study, the ligament parameters utilized were obtained from the literature. These assumptions might affect the prediction of joint dynamics. Likewise, the musculoskeletal geometry was based on the TLEM 2.0 anthropometric database [71]. It can be assumed that changes in the musculoskeletal geometry, e.g., muscle cross-sectional areas or muscle moment arms, of the muscles acting on the knee joint would have an effect on the forces and in turn kinematics of the PF joint. In addition, up to date, MMBS models hardly account for effects related to age or underlying conditions that might affect ligament or muscle structures. However, we believe that our MMBS model captures the physical condition of the examined subject well enough, as was made sure by rigorous model validation and diligent result analysis in cooperation with orthopaedic surgeons. The deployed methods are state-of-the-art and widely used in in silico models [29,33,53,75,80]. As explained above, our findings are based on one patient with one implant design and cannot be applied to all types of knee implants commercially available. Anatomical variability and further implant designs should be analyzed in future studies to support the reported findings. In future studies, different surgical alignment techniques (mechanical vs. kinematic), implant designs, different activities, and more subjects should be considered [37,38].

Orthopedic surgeons attempt to optimize the dynamic behavior of the knee joint by a correct intraoperative alignment of the patellar component. Interindividual morphological joint variation [23–26] is known to influence knee joint dynamics [27] and functionality of TKRs [26,28]. Furthermore, the difference in patellar component designs and the size of the patellar component are factors that influence the PF dynamics [28,37,38]. Regarding patellar resurfacing, the all polyethylene dome-shaped patellar component is currently the most common used design [8,28]. Therefore, our study covers the clinically most relevant TKR type. Our data indicate that the thickness of the analyzed dome-shaped patellar component affects PF contact forces, patellar tilt, and patellar rotation which has also been reported by other research groups [6,8,43]. In this regard, it was found that a higher retropatellar loading due to an increase in the thickness of the patellar component might contribute to anterior knee pain after TKR [6,44]. These higher forces can be explained by the increasing strain of the PF ligaments and can have a similar effect to overstuffing the PF joint which leads to anterior knee pain in some cases [8]. Although reduced patellar component thickness was beneficial due to a decreased PF contact force, it is important to avoid an excessively thin patellar component due to the risk of patellar maltracking and decreased PF ligament forces [46]. Our findings showed that, if the implant thickness increased, the lateral patellar tilt also increased which is in line with previous studies [21,42]. The influence of the thickness was not as high as reported by Bengs et al. [35]. Hsu et al. [44] reported that the PF contact force of the thicker patella was 174% higher compared to the thinner patella; in our study it was 176%. Some biomechanical studies [21,42,44,45] reported that overstuffing the PF joint induced higher lateral tilt of the patella which was also observed in our results. Additionally, a higher thickness might lead to increased polyethylene wear and aseptic loosening of the patellar component [35,42,46,47].

It has been reported that mediolateral positioning of the patellar component affects the PF dynamics [6,41]. We demonstrated the same in our study, i.e., a medialized patellar component reduced the resulting PF contact force during knee flexion in agreement with clinical observations [6,32,36,52,86]. Furthermore, medialization led to a more lateral path and tilt of the patella which is in line with previous findings [36,41,48]. A medialized patellar button decreases the Q-angle [8,41,86] and consequently lateralizes the patella bone, thereby leading to a higher lateral tilt [86]. Our results also indicate that the PF contact force decreased above 60° flexion due to medialization of the patella which might decrease postoperative pain [32,36,48]. In clinical studies, the incidence of lateral release was less for the medially placed patellar component [8,20,48,52]. However, we also showed that medialization caused a higher tendency for lateral tilting. The reported changes in shift, tilt, and PF forces are in good agreement with previous studies [20,48,52,86]. Concerning the patellar tilt, our data are comparable to a clinical study investigating the effects of medialization [50]. The reduction in the PF contact force due to medialization is consistent with a decrease in the Q-angle due to lateral shift of the patella [8,48].

To date, it is not clear whether the superior–inferior position of the patellar component influences knee dynamics after total knee arthroplasty [28,33,49]. We showed that the superior–inferior position of the patellar component affected the PF dynamics which is in line with previous studies [28,32,33]. In another study, Fitzpatrick et al. [28] identified the superior–inferior positioning as the most sensitive parameter using a dynamic finite element model supporting our findings. In this regard, our results, however, favor an inferior placement of the patellar component due to a decreased PF contact force. Nakamura et al. [33] recommended a superior placement of the patellar component, whereas other studies recommended an inferior placement [5,32]. As it concerns the PF contact forces, our results are contradictory to findings presented by Nakamura et al. [33] but consistent with other studies [5,30,32]. One explanation for the divergent data could be differences in the considered load cases and numerical simulation models used. The contradicting consequences of inferior position of the patellar component [33] occur only at high flexion angles (around 130° flexion) which were not reached in our analyzed load case. Furthermore, a higher number of PF ligaments were considered in our study. In general, inferior position of the patellar button might be favorable for daily living activities such as squatting but less desirable in patients who demand to reach higher flexion angles [88]. Based on our findings, the physiological reconstruction of the patellar position is of high importance. The most common conventional surgical techniques for patellar resection, i.e., freehand with a saw, using a saw guide, and a reamer reveal a high variance regarding the accuracy of the patellar cut. Although computer-navigation is used in TKR to improve the accuracy of the bony cut and component placement, currently it is focused on the positioning of the tibial and femoral component. However, studies with navigated patellar resurfacing systems showed an equal or higher accuracy for the patellar cut and the reconstructing of the physiological patellar position in comparison to conventional techniques with good short- and mid-term outcomes [12,13]. Our results emphasize the importance to focus on precise placement of the patellar component.

Another advantage of our study is the use of a detailed MMBS model incorporating relevant muscles of the lower extremity as active force elements compared to previous studies that assumed the muscles to apply only constant forces [15,16,30,51,85], considered only few muscles [28,33], or analyzed passive load cases without active muscle forces [34]. Our study showed the potential of MMBS in very comprehensively investigating the effects of surgical parameters on knee joint dynamics. The presented non-invasive method to simultaneously predict muscle, ligament, and knee joint forces can be used to improve the preclinical testing of TKRs as described by Affatato et al. [47] on the example of knee wear simulators. Computational data can provide additional insight into the influence of patellar component malpositioning on PF dynamics during active knee motions that commonly occur during daily living. Most orthopedic surgeons perform patellar bone preparation and component positioning without a navigation system [12,13]; however, small changes in the positioning leads to crucial changes in PF dynamics as demonstrated in this study. Furthermore, we showed that translational positioning of the patellar component is a relevant factor supporting the need for a navigation-based surgical procedure [12,13]. We pointed out that the patellar component is basically robust to rotational malalignment due to a consistent contact region between the articulating surfaces, which is in agreement with the findings of Fitzpatrick et al. [28]. Therefore, intraoperative placement of patellar components should first focus on translational position rather than rotational orientation.

5. Conclusions

In conclusion, the presented effects of patellar component design and positioning on PF kinematics and kinetics are in good agreement with previous experimental and computational studies. Aiming at an optimal intraoperative patellar component alignment, the most important parameters are component thickness, mediolateral and superior–inferior positioning. In both manual and navigation-based surgical techniques, the patellar thickness, patellar tracking, patellar position relative to the joint line, the orientation of the trochlear, and positioning of the anterior femoral component should be considered in order to prevent PF complications. Our findings will support orthopedic surgeons in intraoperative patellar component positioning from a biomechanical perspective. Regarding the different positioning of the patellar component, close attention should be paid to translational positioning as this might result in poor patient outcome.

Author Contributions: Conceptualization: M.K., T.M.G., R.B. and A.G.; methodology: M.K., M.D., S.K., C.S., R.B. and A.G.; software: M.K. and A.G.; validation: M.K. and A.G.; data curation: M.K., R.B. and A.G.; formal analysis: M.K. and A.G.; investigation: M.K., M.D., S.K., C.S., T.M.G., R.B. and A.G.; resources: R.B.; writing—original draft preparation: M.K.; writing—review and editing: M.K., M.D., S.K., C.S., T.M.G., R.B. and A.G.; visualization: M.K. and A.G.; supervision: M.K., R.B. and A.G. All authors have read and agreed to the published version of the manuscript.

Funding: This research received no external funding.

Conflicts of Interest: Three of the authors (S.K., C.S. and T.M.G.) are employees of Aesculap AG Tuttlingen, a manufacturer of orthopedic implants. The role of the three authors (S.K., C.S. and T.M.G.) had no effect on the collection and processing of the data. There are no personal circumstances or interest that influenced the representation or interpretation of reported results. The other authors (M.K., M.D., R.B. and A.G.) declare no conflict of interest.

Appendix A

Mechanical Material Properties of the Ligaments

Table A1. Mechanical properties (stiffness and reference strain) of the implemented ligaments were adapted from the literature [60,61,74,80]. The tibiofemoral ligament structures included: two bundles posterior cruciate ligament (PCL), two bundles lateral collateral ligament (LCL), three bundle medial collateral ligament (MCL), two bundles oblique popliteal ligament (OPL), one bundle arcuate popliteal ligament (APL), two bundles posterior capsule (pCAP), three bundles lateral patellofemoral ligament (LPFL), three bundles medial patellofemoral ligament (MPFL), and the patellar ligament (PL). Reference strains are the respective ligament strain values for the reference position upright standing. Stiffness is expressed in Newton per unit strain.

Tibiofemoral				Patellofemoral			
Ligament	**Ligament Bundle**	**Stiffness (N)**	**Ref. Strain**	**Ligament**	**Ligament Bundle**	**Stiffness (N)**	**Ref. Strain**
PCL	a	3000	−0.10	MPFL	p	2500	0.12
	p	1500	−0.03		c	2500	0.08
					d	2500	0.08
MCL	a	1500	0.04	LPFL	p	2000	0.06
					c	2000	0.06
					d	2000	0.06
	c	1500	0.04	PL		–	–
	p	1500	0.02		a	anterior	
LCL	a	2250	−0.25		c	central	
	p	2250	0.08		d	distal	
OPL	p	1250	0.06		i	inferior	
	d	1500	0.04		l	lateral	
pCAP	l	2500	0.05		m	medial/middle	
	m	2500	0.05		p	posterior/proximal	
APL		1500	0.04		s	superior	

Appendix B

Investigated joint quantities during active two-leg squat motion for the validation of the musculoskeletal multibody simulation model (Figure A1). The tibiofemoral contact forces (Figure A1A) agreed with the measured contact forces in patients with telemetric TKR [84]. Likewise, the quadriceps force was in good agreement with previously published studies [82,83] (Figure A1B). Patellofemoral contact forces (Figure A1C) calculated at 30°, 60°, and 90° knee flexion were similar to findings by Wallace et al. [81] and Innocenti et al. [82].

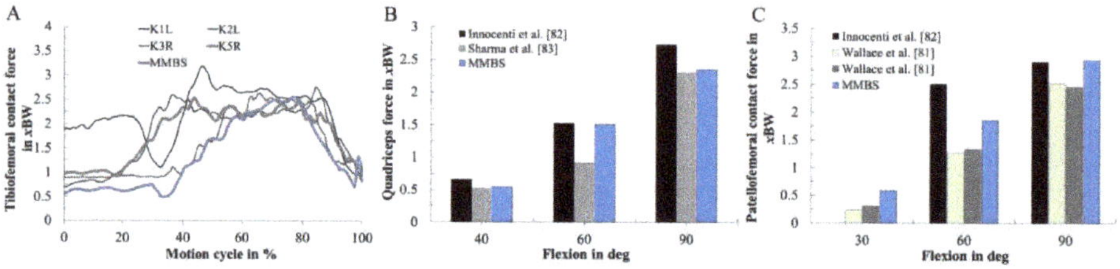

Figure A1. Validation of the musculoskeletal multibody simulation (MMBS) model. The MMBS model has been validated in terms of the tibiofemoral contact force (**A**), quadriceps force (**B**) and the patellofemoral contact force (**C**) during a dynamic squat motion. The tibiofemoral contact force is compared against the in vivo measurements of instrumented total knee replacements for three subjects (K1L, K2L, K3R, and K5R) [84]. The quadriceps force is compared against two simulation studies [82,83]. The resultant patellofemoral contact force is an important validation parameter for patellofemoral joint dynamics: it is compared against studies described in [81,82].

Appendix C

Equations Used for Statistical Evaluation

The computational formulas for mean absolute deviation (*MAD*)

$$MAD = \frac{\sum_{i=1}^{n} |m_i - c_i|}{n}, \tag{A1}$$

root mean square error (*RMSE*),

$$RMSE = \sqrt{\frac{\sum_{i=1}^{n} (m_i - c_i)^2}{n}}, \tag{A2}$$

Pearson correlation coefficient (r^2)

$$r^2 = \frac{n\sum m_i c_i - \sum m_i \sum c_i}{\sqrt{\left[n\sum m_i^2 - (\sum m_i)^2\right]}\sqrt{\left[n\sum c_i^2 - (\sum c_i)^2\right]}}, \tag{A3}$$

and coefficient of determination (R^2)

$$R^2 = \left(r^2\right)^2, \tag{A4}$$

presented here, where m_i and c_i represent the measured and calculated values at each time; step i, n is the total number of steps.

References

1. Kurtz, S.M.; Ong, K.L.; Lau, E.; Bozic, K.J. Impact of the economic downturn on total joint replacement demand in the United States: Updated projections to 2021. *J. Bone Jt. Surg. Am.* **2014**, *96*, 624–630. [CrossRef]
2. Price, A.J.; Alvand, A.; Troelsen, A.; Katz, J.N.; Hooper, G.; Gray, A.; Carr, A.; Beard, D. Knee replacement. *Lancet* **2018**, *392*, 1672–1682. [CrossRef]
3. Bourne, R.B.; Chesworth, B.M.; Davis, A.M.; Mahomed, N.N.; Charron, K.D.J. Patient satisfaction after total knee arthroplasty: Who is satisfied and who is not? *Clin. Orthop. Relat. Res.* **2010**, *468*, 57–63. [CrossRef] [PubMed]
4. Pabinger, C.; Berghold, A.; Boehler, N.; Labek, G. Revision rates after knee replacement. Cumulative results from worldwide clinical studies versus joint registers. *Osteoarthr. Cartil.* **2013**, *21*, 263–268. [CrossRef] [PubMed]

5. Assi, C.; Kheir, N.; Samaha, C.; Chamoun, M.; Yammine, K. Novel anatomical-based surgical technique for positioning of the patellar component in total knee arthroplasty. *SICOT J.* **2017**, *3*, 67. [CrossRef]
6. Gasparini, G.; Familiari, F.; Ranuccio, F. Patellar malalignment treatment in total knee arthroplasty. *Joints* **2013**, *1*, 10–17.
7. He, J.-Y.; Jiang, L.-S.; Dai, L.-Y. Is patellar resurfacing superior than nonresurfacing in total knee arthroplasty? A meta-analysis of randomized trials. *Knee* **2011**, *18*, 137–144. [CrossRef]
8. Matz, J.; Lanting, B.A.; Howard, J.L. Understanding the patellofemoral joint in total knee arthroplasty. *Can. J. Surg.* **2019**, *62*, 57–65. [CrossRef] [PubMed]
9. Huang, A.-B.; Qi, Y.-S.; Song, C.-H.; Zhang, J.-Y.; Yang, Y.-Q.; Yu, J.-K. Novel customized template designing for patellar resurfacing in total knee arthroplasty. *J. Orthop. Res.* **2016**, *34*, 1798–1803. [CrossRef]
10. Shah, J.N.; Howard, J.S.; Flanigan, D.C.; Brophy, R.H.; Carey, J.L.; Lattermann, C. A systematic review of complications and failures associated with medial patellofemoral ligament reconstruction for recurrent patellar dislocation. *Am. J. Sports Med.* **2012**, *40*, 1916–1923. [CrossRef]
11. Fraser, J.F.; Spangehl, M.J. International Rates of Patellar Resurfacing in Primary Total Knee Arthroplasty, 2004–2014. *J. Arthroplast.* **2017**, *32*, 83–86. [CrossRef] [PubMed]
12. Belvedere, C.; Ensini, A.; Tamarri, S.; d'Amato, M.; Barbadoro, P.; Feliciangeli, A.; Rao, A.; Leardini, A. Does navigated patellar resurfacing in total knee arthroplasty result in proper bone cut, motion and clinical outcomes? *Clin. Biomech. (Bristol. Avon.)* **2019**, *69*, 168–177. [CrossRef] [PubMed]
13. Fu, C.K.; Wai, J.; Lee, E.; Hutchison, C.; Myden, C.; Batuyong, E.; Anglin, C. Computer-assisted patellar resection system: Development and insights. *J. Orthop. Res.* **2012**, *30*, 535–540. [CrossRef] [PubMed]
14. Narkbunnam, R.; Electricwala, A.J.; Huddleston, J.I.; Maloney, W.J.; Goodman, S.B.; Amanatullah, D.F. Suboptimal patellofemoral alignment is associated with poor clinical outcome scores after primary total knee arthroplasty. *Arch. Orthop. Trauma Surg.* **2019**, *139*, 249–254. [CrossRef] [PubMed]
15. Steinbrück, A.; Schröder, C.; Woiczinski, M.; Schmidutz, F.; Müller, P.E.; Jansson, V.; Fottner, A. Mediolateral femoral component position in TKA significantly alters patella shift and femoral roll-back. *Knee Surg. Sports Traumatol. Arthrosc.* **2017**, *25*, 3561–3568. [CrossRef]
16. Steinbrück, A.; Fottner, A.; Schröder, C.; Woiczinski, M.; Schmitt-Sody, M.; Müller, T.; Müller, P.E.; Jansson, V. Influence of mediolateral tibial baseplate position in TKA on knee kinematics and retropatellar pressure. *Knee Surg. Sports Traumatol. Arthrosc.* **2017**, *25*, 2602–2608. [CrossRef]
17. Petersen, W.; Rembitzki, I.V.; Brüggemann, G.-P.; Ellermann, A.; Best, R.; Koppenburg, A.G.; Liebau, C. Anterior knee pain after total knee arthroplasty: A narrative review. *Int. Orthop.* **2014**, *38*, 319–328. [CrossRef]
18. Steinbrück, A.; Schröder, C.; Woiczinski, M.; Müller, T.; Müller, P.E.; Jansson, V.; Fottner, A. Influence of tibial rotation in total knee arthroplasty on knee kinematics and retropatellar pressure: An in vitro study. *Knee Surg. Sports Traumatol. Arthrosc.* **2016**, *24*, 2395–2401. [CrossRef]
19. Clement, N.D.; MacDonald, D.; Patton, J.T.; Burnett, R. Post-operative Oxford knee score can be used to indicate whether patient expectations have been achieved after primary total knee arthroplasty. *Knee Surg. Sports Traumatol. Arthrosc.* **2015**, *23*, 1578–1590. [CrossRef]
20. Hofmann, A.A.; Tkach, T.K.; Evanich, C.J.; Camargo, M.P.; Zhang, Y. Patellar component medialization in total knee arthroplasty. *J. Arthroplast.* **1997**, *12*, 155–160. [CrossRef]
21. Merican, A.M.; Ghosh, K.M.; Iranpour, F.; Deehan, D.J.; Amis, A.A. The effect of femoral component rotation on the kinematics of the tibiofemoral and patellofemoral joints after total knee arthroplasty. *Knee Surg. Sports Traumatol. Arthrosc.* **2011**, *19*, 1479–1487. [CrossRef]
22. Schroer, W.C.; Berend, K.R.; Lombardi, A.V.; Barnes, C.L.; Bolognesi, M.P.; Berend, M.E.; Ritter, M.A.; Nunley, R.M. Why are total knees failing today? Etiology of total knee revision in 2010 and 2011. *J. Arthroplast.* **2013**, *28*, 116–119. [CrossRef]
23. Wibeeg, G. Roentgenographs and Anatomic Studies on the Femoropatellar Joint: With Special Reference to Chondromalacia Patellae. *Acta Orthop. Scand.* **1941**, *12*, 319–410. [CrossRef]
24. Mahfouz, M.; Abdel Fatah, E.E.; Bowers, L.S.; Scuderi, G. Three-dimensional morphology of the knee reveals ethnic differences. *Clin. Orthop. Relat. Res.* **2012**, *470*, 172–185. [CrossRef] [PubMed]
25. Kim, T.K.; Phillips, M.; Bhandari, M.; Watson, J.; Malhotra, R. What Differences in Morphologic Features of the Knee Exist Among Patients of Various Races? A Systematic Review. *Clin. Orthop. Relat. Res.* **2017**, *475*, 170–182. [CrossRef] [PubMed]

26. Ali, A.A.; Clary, C.W.; Smoger, L.M.; Dennis, D.A.; Fitzpatrick, C.K.; Rullkoetter, P.J.; Laz, P.J. Computational framework for population-based evaluation of TKR-implanted patellofemoral joint mechanics. *Biomech. Model. Mechanobiol.* **2020**, *2020*, 1–9. [CrossRef]
27. Fitzpatrick, C.K.; Baldwin, M.A.; Laz, P.J.; FitzPatrick, D.P.; Lerner, A.L.; Rullkoetter, P.J. Development of a statistical shape model of the patellofemoral joint for investigating relationships between shape and function. *J. Biomech.* **2011**, *44*, 2446–2452. [CrossRef]
28. Fitzpatrick, C.K.; Baldwin, M.A.; Clary, C.W.; Wright, A.; Laz, P.J.; Rullkoetter, P.J. Identifying alignment parameters affecting implanted patellofemoral mechanics. *J. Orthop. Res.* **2012**, *30*, 1167–1175. [CrossRef]
29. Chen, Z.; Wang, L.; Liu, Y.; He, J.; Lian, Q.; Li, D.; Jin, Z. Effect of component mal-rotation on knee loading in total knee arthroplasty using multi-body dynamics modeling under a simulated walking gait. *J. Orthop. Res.* **2015**, *33*, 1287–1296. [CrossRef]
30. Luyckx, T.; Didden, K.; Vandenneucker, H.; Labey, L.; Innocenti, B.; Bellemans, J. Is there a biomechanical explanation for anterior knee pain in patients with patella alta? Influence of patellar height on patellofemoral contact force, contact area and contact pressure. *J. Bone Jt. Surg. Br.* **2009**, *91*, 344–350. [CrossRef]
31. Kebbach, M.; Grawe, R.; Geier, A.; Winter, E.; Bergschmidt, P.; Kluess, D.; D'Lima, D.; Woernle, C.; Bader, R. Effect of surgical parameters on the biomechanical behaviour of bicondylar total knee endoprostheses—A robot-assisted test method based on a musculoskeletal model. *Sci. Rep.* **2019**, *9*, 14504. [CrossRef] [PubMed]
32. Lee, T.Q.; Budoff, J.E.; Glaser, F.E. Patellar component positioning in total knee arthroplasty. *Clin. Orthop. Relat. Res.* **1999**, *366*, 274–281. [CrossRef] [PubMed]
33. Nakamura, S.; Tanaka, Y.; Kuriyama, S.; Nishitani, K.; Ito, H.; Furu, M.; Matsuda, S. Superior–inferior position of patellar component affects patellofemoral kinematics and contact forces in computer simulation. *Clin. Biomech. (Bristol. Avon.)* **2017**, *45*, 19–24. [CrossRef] [PubMed]
34. Rosso, I.; Surace, C.; Antonaci, P.; Surace, F.M.; Negretto, R.J. Influence of the patellar button thickness on the knee flexion after total knee arthroplasty. *Acta Bioeng. Biomech.* **2018**, *20*, 121–134.
35. Bengs, B.C.; Scott, R.D. The effect of patellar thickness on intraoperative knee flexion and patellar tracking in total knee arthroplasty. *J. Arthroplast.* **2006**, *21*, 650–655. [CrossRef]
36. Miller, M.C.; Zhang, A.X.; Petrella, A.J.; Berger, R.A.; Rubash, H.E. The effect of component placement on knee kinetics after arthroplasty with an unconstrained prosthesis. *J. Orthop. Res.* **2001**, *19*, 614–620. [CrossRef]
37. Mochizuki, T.; Yano, K.; Ikari, K.; Hiroshima, R.; Okazaki, K. Effect on patellar kinematics of the different patellar component designs in total knee arthroplasty: Intraoperative measurement of dome type versus anatomic type. *Eur. J. Orthop. Surg. Traumatol.* **2020**, *30*, 419–424. [CrossRef]
38. Joseph, L.; Batailler, C.; Roger, J.; Swan, J.; Servien, E.; Lustig, S. Patellar component size effects patellar tilt in total knee arthroplasty with patellar resurfacing. *Knee Surg. Sports Traumatol. Arthrosc.* **2020**. [CrossRef]
39. Atzori, F.; Sabatini, L.; Deledda, D.; Schirò, M.; Lo Baido, R.; Baido, R.L.; Massè, A. Evaluation of anterior knee pain in a PS total knee arthroplasty: The role of patella-friendly femoral component and patellar size. *Musculoskelet. Surg.* **2015**, *99*, 75–83. [CrossRef]
40. van de Groes, S.A.W.; Koëter, S.; Waal Malefijt, M.; de Verdonschot, N. Effect of medial–lateral malpositioning of the femoral component in total knee arthroplasty on anterior knee pain at greater than 8years of follow-up. *Knee* **2014**, *21*, 1258–1262. [CrossRef]
41. Yoshii, I.; Whiteside, L.A.; Anouchi, Y.S. The effect of patellar button placement and femoral component design on patellar tracking in total knee arthroplasty. *Clin. Orthop. Relat. Res.* **1992**, *275*, 211–219. [CrossRef]
42. Bracey, D.N.; Brown, M.L.; Beard, H.R.; Mannava, S.; Nazir, O.F.; Seyler, T.M.; Lang, J.E. Effects of patellofemoral overstuffing on knee flexion and patellar kinematics following total knee arthroplasty: A cadaveric study. *Int. Orthop.* **2015**, *39*, 1715–1722. [CrossRef]
43. Abolghasemian, M.; Samiezadeh, S.; Sternheim, A.; Bougherara, H.; Barnes, C.L.; Backstein, D.J. Effect of patellar thickness on knee flexion in total knee arthroplasty: A biomechanical and experimental study. *J. Arthroplast.* **2014**, *29*, 80–84. [CrossRef] [PubMed]
44. Hsu, H.-C.; Luo, Z.-P.; Rand, J.A.; An, K.-N. Influence of patellar thickness on patellar tracking and patellofemoral contact characteristics after total knee arthroplasty. *J. Arthroplast.* **1996**, *11*, 69–80. [CrossRef]
45. Youm, Y.-S.; Cho, W.-S.; Woo, J.-H.; Kim, B.-K. The effect of patellar thickness changes on patellar tilt in total knee arthroplasty. *Knee Surg. Sports Traumatol. Arthrosc.* **2010**, *18*, 923–927. [CrossRef]

46. Reuben, J.D.; McDonald, C.L.; Woodard, P.L.; Hennington, L.J. Effect of patella thickness on patella strain following total knee arthroplasty. *J. Arthroplast.* **1991**, *6*, 251–258. [CrossRef]
47. Oishi, C.S.; Kaufman, K.R.; Irby, S.E.; Colwell, C.W. Effects of patellar thickness on compression and shear forces in total knee arthroplasty. *Clin. Orthop. Relat. Res.* **1996**, 283–290. [CrossRef]
48. Anglin, C.; Brimacombe, J.M.; Wilson, D.R.; Masri, B.A.; Greidanus, N.V.; Tonetti, J.; Hodgson, A.J. Biomechanical consequences of patellar component medialization in total knee arthroplasty. *J. Arthroplast.* **2010**, *25*, 793–802. [CrossRef]
49. Donell, S. Patellar tracking in primary total knee arthroplasty. *EFORT Open Rev.* **2018**, *3*, 106–113. [CrossRef]
50. Kawano, T.; Miura, H.; Nagamine, R.; Urabe, K.; Matsuda, S.; Mawatari, T.; Moro-Oka, T.; Iwamoto, Y. Factors affecting patellar tracking after total knee arthroplasty. *J. Arthroplast.* **2002**, *17*, 942–947. [CrossRef]
51. Didden, K.; Luyckx, T.; Bellemans, J.; Labey, L.; Innocenti, B.; Vandenneucker, H. Anteroposterior positioning of the tibial component and its effect on the mechanics of patellofemoral contact. *J. Bone Jt. Surg. Br.* **2010**, *92*, 1466–1470. [CrossRef] [PubMed]
52. Lewonowski, K.; Dorr, L.D.; McPherson, E.J.; Huber, G.; Wan, Z. Medialization of the patella in total knee arthroplasty. *J. Arthroplast.* **1997**, *12*, 161–167. [CrossRef]
53. Kuriyama, S.; Ishikawa, M.; Furu, M.; Ito, H.; Matsuda, S. Malrotated tibial component increases medial collateral ligament tension in total knee arthroplasty. *J. Orthop. Res.* **2014**, *32*, 1658–1666. [CrossRef] [PubMed]
54. Tischer, T.; Geier, A.; Lenz, R.; Woernle, C.; Bader, R. Impact of the patella height on the strain pattern of the medial patellofemoral ligament after reconstruction: A computer model-based study. *Knee Surg. Sports Traumatol. Arthrosc.* **2017**, *25*, 3123–3133. [CrossRef]
55. Geier, A.; Aschemann, H.; D Lima, D.; Woernle, C.; Bader, R. Force Closure Mechanism Modeling for Musculoskeletal Multibody Simulation. *IEEE Trans. Biomed. Eng.* **2018**, *65*, 2471–2482. [CrossRef] [PubMed]
56. Affatato, S.; Ruggiero, A. A Critical Analysis of TKR In Vitro Wear Tests Considering Predicted Knee Joint Loads. *Materials* **2019**, *12*, 1597. [CrossRef]
57. Fregly, B.J.; Besier, T.F.; Lloyd, D.G.; Delp, S.L.; Banks, S.A.; Pandy, M.G.; D'Lima, D.D. Grand challenge competition to predict in vivo knee loads. *J. Orthop. Res.* **2012**, *30*, 503–513. [CrossRef]
58. Spitzer, V.M.; Whitlock, D.G. The visible human dataset: The anatomical platform for human simulation. *Anat. Rec.* **1998**, *253*, 49–57. [CrossRef]
59. Hippmann, G. An Algorithm for Compliant Contact Between Complexly Shaped Bodies. *Multibody Syst. Dyn.* **2004**, *12*, 345–362. [CrossRef]
60. Wismans, J.; Veldpaus, F.; Janssen, J.; Huson, A.; Struben, P. A three-dimensional mathematical model of the knee-joint. *J. Biomech.* **1980**, *13*, 677–685. [CrossRef]
61. Blankevoort, L.; Kuiper, J.H.; Huiskes, R.; Grootenboer, H.J. Articular contact in a three-dimensional model of the knee. *J. Biomech.* **1991**, *24*, 1019–1031. [CrossRef]
62. Herrmann, S.; Kluess, D.; Kaehler, M.; Grawe, R.; Rachholz, R.; Souffrant, R.; Zierath, J.; Bader, R.; Woernle, C. A Novel Approach for Dynamic Testing of Total Hip Dislocation under Physiological Conditions. *PLoS ONE* **2015**, *10*, e0145798. [CrossRef] [PubMed]
63. Leardini, A.; Chiari, L.; Della Croce, U.; Cappozzo, A. Human movement analysis using stereophotogrammetry. Part 3. Soft tissue artifact assessment and compensation. *Gait Posture* **2005**, *21*, 212–225. [CrossRef]
64. Anderson, F.C.; Pandy, M.G. Static and dynamic optimization solutions for gait are practically equivalent. *J. Biomech.* **2001**, *34*, 153–161. [CrossRef]
65. Geier, A.; Kebbach, M.; Soodmand, E.; Woernle, C.; Kluess, D.; Bader, R. Neuro-musculoskeletal flexible multibody simulation yields a framework for efficient bone failure risk assessment. *Sci. Rep.* **2019**, *9*, 6928. [CrossRef]
66. Thelen, D.G.; Anderson, F.C.; Delp, S.L. Generating dynamic simulations of movement using computed muscle control. *J. Biomech.* **2003**, *36*, 321–328. [CrossRef]
67. Winter, D.A. *Biomechanics and Motor Control of Human Movement*, 4th ed.; Wiley: Hoboken, NJ, USA, 2009; ISBN 9780470398180.
68. Wu, G.; Siegler, S.; Allard, P.; Kirtley, C.; Leardini, A.; Rosenbaum, D.; Whittle, M.; D'Lima, D.D.; Cristofolini, L.; Witte, H.; et al. ISB recommendation on definitions of joint coordinate system of various joints for the reporting of human joint motion–part I: Ankle, hip, and spine. International Society of Biomechanics. *J. Biomech.* **2002**, *35*, 543–548. [CrossRef]

69. Grood, E.S.; Suntay, W.J. A joint coordinate system for the clinical description of three-dimensional motions: Application to the knee. *J. Biomech. Eng.* **1983**, *105*, 136–144. [CrossRef]
70. Bull, A.M.J.; Katchburian, M.V.; Shih, Y.-F.; Amis, A.A. Standardisation of the description of patellofemoral motion and comparison between different techniques. *Knee Surg. Sports Traumatol. Arthrosc.* **2002**, *10*, 184–193. [CrossRef]
71. Carbone, V.; Fluit, R.; Pellikaan, P.; van der Krogt, M.M.; Janssen, D.; Damsgaard, M.; Vigneron, L.; Feilkas, T.; Koopman, H.F.J.M.; Verdonschot, N. TLEM 2.0—A comprehensive musculoskeletal geometry dataset for subject-specific modeling of lower extremity. *J. Biomech.* **2015**, *48*, 734–741. [CrossRef]
72. An, K.N.; Kaufman, K.R.; Chao, E.Y. Physiological considerations of muscle force through the elbow joint. *J. Biomech.* **1989**, *22*, 1249–1256. [CrossRef]
73. Charlton, I.W.; Johnson, G.R. Application of spherical and cylindrical wrapping algorithms in a musculoskeletal model of the upper limb. *J. Biomech.* **2001**, *34*, 1209–1216. [CrossRef]
74. Smith, C.R.; Vignos, M.F.; Lenhart, R.L.; Kaiser, J.; Thelen, D.G. The Influence of Component Alignment and Ligament Properties on Tibiofemoral Contact Forces in Total Knee Replacement. *J. Biomech. Eng.* **2016**, *138*, 21017. [CrossRef] [PubMed]
75. Marra, M.A.; Vanheule, V.; Fluit, R.; Koopman, B.H.F.J.M.; Rasmussen, J.; Verdonschot, N.; Andersen, M.S. A subject-specific musculoskeletal modeling framework to predict in vivo mechanics of total knee arthroplasty. *J. Biomech. Eng.* **2015**, *137*, 20904. [CrossRef]
76. Ortug, A.; Ormeci, T.; Yuzbasioglu, N.; Albay, S.; Seker, M. Evaluation of normal tibial tubercle to trochlear groove distance in adult Turkish population. *Niger. J. Clin. Pract.* **2018**, *21*, 1403–1407. [CrossRef]
77. Schlenzka, D.; Schwesinger, G. The height of the patella: An anatomical study. *Eur. J. Radiol.* **1990**, *11*, 19–21. [CrossRef]
78. Baldwin, J.L.; House, C.K. Anatomic dimensions of the patella measured during total knee arthroplasty. *J. Arthroplast.* **2005**, *20*, 250–257. [CrossRef]
79. Herrmann, S.; Kähler, M.; Grawe, R.; Kluess, D.; Woernle, C.; Bader, R. Physiological-Like Testing of the Dislocation Stability of Artificial Hip Joints. In *New Trends in Mechanism and Machine Science: From Fundamentals to Industrial Applications*; Flores, P., Viadero, F., Eds.; Springer: Cham, Swizerland, 2015; pp. 659–667. ISBN 978-3-319-09410-6.
80. Thelen, D.G.; Won Choi, K.; Schmitz, A.M. Co-simulation of neuromuscular dynamics and knee mechanics during human walking. *J. Biomech. Eng.* **2014**, *136*, 21033. [CrossRef]
81. Wallace, D.A.; Salem, G.J.; Salinas, R.; Powers, C.M. Patellofemoral joint kinetics while squatting with and without an external load. *J. Orthop. Sports Phys. Ther.* **2002**, *32*, 141–148. [CrossRef]
82. Innocenti, B.; Pianigiani, S.; Labey, L.; Victor, J.; Bellemans, J. Contact forces in several TKA designs during squatting: A numerical sensitivity analysis. *J. Biomech.* **2011**, *44*, 1573–1581. [CrossRef]
83. Sharma, A.; Komistek, R.D.; Ranawat, C.S.; Dennis, D.A.; Mahfouz, M.R. In vivo contact pressures in total knee arthroplasty. *J. Arthroplast.* **2007**, *22*, 404–416. [CrossRef] [PubMed]
84. Kutzner, I.; Heinlein, B.; Graichen, F.; Bender, A.; Rohlmann, A.; Halder, A.; Beier, A.; Bergmann, G. Loading of the knee joint during activities of daily living measured in vivo in five subjects. *J. Biomech.* **2010**, *43*, 2164–2173. [CrossRef] [PubMed]
85. Woiczinski, M.; Steinbrück, A.; Weber, P.; Müller, P.E.; Jansson, V.; Schröder, C. Development and validation of a weight-bearing finite element model for total knee replacement. *Comput. Methods Biomech. Biomed. Eng.* **2016**, *19*, 1033–1045. [CrossRef] [PubMed]
86. McPherson, E.J. Patellar tracking in primary total knee arthroplasty. *Instr. Course Lect.* **2006**, *55*, 439–448.
87. Schütz, P.; Postolka, B.; Gerber, H.; Ferguson, S.J.; Taylor, W.R.; List, R. Knee implant kinematics are task-dependent. *J. R. Soc. Interface* **2019**, *16*, 20180678. [CrossRef]
88. Mulholland, S.J.; Wyss, U.P. Activities of daily living in non-Western cultures: Range of motion requirements for hip and knee joint implants. *Int. J. Rehabil. Res.* **2001**, *24*, 191–198. [CrossRef]

materials

MDPI

Article

Corrosion Behavior of Surface-Treated Metallic Implant Materials

Therese Bormann [1,*], Phuong Thao Mai [2], Jens Gibmeier [2], Robert Sonntag [1], Ulrike Müller [1] and J. Philippe Kretzer [1]

1 Laboratory of Biomechanics and Implant Research, Clinic for Orthopedics and Trauma Surgery, Heidelberg University Hospital, 69118 Heidelberg, Germany; robert.sonntag@med.uni-heidelberg.de (R.S.); ulmuel@hotmail.de (U.M.); philippe.kretzer@med.uni-heidelberg.de (J.P.K.)

2 Institute for Applied Materials, Karlsruhe Institute of Technology, 76131 Karlsruhe, Germany; phuong.mai@kit.edu (P.T.M.); jens.gibmeier@kit.edu (J.G.)

* Correspondence: therese.bormann@med.uni-heidelberg.de

Received: 28 February 2020; Accepted: 23 April 2020; Published: 25 April 2020

Abstract: Corrosion of taper connections in total hip arthroplasty remains of concern, as particles and ions generated by corrosive processes can cause clinical problems such as periprosthetic osteolysis or adverse reaction to metallic debris. Mechanical surface treatments that introduce compressive residual stresses (RSs) in metallic materials can lead to a better performance in terms of fretting and fatigue and may lower the susceptibility to corrosion. The study investigates the impact of mechanical surface treatments on the corrosion behavior of metallic biomaterials. Compressive RSs were introduced by deep rolling and microblasting in Ti6Al4V and CoCrMo samples. Polished samples served as reference. Corrosion behavior was characterized by repeated anodic polarization. Residual stresses of up to about −900 MPa were introduced by deep rolling with a reach in depth of approximately 500 µm. Microblasting led to compressive RSs up to approximately −800 and −600 MPa for Ti6Al4V and CoCrMo, respectively, in the immediate vicinity of the surface. For Ti6Al4V, microblasting resulted in decreased corrosion resistance with lower breakdown potentials and/or increased passive current densities in comparison to the polished and deep-rolled samples. The corrosion behavior of CoCrMo on the other hand was not affected by the mechanical surface treatments.

Keywords: implant; biomaterial; corrosion; residual stress; total hip replacement; taper connection; anodic polarization; surface treatment

1. Introduction

Corrosion of metallic parts of artificial joints remains of concern as ions and particles that are generated by the corrosive processes can lead to clinical problems [1,2]. Alloys based on titanium and cobalt-chromium or stainless steels are usually applied as implant materials. The alloys themselves are highly corrosion resistant as they are passive metals (i.e., a protective oxide layer insulates the metal from the environment preventing corrosion). However, if the passive layer is continuously damaged, corrosion can—and does—occur.

In endoprosthetic implants, modularity enables intraoperative flexibility, which is especially beneficial in case of revision operations. In total hip arthroplasty (THA), the most prominent modular example is the taper connection between hip stem and femoral head. The material combination Ti6Al4V–CoCrMo is commonly used in these taper connections, because hip stems are often manufactured from Ti6Al4V due to its relatively low elastic modulus and the good osseointegrative properties, while metallic femoral heads are usually made of CoCrMo because of its high abrasion resistance. Corrosive processes at this interface are often referred to as *mechanically assisted crevice corrosion*

(MACC) [3]. This means that micromotion caused by cyclic implant loading leads to passive layer damage, which, in turn, promotes fretting wear as well as galvanic and crevice corrosion. In extreme cases, almost the complete trunnion of the taper connection was found to be worn off (trunnionosis) [4–6]. However, even in less dramatic cases, a considerable amount of metal loss can be generated at the head–neck interface, which can cause clinical problems, such as periprosthetic osteolysis and adverse reaction to metallic debris [7]. Furthermore, elevated blood metal ion levels can be provoked by corroding taper connections [7]. Using ceramic heads instead of heads from CoCrMo is a possibility to reduce corrosion at this particular interface. However, also for ceramic heads on metallic bone stems, taper corrosion has been reported, albeit to a lower extent [8]. Independent of head material, state-of-the-art to prevent taper corrosion is a proper head impaction during surgery. It has been shown that an impact power of at least 4 kN should be applied to ensure good interlocking of head and stem and reduce micromotion as far as possible [9–12].

Another approach in increasing fatigue and corrosion resistance is to alter the surface properties of the metals. The induction of compressive residual stresses in the surface and subsurface region reduces the susceptibility to crack initiation, for example by nanocrystal formation and strain hardening, which, in turn, enhances the resistance towards fatigue and fretting fatigue [13–15]. For Ti6Al4V, a superior fatigue and fretting fatigue performance after shot peening processes has been shown, for example, by Sonntag et al., Altenberger et al. and Liu et al. [14,16,17]. Aside from the mechanical behavior, corrosion characteristics of metals are influenced by residual stresses. It is well known that tensile RSs increase the susceptibility of stress corrosion cracking. Compressive RSs, on the other hand, can stop crack propagation and decrease the susceptibility of pitting corrosion [18–20]. The beneficial effects of compressive RSs on the corrosion resistance were shown for example for AA2024-T3—an aluminum alloy with poor corrosion properties—and a magnesium alloy for biomedical applications [19,20]. Lee at al. investigated the effect of shot peening on Ti6Al4V, which is a well-known technique to induce compressive RSs, and found increased high-cycle fatigue resistance even under seawater environment [21]. Both by shot peening and deep rolling, compressive RSs can be induced for depth ranges of some 100 μm depending on the process parameters used. However, techniques like deep rolling or laser shock peening have likewise been applied to induce compressive RSs to depths of up to a millimeter and even more depending on the intensity of the process [16].

This study aims at investigating the corrosion properties of Ti6Al4V and CoCrMo after inducing compressive RSs by the techniques of deep rolling and microblasting. The corrosion behavior of the mechanically surface-treated materials is investigated by means of anodic polarization. The two investigated alloys are commonly used as pairing in head–neck taper connections in total hip arthroplasty. As corrosion at this interface is complex, for separation of the individual influencing factors in this first approach, only the material–electrolyte interaction was investigated. We are aiming to assess the potential of mechanical surface treatments with regard to improved corrosion resistance of modular taper junctions in THA.

2. Materials and Methods

Surface treatments and electrochemical corrosion experiments were performed on discs made of Ti6Al4V ELI (Ti) as defined in DIN ISO 5832-3 and the low carbon CoCrMo-alloy (CoCr) defined in DIN ISO 5832-12, respectively. Ti6Al4V was received as rod material with a diameter of 15 mm. To provide a defined microstructure and to eliminate all residual stresses, the rod material was heat treated at 900 °C for 10 min under vacuum followed by oil quenching. Finally, the material was annealed at 500 °C for 4 h followed by a slow furnace cooling. CoCrMo was received as rod material with a diameter of 28 mm. This material was used in the as-received state. For surface treatments and electrochemical tests, discs with a diameter of 15 mm (Ti) and 28 mm (CoCr) and a thickness of 3 mm were manufactured.

2.1. Surface Treatments

The corrosion behavior of the two materials was investigated for the three different surface conditions of (i) metallographical polishing (PO), (ii) deep rolling (DR) and (iii) microblasting (MB).

A metallographic preparation (i.e., grinding, fine grinding and polishing by diamond and/or oxide suspension) removes the residuals from previous processing. Therefore, the metallographically-prepared samples served as reference. The mechanical surface treatments of deep rolling and microblasting on the other hand induced significant compressive residual stresses. Microblasting is a shot peening process with grit sizes in the micrometer range and a commonly used technique for surface treatments of metallic implant materials [22,23].

All samples were at first on one face ground and fine ground using SiC-paper from a grit of P320 to finally P2500. For the polished samples, the final surface treatment comprised diamond polishing applying diamond suspension down to a grain size of 3 µm.

Deep rolling and microblasting were carried out according to the process parameters listed in Table 1.

Table 1. Processing parameters applied for deep rolling and microblasting.

	Parameters Deep Rolling	Parameters Microblasting
Ball material / diameter	Ceramic / 6 mm	Glass beads MS 550B / 20–30 µm
Pressure	180 bar	3 bar
Feed rate	2000 mm/min	8 mm/s
Line spacing between traces	0.04 mm	1 mm
Initial distance to surface	–	10 mm
Coverage ratio	–	100%
Flow rate	–	1.5 kg/min

For the deep rolling, the rolling tool is directed using a meandering pattern, as schematically shown in Figure 1. Here, the compressive residual stresses at the material's surface are generated through continuous plastic deformation induced through the rolling contact (Hertzian pressure) between the tool and the workpiece material. The overlapping of adjacent rolling tracks induces characteristic non-axisymmetric residual stress distributions in lateral direction, i.e., maximum compressive residual stresses occur in feed direction, while in rolling direction, typically, much lower compressive residual stresses develop.

Microblasting is carried out by moving the blast nozzle also following a meandering pattern. Each blasting grain induces a residual imprint due to plastic deformation of the interacting material. Due to the high coverage the resulting lateral compressive residual stress distribution in the near-surface region is generally independent of direction (i.e., an axisymmetric residual stress state is generated).

The results of the surface treatments were assessed by X-ray residual stress analysis.

Scanning electron microscopy (LEO EVO 50, Zeiss, Oberkochen, Germany) in secondary electron contrast mode was applied in order to visualize the sample surfaces after the final surface treatment.

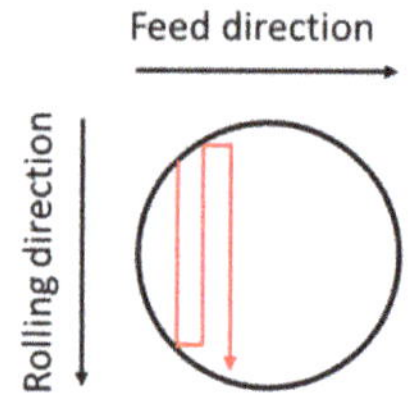

Figure 1. Scheme of the meandering pattern used for deep rolling and microblasting with indication of the feed and the rolling direction.

2.2. X-ray Residual Stress Analysis

Residual stresses in the two principal directions were determined using a custom-made 3-axis X-ray diffractometer in ψ-configuration according to the well-known $\sin^2\psi$-technique [24]. X-ray diffraction analysis was performed with Ni-filtered Cu Kα radiation for the {213}-lattice planes in the 2Θ-range $136° \leq 2\Theta \leq 147.4°$. Fifteen sample tilts were considered in the range $-60° \leq \psi \leq 60°$. As primary aperture, a ø 1 mm pin hole collimator was applied. On the secondary side, a 4 mm symmetrizing slit was used in front of the scintillation counter. Data post-processing was carried out using a Pearson VII fit after background subtraction. The diffraction elastic constants (DEC) $E_{\{213\}}$ = 113 GPa and $\nu_{\{213\}}$ = 0.32 for Ti6Al4V and $E_{\{220\}}$ = 227 GPa and $\nu_{\{220\}}$ = 0.3 for the CoCr samples were applied. Depth distributions of residual stresses were determined by means of electrochemical sublayer removal and reapplication of the $\sin^2\psi$-measurement at the newly created surface.

2.3. Electrochemical Testing

Electrochemical tests were accomplished using a potentiostat (Wenking MLab 200, Bank Elektronik – Intelligent Controls, Pohlheim, Germany) in combination with an electrochemical cell. The standard three-electrode system was applied for electrochemical testing with the Ti or CoCr discs serving as working electrode. Discs were covered by a silicone sealing mask which exposed the metal to the electrolyte through an orifice of 11.3 mm diameter, resulting in a working electrode area of 1 cm^2. The KCl saturated Ag/AgCl reference electrode (Sensortechnik Meinsberg, Waldheim, Germany) was placed in a Luggin capillary. The distance between capillary tip and working electrode was set to 1 mm. A platinized titanium bolt of 6 mm diameter served as counter electrode. Anodic polarization curves were recorded between −800 and 2000 mV for Ti and between −800 and 1000 mV for CoCr samples. Anodic polarization was repeated four times for each sample. The scan rate was set to 1 mV/s as proposed in the ASTM standard F2129-15 [25]. Open-circuit potential (OCP) measurements were conducted before, in between and after the polarization scans for one hour in each case. Dulbecco's phosphate buffered saline (PBS) (Biochrom, Berlin, Germany) with a pH of 7.4 was used as electrolyte for electrochemical tests of the three generated surface conditions for Ti and CoCr samples. In addition, a second electrolyte containing 10 g/l $FeCl_3$ in Ringer's solution ($FeCl_3$) (B. Braun AG, Melsungen, Germany) was used for electrochemical tests of polished, deep-rolled and microblasted Ti discs. The $FeCl_3$-containing electrolyte has a low pH of 1.8 and a higher concentration of Cl^- ions and was used as corrosion accelerating environment [26,27]. Electrochemical testing was done at 37 °C and by using aerated electrolytes. Before electrochemical testing, all samples were ultrasonically cleaned in ethanol for 10 min. The current density of the passive plateau (I_p) was determined from the anodic polarization curves as qualitative measurement of material degradation processes. The breakdown potential (E_b), which characterizes the onset of pitting corrosion, was determined from the intersection between passive current density and a linear fit of the polarization curve in the transpassive regime (i.e., the part of the curve leaving the passive plateau abruptly towards higher current density values). All curves were median filtered prior to the determination of I_p and E_b.

2.4. *Surface Roughness Measurements*

Surface roughness measurements were carried out after electrochemical testing using a perthometer (M2, Mahr, Göttingen, Germany). For each sample, four measurements were done in the area exposed to the electrolyte as well as in sample regions that were not exposed to the electrolyte during electrochemical testing but were covered by a silicone sealing sheet. The direction of the surface roughness profiles was changed about 90° for each measurement. The first roughness profile was placed arbitrarily on the sample surface. We consider the unexposed surface region to reflect the initial surface roughness of the samples before electrochemical testing. The length of the measuring line was 1.75 mm.

All results are given as mean ± standard deviation from three samples (n = 3).

3. Results

3.1. *Mechanical Surface Treatment*

The scanning electron microscope (SEM) images presented in Figure 2 emphasize the different characteristics of the mechanical surface treatments. Compared to the polishing and deep rolling process, microblasting results in a more rugged surface due to the stochastic impact of individual grits.

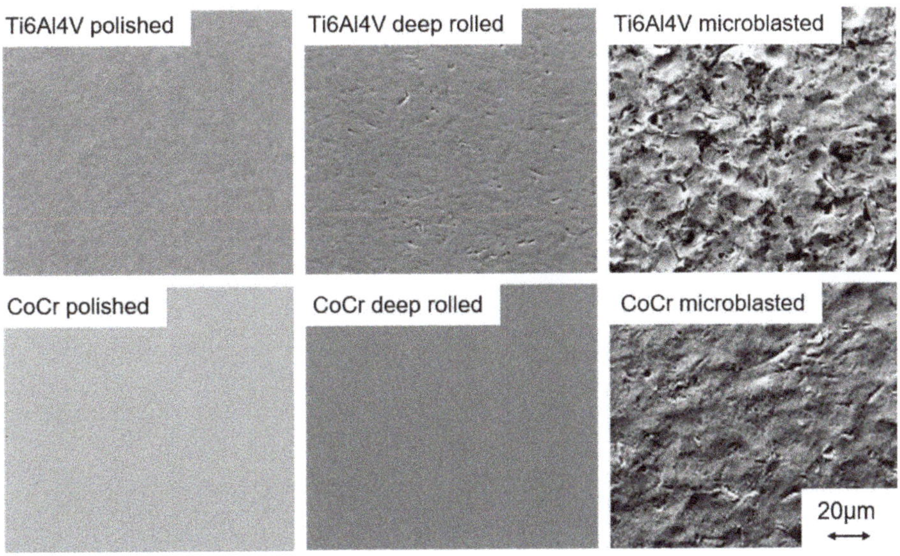

Figure 2. Scanning electron microscope images of the polished, deep-rolled and microblasted Ti6Al4V and CoCrMo (CoCr) samples.

The residual stress depth distributions for the Ti6Al4V samples are presented in Figure 3 together with the depth distributions of the mean integral widths of the recorded diffraction lines. In comparative studies, using the same measuring parameters, the change in the integral widths values corresponds to the degree of cold working (i.e., higher values indicate a larger degree of cold working). As expected, the polished samples can be treated as stress-free. The integral widths of the diffraction lines also indicate that almost no plastic deformation was induced by the polishing process, since the values are similar to the ones of the bulk material, which is unaffected by the surface treatment. Deep-rolled and microblasted samples show maximum RSs of about −800 MPa. While the microblasted sample shows its maximum directly at the surface, deep rolling led to maximum compressive RSs at a depth of about 200 µm. It should be noted that the compressive RSs resulting from deep rolling are directional (i.e., the RSs are significantly higher in the feed direction compared to the rolling direction (Figure 1)). Compressive RSs due to the deep rolling process were introduced up to a depth of approximately 600 μm, whereas for the microblasted sample the range of influence extents into approximately 50 μm depth. Regarding the mean integral width, for the microblasted sample the value dropped from 3.7° to 1.8° within the first 50 μm below the surface, indicating a rather steep cold working gradient. For the deep-rolled sample, the value directly at the surface referred to about 2.4° and decreased continuously with increasing distance to the surface.

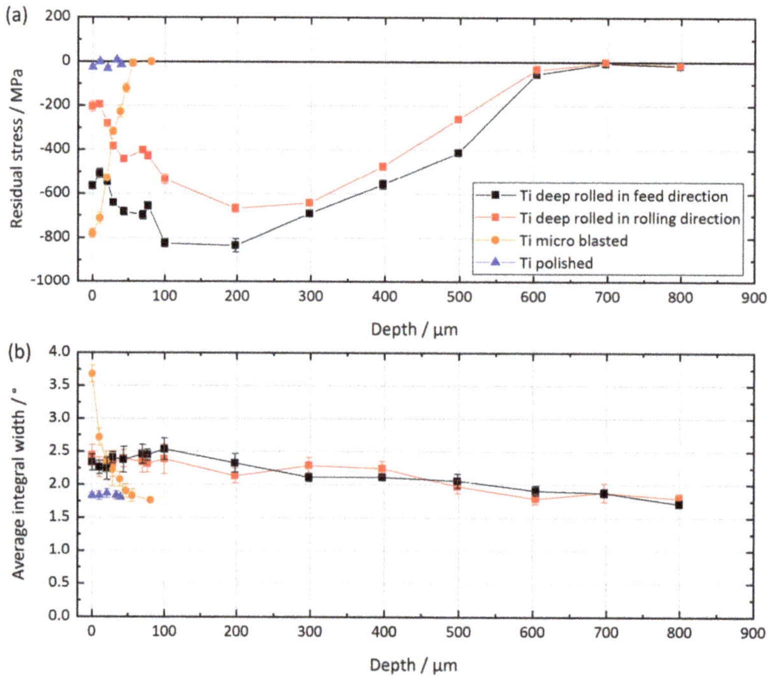

Figure 3. Residual stress depth distributions (**a**) and depth distributions of the average integral widths of the diffraction lines (**b**) for the three differently mechanically surface-treated Ti6Al4V (Ti) samples.

Figure 4 shows the depth distributions of RSs and mean integral widths for the CoCrMo samples. The courses show similarities to the ones of the Ti6Al4V samples. The polished sample is almost unaffected by the surface treatment, it shows only a small amount of tensile residual stresses close to the surface. Deep rolling and microblasting on the other hand induced characteristic compressive residual stress distributions. Microblasting resulted in compressive RSs of approximately −600 MPa directly at the surface. In addition, a steep gradient with a zero crossing and a change of sign towards tensile RSs in a depth of about 30 μm was found. As for the Ti6Al4V samples, the RSs are higher in the feed direction than in the rolling direction. Directly at the surface, compressive RSs in the feed direction of about −470 MPa and in the rolling direction of about −120 MPa were determined. At a depth of approximately 100 μm, the maximum of the compressive RSs reached about −900 MPa in the feed direction and about −580 MPa in the rolling direction. Zero crossing is observed at a depth of about 500 μm. At depths of more than 500 μm, tensile RSs were determined.

Regarding mean integral widths, the polished sample exhibited values in the range of the unaffected bulk material (see Figure 4). The microblasted sample showed significantly higher values of about 1.9° mean integral width directly at the surface. The value dropped within the first 30 μm below the surface to the values of the unaffected bulk material. The deep-rolled CoCrMo samples exhibited mean integral widths of about 1.2° at the very surface, which is slightly higher than for the unaffected bulk material. The depth profile shows a maximum of about 1.5° in approximately 100 µm depth, which is related to the location of maximal Hertzian stresses. With increasing depth, the integral widths continuously decreased to the value of the unaffected bulk material (1.05°).

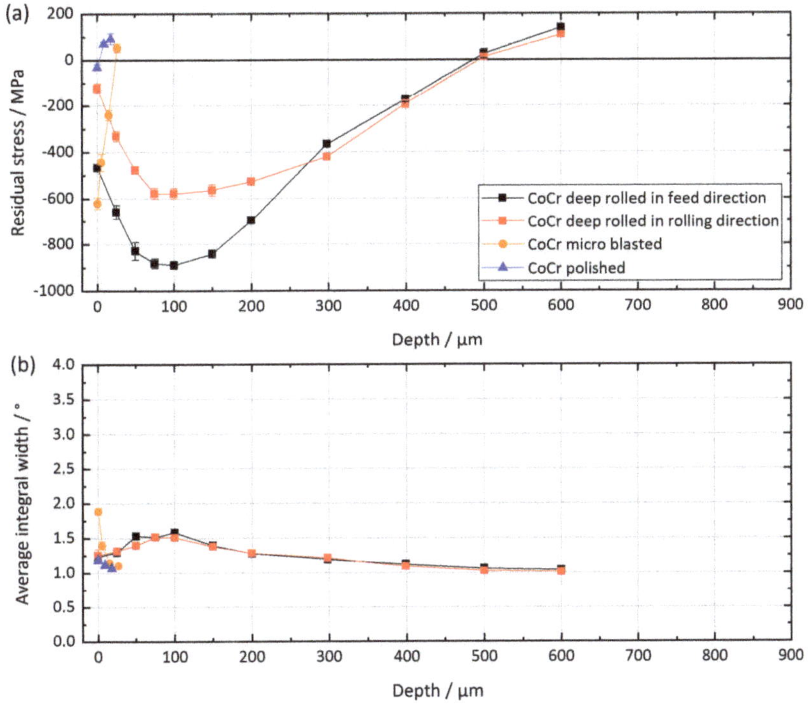

Figure 4. Residual stress depth distributions (**a**) and depths distributions of the average integral widths of the diffraction lines (**b**) for the three differently surface-treated CoCrMo (CoCr) samples.

3.2. Electrochemical Tests

Figure 5 represents the breakdown potentials (E_b) obtained from anodic polarization curves of the two materials in different surface conditions. For the polished and deep-rolled Ti discs, the breakdown potential could not be reached within the measuring range of up to 2000 mV if the used electrolyte was PBS. The microblasted samples; however, showed a distinct passive layer breakdown potential of 1536 ± 105 mV. If tested in the $FeCl_3$-containing electrolyte, the breakdown potential of all three investigated surface conditions was below 2000 mV. Polished samples had the lowest breakdown potential of 1504 ± 17 mV, while E_b of deep-rolled and microblasted samples referred to 1688 ± 115 and 1704 ± 100 mV, respectively. The surface treatments did not influence the breakdown potentials of CoCr discs, E_b accounted to 589 ± 10, 591 ± 10 and 593 ± 9 mV for polished, deep-rolled and microblasted discs, respectively. The breakdown potentials were hardly influenced by the scan number of the cyclic repeated polarization.

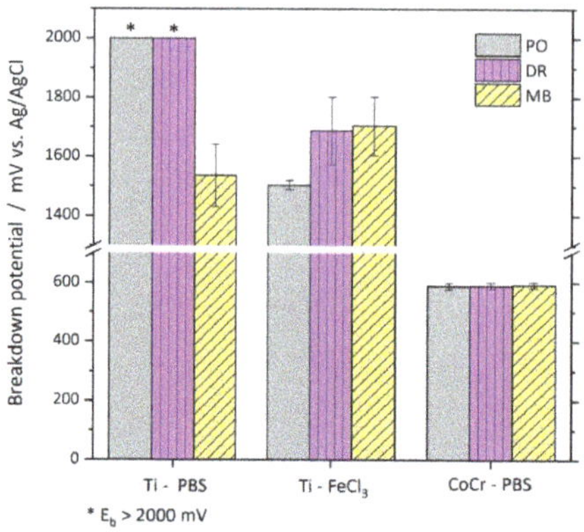

Figure 5. Breakdown potential (E_b) of polished (PO), deep-rolled (DR) and microblasted (MB) Ti6Al4V samples tested in PBS and in $FeCl_3$-containing electrolyte ($FeCl_3$) and of polished, deep-rolled and microblasted CoCrMo samples tested in PBS. Values comprise all four anodic polarization scans of three tested samples. Values marked with * refer to breakdown potentials above 2000 mV.

Figure 6 presents the passive current densities from the anodic polarization curves. Passive current densities of the first and the following scans, respectively, are displayed separately because Ti shows clearly increased passive current densities during the first anodic polarization cycle compared to all following scans (cp. Table 2). The difference amounts to about one order of magnitude. Passive current densities of the microblasted Ti samples are about two times higher than those of polished and deep-rolled samples in both tested electrolytes. In PBS, polished and deep-rolled surfaces exhibited comparable passive current densities. If tested in $FeCl_3$-containing electrolyte, the polished and deep-rolled samples showed comparable passive current densities in the first polarization cycle, while in the following cycles the deep-rolled surface show a tendency towards lower I_p. Passive current density of MB Ti samples was raised during the first anodization cycle in the $FeCl_3$-containing electrolyte in comparison to PBS. This tendency diminished during repeated polarization.

Passive current densities of CoCr are less affected by cycle number and surface condition, respectively (cp. Table 2). However, there are slight tendencies towards higher passive current densities during the first anodization and for the microblasted surface, respectively.

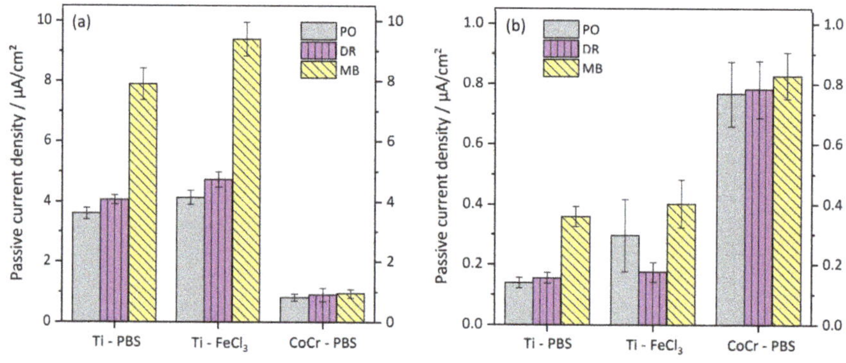

Figure 6. Passive current densities of polished, deep-rolled and microblasted Ti6Al4V samples tested in PBS and in $FeCl_3$ and of polished, deep-rolled and microblasted CoCrMo samples tested in PBS. (**a**) Passive currents from the first anodic polarization cycle. (**b**) Passive currents from polarization cycles no. 2–4.

Table 2. Passive current densities (mean (SD)) during first anodic polarization cycles and anodic polarization cycles no. 2–4 of polished, deep-rolled and microblasted Ti6Al4V and CoCrMo samples.

	Cycle 1 / nAcm^{-2}			**Cycles no. 2–4 / nAcm^{-2}**		
	PO	**DR**	**MB**	**PO**	**DR**	**MB**
Ti-PBS	3600 (200)	4067 (153)	7900 (529)	140 (17)	156 (17)	359 (34)
Ti-$FeCl_3$	4133 (231)	4733 (252)	9400 (557)	297 (11)	175 (6)	403 (79)
CoCr-PBS	813 (118)	908 (225)	950 (145)	767 (107)	783 (95)	828 (78)

The anodic current density–potential plots shown in Figure 7 illustrate the distinct first polarization cycle of Ti samples. The passive region subdivides in two parts of a less distinct plateau at passive current densities in the range of several hundred nA, and a second, more pronounced plateau in the μA range. In scans no. 2–4, a stable plateau region at the lower passive current densities developed. The $FeCl_3$-containing electrolyte resulted further in a displacement of the corrosion potential (i.e., the potential at zero crossing of the current density, which refers to the minimal value in logarithmic representation), from about −600 to about 500 mV (cp. Table 3), which means that the cathodic reaction is enhanced if $FeCl_3$-containing electrolyte is used.

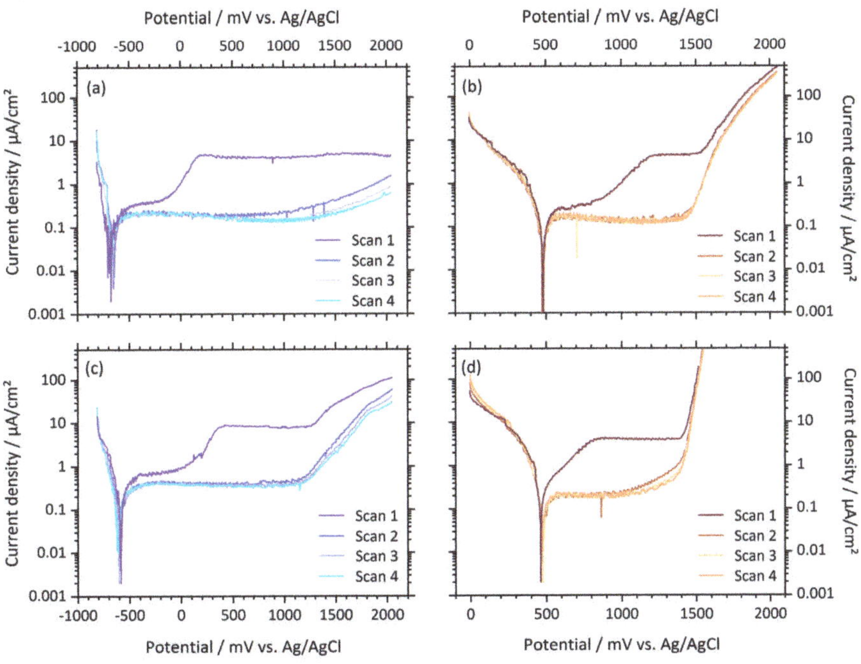

Figure 7. Anodic polarization curves of Ti6Al4V samples. (**a**) Deep-rolled sample tested in PBS, (**b**) deep-rolled sample tested in $FeCl_3$, (**c**) microblasted sample tested in PBS and (**d**) polished sample tested in $FeCl_3$.

Table 3. Corrosion potential (mean (SD)) of polished, deep-rolled and microblasted Ti6Al4V specimens tested in PBS and $FeCl_3$-containing electrolyte, respectively.

	PO / mV	**DR / mV**	**MB / mV**
PBS	−615 (94)	−642 (71)	−566 (30)
$FeCl_3$	498 (27)	469 (25)	461 (31)

For CoCr discs, again the curve progression in the first polarization cycle differs from the following cycles, as illustrated for a polished and a microblasted CoCr sample, respectively, in Figure 8. For polished and deep-rolled samples, during the first cycle, a passivation peak at approximately −700 mV can be noticed (cp. Figure 8a). After reaching the local maximum, the current density lowers to the passive current densities displayed in Table 2. With further increasing potential starting at approximately 0 mV, the passive current rised again slowly to a shoulder before the breakdown potential—characterized by the abruptly rising current density—is reached. This progression was observed for all mechanical surface treatments.

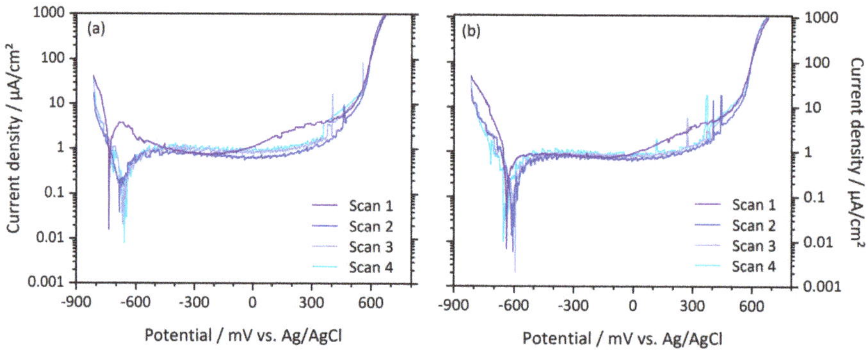

Figure 8. Anodic polarization curves of (**a**) a polished and (**b**) a microblasted CoCrMo sample tested in PBS.

3.3. *Surface Roughness*

Figure 9 presents the surface roughness value R_a of all tested samples in the as-mechanically-treated condition as well as after electrochemical testing. In Table 4, the R_a and R_z values are displayed. Generally, Ti samples comprise a higher surface roughness than CoCr samples. The roughness values of polished and deep-rolled samples are comparable, as shown in Table 4. For Ti samples only, deep rolling leads to slightly increased R_z values in comparison to polished samples. Microblasting results in a surfaces roughness of about one order of magnitude higher than in case of polishing or deep rolling. These findings are reflected by the SEM images in Figure 2.

During electrochemical testing, the samples were exposed to the electrolyte in combination with cyclic anodic polarization. Electrochemical testing of Ti samples in PBS did not result in significant changes of the surface roughness of the exposed areas. Polished Ti samples showed; however, a tendency towards decreasing surface roughness values in the regions that have been exposed to the $FeCl_3$-containing electrolyte.

CoCr samples were solely tested in PBS. After electrochemical testing, a slightly increased surface roughness of the polished and deep-rolled surfaces was detected. The microblasted surfaces; however, were not altered by the cyclic anodic polarization, regardless of the sample material or electrolyte.

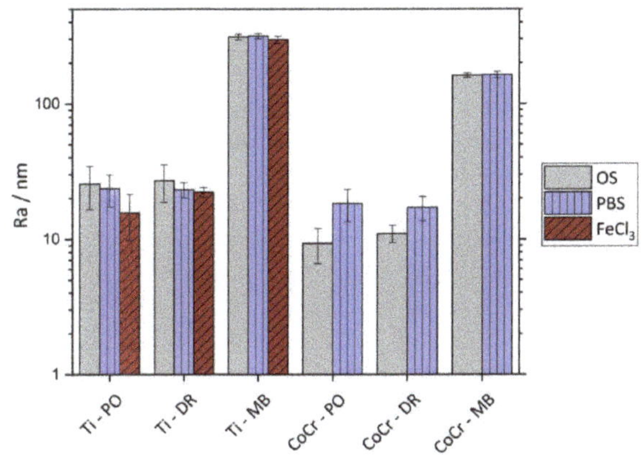

Figure 9. R_a values (nm) for Ti6Al4V and CoCrMo in polished, deep-rolled and microblasted surface condition. "Original surface" (OS) refers to the region which had not been in contact with the electrolyte during electrochemical testing; PBS refers to the surface after being electrochemically tested in PBS; $FeCl_3$ refers to the surface after being electrochemically tested in $FeCl_3$-containing electrolyte.

Table 4. R_a and R_z values in nm for all tested materials, surface conditions and electrolytes.

	Ti6Al4V / nm			CoCrMo / nm		
	PO	DR	MB	PO	DR	MB
R_a / OS	25.5 (8.9)	27.0 (8.3)	311.9 (15.1)	9.3 (2.7)	11.0 (1.6)	162.2 (6.0)
R_a / PBS	23.6 (6.3)	23.1 (3.0)	315.9 (14.2)	18.3 (4.9)	17.1 (3.5)	163.7 (8.9)
R_a / $FeCl_3$	15.6 (5.8)	22.2 (1.8)	297.5 (17.6)	-	-	-
R_z / OS	177.9 (63.3)	210.1 (53.4)	2004 (131)	68.5 (20.4)	69.1 (17.5)	1077 (91)
R_z / PBS	178.3 (61.6)	200.5 (28.7)	2010 (160)	122.7 (23.9)	106.4 (15.4)	1097 (89)
R_z / $FeCl_3$	109.5 (32.4)	192.8 (22.5)	1883 (115)	-	-	-

4. Discussion

Generally, Ti6Al4V showed higher breakdown potentials than CoCrMo, meaning that Ti has a higher resistance to pitting corrosion. At polished and deep-rolled samples, breakdown of the passive layer could not be accomplished in PBS up to a potential of 2000 mV. This is consistent with the literature reporting breakdown potentials between 2500 and 3500 mV for Ti6Al4V in chloride-containing electrolytes [28,29]. The microblasted sample, on the other hand, exhibited a clearly lower breakdown potential of 1536 mV in PBS, which indicates a greater susceptibility to pitting corrosion of microblasted surfaces in comparison to polished or deep-rolled surfaces. Polarization of Ti samples in the $FeCl_3$-containing electrolyte led to breakdown of the passive layer for all tested surface conditions well below 2000 mV, while in this corrosion-accelerating electrolyte, the polished samples exhibited passive layer breakdown at lower potential than microblasted and deep-rolled samples. Interestingly, the addition of $FeCl_3$ only resulted in decreased breakdown potentials for PO and DR surfaces, the breakdown potential of the microblasted surface was not further lowered by the $FeCl_3$-containing environment.

We consider passive current densities as qualitative measurement for material degradation processes, as these involve exchange and transport of ions and charged molecules. Higher passive current densities would therefore represent increased material degradation rates, while lower passive current densities would refer to smaller corrosion rates. Polished and deep-rolled Ti6Al4V samples showed substantially lower passive current densities than samples with microblasted surfaces, which confirms that the corrosion resistance of Ti6Al4V is higher after metallographical polishing and deep rolling than after microblasting of the surface.

However, during the first anodic polarization cycle, passive current densities of Ti are substantially higher than passive current densities of CoCr. In addition, we observed two distinct regions in the passive potential range. It is well known that Ti6Al4V spontaneously forms a stable oxide layer when exposed to air. The oxide layer has a thickness of approximately 5 nm and consists mainly of TiO_2 and a small quantity of TiO and Ti_2O_3 at the metal–oxide interface [28]. Milosev et al. showed that upon polarization of Ti6Al4V, the thickness of the TiO_2 layer increases and Al_2O_3 is introduced in the passive layer already at low potentials [28]. We; therefore, attribute the distinctive passive region during the first polarization to growth and chemical transformation of the oxide layer. In all following polarization cycles, the oxide layer stabilized and does not undergo further changes, as can be seen from the constant current density within the passive region and the almost identical current density–potential plots.

As illustrated in Figure 8, the first polarization cycle of CoCrMo samples differed from all following cycles. For polished and deep-rolled samples, a passivation peak (i.e., an active–passive transition) was observed during the first polarization scan. Hence, we suppose a thickening or healing of the existing oxide layer at low potentials. All surface treatments exhibited increased passive current densities over a potential range of approximately 300 mV just before passive–transpassive transition. This might be attributed to chemical changes, as the composition of the oxide layer is highly dependent on the applied potential [30,31]. The spontaneously formed oxide layer mostly consists of Cr_2O_3 and small quantities of Co- and Mo-oxides and—according to Milosev et al.—has a thickness of about 1.8 nm [31]. Chemical modifications caused by polarization in intermediate potential ranges include the incorporation of Co- and Mo-oxides in the passive layer, which is accompanied by an increase in thickness [31]. Milosev et al. found the increase in thickness of the passive layer to account for about a factor 3, if CoCrMo is polarized up to 800 mV (against a saturated calomel electrode) in simulated physiological solution [31]. However, after the first polarization cycle, the passive layer of CoCrMo stabilized, so that, in all following cycles, neither passivation peaks nor increasing passive current densities were observed before passive–transpassive transition. Among the different surface treatments, the passive current densities of CoCrMo hardly varied. The changes within the oxide layer on CoCrMo caused by polarization can; therefore, be considered to be independent of the mechanical surface treatments.

The decreased corrosion behavior of microblasted Ti6Al4V samples can be attributed to the higher roughness of the surface. A higher corrosion resistance related to lower surface roughness has been reported for Ti6Al4V but also for other metals and alloys, such as cp-Ti and NiTi [29,32–34]. The higher passive current densities in microblasted Ti samples can be attributed to the true surface area, which increases with roughness. The irregular grooves that are formed upon microblasting might, in addition, lead to small, enclosed areas with reduced diffusion in the electrolyte. This could lead to a locally-restricted oxygen depletion in the electrolyte, which may result in localized corrosion similar to the process of crevice corrosion [29]. Additionally, the oxide layer of the rough surface may possess a higher degree of imperfections, which would make the surface more susceptible to pitting corrosion [29]. The lower breakdown potential of the microblasted Ti6Al4V samples tested in PBS can be attributed to these effects. In a more aggressive corrosive environment; however, these effects seem to play a minor role, as in the $FeCl_3$-containing electrolyte all three Ti6Al4V surface modifications exhibited breakdown potentials in the same range.

For CoCrMo; however, neither influence of the near-surface RS state nor influence of roughness on corrosion behavior could be observed. The roughness values of microblasted CoCrMo are about two times lower than the roughness values of microblasted Ti. There might be a critical roughness, above which the corrosion behavior is affected. It is possible; however, that the influence of roughness on breakdown potential and passive current densities is alloy dependent. Evidence can be found in literature of metals and alloys that do not show a clear relation of corrosion resistance and surface roughness (e.g., β-Ti alloy and stainless steel) [33,35].

In addition to the analysis of the potentiodynamic polarization curves, the roughness of the materials before and after polarization might be an indicator for the extent of the corrosional attack during cyclic polarization. The deep-rolled and polished CoCrMo samples showed a tendency towards increased surface roughness after the four conducted polarization cycles. This indicates material loss due to pitting corrosion. Deep-rolled and microblasted Ti samples did not show altered roughness values after cyclic polarization in PBS and $FeCl_3$-containing electrolyte. The polished samples polarized in $FeCl_3$-containing electrolyte showed; however, a slight tendency towards decreasing surface roughness after the electrochemical tests. Roughness decrease due to electrochemical material degradation can be explained by the difference in Volta potential between valley and peaks, as electrons in peak areas are more likely to escape the material than electrons in valley areas [36]. This would mean that roughness progression due to galvanic corrosion would pass a minimum before a roughness increase due to pitting becomes evident. Furthermore, in the $FeCl_3$-containing environment, polished Ti samples exhibited lower breakdown potential than deep-rolled Ti samples. This indicates that the corrosion resistance of deep-rolled Ti6Al4V might be superior to that of polished Ti6Al4V, which might be correlated to the high compressive RSs in the near-surface region. Thus, for the three investigated surface conditions of Ti6Al4V, deep rolling resulted in relatively higher corrosion resistance.

This work investigated the potential of surface treatments for enhancing the corrosion resistance of passive metals, which are applied in the field of surgical implants. It has been shown that the surface treatments did impact the corrosion behavior of the metals, even though the observed differences on the breakdown potentials and passive current densities between polished and deep-rolled samples were rather small. The corrosion of taper connections is; however, mainly initiated by occurring micromotion, which can damage the passive layer leading, in turn, to fretting and galvanic corrosion [3,37,38]. The mechanical resistance of the oxide layer is; therefore, crucial for the corrosion resistance of the applied alloys. Assessment of corrosion in this study was somehow simplified, as the mechanical component of taper corrosion was not taken into account. This was done in order to separate the effects of occurring micromotion from the effects caused by galvanic corrosion. In this setup, the corrosion behavior is; therefore, rather governed by sample roughness than by the induced compressive RSs, especially since the investigated alloys are passive metals with a naturally high corrosion resistance. However, apart from the slightly improved corrosion resistance for the deep-rolled surface of Ti6Al4V observed in this study, the induction of compressive RSs by means of deep rolling have moreover the potential of combining a mirror smooth sample surface with the beneficial effects in terms of fretting and fatigue resistance.

Further studies are needed that involve actually occurring forces and resulting micromotion in order to evaluate the impact of compressive RSs onto the corrosion resistance of modular taper connections.

5. Conclusions

Compressive RSs were introduced in Ti6Al4V and CoCrMo by means of deep rolling and microblasting, while metallographically-prepared samples served as reference. For CoCrMo, the surface treatments did not alter the corrosion behavior. For Ti6Al4V, microblasting led to a lowered breakdown potential and/or increased passive current densities, which indicates decreased corrosion resistance. Further studies comprising mechanical loads are needed in order to assess the impact of mechanical surface treatments with respect to corrosion of modular taper connections in total joint arthroplasty.

Author Contributions: Conceptualization, J.P.K., J.G. and R.S.; methodology, J.P.K., U.M. and J.G.; validation, all authors; formal analysis, T.B. and P.T.M.; investigation, T.B. and P.T.M.; resources, J.P.K. and J.G.; data curation, T.B. and P.T.M.; writing—original draft preparation, T.B. and P.T.M.; writing—review and editing, J.P.K., R.S., U.M. and J.G.; visualization, T.B. and P.T.M.; supervision, J.P.K. and J.G.; project administration, J.P.K and J.G.; funding acquisition, J.P.K., R.S. and J.G. All authors have read and agreed to the published version of the manuscript.

Funding: This research was funded by the *Deutsche Forschungsgemeinschaft* (DFG) within the framework of the project no. 382919963.

Conflicts of Interest: The authors declare no conflict of interest related to this study.

References

1. Hussey, D.K.; McGrory, B.J. Ten-Year Cross-Sectional Study of Mechanically Assisted Crevice Corrosion in 1352 Consecutive Patients With Metal-on-Polyethylene Total Hip Arthroplasty. *J. Arthroplast.* **2017**, *32*, 2546–2551. [CrossRef] [PubMed]
2. Hall, D.J.; Pourzal, R.; Jacobs, J.J. What Surgeons Need to Know About Adverse Local Tissue Reaction in Total Hip Arthroplasty. *J. Arthroplast.* **2020**. [CrossRef] [PubMed]
3. Gilbert, J.L.; Buckley, C.A.; Jacobs, J.J. In vivo corrosion of modular hip prosthesis components in mixed and similar metal combinations. The effect of crevice, stress, motion, and alloy coupling. *J. Biomed. Mater. Res.* **1993**, *27*, 1533–1544. [CrossRef] [PubMed]
4. Wood, T.J.; Alzahrani, M.; Langohr, D.; Teeter, M.G.; Howard, J.L.; Schemitsch, E.H.; Lanting, B.A. Catastrophic femoral head trunnion dissociation: A case series with surface wear analysis. *HIP Int.* **2019**, *29*, NP1–NP5. [CrossRef]
5. Hunter, A.M.; Hsu, A.; Shah, A.; Naranje, S.M. A case example and literature review of catastrophic wear before catastrophic failure: Identification of trunnionosis and metallosis in metal-on-polyethylene hip arthroplasty prior to frank failure or fracture. *Eur. J. Orthop. Surg.Traumatol.* **2019**, *29*, 711–715. [CrossRef]
6. Martin, A.J.; Jenkins, D.R.; Van Citters, D.W. Role of corrosion in taper failure and head disassociation in total hip arthroplasty of a single design. *J. Orthop. Res.* **2018**, *36*, 2996–3003. [CrossRef]
7. Gascoyne, T.C.; Turgeon, T.R.; Burnell, C.D. Retrieval Analysis of Large-Head Modular Metal-on-Metal Hip Replacements of a Single Design. *J. Arthroplast.* **2018**, *33*, 1945–1952. [CrossRef]
8. Kurtz, S.M.; Kocagöz, S.B.; Hanzlik, J.A.; Underwood, R.J.; Gilbert, J.L.; MacDonald, D.W.; Lee, G.-C.; Mont, M.A.; Kraay, M.J.; Klein, G.R.; et al. Do Ceramic Femoral Heads Reduce Taper Fretting Corrosion in Hip Arthroplasty? A Retrieval Study. *Clin. Orthop. Relat. Res.* **2013**, *471*, 3270–3282. [CrossRef]
9. Rehmer, A.; Bishop, N.E.; Morlock, M.M. Influence of assembly procedure and material combination on the strength of the taper connection at the head–neck junction of modular hip endoprostheses. *Clin. Biomech.* **2012**, *27*, 77–83. [CrossRef]
10. Falkenberg, A.; Biller, S.; Morlock, M.M.; Huber, G. Micromotion at the head-stem taper junction of total hip prostheses is influenced by prosthesis design-, patient- and surgeon-related factors. *J. Biomech.* **2020**, *98*, 109424. [CrossRef]
11. Panagiotidou, A.; Cobb, T.; Meswania, J.; Skinner, J.; Hart, A.; Haddad, F.; Blunn, G. Effect of impact assembly on the interface deformation and fretting corrosion of modular hip tapers: An in vitro study. *J. Orthop. Res.* **2018**, *36*, 405–416. [CrossRef] [PubMed]
12. Ramoutar, D.N.; Crosnier, E.A.; Shivji, F.; Miles, A.W.; Gill, H.S. Assessment of Head Displacement and Disassembly Force With Increasing Assembly Load at the Head/Trunnion Junction of a Total Hip Arthroplasty Prosthesis. *J. Arthroplast.* **2017**, *32*, 1675–1678. [CrossRef] [PubMed]
13. Majzoobi, G.H.; Abbasi, F. On the effect of shot-peening on fretting fatigue of Al7075-T6 under cyclic normal contact loading. *Surf. Coat. Technol.* **2017**, *328*, 292–303. [CrossRef]
14. Liu, K.K.; Hill, M.R. The effects of laser peening and shot peening on fretting fatigue in Ti–6Al–4V coupons. *Tribol. Int.* **2009**, *42*, 1250–1262. [CrossRef]
15. Zhao, X.; Zhou, H.; Liu, Y. Effect of shot peening on the fatigue properties of nickel-based superalloy GH4169 at high temperature. *Results Phys.* **2018**, *11*, 452–460. [CrossRef]
16. Sonntag, R.; Reinders, J.; Gibmeier, J.; Kretzer, J.P. Fatigue performance of medical Ti6Al4V alloy after mechanical surface treatments. *PLoS ONE* **2015**, *10*, e0121963. [CrossRef] [PubMed]
17. Altenberger, I.; Nalla, R.K.; Sano, Y.; Wagner, L.; Ritchie, R.O. On the effect of deep-rolling and laser-peening on the stress-controlled low- and high-cycle fatigue behavior of Ti–6Al–4V at elevated temperatures up to 550 °C. *Int. J. Fatigue* **2012**, *44*, 292–302. [CrossRef]
18. Eto, S.; Miura, Y.; Tani, J.; Fujii, T. Effect of residual stress induced by pulsed-laser irradiation on initiation of chloride stress corrosion cracking in stainless steel. *Mater. Sci. Eng. A* **2014**, *590*, 433–439. [CrossRef]
19. Wu, S.-X.; Wang, S.-R.; Wang, G.-Q.; Yu, X.-C.; Liu, W.-T.; Chang, Z.-Q.; Wen, D.-S. Microstructure, mechanical and corrosion properties of magnesium alloy bone plate treated by high-energy shot peening. *Trans. Nonferrous Met. Soc. China* **2019**, *29*, 1641–1652. [CrossRef]

20. Trdan, U.; Sano, T.; Klobčar, D.; Sano, Y.; Grum, J.; Šturm, R. Improvement of corrosion resistance of AA2024-T3 using femtosecond laser peening without protective and confining medium. *Corros. Sci.* **2018**, *143*, 46–55. [CrossRef]
21. Lee, H.; Mall, S.; Allen, W.Y. Fretting fatigue behavior of shot-peened Ti–6Al–4V under seawater environment. *Mater. Sci. Eng. A* **2006**, *420*, 72–78. [CrossRef]
22. Weingärtner, R.; Hoffmeister, J.; Schulze, V. Mechanical Surface Treatment by Micro Peening. *HTM J. Heat Treat. Mater.* **2015**, *70*, 59–65. [CrossRef]
23. Pypen, C.M.J.M.; Plenk, H., Jr.; Ebel, M.F.; Svagera, R.; Wernisch, J. Characterization of microblasted and reactive ion etched surfaces on the commercially pure metals niobium, tantalum and titanium. *J. Mater. Sci. Mater. Med.* **1997**, *8*, 781–784. [CrossRef] [PubMed]
24. Eigenmann, B.; Macherauch, E. Röntgenographische Untersuchung von Spannungszuständen in Werkstoffen. *Materialwissenschaft Werkst.* **1995**, *26*, 148–160. [CrossRef]
25. ASTM. *F2129-15, Standard Test Method for Conducting Cyclic Poentiodynamic Polarization Measurements to Determine the Corrosion Susceptibility of Small Implant Devices*; ASTM International: West Conshohocken, PA, USA, 2015.
26. Baxmann, M.; Pfaff, A.M.; Schilling, C.; Grupp, T.M.; Morlock, M.M. Biomechanical Evaluation of the Fatigue Performance, the Taper Corrosion and the Metal Ion Release of a Dual Taper Hip Prosthesis under Physiological Environmental Conditions. *Biotribology* **2017**, *12*, 1–7. [CrossRef]
27. ASTM. *G48-03, Standard Test Methods for Pitting and Crevice Corrosion Resistance of Stainless Steels and Related Alloys by Use of Ferric Chloride Solution*; ASTM International: West Conshohocken, PA, USA, 2003.
28. Milošev, I.; Metikoš-Huković, M.; Strehblow, H.H. Passive film on orthopaedic TiAlV alloy formed in physiological solution investigated by X-ray photoelectron spectroscopy. *Biomaterials* **2000**, *21*, 2103. [CrossRef]
29. Sivakumar, B.; Pathak, L.C.; Singh, R. Role of surface roughness on corrosion and fretting corrosion behaviour of commercially pure titanium in Ringer's solution for bio-implant application. *Appl. Surf. Sci.* **2017**, *401*, 385–398. [CrossRef]
30. Karimi, S.; Nickchi, T.; Alfantazi, A. Effects of bovine serum albumin on the corrosion behaviour of AISI 316L, Co–28Cr–6Mo, and Ti–6Al–4V alloys in phosphate buffered saline solutions. *Corros. Sci.* **2011**, *53*, 3262–3272. [CrossRef]
31. Milošev, I.; Strehblow, H.H. The composition of the surface passive film formed on CoCrMo alloy in simulated physiological solution. *Electrochim. Acta* **2003**, *48*, 2767–2774. [CrossRef]
32. Milošev, I.; Kapun, B. The corrosion resistance of Nitinol alloy in simulated physiological solutions: Part 1: The effect of surface preparation. *Mater. Sci. Eng. C* **2012**, *32*, 1087–1096. [CrossRef]
33. Hunt, N.P.; Cunningham, S.J.; Golden, C.G.; Sheriff, M. An investigation into the effects of polishing on surface hardness and corrosion of orthodontic archwires. *Angle Orthod.* **1999**, *69*, 433–440. [PubMed]
34. Chi, G.; Yi, D.; Liu, H. Effect of roughness on electrochemical and pitting corrosion of Ti-6Al-4V alloy in 12 wt.% HCl solution at 35 °C. *J. Mater. Res. Technol.* **2019**. [CrossRef]
35. Azar, V.; Hashemi, B.; Rezaee Yazdi, M. The effect of shot peening on fatigue and corrosion behavior of 316L stainless steel in Ringer's solution. *Surf. Coat. Technol.* **2010**, *204*, 3546–3551. [CrossRef]
36. Kim, S.K.; Park, I.J.; Lee, D.Y.; Kim, J.G. Influence of surface roughness on the electrochemical behavior of carbon steel. *J. Appl. Electrochem.* **2013**, *43*, 507–514. [CrossRef]
37. Goldberg, J.R.; Gilbert, J.L.; Jacobs, J.J.; Bauer, T.W.; Paprosky, W.; Leurgans, S. A multicenter retrieval study of the taper interfaces of modular hip prostheses. *Clin. Orthop. Relat. Res.* **2002**, *401*, 149–161. [CrossRef] [PubMed]
38. Osman, K.; Panagiotidou, A.P.; Khan, M.; Blunn, G.; Haddad, F.S. Corrosion at the head-neck interface of current designs of modular femoral components. *Bone Joint J.* **2016**, *98*, 579–584. [CrossRef]

materials

MDPI

Review

What the Surgeon Can Do to Reduce the Risk of Trunnionosis in Hip Arthroplasty: Recommendations from the Literature

Claude B. Rieker [1,*] and Peter Wahl [2]

1 Scientific Affairs, Zimmer Biomet EMEA (Europa, Middle East and Africa), Sulzerallee 8, 8404 Winterthur, Switzerland

2 Division of Orthopaedics and Traumatology, Cantonal Hospital Winterthur, Brauerstrasse 15, 8400 Winterthur, Switzerland; peter.wahl@ksw.ch

* Correspondence: claude.rieker@zimmerbiomet.com

Received: 29 February 2020; Accepted: 17 April 2020; Published: 21 April 2020

Abstract: Trunnionosis, defined as wear and corrosion at the head–neck taper connection, is a cause of failure in hip arthroplasty. Trunnionosis is linked to a synergistic combination of factors related to the prosthesis, the patient, and the surgeon. This review presents analytical models that allow for the quantification of the impact of these factors, with the aim of providing practical recommendations to help surgeons minimize the occurrence of this failure mode. A tighter fit reduces micromotion and, consequently, fretting of the taper connection. The paramount parameters controlling the fixation force are the coefficient of friction and the impaction force. The influence of the head diameter, as well as of the diameter and angle of the taper, is comparatively small, but varus alignment of the taper and heads with longer necks are unfavourable under physiologic loads. The trunnion should be rinsed, cleaned, and dried carefully, while avoiding any contamination of the bore—the female counterpart within the head—prior to assembly. Biological debris, and even residual water, might critically reduce the fixation of the taper connection between the head and the neck. The impaction force applied to the components should correspond to at least two strong blows with a 500 g hammer, striking the head with an ad hoc impactor aligned with the axis of the taper. These strong blows should correspond to a minimum impaction force of 4000 N.

Keywords: hip arthroplasty; trunnionosis; trunnion failure; fretting corrosion; head–neck junction; mechanically assisted crevice corrosion

1. Introduction

Total hip arthroplasty (THA) is so successful in restoring mobility and relieving pain in patients with degenerated hip joints [1] that it has been nominated as the operation of the 20th century [2]. Failure, however, remains an issue, with between one-third and nearly half of THA procedures requiring postoperative revision within 30 years [3,4]. One of the possible causes of failure is trunnionosis [5]. Trunnionosis is defined as wear and corrosion at the head–neck taper connection [6]. Hence, it is associated with the modularity of the head–stem construct. Modularity gives the surgeon the flexibility to choose femoral heads of varying materials and diameters, with variable neck lengths, so that the joint replacement can be adjusted according to the patient's anatomy [7]. It can also reduce the inventory and consecutive storage costs [8,9]. In THA, the use of modular heads began in the early 1970s and has almost completely supplanted monobloc femoral components [6,10].

Epidemiological data on the incidence of clinically relevant trunnionosis are scarce. Up to 4.7% of revisions are reported to be attributable to taper corrosion [11–13]. Rates of up to 10.5% are even reported for certain subgroups [12]. While the latter number seems rather high, the occurrence of trunnionosis is certainly underreported, as the taper is seldom analysed. In some revisions, only the femoral head is revised and the stem is retained, making a full analysis impossible. Sometimes, the stem may be revised without disconnecting the head. Additionally, it remains quite difficult to determine the clinical relevance of this issue when other reasons for revision are present concomitantly. While taper corrosion might be observed frequently on retrievals [14–16], the clinical relevance might remain difficult to determine [11], particularly because approximately one-third of the adverse local tissue reaction (ALTR) pseudotumours related to taper corrosion identified on magnetic resonance imaging are asymptomatic [17]. A determination of the levels of cobalt and chromium in the synovial fluid might help identify taper corrosion-related issues [18].

The aetiology of trunnionosis is believed to be a synergistic combination of factors related to the prosthesis, the patient, and the surgeon [19–25]. Trunnionosis involves both fretting corrosion as well as crevice corrosion [26,27]. This process is called mechanically assisted crevice corrosion (MACC) [19]. Taper corrosion can lead to elevated metal ion levels in the synovial fluid of the affected joint as well as in the serum, and may cause ALTR [11,13,18,28,29]. ALTR includes lymphocyte-dominated inflammatory reactions and macrophage infiltrates reacting to particulate corrosion products [11,30]. These can lead to synovitis, local osteolysis, the necrosis of periprosthetic tissues and, finally, component loosening [11,13,31]. Long-term MACC leads to material loss at the taper junction, which can, in rare cases, lead to the frank dissociation of the connection, as well as marked taper deformity [13,20,32–34].

This manuscript aims to quantify the impact of these factors according to published analytical models, as well as clinical and in vitro studies, and provide practical recommendations to help surgeons minimize the occurrence of trunnionosis. We conducted a thorough non-systematic review of the literature using two search engines (PubMed and Google Scholar, using the following keywords: trunnionosis, fretting corrosion, taper connection, taper corrosion, taper failure, MACC, modularity, assembly force, disassembly force, micromotion) and cross-referenced related studies to identify the relevant literature.

2. Technical Aspects of Taper Connections in Hip Arthroplasty

A taper connection is a means of reliably joining two mechanical components, by tightly fitting a cone into a negative cone-shaped counterpart [35]. The male component is referred to as the trunnion, while the female counterpart is a bore [24,35]. A taper is defined by three parameters: the largest diameter at its base, the smallest diameter at its opening or tip, and its angle [36].

MACC of a Morse taper connection is caused by fretting and crevice corrosion [26,27]. Fretting first disrupts the protective oxide layer on the surfaces of the taper and causes wear. Changes in local chemistry within crevices then lead to the complex interactions of crevice corrosion [26,27]. Although repassivation (i.e., reformation of the protective oxide layers) occurs naturally, fretting alters the repassivation of the exposed metals [24,25,37].

A recently published study indicates that trunnionosis is mainly determined by fretting corrosion, rather than by crevice corrosion [38]. Therefore, minimizing the micromotions at the head–neck taper interface would mitigate the starting conditions of trunnionosis. A strong press-fit fixation of the taper interface will logically lower these micromotions [21,39]. This underlines the importance of a stable fixation between the femoral head and the stem's trunnion [6,20,21]. Fretting corrosion at the taper interface is linked to micromotions of 5 μm to 12 μm [21,40]. Time in vivo (i.e., exposure to repeated loads) is also linked to the degree of corrosion of tapers [12,26]. Thus, a stable taper connection with lower micromotions under physiological loads will produce a reduced risk of trunnionosis. The force required to remove the head from the taper is a measure of taper stability, denoted as the fixation force or the pull-off force. Given the association between fixation force (i.e., the force necessary to dissociate the taper connection) and micromotion at the taper's interface [21,39], the fixation force is a surrogate parameter for the rest of this analysis. Fessler et al. [41,42] and MacLeod et al. [43] have both provided analytical models to estimate the fixation force between the neck and the head. These two analytic models are presented in Figure 1.

Fessler's model:

$$F_{fixation} = F_{impact} \cdot \frac{\mu\cos(0.5\alpha) - \sin(0.5\alpha)}{\mu\cos(0.5\alpha) + \sin(0.5\alpha)}$$

MacLeod's model:

$$F_{fixation} = F_{impact} \cdot \frac{\mu\cos(0.5\alpha) - \sin(0.5\alpha)}{\mu\cos(0.5\alpha) + \sin(0.5\alpha)} \cdot \frac{E_h r_f}{2 r_h^2} \cdot \left(1 - \frac{r_h^2}{r_f^2}\right) \cdot \left[\frac{r_f}{E_h}\left(\left\{\frac{r_h^2 + r_f^2}{r_h^2 - r_f^2}\right\} + \vartheta_h\right) + \frac{r_f}{E_t}(1 - \vartheta_t)\right]$$

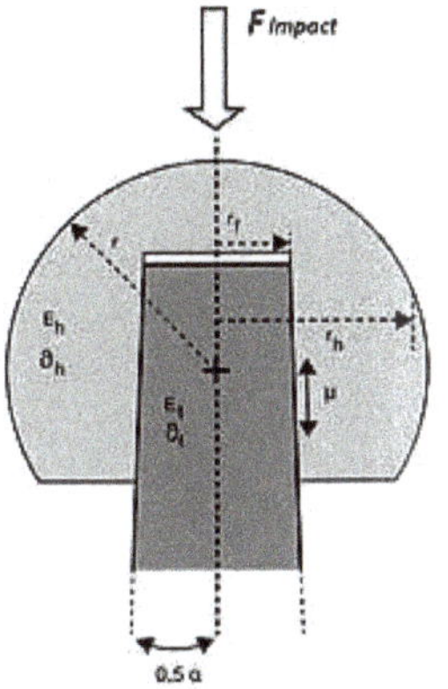

- F_{impact}: impaction force
- μ: coefficient of friction
- α: angle of the taper
- r: radius of the femoral head
- r_h: radius of the surrounding material
- r_f: proximal radius of the taper
- E_t: modulus of elasticity of the stem
- E_h: modulus of elasticity of the head
- ϑ_h: Poisson's ratio of the head
- ϑ_t: Poisson's ratio of the stem

Figure 1. Fessler and MacLeod's formulas illustrated. MacLeod's model uses the same fundamental formula as Fessler's model, as illustrated in the figure, but adds three factors that consider the geometry and the material properties of the system. In both models, the fixation force correlates linearly with the force applied to impact the taper. The effects of the other parameters are illustrated in Figures 2–4.

MacLeod's model is an extension of Fessler's model and adds three multiplying factors that consider the geometry and the material properties of the system. In both models, the fixation force correlates linearly with the force applied to impact the taper. The fixation force, therefore, has a major effect on the stability of the taper connection [6,44–52]. Because the taper angle α is determined by the chosen prosthesis design (typically between 5°30′ and 6° [36]), this parameter has to be considered as fixed. The effect of the taper angle α on the fixation force is negligible, as illustrated in Figure 2. The taper angle α should not be confused with the slope of the taper, which is α/2 [43]. MacLeod's equation demonstrates that the influence of the diameter of the head (28 mm to 60 mm) is also relatively small, accounting for a maximum variation of 6.4%, as demonstrated in Figure 3. Both models predict that the fixation force is nil for a coefficient of friction of about 0.05, and this fixation force increases as the coefficient of friction increases (Figure 4).

When all other parameters are equal, MacLeod's model estimates a higher fixation force than Fessler's model by about a third (Figure 2). However, when considering the technical aspects of the different experimental setups, nearly all fixation force values described in the literature correspond to the values estimated by Fessler's formula, ranging 40% to 55% of the impaction force [6,21,44,47–52]. Surprisingly, the values measured by MacLeod et al. do not correspond to those estimated by their own model [43].

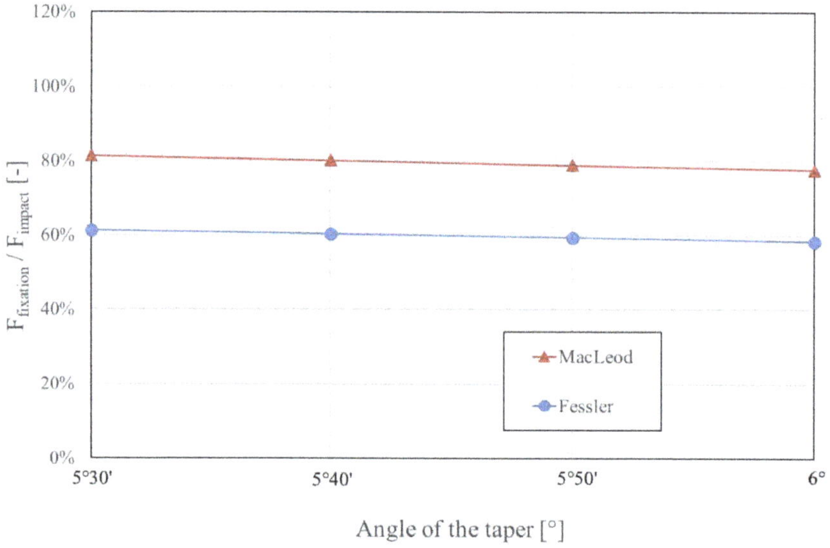

Figure 2. Influence of the angle of the taper on fixation force, as estimated by Fessler and MacLeod's models. With the impaction force having a linear effect in both models, the fixation force is represented as a proportion of the impaction force. The range of taper angles illustrated covers the range of tapers available commercially in hip arthroplasty. For both models, a coefficient of friction μ of 0.2 was considered. For MacLeod's estimate, a 32-mm head was considered.

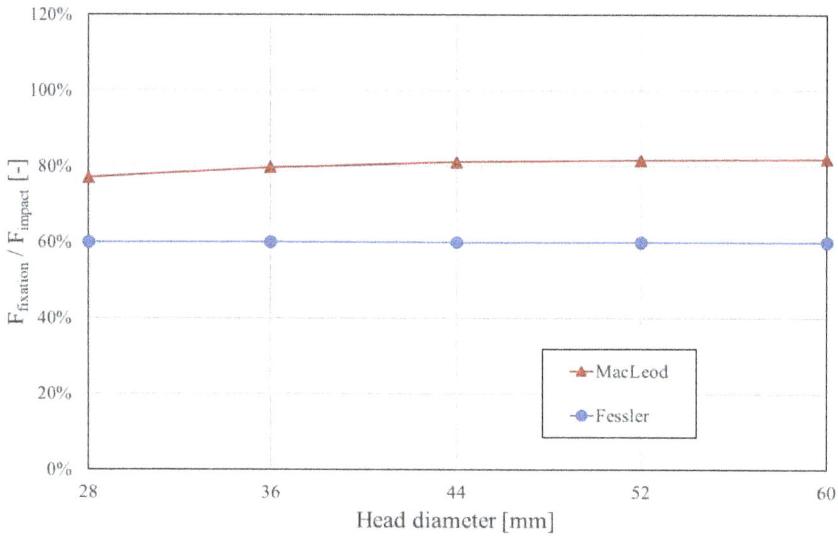

Figure 3. Influence of the diameter of the head on fixation force, as estimated by Fessler and MacLeod's models. As this parameter is not considered in Fessler's model, only MacLeod's model shows a variability depending on this factor. The influence on fixation force, however, remains relatively small and negligible. Notably, MacLeod's formula provides results with a positive influence for an increasing head size, whereas the measurements describe a decrease of approximately 20% in the fixation force when the head size is increased from 28 to 36 mm. With the impaction force having a linear effect in both models, the fixation force is represented as a proportion of the impaction force.

MacLeod's model is not completely trustworthy for two additional reasons. First, it predicts a higher fixation force than the impaction force, starting from a coefficient of friction of 0.372 (for a 32-mm head) and upward. This would imply the creation of energy within the system, which is clearly impossible. Secondly, while the experimental observations indicated a reduction in the fixation force with an increasing head diameter (with the fixation force of 36 mm heads being 20% less than the fixation force for 28 mm heads [43]), the model predicts the contrary (Figure 3). Considering all these elements, MacLeod's equation will be omitted for further analysis in this review, and only Fessler's equation will be considered.

Therefore, the two main parameters controlling the fixation force are the impaction force, which has a linear effect, and the coefficient of friction μ, which should be as high as possible. Under ideal conditions, the value of μ is approximately 0.15 to 0.25 [41,53–55]. However, the relationship between the fixation force and the coefficient of friction is not linear. This force becomes nil when the coefficient of friction approaches 0.05 (Figure 4).

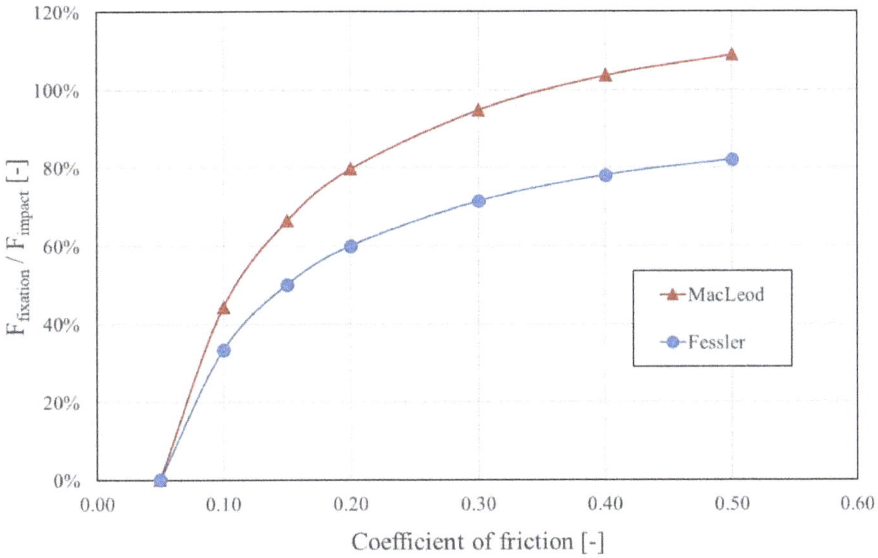

Figure 4. Influence of the coefficient of friction μ on fixation force, as estimated by Fessler and MacLeod's models. For MacLeod's model, a head diameter of 32 mm was considered. With the impaction force having a linear effect in both models, the fixation force is represented as a proportion of the impaction force. The effect of the coefficient of friction μ is not linear. With a coefficient of friction < 0.05, stable fixation is not possible. MacLeod's model not only predicts higher fixation forces than Fessler's model, it also predicts a fixation force higher than the impaction force starting from a coefficient of friction of 0.372 (for a 32-mm head) and upward, which is physically impossible.

Symptomatic trunnionosis appears to be associated with increased head offsets and longer neck lever arms [21,22,56,57]. Micro-grooved tapers, designed to improve stress distributions in ceramic heads, may increase the likelihood of corrosion when combined with metal heads [21,58,59]. However, retrieval studies have not been able to confirm the impact of surface topography, because they have shown comparable fretting corrosion in vivo [21,60]. The variability induced by surface topography is obviously far less important than the other relevant factors in vivo. The association of head diameter with trunnionosis is not always consistent in the literature [12,22,26,57,61,62], but confounding parameters such as time in vivo were not always considered in these studies. Based on some of these publications, large-diameter heads induce force transmission on the taper with a greater lever arm, and this will be mechanically unfavourable when micromotions have to be minimized [43,57,63]. Interestingly, the influence of the head size is not always seen on retrievals [62]. In addition, and perhaps somewhat counterintuitively, the length of the trunnion does not appear to have a significant impact on micromotion or fixation force [47,64]. Increased trunnionosis has also been linked with low flexural rigidity necks (i.e., those with increased elasticity) [65] and small taper angle differences [66]. However, all these parameters are associated with the chosen design of the prosthesis and may not be modifiable intraoperatively. Taper incongruences caused by mixing heads and tapers from different manufacturers might affect the strength of the taper connection and must be avoided, given the variability of the effect associated with cobalt–chromium (CoCr) heads, because taper incongruences may critically reduce the fracture load of ceramic heads [67,68], and because of certification issues. Gross trunnion failures are reportedly above average for tapers made of beta titanium alloy (titanium–molybdenum–zirconium–iron (TMZF)) [34]. This specific beta titanium alloy might have fretting corrosion characteristics that are unfavourable for taper connections.

3. Patient-Related Factors for Trunnionosis

Patient-related factors that affect corrosion may include excess body weight [69] and high-impact activities with a resultant increase in demand on the prosthesis [70]. Bergmann et al. have shown that prosthetic loads are several times the patient's body weight, and the exact load level greatly depends on the patient's activity [71]. Impact activities greatly increase prosthetic loads; for example, jogging and brisk walking increased the prosthetic load by 3.9 times a patient's body weight, while stumbling increased the prosthetic load by 11 times a patient's body weight [71]. These larger loads obviously amplify micromotions at the taper interface. However, the surgeon cannot influence these parameters, as they depend on the patient's activities in daily life.

The risk of trunnionosis should be considered when planning for THA in patients with larger femoral neck offsets. Varus orientation of the taper and the use of heads with +4 mm necks or longer significantly increase the risk of micromotion and MACC in the taper interface [21]. Given the risk of trunnionosis, implanting the stem in a varus axis to increase the offset might not be mechanically sound. This might well become an issue, particularly with the so-called short stems, which are often recommended for implantation in varus to reconstruct larger femoral offsets [72]. As heads with longer necks might also be detrimental [21], stem designs with larger offsets might have to be favoured. In general, older designs have smaller offsets. When choosing stems with increased offsets, the reduced flexural rigidity of the neck or the taper should be avoided [65,73]. In our opinion, the traditional solution in THA in compensating offset loss with leg-lengthening should be re-evaluated and updated in light of the increasing awareness of trunnionosis.

4. Surgeon-Related Factors Determining Taper Fixation

Micromotions in the taper interface, and thus the risk of trunnionosis, are determined by factors under the direct control of the surgeon. These factors are the coefficient of friction and the impaction force, which also appear to be the main determinants of the fixation force, once the prosthetic design has been chosen.

The coefficient of friction is significantly affected by the condition of the trunnion at the time of assembly. This has a major influence on the fixation force, as illustrated in Figure 4. According to several in vitro studies, fluid or fat left on the trunnion at the time of head assembly negatively affect fixation force [10,45,74,75]. By cleaning the trunnion with saline solution and drying it with gauze directly before the assembly of the head, the disassembly forces increase to values observed on pristine control trunnions [10,44,75]. The effect of contamination on the coefficient of friction is independent of the head material (CoCr versus ceramic) [44]. The contamination of the female taper should be avoided carefully while manipulating the head before seating it, as adequate cleaning of the bore would be particularly difficult. A lower coefficient of friction reduces the fixation force and causes higher hoop stresses, increasing the fracture risk for ceramic heads [75]. Drying and cleaning should be done with gauze only, as metallic brushes or pads damage the surface of the taper [76].

The head should never be struck directly with the hammer, but instead with an adequate impactor to avoid damaging the bearing surface of metal heads or fracturing ceramic heads due to point loading. The characteristics of the impactor greatly influence the force transmitted from the hammer blow to the taper connection. Hard plastic tips found on commercially available impactors avoid damage to the head, but they reduce the assembly force by approximately 20% compared to a metallic tip [77]. Older impactors (with questionable stiffness) and impactors with rubber tips should be abandoned [77].

Practical recommendations regarding the impaction force require a discussion of the technical aspects of the studies found in the literature. Quasi-static assembly procedures ensure the precise measurement of the impaction force but do not correspond to the technical solutions available intraoperatively. As the rate of impaction has no relevant effect on the fixation force [49,78], impaction with a hammer might be considered equivalent, even if minor mechanical differences might be identified [55]. While force sensors at the tip of the impactor might approximate the assembly force applied to the taper, the energy dissipated by the impactor must be subtracted from the values measured by the sensing hammers [10,43,50,78,79]. Some studies do not even adequately describe the impaction force used or the fixation force measured [45,74].

The number of hammer blows does not play a significant role, as the impaction force is controlled by the impact with the highest energy [6,45,79]. To ensure the optimal impaction of the taper, two hammer blows are recommended [74]. The first blow could be seen as being the alignment blow and the second blow as being the definitive impaction blow. Many lighter impactions are not useful, as the effect is not cumulative. We recommend a good alignment between the impactor and the neck of the femoral stem. Based on a trigonometric model, a misalignment up to 20° could be tolerated, as 94% of the impacting force would still be maintained in the correct direction. This is confirmed experimentally, with fixation forces not altered significantly by seating loads applied at 20° [52]. Asymmetries of the seating load displacement could, however, be observed at the taper [51,52], and this might explain notable differences in the fixation force observed with off-axis impactions of only 10°, when combined in different planes [80]. Axis deviations of more than 20° are common during THA, at least when using a posterior approach to the hip [81]. Due to the off-axis orientation of physiologic loads, the manual assembly of a taper, relying on later impaction under the patient's weight, is inadequate [49,70,71].

Impaction with a hammer has its advantages, in that a short impulse causes a lower transmission of energy to the tissues distal to the taper than a slower application of the force [44,55,77]. It might be expected that the impaction of the taper with a hammer is associated with a low risk of femoral fracture, even if high forces are applied. An increased impaction force increases the contact surface, and this has a favourable effect on the stability of the taper connection [39]. However, there is a clinically relevant upper limit to the amount of force applicable. However, to the best of our knowledge, no study has established a maximum recommended force.

Impaction with at least 4000 N is recommended. An adequate impaction force can be reached with a strong blow from a 500 g hammer [46,50]. Surgeons should become familiar and proficient with force-measuring instruments, especially those surgeons with low levels of experience in arthroplasty, to avoid the application of insufficient or off-axis impaction force [46]. The settings of the instrumentation should be checked carefully to ensure that the force measured corresponds to the potential assembly force, and not the force applied with the hammer.

5. Discussion

Contemporary hip arthroplasty includes modular heads with variable neck lengths, as this increases the surgical options to tailor the implant to the patient's individual anatomy and allows for the use of heads made of materials that differ from that of the stem [7]. However, because of the additional interface, modularity may result in additional specific failures. Trunnionosis, defined as fretting and corrosion of the taper connection [6], may result in metal ions and metal particles entering the joint, causing ALTR. Since the first report of a pseudotumour related to taper corrosion was published as early as 1988 [82], corrosion and fretting at the taper junction have become increasingly linked to implant failure [11–13,20,22,83]. Surgeons should be aware of all controllable factors to minimize the risk of trunnionosis. This review highlights the important controllable factors that determine the stability of the head–neck taper connection: impaction force and the coefficient of friction at the taper interface.

Fessler's model quantifies the fixation force of a taper connection [6]. A higher fixation force leads to a tighter fit between components. This will reduce micromotions in the interface between the bore in the head and the stem's trunnion during physiologic loads, thereby preventing fretting corrosion [10,20,21,40]. The head size is not considered in Fessler's analytical model of fixation force (Figure 1). Since trunnionosis came into focus as clinically relevant, around ten years ago, multiple factors with a possible impact on the lever arm to the centre of rotation (e.g., large-diameter heads, high mediolateral offset, large neck length, and bearing type) have been discussed as factors that may exacerbate the fretting corrosion process [23,29,58,70]. However, the available data remain inconclusive as to their relevance, and their actual impact remains unclear even to date. Other factors excluded from the analytical model, but under the direct influence of the surgeon, are the impaction technique [6] and the avoidance of mismatch by combining heads and stems with different taper designs (i.e., from different manufacturers) [25,67,68,84].

Modern hip arthroplasty is prone to fretting and corrosion at the taper junction for three reasons. First, procedures are increasingly being performed through small incisions [85], which may impair the proper cleaning and drying of the taper due to reduced exposure. Secondly, driven by the trend to increase the range of motion, there has been a reduction in the diameter of the neck and of the length of the tapers, which reduces flexural rigidity [65,73]. At the same time, head diameters have increased with the aim of reducing the risk of dislocation, along with an increasing range of motion [24,86,87], a change justified by the improved wear characteristics of highly cross-linked polyethylene and ceramic bearings [88–90]. Thirdly, these developments have coincided with an increase in obesity rates in most patient populations, with the prevalence of adult obesity exceeding 50% in numerous countries [91]. Taken together, these three developments have created challenges for proper taper fixation and have resulted in corrosion issues.

Finally, there is an increasing understanding of the inter-subject variation in terms of the biologic response to wear particles. Macrophages are activated by wear particles from CoCr alloys [92]. This may lead to the cell-induced corrosion of the taper interface [93]. Improperly seated heads may develop micromotions under physiological loads large enough for macrophages to penetrate the interface and contribute to taper corrosion and failure [52]. This may also be the case with heads engaging the taper proximally, which is the rule for ceramic heads [24]. Differences in alleles are strongly associated with the development of pseudotumours after THA with metal-on-metal bearings [94]. Variations in the genetic signal on the seventh chromosome can influence the probability of developing osteolysis after THA [95].

Trunnionosis develops due to the insufficient fixation of the taper connection. Wear and corrosion alter the surface and the geometry of the trunnion. Despite this, well-fixed stems do not necessarily need to be revised when the trunnion shows severe corrosion. The correct application of a new head does not lead to increased revision rates for corrosion-related issues [11,96,97]. Macroscopic material loss on the trunnion obviously does not allow the proper seating of a new head, and the stem should then be revised. Ceramic heads with titanium alloy inner sleeves may reduce the risk of the recurrence of corrosion issues at the level of the taper compared to CoCr heads [11,97]. However, this remains unconfirmed, considering the difficulties in identifying relevant material issues and the small sample sizes in the studies cited.

In conclusion, trunnionosis is a multifactorial phenomenon related to wear and corrosion in the modular links in hip arthroplasty. A tighter fit decreases micromotion and fretting of the taper interface in the long term and thus reduces the risk of trunnionosis. To minimize micromotions leading to wear and corrosion of the taper connection, surgeons should be aware of the factors directly under their control. Analytical models and empirical investigations reflect the critical significance of carefully cleaning, rinsing, and drying the taper before assembly. The absence of biological residues will yield a higher fixation force between the taper and the femoral head. Fixation strength increases linearly with impaction force. Based on the literature, the adequate impaction of the taper connection can be achieved with at least two strong blows from a 500 g hammer. Surgeons are encouraged to undertake training on force-measuring machines to ensure adequate impaction.

Author Contributions: Conceptualization, C.B.R. and P.W.; methodology, C.B.R. and P.W.; formal analysis, C.B.R. and P.W.; writing—original draft preparation, C.B.R. and P.W.; writing—review and editing, C.B.R. and P.W.; visualization, C.B.R. and P.W. All authors have read and agreed to the published version of the manuscript.

Funding: This research received no external funding.

Conflicts of Interest: The corresponding author (C.B.R.) is a salaried employee and owns stock of Zimmer Biomet, a manufacturer of orthopaedic implants. P.W. declares no conflict of interest.

References

1. Lindgren, J.V.; Wretenberg, P.; Karrholm, J.; Garellick, G.; Rolfson, O. Patient-reported outcome is influenced by surgical approach in total hip replacement: A study of the Swedish Hip Arthroplasty Register including 42,233 patients. *Bone Jt. J.* **2014**, *96-B*, 590–596. [CrossRef] [PubMed]
2. Learmonth, I.D.; Young, C.; Rorabeck, C. The operation of the century: Total hip replacement. *Lancet (Lond. Engl.)* **2007**, *370*, 1508–1519. [CrossRef]
3. Sadoghi, P.; Schroder, C.; Fottner, A.; Steinbruck, A.; Betz, O.; Muller, P.E.; Jansson, V.; Holzer, A. Application and survival curve of total hip arthroplasties: A systematic comparative analysis using worldwide hip arthroplasty registers. *Int. Orthop.* **2012**, *36*, 2197–2203. [CrossRef] [PubMed]
4. Evans, J.T.; Evans, J.P.; Walker, R.W.; Blom, A.W.; Whitehouse, M.R.; Sayers, A. How long does a hip replacement last? A systematic review and meta-analysis of case series and national registry reports with more than 15 years of follow-up. *Lancet (Lond. Engl.)* **2019**, *393*, 647–654. [CrossRef]
5. Esposito, C.I.; Wright, T.M.; Goodman, S.B.; Berry, D.J.; Clinical, B.; Bioengineering Study Groups from Carl, T.B.W. What is the trouble with trunnions? *Clin. Orthop. Relat. Res.* **2014**, *472*, 3652–3658. [CrossRef]
6. Rehmer, A.; Bishop, N.E.; Morlock, M.M. Influence of assembly procedure and material combination on the strength of the taper connection at the head-neck junction of modular hip endoprostheses. *Clin. Biomech. (Bristolavon)* **2012**, *27*, 77–83. [CrossRef]
7. Barrack, R.L. Modularity of Prosthetic Implants. *J. Am. Acad. Orthop. Surg.* **1994**, *2*, 16–25. [CrossRef]
8. Cameron, H.U. Modularity in primary total hip arthroplasty. *J. Arthroplast.* **1996**, *11*, 332–334; discussion 337–338. [CrossRef]
9. McCarthy, J.C.; Bono, J.V.; O'Donnell, P.J. Custom and modular components in primary total hip replacement. *Clin. Orthop. Relat. Res.* **1997**, *344*, 162–171. [CrossRef]
10. Lavernia, C.J.; Baerga, L.; Barrack, R.L.; Tozakoglou, E.; Cook, S.D.; Lata, L.; Rossi, M.D. The effects of blood and fat on Morse taper disassembly forces. *Am. J. Orthop. (Belle Mead Nj)* **2009**, *38*, 187–190.
11. Cooper, H.J.; Della Valle, C.J.; Berger, R.A.; Tetreault, M.; Paprosky, W.G.; Sporer, S.M.; Jacobs, J.J. Corrosion at the head-neck taper as a cause for adverse local tissue reactions after total hip arthroplasty. *J. Bone Jt. Surg. Am.* **2012**, *94*, 1655–1661. [CrossRef] [PubMed]
12. Hussey, D.K.; McGrory, B.J. Ten-Year Cross-Sectional Study of Mechanically Assisted Crevice Corrosion in 1352 Consecutive Patients with Metal-on-Polyethylene Total Hip Arthroplasty. *J. Arthroplast.* **2017**, *32*, 2546–2551. [CrossRef] [PubMed]
13. Urish, K.L.; Hamlin, B.R.; Plakseychuk, A.Y.; Levison, T.J.; Higgs, G.B.; Kurtz, S.M.; DiGioia, A.M. Trunnion Failure of the Recalled Low Friction Ion Treatment Cobalt Chromium Alloy Femoral Head. *J. Arthroplast.* **2017**, *32*, 2857–2863. [CrossRef] [PubMed]
14. Kurtz, S.M.; Kocagoz, S.B.; Hanzlik, J.A.; Underwood, R.J.; Gilbert, J.L.; MacDonald, D.W.; Lee, G.C.; Mont, M.A.; Kraay, M.J.; Klein, G.R.; et al. Do ceramic femoral heads reduce taper fretting corrosion in hip arthroplasty? A retrieval study. *Clin. Orthop. Relat. Res.* **2013**, *471*, 3270–3282. [CrossRef] [PubMed]

15. Witt, F.; Bosker, B.H.; Bishop, N.E.; Ettema, H.B.; Verheyen, C.C.; Morlock, M.M. The relation between titanium taper corrosion and cobalt-chromium bearing wear in large-head metal-on-metal total hip prostheses: A retrieval study. *J. Bone Jt. Surg. Am.* **2014**, *96*, e157. [CrossRef] [PubMed]
16. McGrory, B.J.; MacKenzie, J.; Babikian, G. A High Prevalence of Corrosion at the Head-Neck Taper with Contemporary Zimmer Non-Cemented Femoral Hip Components. *J. Arthroplast.* **2015**, *30*, 1265–1268. [CrossRef]
17. Kwon, Y.M.; Khormaee, S.; Liow, M.H.; Tsai, T.Y.; Freiberg, A.A.; Rubash, H.E. Asymptomatic Pseudotumors in Patients with Taper Corrosion of a Dual-Taper Modular Femoral Stem: MARS-MRI and Metal Ion Study. *J. Bone Jt. Surg. Am.* **2016**, *98*, 1735–1740. [CrossRef]
18. McGrory, B.J.; Payson, A.M.; MacKenzie, J.A. Elevated Intra-Articular Cobalt and Chromium Levels in Mechanically Assisted Crevice Corrosion in Metal-on-Polyethylene Total Hip Arthroplasty. *J. Arthroplast.* **2017**, *32*, 1654–1658. [CrossRef]
19. Goldberg, J.R.; Gilbert, J.L.; Jacobs, J.J.; Bauer, T.W.; Paprosky, W.; Leurgans, S. A multicenter retrieval study of the taper interfaces of modular hip prostheses. *Clin. Orthop. Relat. Res.* **2002**, 149–161. [CrossRef]
20. Morlock, M.M. The taper disaster–how could it happen? *Hip Int.* **2015**, *25*, 339–346. [CrossRef] [PubMed]
21. Falkenberg, A.; Biller, S.; Morlock, M.M.; Huber, G. Micromotion at the head-stem taper junction of total hip prostheses is influenced by prosthesis design-, patient- and surgeon-related factors. *J. Biomech.* **2019**, in press. [CrossRef] [PubMed]
22. Higgs, G.B.; Hanzlik, J.A.; MacDonald, D.W.; Gilbert, J.L.; Rimnac, C.M.; Kurtz, S.M.; Implant Research Center Writing, C. Is increased modularity associated with increased fretting and corrosion damage in metal-on-metal total hip arthroplasty devices?: A retrieval study. *J. Arthroplast.* **2013**, *28*, 2–6. [CrossRef]
23. Morlock, M.; Bunte, D.; Guhrs, J.; Bishop, N. Corrosion of the Head-Stem Taper Junction-Are We on the Verge of an Epidemic?: Review Article. *Hss J.* **2017**, *13*, 42–49. [CrossRef] [PubMed]
24. Hussenbocus, S.; Kosuge, D.; Solomon, L.B.; Howie, D.W.; Oskouei, R.H. Head-neck taper corrosion in hip arthroplasty. *Biomed. Res. Int.* **2015**, *2015*, 758123. [CrossRef] [PubMed]
25. Pourzal, R.; Lundberg, H.J.; Hall, D.J.; Jacobs, J.J. What Factors Drive Taper Corrosion? *J. Arthroplast.* **2018**, *33*, 2707–2711. [CrossRef] [PubMed]
26. Gilbert, J.L.; Buckley, C.A.; Jacobs, J.J. In vivo corrosion of modular hip prosthesis components in mixed and similar metal combinations. The effect of crevice, stress, motion, and alloy coupling. *J. Biomed. Mater. Res.* **1993**, *27*, 1533–1544. [CrossRef]
27. Jacobs, J.J.; Gilbert, J.L.; Urban, R.M. Corrosion of metal orthopaedic implants. *J. Bone Jt. Surg. Am.* **1998**, *80*, 268–282. [CrossRef]
28. Chalmers, B.P.; Perry, K.I.; Taunton, M.J.; Mabry, T.M.; Abdel, M.P. Diagnosis of adverse local tissue reactions following metal-on-metal hip arthroplasty. *Curr. Rev. Musculoskelet Med.* **2016**, *9*, 67–74. [CrossRef]
29. Kwon, Y.M.; Fehring, T.K.; Lombardi, A.V.; Barnes, C.L.; Cabanela, M.E.; Jacobs, J.J. Risk stratification algorithm for management of patients with dual modular taper total hip arthroplasty: Consensus statement of the American Association of Hip and Knee Surgeons, the American Academy of Orthopaedic Surgeons and the Hip Society. *J. Arthroplast.* **2014**, *29*, 2060–2064. [CrossRef]
30. Krenn, V.; Morawietz, L.; Perino, G.; Kienapfel, H.; Ascherl, R.; Hassenpflug, G.J.; Thomsen, M.; Thomas, P.; Huber, M.; Kendoff, D.; et al. Revised histopathological consensus classification of joint implant related pathology. *Pathol. Res. Pract.* **2014**, *210*, 779–786. [CrossRef] [PubMed]
31. Abu-Amer, Y.; Darwech, I.; Clohisy, J.C. Aseptic loosening of total joint replacements: Mechanisms underlying osteolysis and potential therapies. *Arthritis Res. Ther.* **2007**, *9* (Suppl. 1), S6. [CrossRef] [PubMed]
32. Banerjee, S.; Cherian, J.J.; Bono, J.V.; Kurtz, S.M.; Geesink, R.; Meneghini, R.M.; Delanois, R.E.; Mont, M.A. Gross trunnion failure after primary total hip arthroplasty. *J. Arthroplast.* **2015**, *30*, 641–648. [CrossRef] [PubMed]
33. Bolarinwa, S.A.; Martino, J.M.; Moskal, J.T.; Wolfe, M.W.; Shuler, T.E. Gross trunnion failure after a metal-on-polyethylene total hip arthroplasty leading to dissociation at the femoral head-trunnion interface. *Arthroplast. Today* **2019**, *5*, 5–10. [CrossRef] [PubMed]
34. Bansal, T.; Aggarwal, S.; Dhillon, M.S.; Patel, S. Gross trunnion failure in metal on polyethylene total hip arthroplasty-a systematic review of literature. *Int. Orthop.* **2020**. [CrossRef]

35. Hernigou, P.; Queinnec, S.; Flouzat Lachaniette, C.H. One hundred and fifty years of history of the Morse taper: From Stephen A. Morse in 1864 to complications related to modularity in hip arthroplasty. *Int. Orthop.* **2013**, *37*, 2081–2088. [CrossRef] [PubMed]
36. Mueller, U.; Braun, S.; Schroeder, S.; Sonntag, R.; Kretzer, J.P. Same Same but Different? 12/14 Stem and Head Tapers in Total Hip Arthroplasty. *J. Arthroplast.* **2017**, *32*, 3191–3199. [CrossRef] [PubMed]
37. Virtanen, S.; Milosev, I.; Gomez-Barrena, E.; Trebse, R.; Salo, J.; Konttinen, Y.T. Special modes of corrosion under physiological and simulated physiological conditions. *Acta Biomater.* **2008**, *4*, 468–476. [CrossRef]
38. Bermudez Castefieda, A. Degradation of Modular Hip Joint Implants, a Corrosion and Tribocorrosion Approach. Ph.D. Thesis, Ecole Polytechnique de Lausanne, Lausanne, Switzerland, 2018.
39. Panagiotidou, A.; Cobb, T.; Meswania, J.; Skinner, J.; Hart, A.; Haddad, F.; Blunn, G. Effect of impact assembly on the interface deformation and fretting corrosion of modular hip tapers: An in vitro study. *J. Orthop. Res. Off. Publ. Orthop. Res. Soc.* **2018**, *36*, 405–416. [CrossRef]
40. Mali, S.A.; Gilbert, J.L. Correlating Fretting Corrosion and Micromotions in Modular Tapers: Test Method Development and Assessment. In *Modularity and Tapers in Total Joint Replacement Devices*; Greenwald, A., Kurtz, S., Lemons, J., Mihalko, W., Eds.; ASTM International: West Conshohocken, PA, USA, 2015; pp. 259–282. [CrossRef]
41. Fessler, H.; Fricker, D.C. Friction in femoral prosthesis and photoelastic model cone taper joints. *Proc. Inst. Mech. Eng. H* **1989**, *203*, 1–14. [CrossRef]
42. Fessler, H.; Fricker, D.C. A study of stresses in alumina universal heads of femoral prostheses. *Proc. Inst. Mech. Eng. H* **1989**, *203*, 15–34. [CrossRef]
43. MacLeod, A.R.; Sullivan, N.P.; Whitehouse, M.R.; Gill, H.S. Large-diameter total hip arthroplasty modular heads require greater assembly forces for initial stability. *Bone Jt. Res.* **2016**, *5*, 338–346. [CrossRef] [PubMed]
44. Krull, A.; Morlock, M.M.; Bishop, N.E. The Influence of Contamination and Cleaning on the Strength of Modular Head Taper Fixation in Total Hip Arthroplasty. *J. Arthroplast.* **2017**, *32*, 3200–3205. [CrossRef] [PubMed]
45. Pennock, A.T.; Schmidt, A.H.; Bourgeault, C.A. Morse-type tapers: Factors that may influence taper strength during total hip arthroplasty. *J. Arthroplast.* **2002**, *17*, 773–778. [CrossRef]
46. Nassutt, R.; Mollenhauer, I.; Klingbeil, K.; Henning, O.; Grundei, H. Relevance of the insertion force for the taper lock reliability of a hip stem and a ceramic femoral head. *Biomed. Tech. (Berl)* **2006**, *51*, 103–109. [CrossRef] [PubMed]
47. Jauch-Matt, S.Y.; Miles, A.W.; Gill, H.S. Effect of trunnion roughness and length on the modular taper junction strength under typical intraoperative assembly forces. *Med. Eng. Phys.* **2017**, *39*, 94–101. [CrossRef] [PubMed]
48. Ramoutar, D.N.; Crosnier, E.A.; Shivji, F.; Miles, A.W.; Gill, H.S. Assessment of Head Displacement and Disassembly Force with Increasing Assembly Load at the Head/Trunnion Junction of a Total Hip Arthroplasty Prosthesis. *J. Arthroplast.* **2017**, *32*, 1675–1678. [CrossRef] [PubMed]
49. Ouellette, E.S.; Shenoy, A.A.; Gilbert, J.L. The seating mechanics of head-neck modular tapers in vitro: Load-displacement measurements, moisture, and rate effects. *J. Orthop. Res.* **2018**, *36*, 1164–1172. [CrossRef]
50. Scholl, L.; Schmidig, G.; Faizan, A.; TenHuisen, K.; Nevelos, J. Evaluation of surgical impaction technique and how it affects locking strength of the head-stem taper junction. *Proc. Inst. Mech. Eng. H* **2016**, *230*, 661–667. [CrossRef]
51. Pierre, D.; Swaminathan, V.; Scholl, L.Y.; TenHuisen, K.; Gilbert, J.L. Effects of Seating Load Magnitude on Incremental Cyclic Fretting Corrosion in 5 degrees 40′ Mixed Alloy Modular Taper Junctions. *J. Arthroplast.* **2018**, *33*, 1953–1961. [CrossRef]
52. Pierre, D.; Swaminathan, V.; Scholl, L.; TenHuisen, K.; Gilbert, J.L. Effects of seating load magnitude and load orientation on seating mechanics in 5 degrees 40′ mixed-alloy modular taper junctions. *J. Biomech.* **2019**, *82*, 251–258. [CrossRef]
53. Bitter, T.; Khan, I.; Marriott, T.; Schreurs, B.W.; Verdonschot, N.; Janssen, D. Experimental Measurement of the Static Coefficient of Friction at the Ti-Ti Taper Connection in Total Hip Arthroplasty. *J. Biomech. Eng.* **2016**, *138*, 4032446. [CrossRef] [PubMed]
54. English, R.; Ashkanfar, A.; Rothwell, G. The effect of different assembly loads on taper junction fretting wear in total hip replacements. *Tribol. Int.* **2016**, *95*, 199–210. [CrossRef]

55. Swaminathan, V.; Gilbert, J.L. Fretting corrosion of CoCrMo and Ti6Al4V interfaces. *Biomaterials* **2012**, *33*, 5487–5503. [CrossRef] [PubMed]
56. Cartner, J.; Aldinger, P.; Li, C.; Collins, D. Characterization of Femoral Head Taper Corrosion Features Using a 22-Year Retrieval Database. *Hss J.* **2017**, *13*, 35–41. [CrossRef] [PubMed]
57. Langton, D.J.; Sidaginamale, R.; Lord, J.K.; Nargol, A.V.; Joyce, T.J. Taper junction failure in large-diameter metal-on-metal bearings. *Bone Jt. Res.* **2012**, *1*, 56–63. [CrossRef]
58. Panagiotidou, A.; Meswania, J.; Hua, J.; Muirhead-Allwood, S.; Hart, A.; Blunn, G. Enhanced wear and corrosion in modular tapers in total hip replacement is associated with the contact area and surface topography. *J. Orthop. Res.* **2013**, *31*, 2032–2039. [CrossRef]
59. Pourzal, R.; Hall, D.J.; Ha, N.Q.; Urban, R.M.; Levine, B.R.; Jacobs, J.J.; Lundberg, H.J. Does Surface Topography Play a Role in Taper Damage in Head-neck Modular Junctions? *Clin. Orthop. Relat. Res.* **2016**, *474*, 2232–2242. [CrossRef]
60. Arnholt, C.M.; MacDonald, D.W.; Underwood, R.J.; Guyer, E.P.; Rimnac, C.M.; Kurtz, S.M.; Implant Research Center Writing, C.; Mont, M.A.; Klein, G.R.; Lee, G.C.; et al. Do Stem Taper Microgrooves Influence Taper Corrosion in Total Hip Arthroplasty? A Matched Cohort Retrieval Study. *J. Arthroplast.* **2017**, *32*, 1363–1373. [CrossRef]
61. Dyrkacz, R.M.; Brandt, J.M.; Ojo, O.A.; Turgeon, T.R.; Wyss, U.P. The influence of head size on corrosion and fretting behaviour at the head-neck interface of artificial hip joints. *J. Arthroplast.* **2013**, *28*, 1036–1040. [CrossRef]
62. Triantafyllopoulos, G.K.; Elpers, M.E.; Burket, J.C.; Esposito, C.I.; Padgett, D.E.; Wright, T.M. Otto Aufranc Award: Large Heads Do Not Increase Damage at the Head-neck Taper of Metal-on-polyethylene Total Hip Arthroplasties. *Clin. Orthop. Relat. Res.* **2016**, *474*, 330–338. [CrossRef]
63. Damm, P.; Dymke, J.; Ackermann, R.; Bender, A.; Graichen, F.; Halder, A.; Beier, A.; Bergmann, G. Friction in total hip joint prosthesis measured in vivo during walking. *PLoS ONE* **2013**, *8*, e78373. [CrossRef] [PubMed]
64. Kleinschnittger, P. Einfluss der Kontaktflächen auf die Höhe der Relativbewegung zwischen Prothesenkonus und -kopf. Bachelor's Thesis, Technical University Hamburg - Harburg, Hamburg, Germany, November 2016.
65. Haschke, H.; Konow, T.; Huber, G.; Morlock, M.M. Influence of flexural rigidity on micromotion at the head-stem taper interface of modular hip prostheses. *Med. Eng. Phys.* **2019**, *68*, 1–10. [CrossRef] [PubMed]
66. Nassif, N.A.; Nawabi, D.H.; Stoner, K.; Elpers, M.; Wright, T.; Padgett, D.E. Taper design affects failure of large-head metal-on-metal total hip replacements. *Clin. Orthop. Relat. Res.* **2014**, *472*, 564–571. [CrossRef]
67. Guhrs, J.; Krull, A.; Witt, F.; Morlock, M.M. The influence of stem taper re-use upon the failure load of ceramic heads. *Med. Eng. Phys.* **2015**, *37*, 545–552. [CrossRef] [PubMed]
68. Mueller, U.; Panzram, B.; Braun, S.; Sonntag, R.; Kretzer, J.P. Mixing of Head-Stem Components in Total Hip Arthroplasty. *J. Arthroplast.* **2018**, *33*, 945–951. [CrossRef]
69. Higgs, G.B.; MacDonald, D.W.; Gilbert, J.L.; Rimnac, C.M.; Kurtz, S.M.; Implant Research Center Writing, C. Does Taper Size Have an Effect on Taper Damage in Retrieved Metal-on-Polyethylene Total Hip Devices? *J. Arthroplast.* **2016**, *31*, 277–281. [CrossRef]
70. Bergmann, G.; Bender, A.; Dymke, J.; Duda, G.; Damm, P. Standardized Loads Acting in Hip Implants. *PLoS ONE* **2016**, *11*, e0155612. [CrossRef]
71. Bergmann, G.; Graichen, F.; Rohlmann, A.; Bender, A.; Heinlein, B.; Duda, G.N.; Heller, M.O.; Morlock, M.M. Realistic loads for testing hip implants. *Biomed. Mater. Eng.* **2010**, *20*, 65–75. [CrossRef]
72. Lidder, S.; Epstein, D.J.; Scott, G. A systematic review of short metaphyseal loading cementless stems in hip arthroplasty. *Bone Jt. Res.* **2019**, *5*. [CrossRef]
73. Porter, D.A.; Urban, R.M.; Jacobs, J.J.; Gilbert, J.L.; Rodriguez, J.A.; Cooper, H.J. Modern trunnions are more flexible: A mechanical analysis of THA taper designs. *Clin. Orthop. Relat. Res.* **2014**, *472*, 3963–3970. [CrossRef] [PubMed]
74. Heiney, J.P.; Battula, S.; Vrabec, G.A.; Parikh, A.; Blice, R.; Schoenfeld, A.J.; Njus, G.O. Impact magnitudes applied by surgeons and their importance when applying the femoral head onto the Morse taper for total hip arthroplasty. *Arch. Orthop. Trauma Surg.* **2009**, *129*, 793–796. [CrossRef]
75. Weisse, B.; Affolter, C.; Stutz, A.; Terrasi, G.P.; Kobel, S.; Weber, W. Influence of contaminants in the stem-ball interface on the static fracture load of ceramic hip joint ball heads. *Proc. Inst. Mech. Eng. H* **2008**, *222*, 829–835. [CrossRef]

76. Faizan, A.; Longaray, J.; Nevelos, J.E.; Westrich, G.H. Effects of corrosion and cleaning method on taper dimensions: An in vitro investigation. *Hip Int.* **2020**, *30*, 87–92. [CrossRef]
77. Krull, A.; Morlock, M.M.; Bishop, N.E. Maximizing the fixation strength of modular components by impaction without tissue damage. *Bone Jt. Res.* **2018**, *7*, 196–204. [CrossRef]
78. Scholl, L.; Pierre, D.; Rajaravivarma, R.; Lee, R.; Faizan, A.; Swaminathan, V.; TenHuisen, K.; Gilbert, J.L.; Nevelos, J. Effect of the support systems' compliance on total hip modular taper seating stability. *Proc. Inst. Mech. Eng. H* **2018**, *232*, 862–870. [CrossRef]
79. Danoff, J.R.; Longaray, J.; Rajaravivarma, R.; Gopalakrishnan, A.; Chen, A.F.; Hozack, W.J. Impaction Force Influences Taper-Trunnion Stability in Total Hip Arthroplasty. *J. Arthroplast.* **2018**, *33*, S270–S274. [CrossRef]
80. Frisch, N.B.; Lynch, J.R.; Banglmaier, R.F.; Silverton, C.D. The stability of dual-taper modular hip implants: a biomechanical analysis examining the effect of impact location on component stability. *Arthroplast. Today* **2017**, *3*, 119–124. [CrossRef]
81. Noble, P.; Foley, E.; Simpson, J.; Gold, J.; Choi, J.; Ismaily, S.; Mathis, K.; Incavo, S. Inter-surgeon variability in the assembly of modular head-neck tapers in THA via the posterior approach. *Proc. Orthop. Proc.* **2017**, *99B*, 59.
82. Svensson, O.; Mathiesen, E.B.; Reinholt, F.P.; Blomgren, G. Formation of a fulminant soft-tissue pseudotumor after uncemented hip arthroplasty. A case report. *J. Bone Jt. Surg. Am.* **1988**, *70*, 1238–1242. [CrossRef]
83. Weiser, M.C.; Lavernia, C.J. Trunnionosis in Total Hip Arthroplasty. *J. Bone Jt. Surg. Am.* **2017**, *99*, 1489–1501. [CrossRef]
84. Chana, R.; Esposito, C.; Campbell, P.A.; Walter, W.K.; Walter, W.L. Mixing and matching causing taper wear: Corrosion associated with pseudotumour formation. *J. Bone Jt. Surg. Br.* **2012**, *94*, 281–286. [CrossRef]
85. Klug, A.; Gramlich, Y.; Hoffmann, R.; Pfeil, J.; Drees, P.; Kutzner, K.P. Trends in Total Hip Arthroplasty in Germany from 2007 to 2016: What Has Changed and Where Are We Now? *Z Orthop. Unf.* **2019**. [CrossRef]
86. Sariali, E.; Lazennec, J.Y.; Khiami, F.; Catonné, Y. Mathematical evaluation of jumping distance in total hip arthroplasty: Influence of abduction angle, femoral head offset, and head diameter. *Acta Orthop.* **2009**, *80*, 277–282. [CrossRef]
87. Zijlstra, W.P.; De Hartog, B.; Van Steenbergen, L.N.; Scheurs, B.W.; Nelissen, R. Effect of femoral head size and surgical approach on risk of revision for dislocation after total hip arthroplasty. *Acta Orthop.* **2017**, *88*, 395–401. [CrossRef]
88. Jameson, S.S.; Lees, D.; James, P.; Serrano-Pedraza, I.; Partington, P.F.; Muller, S.D.; Meek, R.M.; Reed, M.R. Lower rates of dislocation with increased femoral head size after primary total hip replacement: A five-year analysis of NHS patients in England. *J. Bone Jt. Surg. Br.* **2011**, *93*, 876–880. [CrossRef]
89. Hanna, S.A.; Somerville, L.; McCalden, R.W.; Naudie, D.D.; MacDonald, S.J. Highly cross-linked polyethylene decreases the rate of revision of total hip arthroplasty compared with conventional polyethylene at 13 years' follow-up. *Bone Jt. J.* **2016**, *98-B*, 28–32. [CrossRef]
90. Elke, R.; Rieker, C.B. Estimating the osteolysis-free life of a total hip prosthesis depending on the linear wear rate and head size. *Proc. Inst. Mech. Eng. H* **2018**, *232*, 753–758. [CrossRef]
91. Hruby, A.; Hu, F.B. The Epidemiology of Obesity: A Big Picture. *Pharmacoeconomics* **2015**, *33*, 673–689. [CrossRef]
92. Bijukumar, D.R.; Salunkhe, S.; Zheng, G.; Barba, M.; Hall, D.J.; Pourzal, R.; Mathew, M.T. Wear particles induce a new macrophage phenotype with the potential to accelerate material corrosion within total hip replacement interfaces. *Acta Biomater.* **2020**, *101*, 586–597. [CrossRef]
93. Hall, D.J.; Pourzal, R.; Lundberg, H.J.; Mathew, M.T.; Jacobs, J.J.; Urban, R.M. Mechanical, chemical and biological damage modes within head-neck tapers of CoCrMo and Ti6Al4V contemporary hip replacements. *J. Biomed. Mater. Res. B Appl. Biomater.* **2018**, *106*, 1672–1685. [CrossRef]
94. Malik, M.H.; Jury, F.; Bayat, A.; Ollier, W.E.; Kay, P.R. Genetic susceptibility to total hip arthroplasty failure: a preliminary study on the influence of matrix metalloproteinase 1, interleukin 6 polymorphisms and vitamin D receptor. *Ann. Rheum. Dis.* **2007**, *66*, 1116–1120. [CrossRef]
95. Kilb, B.K.J.; Kurmis, A.P.; Parry, M.; Sherwood, K.; Keown, P.; Masri, B.A.; Duncan, C.P.; Garbuz, D.S. Frank Stinchfield Award: Identification of the At-risk Genotype for Development of Pseudotumors Around Metal-on-metal THAs. *Clin. Orthop. Relat. Res.* **2018**, *476*, 230–241. [CrossRef]

96. Goyal, N.; Ho, H.; Fricka, K.B.; Engh, C.A., Jr. Do you have to remove a corroded femoral stem? *J. Arthroplast.* **2014**, *29*, 139–142. [CrossRef]
97. Whitehouse, M.R.; Endo, M.; Zachara, S.; Nielsen, T.O.; Greidanus, N.V.; Masri, B.A.; Garbuz, D.S.; Duncan, C.P. Adverse local tissue reactions in metal-on-polyethylene total hip arthroplasty due to trunnion corrosion: The risk of misdiagnosis. *Bone Jt. J.* **2015**, *97-B*, 1024–1030. [CrossRef]

materials

MDPI

Article

The Effect of N, C, Cr, and Nb Content on Silicon Nitride Coatings for Joint Applications

Luimar Correa Filho [1], Susann Schmidt [2], Cecilia Goyenola [3], Charlotte Skjöldebrand [1], Håkan Engqvist [1], Hans Högberg [3], Markus Tobler [2] and Cecilia Persson [1,*]

1 Division of Applied Materials Science, Department of Materials Science and Engineering, Uppsala University, 751 21 Uppsala, Sweden; luimar.filho@angstrom.uu.se (L.C.F.); charlotte.skjoldebrand@angstrom.uu.se (C.S.); hakan.engqvist@angstrom.uu.se (H.E.)

2 Ionbond AG, Industriestrasse 9, Dulliken 4657, Switzerland; susann_Schmidt@gmx.de (S.S.); markus.tobler@ionbond.com (M.T.)

3 Thin Film Physics Division, Department of Physics, Chemistry and Biology (IFM), Linköping University, 58183 Linköping, Sweden; cecigoyenola@hotmail.com (C.G.); hans.hogberg@liu.se (H.H.)

* Correspondence: cecilia.persson@angstrom.uu.se; Tel.: +46-0702-489-624

Received: 29 February 2020; Accepted: 14 April 2020; Published: 17 April 2020

check for updates

Abstract: Ceramic coatings deposited on orthopedic implants are an alternative to achieve and maintain high wear resistance of the metallic device, and simultaneously allow for a reduction in metal ion release. Silicon nitride based (SiN_x) coatings deposited by high power impulse magnetron sputtering (HiPIMS) have shown potential for use in joint replacements, as a result of an improved chemical stability in combination with a good adhesion. This study investigated the effect of N, C, Cr, and Nb content on the tribocorrosive performance of 3.7 to 8.8 μm thick SiN_x coatings deposited by HiPIMS onto CoCrMo discs. The coating composition was assessed from X-ray photoelectron spectroscopy and the surface roughness by vertical scanning interferometry. Hardness and Young's modulus were measured by nanoindentation and coating adhesion was investigated by scratch tests. Multidirectional wear tests against ultrahigh molecular weight polyethylene pins were performed for 2 million cycles in bovine serum solution (25%) at 37 °C, at an estimated contact pressure of 2.1 MPa. Coatings with a relatively low hardness tended to fail earlier in the wear test, due to chemical reactions and eventually dissolution, accelerated by the tribological contact. In fact, while no definite correlation could be observed between coating composition (N: 42.6–55.5 at %, C: 0–25.7 at %, Cr: 0 or 12.8 at %, and Nb: 0–24.5 at %) and wear performance, it was apparent that high-purity and/or -density coatings (i.e., low oxygen content and high nitrogen content) were desirable to prevent coating and/or counter surface wear or failure. Coatings deposited with a higher energy fulfilled the target profile in terms of low surface roughness (Ra < 20 nm), adequate adhesion (L_{c2} > 30 N), chemical stability over time in the tribocorrosive environment, as well as low polymer wear, presenting potential for a future application in joint bearings.

Keywords: silicon nitride; coating; joint replacement; wear; adhesion

1. Introduction

Total joint replacements (TJRs) are surgical procedures carried out most frequently on patients suffering from arthritic pain or bone fractures [1–5]. These procedures are largely considered successful, with success rates up to 90% at 10 years follow-up for total hip replacements (THRs) and total knee replacements (TKRs) [6,7]. However, the aging and more active population places higher demands on these implants.

Typically, for THRs the femoral head is replaced by a metal alloy (CoCrMo) or ceramic (zirconia-toughened alumina (ZTA)), and the acetabulum by a ceramic or polymer (ultrahigh molecular weight polyethylene (UHMWPE) or cross-linked polyethylene (XLPE)) [8–12]. TKRs are composed of a metallic femoral component (CoCrMo) as well as a polyethylene XLPE insert attached in the tray [3,13–16]. Ceramic coatings on metallic substrates can be used to reduce wear of structural materials, including manufacturing tools as well as joint implants [17,18]. Different types of ceramic coatings are under investigation for hip joint applications (e.g., TiN, DLC, ZrO_2, ZrN, CrN, and Si_3N_4), while TiN and ZrN coatings are already in clinical use in knee implants [19–23]. These transition metal nitride coatings have the expressed purpose of extending the implant's life time, by either preventing or minimizing the body's immune reaction, which might result in osteolysis, aseptic loosening, and ultimately implant revision [24–35].

According to our previous work, SiN_x based coatings have shown potential as an alternative for joint bearings due to their biocompatibility, high wear resistance and hardness, and reduced metal ion release [36–39]. However, it is challenging to achieve an optimal combination of adhesion, coating density, and reactivity in SiN_x coatings; a high coating density resulting in a lower reactivity may give an insufficient adhesion to the substrate due to high residual stresses [40]. Alloying with a third element may be an option to improve the chemical stability while maintaining a balance in coating density and adhesion. For silicon nitride, previous studies have shown that the addition of Cr increases oxidation resistance and mechanical properties [41,42], while Nb improves the wear resistance and increases the hardness [43]. In previous studies, we reported that the addition of C altered the surface reactivity of silicon nitride and influenced the coating density and surface morphology [44,45]. In addition, we have shown that an increased N content results in a higher hardness and density [46–48]. In this study, we investigated the effect of N, C, Cr, and Nb content, as well as ion energy, on the properties of silicon nitride (SiN_x)-based coatings for joint applications, with a focus on their wear performance in a hard-on-soft contact, since, as mentioned above, the counter surface in a joint implant is usually a polyethylene polymer. The coatings were deposited on top of Cr-based interlayers, and were evaluated in terms of chemical composition, surface roughness, mechanical properties, adhesion, and wear resistance in a hard-on-soft contact.

2. Materials and Methods

2.1. Coating Deposition

Coating deposition was conducted in an industrial coating system, with a chamber volume of about 1 m^3, equipped with four magnetrons, of which two were operated in unbalanced magnetron sputtering (UBM) and two in high power impulse magnetron sputtering (HiPIMS) mode. The coatings were deposited using 2-fold substrate rotation. The Si targets were operated at average powers of 5 kW and 8 kW in HiPIMS mode, while the Cr and Nb targets were operated in UBM mode at a sputtering power 1 kW for Cr and sputter powers of 1 kW, 2 kW, or 5 kW for Nb. SiN_x coatings with thicknesses ranging from 3.7 to 8.8 μm were deposited on mirror polished CoCrMo discs. Ion energies were controlled using three different bias voltages (low, medium, and high) as well as average target power settings. The sputter atmosphere was controlled at a pressure of 600 mPa, with N_2-to-Ar ratios ranging between 17% and 40% and the remaining percentage to reach 100% was Ar. Detailed information can be found in Table 1.

Table 1. Description of the coatings and deposition processes used in this study. A pressure of 600 mPa was used for all deposition runs.

Analysis Aim	Coating Designation	Magnetrons				Bias	Gas	
		Si (HIPIMS) (kW)	Cr/Nb (UBM) (kW)	Si (HIPIMS) (kW)	Cr/Nb (UBM) (kW)		N_2 Content (%)	C_2H_2 Content (%)
Standard coating	Standard	5.0	-	5.0	-	low	40.0	-
Effect of N content in top layer	N-low	5.0	-	5.0	-	low	17.0	-
	N-medium	5.0	-	5.0	-	low	25.0	-
	N-high	5.0	-	5.0	-	low	40.0	-
Effect of C content	C-low	5.0	-	5.0	-	medium	38.0	2.5
	C-high	5.0	-	5.0	-	medium	36.0	4.0
Effect of Nb content	Nb-low	5.0	1.0	5.0	1.0	low	40.0	-
	Nb-medium	5.0	2.0	5.0	2.0	low	40.0	-
	Nb-high	5.0	5.0	5.0	5.0	low	40.0	-
Effect of Cr content	Cr	5.0	1.0	5.0	1.0	low	40.0	-
Effect of deposition energy	Bias-medium	5.0	-	5.0	-	medium	40.0	-
	Bias-high	5.0	-	5.0	-	high	40.0	-
	Si Power-high	8.0	-	8.0	-	low	40.0	-

2.2. Compositional Analysis

The composition of the SiN_x coatings was investigated by X-ray photoelectron spectroscopy (XPS, Axis UltraDLD, Kratos Analytical, Manchester, UK) using monochromatic Al(K_α) X-ray radiation ($h\nu$ = 1486.6 eV). The base pressure in the analysis chamber during acquisition was $< 1 \times 10^{-7}$ Pa. The experimental conditions were such that the full width at half maximum (FWHM) of the $Ag3d_{5/2}$ peak from the reference Ag sample was 0.45 eV. For all coatings, XPS survey spectra and core levels were recorded on as-received samples and after sputter cleaning. Sputter cleaning consisted of an initial step of 900 s at an Ar^+ beam energy of 2 keV, followed by a second step for 900 s at an Ar^+ beam energy of 4 keV. During sputter cleaning the Ar^+ beam was rastered over an area of 3×3 mm^2 at an incidence angle of 20°. Automatic charge compensation was applied throughout the acquisition, using low energy electrons provided by a flood gun. The composition of the coatings was assessed from XPS high-resolution core level spectra recorded from the Si 2p, Ar 2p, N 1s, C 1s, and O 1s regions after sputter cleaning. Core level spectra were analyzed with CasaXPS (v2.3.15, Casa Software Ltd, Teignmouth, UK). A Shirley-type background was subtracted, and the spectra were calibrated using adventitious surface carbon at 284.8 eV as a charge reference. For quantitative analysis of the metal-containing coatings the core levels of the Cr 2p and Nb 3d were applied for determination. The measurement precision for XPS analysis was ±5% for compositions below 10 at % and ±2–3% for compositions above 10 at % [49].

During wear tests and exposure to fetal bovine serum (FBS) solution a reaction occurred, and a white layer was formed on the surface of some coatings. This layer was examined using monochromatic Al (K_α) X-ray photoelectron spectroscopy (XPS, Quantera II, Physical Electronics (PHI), Eden Prairie, MN, USA). Measurements were conducted on the surface after 2 min of sputtering Ar^+ ions at 500 V and after an additional 20 min at 1 kV, to investigate the coating surface and further down in the coating, respectively. The sample was mounted on a glass slide in order to float the sample and automatic charge compensation was used throughout the measurement. Core level spectra were analyzed in CasaXPS, a Shirley-type background was subtracted, and the spectra were calibrated using adventitious surface carbon at 284.8 eV as a charge reference.

2.3. Surface Roughness

The coating roughness was measured before wear testing using optical profilometry, specifically vertical scanning interferometry (VSI) at 10× and a field of view (FOV) of 1.0. Each measurement corresponded to an area of 451×594 µm^2. Typically, four measurements were performed on each sample to obtain R_a (arithmetic average).

2.4. Nanoindentation

The hardness and elastic modulus of the coatings were measured in a CSIRO UMIS nanoindenter (Fischer-Cripps Laboratories, New South Wales, Australia) equipped with a three-sided Berkovich tip. All films were tested in the load-controlled mode and for calculations a Poisson's ratio of 0.3 was used. For the tests, at least 30 indents with a load of 20 mN were performed [50].

2.5. Scratch Testing

In order to investigate coating adhesion, a scratch test was performed at different time points [51] using a scratch tester (CSEM-Revetest (CSEM, Neuchatel, Switzerland)) with a Rockwell C tip (apex 120°, tip radius 200 μm). A progressive load up to 100 N, at a loading rate of 120 N/min and a horizontal displacement rate of 6 mm/min were applied. This resulted in a scratch length of 5 mm, which was evaluated in a light optical microscope to determine the critical load L_{C2} indicating where the adhesion failure occurred [52,53]. Each sample was scratched three times at each time point.

2.6. Wear Resistance (2D) in a Hard-on-Soft Contact

Multidirectional wear tests (MWT) were carried out to evaluate the response of the coatings against polyethylene using cylindrical pins with a nominal length of 19.1 mm and diameter of 9.5 mm. The pins were made of UHMWPE GUR1020 (one of the two most commonly used grades of UHMWPE in orthopaedics, defined as per BS ISO 5834-2 2019 [54]), provided by the collaborating industrial partners Peter Brehm GmbH (Weisendorf, Germany). MWT tests were performed in 0.2 μm filtered bovine serum solution (25%) at 37 °C. Prior to testing, the pins were presoaked in serum and cleaned according to standard [55]. The test was carried out with a nominal load of 150 N resulting in an estimated contact pressure of 2.1 MPa within the guidelines of [56], frequency 2 Hz, and sliding velocity 56 mm/s for 2,000,000 cycles (2.0 MC) using a 7 mm × 7 mm square path for a sliding distance of 28 mm/cycle.

2.7. Statistical Analysis

IBM SPSS Statistics v 26 (New York, NY, USA) was used for all statistical analyses. An analysis of variance (ANOVA) was performed, followed by a Scheffe's post hoc test. When Levene's test for homogeneity of variances was significant, Welch's robust test followed by a Tamhane post hoc test was used instead. The Pearson correlation coefficient was used to evaluate potential correlations. A critical level of $\alpha = 0.05$ was used to determine significance.

3. Results and Discussion

3.1. Coating Thickness and Composition

The growth rate for the SiN_x coatings depended on applied target power settings and bias voltages, as well as the N_2 and C_2H_2 gas flows. Increased SiN_x and $SiMeN_x$ growth rates resulted from more material being removed from the target due to elevated target potentials [57]. Additionally, as Nb or Cr were added to the process, the number of operated targets increased and contributed to the $SiMeN_x$ growth. The growth rate for the different bias settings showed a maximum at a medium level, indicating that the flux of film-forming species at low bias voltages was not optimally directed to the substrate table and resputtering occurred at high bias voltage settings. Increasing amounts of N_2 led to decreased growth rates due to target poisoning [47] while an increased C_2H_2 gas flow resulted in an increased growth rate. This was attributed to a reduction in coating mass density and morphological density, specifically a pronounced growth of columns [45].

Following the trend of the coating growth rate, the coating composition changed by increasing N_2 and C_2H_2 gas flows, leading to increased amounts of N and C in the coating. XPS results showed a nitrogen content close to 50 at % for all coatings (Table 2). SiN_x coatings with a higher ion energy and 40% of N_2 during deposition yielded a nitrogen content in the coatings exceeding 51 at % and a N/Si ratio ≥1. This ratio has previously been shown to be beneficial to a lower dissolution rate, which could be advantageous to the coating lifespan [39]. When C_2H_2 was added to the process the coating showed higher O contents, which in turn supported the interpretation for the growth rate of $SiCN_x$ at elevated C_2H_2 flows. Here, the reduction in coating density and the pronounced growth of columns led to incorporation of O as the coatings were exposed to air prior to analysis. Further, a reduced morphological density was observed as Cr and Nb were added to the process [40]. This was mirrored in higher O and C contents in the corresponding coatings.

Table 2. Deposition settings, coating thicknesses, growth rates, and composition.

Coating Designation	Si Power (kW)	Bias	Settings SiN/SiMeN/SiCN Thickness (μm)	Growth Rate (nm/s)	Elemental Composition—XPS Si (at %)	N (at %)	O (at %)	C (at %)	Nb/Cr (at %)	N/Si -
Standard	5.0	low	5.5	0.13	42.8	49.7	5.5	2.0	-	1.16
N2-low	5.0	low	3.8	0.21	51.8	47.9	-	0.3	-	0.92
N2-medium	5.0	low	3.5	0.19	43.1	46.3	7.4	3.2	-	1.07
N2-high/Bias low/Standard	5.0	low	2.1	0.12	44.7	54.7	0.2	0.4	-	1.22
Bias-medium	5.0	medium	6.4	0.15	45.6	54.4	-	-	-	1.19
Bias-high	5.0	high	6.2	0.14	44.5	55.5	-	-	-	1.25
C-low, 2.5%	5	medium	5.4	0.17	33.1	47.2	2.9	16.9	-	1.43
C-high, 4%	5	medium	6.4	0.2	27.9	42.6	3.9	25.7	-	1.53
Nb-low, 2 × 1 kW Nb	5.0	low	3.1	0.17	39.9	46.8	3.6	1	7	1.17
Nb-medium, 2 × 2 kW Nb	5.0	low	3.1	0.17	33.9	45.2	4.2	1.3	13.7	1.33
Nb-high, 2 × 5 kW Nb	5.0	low	5.8	0.32	25.1	43.1	4.1	1.5	24.5	1.72
Cr-medium, 2 × 1 kW Cr	5.0	low	3.2	0.18	33.8	44.6	4.4	3.6	12.8	1.32
Si Power-high	8.0	low	4	0.24	46.5	52.3	1.2	-	-	1.12

The microstructure of similar coatings has been published earlier, for a range of coating parameters [47]. The SiN_x coatings that performed well were very dense and, thus, had low O and C contents, but also displayed more residual stresses. Likewise the coatings that did not perform well were less dense and contained more O and C (by adsorption) [48].

3.2. *Surface Roughness*

The average surface roughness (R_a) determined for the as-deposited coatings was <50 nm (Table 3), thus fulfilling the standard for biomedical implants (ASTM F2033-12). Coatings *C-low, Standard, Nb-medium*, and *Nb-high* displayed the lowest values (7.69–12.97 nm), followed by coatings *Nb-low, Si Power-high, Bias-medium, C-high*, and i (14.71–19.97 nm). The highest values of R_a were obtained for coatings *Bias-high, N-medium, N-low*, and *N-high* (22.2–42.05 nm). As shown in Table 3, the coatings with higher Nb and C content presented relatively low surface roughness values. A lower surface roughness could possibly be attributed to the ionization energy of N being higher than C, which resulted in more carbon atoms being deposited. On the other hand more of the amorphous phase was being created, resulting in a smoother surface [58–62]. No statistically significant correlation could be found between surface roughness and coating thickness, nor between surface roughness and deposition rate.

Table 3. Average surface roughness of SiNx coatings, as measured by interferometry. Coatings attributed with the same letters from a–e were not statistically significantly different (i.e., $p > 0.05$).

Coating Designation	R_a (nm)	Statistical Differences
Uncoated CoCr	3.5 ± 0.2	*a*
Standard	10 ± 0.9	*b, c*
N-low	33.2 ± 2.9	-
N-medium	33.1 ± 12.4	*e*
N-high	42.0 ± 6.0	-
C-low	7.6 ± 0.5	*a, b*
C-high	18.8 ± 3.0	*d, e*
Nb-low	14.7 ± 0.7	*c, d*
Nb-medium	12.9 ± 0.4	*c*
Nb-high	10.1 ± 3.4	*b, c*
Cr	19.9 ± 0.8	*e*
Bias-medium	18.0 ± 1.1	*d, e*
Bias-high	22.2 ± 1.1	-
Si Power-high	16.8 ± 0.5	-

3.3. Mechanical Properties (Nanoindentation)

The coating hardness varied from 13–25.4 GPa, with coatings *Standard*, *N-high*, and *N-low* exhibiting higher values (Figure 1). A similar tendency could be observed for the Young's modulus.

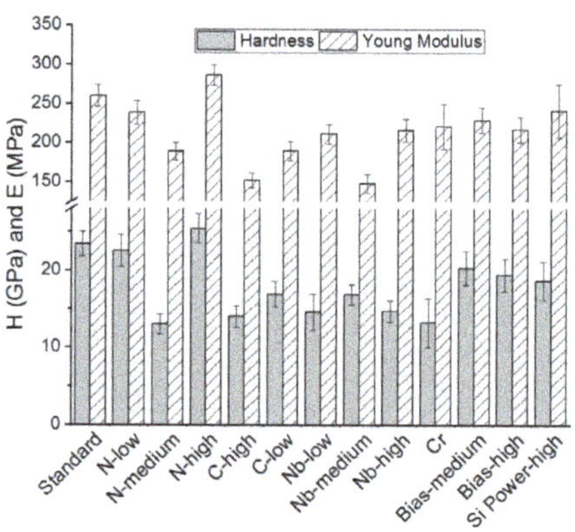

Figure 1. Hardness and Young's modulus for SiN_x based coatings.

Earlier studies on SiN_x coatings have determined similar values for hardness and Young's modulus, although different deposition methods were applied [39,44,47,63,64]. A higher hardness suggests a higher coating density. Hardness values reported for other coatings for joint implants such as ZrN, TiNbN, Ox-Zr, and TiN coatings resided in a similar range, namely 14.0–31.0, 14.0–24.5 and 12.0–14.0 and 33–56 GPa, respectively [65].

3.4. Adhesion

The scratch test results in terms of Lc_2 values are shown in Figure 2. As can be seen, coatings deposited with a higher target power showed lower L_{C2} values. This was due to higher residual stresses resulting from a higher N content and the increase in Si-N bonds [45,48]. Moreover, these coatings showed a generally denser morphology (data not shown), which in turn contributed to increased residual stresses [66], as demonstrated previously [48]. Furthermore, coatings with higher O and Cr contents displayed higher Lc_2 values, which may be related to a lower amount of N-bonds, i.e., an opposite trend to that previously mentioned and/or a decreased coating density and, hence, residual stresses. The following coatings showed statistical differences: *Standard* vs. *C-high*, *Nb-medium*, *Nb-high*, *Bias-medium*, *Bias-high*, and *Si power high* as well as *C-high* vs. *Nb-medium*, *Nb-high*, *Bias-high*, and *Si power high*.

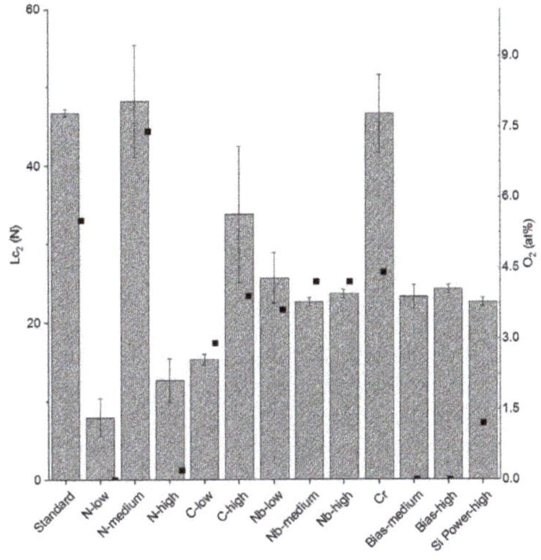

Figure 2. On the left axis the results for adhesion (Lc_2) are shown for the coatings tested in this study, with bars and standard deviations. On the right axis the O_2 content of the coatings is shown, represented by square dots.

3.5. Multidirectional Wear Tests

3.5.1. Macroscopic Appearance and Surface Analysis

The macroscopic surface structure of the coatings after the wear tests is shown in Figure 3. The formation of an opaque layer on the surface could be observed during testing on some of the coatings (Figure 3). XPS measurements were, therefore, performed on coating *N-medium*, at a region that still displayed a reflective surface (assumed to be unworn) and a region that had formed an opaque layer on the surface. Previous work showed a tribofilm formation in aqueous environments for Si_3N_4 materials, and in those conditions a SiO_2 and $Si(OH)_2$ layer could be found, improving the wear resistance and reducing the coefficient of friction by acting as a self-lubricating layer [67–69]. However, in the XPS measurements herein the use of charge neutralizers and lack of a good charge reference made the positions of the peaks uncertain. To determine whether the Si2p and O1s peaks originated from Si-O bonds, the distance between the peaks, ΔE_b, was determined and compared to the distance (ΔE_b) from literature according to Briggs et al. [70]. The deconvoluted Si peaks were fitted with the smallest number of curves possible. The spectrum obtained at the surface revealed contributions attributed to Si-C (100.8 eV), Si-N (101.4 eV), and Si-O (102.8 eV), which correlated well with findings from similar materials. After 2 min sputtering at 500 V the Si-O contribution was no longer detected, while there were still contributions attributed to Si-N and Si-C, and after additional sputtering for 20 min at 1 kV only two contributions were identified, Si-N and Si-Si. These results indicated that the outer layer contained more O and C compared to the bulk of the coating, which could be due to the formation of a tribofilm during wear testing.

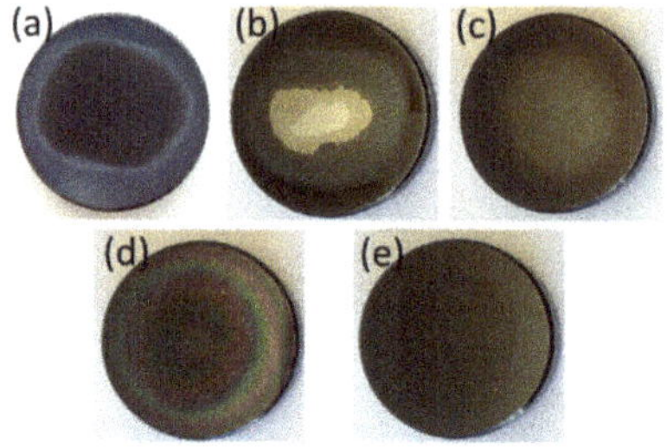

Figure 3. Typical macroscopic appearances of (**a**) a reacted surface (Standard), (**b**) a failed coating (Nb-medium), and (**c**,**d**) coatings with a surface layer: (**c**) Coating Cr and (**d**) coating Si Power-high. In (**e**) a Bias-high coating is shown, which did not present any layer formation or upcoming failure up to 2 MC.

3.5.2. Coefficient of Friction

Low coefficients of friction were observed for the first 0.5 million cycles for *N-high*, *Standard*, *Bias-medium*, and *Bias-high* (0.051–0.067). Coatings *N-low*, *Cr*, and *Si Power-high* showed somewhat higher values, from 0.103–0.108, with little variation. Coefficients of friction did not change markedly throughout the tests, except for coatings that reacted or failed during the test (Figure 4). This work generally showed lower coefficients of friction compared to previous work on Nb-Ti-N coatings (ranging from 0.11 to 0.12) and on TiN (0.14) [66].

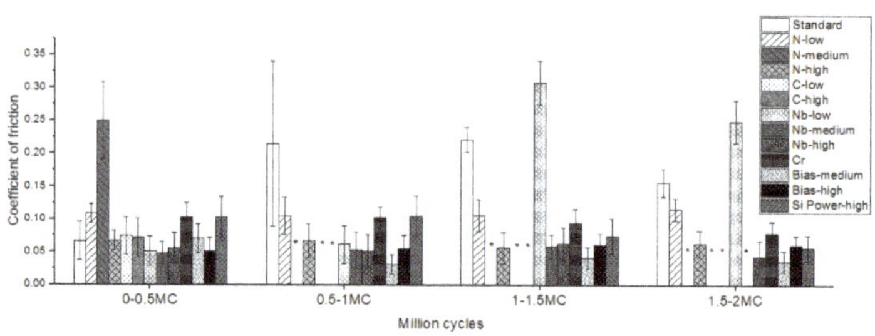

Figure 4. Coefficient of friction up to 2.0 MC for the tested coatings. During wear tests the following coatings wore through: (*) N-medium, C-high, and C-low at 0.5 MC; Nb-medium at 1.5 MC; Nb-low and Nb-high at 2.0 MC.

3.5.3. Volumetric Wear Rate

While *N-low* and *N-high* gave the lowest wear rates for the UHMWPE pins (< 0.37 mm^3/MC, Figure 4), N-medium failed already in the first 0.5 MC (Figure 3), giving a high wear rate due to the increased surface roughness from the reacted surface. The *Standard* coating also gave a high wear rate, due to a reacted surface (Figure 3). The coatings with a higher C content all failed at 0.5 MC. Nb coatings failed at different time points, for example, *Nb-low* and *Nb-high* had failed at 2.0 MC and *Nb-medium* at 1.5 MC. The remaining coatings did not fail and presented low wear rates (0.74–3.63 mm^3/MC). Figures 3 and 5 show that coatings with no apparent reaction or coating failure, and that gave low pin wear rates, were those with an initially high hardness (22.5–28.4 GPa), and, hence, presumably higher density and lower reactivity and/or a high N content (*N-low*, *N-high*, *Bias-medium*, *Bias-high*, and *Si Power-high*), with the exception being the *Standard* (H = 23.4 GPa) coating, which, however, contained more oxygen than the well-performing coatings (Table 2), suggesting a higher reactivity. Coating wear through contact with UHMWPE during the tests was not expected, and coatings failing would rather be associated to a higher reactivity and subsequent dissolution [40].

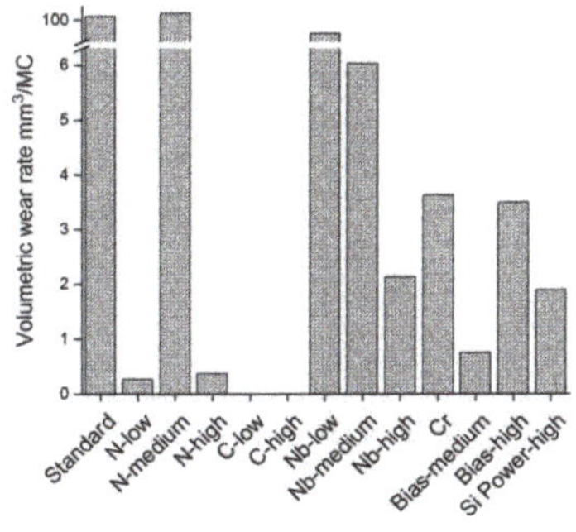

Figure 5. Volumetric wear rate for the UHMWPE pins ran against all coatings in this study. Coatings C-high and C-low showed negative values of −2.90 and −8.40 mm^3/MC, respectively.

4. Conclusions

Based on the results of the coatings tested in this work, some important conclusions were drawn. First, the low-ion energy coatings generally exhibited a lower hardness and initially higher critical load in scratch testing. High concentrations of impurities (higher O content and lower N content) were associated with early reactions and/or dissolution of the coating, as shown by XPS compositional analysis as well as multidirectional wear tests. During the wear tests coatings with lower or no apparent O content did not fail and showed a low volumetric wear rate of UHMWPE pins. SiN_x coatings of high N content, low O content (e.g., *N-high*, *Bias-medium*, *Bias-high*, and *Si Power-high*) are needed for the target–joint implant applications. Promising low wear rates were found for UHWMPE pins sliding against these latter coatings in a multidirectional wear test.

5. Patents

Ionbond AG, where S.S. and M.T. are employees, owns patents related to similar coatings.

Author Contributions: Conceptualization, C.G., M.T., S.S., H.E., H.H., and C.P.; methodology, L.C.F., C.S., M.T., S.S., and C.P.; validation, L.C.F., C.G., C.S., M.T., and S.S.; formal analysis, L.C.F., C.G., C.S., M.T., S.S., and C.P.; investigation, L.C.F., C.G., C.S., M.T., and S.S.; resources, C.P., H.H., and H.E.; writing—original draft preparation, L.C.F., M.T., S.S., and C.P.; writing—review and editing, L.C.F., S.S., C.G., C.S., H.E., M.T., H.H., and C.P.; visualization, L.C.F., S.S., and C.P.; supervision, H.E., H.H., and C.P.; project administration, H.H., H.E., and C.P.; funding acquisition, C.P., H.E., and H.H. All authors have read and agreed to the published version of the manuscript.

Funding: This research was funded by the European Union, grant number FP7-NMP-2012-310477 (Life Long Joints project); EBW+ Project Erasmus Mundus Programme, Action 2–STRAND 1, Lot 9 (Latin America), Brazil, Grant number 2014-0982. H.H. acknowledges financial support from the Swedish Government Strategic Research Area in Materials Science on Advanced Functional Materials at Linköping University (Faculty Grant SFO Mat LiU No. 2009 00971).

Acknowledgments: The authors are grateful for assistance from Mathilde Cogrel and Alejandro López for assistance with MWT measurements.

Conflicts of Interest: The authors declare no conflict of interest, and Ionbond AG, where S.S. and M.T. are employees, owns patents related to similar coatings.

References

1. De Bellis, U.G.; Legnani, C.; Calori, G.M. Acute total hip replacement for acetabular fractures: A systematic review of the literature. *Injury* **2014**, *45*, 356–361. [CrossRef] [PubMed]
2. Katz, J.N. Total joint replacement in osteoarthritis. *Best Pract. Res. Clin. Rheumatol.* **2006**, *20*, 145–153. [CrossRef] [PubMed]
3. Salih, S.; Hamer, A. Hip and knee replacement. *Surgery* **2013**, *31*, 482–487. [CrossRef]
4. Shan, L.; Shan, B.; Graham, D.; Saxena, A. Total hip replacement: A systematic review and meta-analysis on mid-term quality of life. *Osteoarthr. Cartil.* **2014**, *i22*, 389–406. [CrossRef] [PubMed]

5. Martinez-Cano, J.P.; Herrera-Escobar, J.P.; Arango Gutierrez, A.S.; Sanchez Vergel, A.; Martinez-Rondanelli, A. Prospective quality of life assessment after hip and knee arthroplasty: short- and mid-term follow-up results. *Arthroplast. Today* **2017**. [CrossRef] [PubMed]
6. Kärrholm, J.; Mohaddes, M.; Odin, D.; Vinblad, J.; Rogmark, C.; Rolfson, O. *Swedish Hip Arthroplasty Register Annual Report 2017*; Swedish Hip Arthroplasty Register: Göteborg, Swedenm, 2017; ISBN 9789188017208.
7. Robertsson, O.; W-Dahl, A.; Lidgren, L.; Sundberg, M. *Swedish Knee Arthroplasty Register Annual Report 2018*; Swedish Hip Arthroplasty Register: Göteborg, Sweden, 2018.
8. Sharma, V.; Chang, J.-D.; Kim, I.-S.; Yoo, J.-H.; Lee, S.-S.; Mansukhani, S.A. Midterm outcome of fourth-generation ceramic-on-ceramic bearing surfaces in revision total hip arthroplasty. *J. Orthop. Surg.* **2018**, *26*, 230949901878391. [CrossRef]
9. Kurtz, S.M.; Kocagöz, S.; Arnholt, C.; Huet, R.; Ueno, M.; Walter, W.L. Advances in zirconia toughened alumina biomaterials for total joint replacement. *J. Mech. Behav. Biomed. Mater.* **2014**. [CrossRef]
10. Gamble, D.; Jaiswal, P.K.; Lutz, I.; Johnston, K.D. The Use of Ceramics in Total Hip Arthroplasty. *Orthop. Rheumatol.* **2017**, *4*, 1–7. [CrossRef]
11. Maheshwari, A.V.; Shah, N.V.; Newman, J.M.; Pascal, S.; Sheth, N.P.; Grieco, P.W.; Stroud, S.G. New alternate bearing surfaces in total hip arthroplasty: A review of the current literature. *J. Clin. Orthop. Trauma* **2017**, *9*, 7–16. [CrossRef]
12. Kumar, N.; Arora, N.C.; Datta, B. Bearing surfaces in hip replacement—Evolution and likely future. *Med. J. Armed Forces India* **2014**, *70*, 371–376. [CrossRef]
13. Garrett, S.; Jacobs, N.; Yates, P.; Smith, A.; Wood, D. Differences in metal ion release following cobalt-chromium and oxidized zirconium total knee arthroplasty. *Acta Orthop. Belg.* **2010**, *76*, 513–520. [PubMed]
14. Hui, C.; Salmon, L.; Maeno, S.; Roe, J.; Walsh, W.; Pinczewski, L. Five-Year Comparison of Oxidized Zirconium and Cobalt-Chromium Femoral Components in Total Knee Arthroplasty: A Randomized Controlled Trial. *JBJS* **2011**, *93*, 624–630. [CrossRef] [PubMed]
15. Nowakowski, A.M.; Vavken, P.; Pagenstert, G.; Valderrabano, V. Design, Shape, and Materials of Total Knee Replacement. In *The Unhappy Total Knee Replacement: A Comprehensive Review and Management Guide*; Hirschmann, M.T., Becker, R., Eds.; Springer International Publishing: Cham, Switzerland, 2015; ISBN 978-3-319-08099-4.
16. Meier, E.; Gelse, K.; Trieb, K.; Pachowsky, M.; Hennig, F.F.; Mauerer, A. First clinical study of a novel complete metal-free ceramic total knee replacement system. *J. Orthop. Surg. Res.* **2016**, *11*, 21. [CrossRef]
17. Haider, H.; Weisenburger, J.N.; Namavar, F.; Garvin, K.L. Why Coating Technologies for Hip Replacement Systems, and the Importance of Testing Them In Vitro. *Oper. Tech. Orthop.* **2017**, *27*, 152–160. [CrossRef]
18. Wang, Y.X.; Zhang, S. Toward hard yet tough ceramic coatings. *Surf. Coat. Technol.* **2014**, *258*, 1–16. [CrossRef]
19. Breugem, S.J.M.; Linnartz, J.; Sierevelt, I.; Bruijn, J.D.; Driessen, M.J.M. Evaluation of 1031 primary titanium nitride coated mobile bearing total knee arthroplasties in an orthopedic clinic. *World J. Orthop.* **2017**, *8*, 922–928. [CrossRef]
20. Mohammed, A.; Metcalfe, A.; Woodnutt, D. Medium-term outcome of titanium nitride, mobile bearing total knee replacement. *Acta Orthop. Belg.* **2014**, *80*, 269–275.
21. Fabry, C.; Zietz, C.; Baumann, A.; Ehall, R.; Bader, R. High wear resistance of femoral components coated with titanium nitride: a retrieval analysis. *Knee Surg. Sport. Traumatol. Arthrosc.* **2018**, *26*, 2630–2639. [CrossRef]
22. Thomsen, M.; Rozak, M.; Thomas, P. Pain in a chromium-allergic patient with total knee arthroplasty: disappearance of symptoms after revision with a special surface-coated TKA—A case report. *Acta Orthop.* **2011**, *82*, 386–388. [CrossRef]
23. Lützner, J.; Dinnebier, G.; Hartmann, A.; Günther, K.-P.; Kirschner, S. Study rationale and protocol: prospective randomized comparison of metal ion concentrations in the patient's plasma after implantation of coated and uncoated total knee prostheses. *BMC Musculoskelet. Disord.* **2009**, *10*, 128. [CrossRef]
24. Spector, M.; Ries, M.D.; Bourne, R.B.; Sauer, W.S.; Long, M.; Hunter, G. Wear performance of ultra-high molecular weight polyethylene, on oxidized zirconium total knee femoral components. *JBJS* **2001**, *83*, S80–S86. [CrossRef] [PubMed]
25. Hauert, R.; Thorwarth, K.; Thorwarth, G. An overview on diamond-like carbon coatings in medical applications. *Surf. Coat. Technol.* **2013**. [CrossRef]
26. Hendry, J.A.; Pilliar, R.M. The fretting corrosion resistance of PVD surface-modified orthopedic implant alloys. *J. Biomed. Mater. Res.* **2001**, *58*, 156–166. [CrossRef]

27. Bader, R.; Bergschmidt, P.; Fritsche, A.; Ansorge, S.; Thomas, P.; Mittelmeier, W. Alternative Werkstoffe und Lösungen in der Knieendoprothetik für Patienten mit Metallallergie. *Orthopade* **2008**, *37*, 136–142. [CrossRef] [PubMed]
28. Alakoski, E.; Tiainen, V.-M.; Soininen, A.; Konttinen, Y.T. Load-Bearing Biomedical Applications of Diamond-Like Carbon Coatings - Current Status. *Open Orthop. J.* **2008**. [CrossRef]
29. Dearnaley, G.; Arps, J.H. Biomedical applications of diamond-like carbon (DLC) coatings: A review. *Surf. Coat. Technol.* **2005**. [CrossRef]
30. Maru, M.M.; Amaral, M.; Rodrigues, S.P.; Santos, R.; Gouvea, C.P.; Archanjo, B.S.; Trommer, R.M.; Oliveira, F.J.; Silva, R.F.; Achete, C.A. The High performance of nanocrystalline CVD diamond coated hip joints in wear simulator test. *J. Mech. Behav. Biomed. Mater.* **2015**. [CrossRef]
31. Pappas, M.J.; Makris, G.; Buechel, F.F. Titanium nitride ceramic film against polyethylene. A 48 million cycle wear test. *Clin. Orthop. Relat. Res.* **1995**, *317*, 64–70.
32. Ching, H.A.; Choudhury, D.; Nine, M.J.; Abu Osman, N.A. Effects of surface coating on reducing friction and wear of orthopaedic implants. *Sci. Technol. Adv. Mater.* **2014**, *15*. [CrossRef]
33. Leslie, I.; Williams, S.; Isaac, G.; Hatto, P.; Ingham, E.; Fisher, J. Wear of surface-engineered metal-onmetal bearings for hip prostheses under adverse conditions with the head loading on the rim of the cup. *Proc. Inst. Mech. Eng. Part H J. Eng. Med.* **2013**. [CrossRef]
34. Van Hove, R.P.; Sierevelt, I.N.; van Royen, B.J.; Nolte, P.A. Titanium-nitride coating of orthopaedic implants: a review of the literature. *Biomed. Res. Int.* **2015**, *2015*, 9. [CrossRef] [PubMed]
35. Nelissen, R.G.H.H.; Valstar, E.R.; Rozing, P.M. The effect of hydroxyapatite on the micromotion of total knee prostheses. A prospective, randomized, double-blind study. *JBJS* **1998**, *80*, 1665–1672. [CrossRef] [PubMed]
36. Bal, B.S.; Rahaman, M.N. Orthopedic applications of silicon nitride ceramics. *Acta Biomater.* **2012**, *8*, 2889–2898. [CrossRef] [PubMed]
37. Bock, R.M.; McEntire, B.J.; Bal, B.S.; Rahaman, M.N.; Boffelli, M.; Pezzotti, G. Surface modulation of silicon nitride ceramics for orthopaedic applications. *Acta Biomater.* **2015**. [CrossRef]
38. Pettersson, M.; Bryant, M.; Schmidt, S.; Engqvist, H.; Hall, R.M.; Neville, A.; Persson, C. Dissolution behaviour of silicon nitride coatings for joint replacements. *Mater. Sci. Eng. C* **2016**, *62*, 497–505. [CrossRef]
39. Pettersson, M.; Tkachenko, S.; Schmidt, S.; Berlind, T.; Jacobson, S.; Hultman, L.; Engqvist, H.; Persson, C. Mechanical and tribological behavior of silicon nitride and silicon carbon nitride coatings for total joint replacements. *J. Mech. Behav. Biomed. Mater.* **2013**, *25*, 41–47. [CrossRef]
40. Filho, L.C.; Schmidt, S.; López, A.; Cogrel, M.; Leifer, K.; Engqvist, H.; Högberg, H.; Persson, C. The Effect of Coating Density on Functional Properties of SiN(x) Coated Implants. *Materials* **2019**, *12*, 3370. [CrossRef]
41. Lee, J.-W.; Chang, Y.-C. A study on the microstructures and mechanical properties of pulsed DC reactive magnetron sputtered Cr-Si-N nanocomposite coatings. *Surf. Coat. Technol.* **2007**, *202*, 831–836. [CrossRef]
42. Park, J.H.; Chung, W.S.; Cho, Y.-R.; Kim, K.H. Synthesis and mechanical properties of Cr-Si-N coatings deposited by a hybrid system of arc ion plating and sputtering techniques. *Surf. Coat. Technol.* **2004**, *188*, 425–430. [CrossRef]
43. Hones, P.; Sanjines, R.; Lévy, F. Sputter deposited chromium nitride based ternary compounds for hard coatings. *Thin Solid Films* **1998**, *332*, 240–246. [CrossRef]
44. Olofsson, J.; Pettersson, M.; Teuscher, N.; Heilmann, A.; Larsson, K.; Grandfield, K.; Persson, C.; Jacobson, S.; Engqvist, H. Fabrication and evaluation of SixNy coatings for total joint replacements. *J. Mater. Sci. Mater. Med.* **2012**, *23*, 1879–1889. [CrossRef] [PubMed]
45. Hänninen, T.; Schmidt, S.; Ivanov, I.G.; Jensen, J.; Hultman, L.; Högberg, H. Silicon carbonitride thin films deposited by reactive high power impulse magnetron sputtering. *Surf. Coat. Technol.* **2018**, *335*, 248–256. [CrossRef]
46. Skjöldebrand, C.; Schmidt, S.; Vuong, V.; Pettersson, M.; Grandfield, K.; Högberg, H.; Engqvist, H.; Persson, C. Influence of substrate heating and nitrogen flow on the composition, morphological and mechanical properties of SiNx coatings aimed for joint replacements. *Materials* **2017**, *10*, 173. [CrossRef] [PubMed]
47. Schmidt, S.; Hänninen, T.; Goyenola, C.; Wissting, J.; Jensen, J.; Hultman, L.; Goebbels, N.; Tobler, M.; Högberg, H. SiN x Coatings Deposited by Reactive High Power Impulse Magnetron Sputtering: Process Parameters Influencing the Nitrogen Content. *ACS Appl. Mater. Interfaces* **2016**. [CrossRef] [PubMed]

48. Schmidt, S.; Hänninen, T.; Wissting, J.; Hultman, L.; Goebbels, N.; Santana, A.; Tobler, M.; Högberg, H. SiNx coatings deposited by reactive high power impulse magnetron sputtering: Process parameters influencing the residual coating stress. *J. Appl. Phys.* **2017**, *121*, 171904. [CrossRef]
49. Hofmann, S. *Auger- and X-ray Photoelectron Spectroscopy in Materials Science: A User-Oriented Guide*; Springer Science & Business Media: Berlin/Heidelberg, Germany, 2012; Volume 49.
50. Oliver, W. Improved technique for determining hardness and elastic modulus using load and displacement sensing indentation experiments. *J. Mater. Res.* **1992**. [CrossRef]
51. ISO 20502:2005. *Fine Ceramics (Advanced Ceramics, Advanced Technical Ceramics)—Determination of Adhesion of Ceramic Coatings by Scratch Testing*; ISO: Geneva, Switzerland, 2005.
52. Hassan, M.A.; Bushroa, A.R.; Mahmoodian, R. Identification of critical load for scratch adhesion strength of nitride-based thin films using wavelet analysis and a proposed analytical model. *Surf. Coat. Technol.* **2015**, *277*, 216–221. [CrossRef]
53. Beake, B.D.; Vishnyakov, V.M.; Valizadeh, R.; Colligon, J.S. Influence of mechanical properties on the nanoscratch behaviour of hard nanocomposite TiN/Si3N4 coatings on Si. *J. Phys. D. Appl. Phys.* **2006**, *39*, 1392. [CrossRef]
54. ISO 5834-2:2019. *Implants for Surgery—Ultra-High-Molecular-Weight Polyethylene—Part 2: Moulded Forms*; ISO: Geneva, Switzerland, 2019.
55. ASTM. *ASTM F2025-06(2018), Standard Practice for Gravimetric Measurement of Polymeric Components for Wear Assessment*; ASTM International: West Conshohocken, PA, USA, 2018. [CrossRef]
56. ASTM. *ASTM F732-17, Standard Test Method for Wear Testing of Polymeric Materials Used in Total Joint Prostheses*; ASTM International: West Conshohocken, PA, USA, 2017.
57. Wasa, K.; Hayakawa, S. *Handbook of Sputter Deposition Technology*; IAEA: Vienna, Austria, 1992.
58. Gu, J.; Li, L.; Miao, H.; Xu, Y.; Xu, Y.; Sun, J.; Wang, X.; He, Z. Effect of C_2H_2/N_2 partial pressure ratio on microstructure and mechanical properties of Ti-Al-Si-C-N coatings. *Surf. Coat. Technol.* **2019**, *365*, 200–207. [CrossRef]
59. Huang, S.W.; Ng, M.W.; Samandi, M.; Brandt, M. Tribological behaviour and microstructure of TiCxN(1−x) coatings deposited by filtered arc. *Wear* **2002**, *252*, 566–579. [CrossRef]
60. Guo, C.; Li, M.; Nibarger, J.P.; Gibson, G.N. Single and double ionization of diatomic molecules in strong laser fields. *Phys. Rev. A* **1998**, *58*, R4271–R4274. [CrossRef]
61. Sorensen, S.L.; Björneholm, O.; Hjelte, I.; Kihlgren, T.; Öhrwall, G.; Sundin, S.; Svensson, S.; Buil, S.; Descamps, D.; L'Huillier, A.; et al. Femtosecond pump-probe photoelectron spectroscopy of predissociative Rydberg states in acetylene. *J. Chem. Phys.* **2000**, *112*, 8038–8042. [CrossRef]
62. Reutt, J.E.; Wang, L.S.; Pollard, J.E.; Trevor, D.J.; Lee, Y.T.; Shirley, D.A. Photoelectron spectroscopy and inferred femtosecond intramolecular dynamics of $C_2H_2^+$ and $C_2D_2^+$. *J. Chem. Phys.* **1986**, *84*, 3022–3031. [CrossRef]
63. Pusch, C.; Hoche, H.; Berger, C.; Riedel, R.; Ionescu, E.; Klein, A. Influence of the PVD sputtering method on structural characteristics of SiCN-coatings—Comparison of RF, DC and HiPIMS sputtering and target configurations. *Surf. Coat. Technol.* **2011**, *205*, S119–S123.
64. Shi, Z.; Wang, Y.; Du, C.; Huang, N.; Wang, L.; Ning, C. Silicon nitride films for the protective functional coating: blood compatibility and biomechanical property study. *J. Mech. Behav. Biomed. Mater.* **2012**, *16*, 9–20. [PubMed]
65. McEntire, B.J.; Bal, B.S.; Rahaman, M.N.; Chevalier, J.; Pezzotti, G. Ceramics and ceramic coatings in orthopaedics. *J. Eur. Ceram. Soc.* **2015**, *35*, 4327–4369.
66. Dumitriu La Grange, D.; Goebbels, N.; Santana, A.; Heuberger, R.; Imwinkelried, T.; Eschbach, L.; Karimi, A. Effect of niobium onto the tribological behavior of cathodic arc deposited Nb-Ti-N coatings. *Wear* **2016**, *368–369*. [CrossRef]
67. Park, I.-W.; Kang, D.S.; Moore, J.J.; Kwon, S.C.; Rha, J.J.; Kim, K.H. Microstructures, mechanical properties, and tribological behaviors of Cr–Al–N, Cr–Si–N, and Cr–Al–Si–N coatings by a hybrid coating system. *Surf. Coat. Technol.* **2007**, *201*, 5223–5227. [CrossRef]
68. Xu, J.; Kato, K. Formation of tribochemical layer of ceramics sliding in water and its role for low friction. *Wear* **2000**, *245*, 61–75. [CrossRef]

69. Dante, R.C.; Kajdas, C.K. A review and a fundamental theory of silicon nitride tribochemistry. *Wear* **2012**. [CrossRef]
70. Briggs, D. *Handbook of X-ray Photoelectron Spectroscopy*; Wanger, C.D., Riggs, W.M., Davis, L.E., Moulder, J.F., Muilenberg, G.E., Eds.; Perkin-Elmer Corp., Physical Electronics Division: Eden Prairie, MN, USA, 1979; pp. 190–195.

materials

MDPI

Article

Functional Characteristics and Mechanical Performance of PCU Composites for Knee Meniscus Replacement

Adijat Omowumi Inyang * and Christopher Leonard Vaughan

Division of Biomedical Engineering, Human Biology Department, Faculty of Health Sciences, University of Cape Town, Anzio Road, Observatory, Cape Town 7925, South Africa; kit.vaughan@uct.ac.za

* Correspondence: wumi.inyang@uct.ac.za; Tel.: +27-62-207-6069

Received: 27 February 2020; Accepted: 30 March 2020; Published: 17 April 2020

Abstract: The potential use of fiber-reinforced based polycarbonate-urethanes (PCUs) as candidate meniscal substitutes was investigated in this study. Mechanical test pieces were designed and fabricated using a compression molding technique. Ultra-High Molecular Weight Polyethylene (UHMWPE) fibers were impregnated into PCU matrices, and their mechanical and microstructural properties evaluated. In particular, the tensile moduli of the PCUs were found unsuitable, since they were comparatively lower than that of the meniscus, and may not be able to replicate the inherent role of the meniscus effectively. However, the inclusion of fibers produced a substantial increment in the tensile modulus, to a value within a close range measured for meniscus tissues. Increments of up to 227% were calculated with a PCU fiber reinforcement composite. The embedded fibers in the PCU composites enhanced the fracture mechanisms by preventing the brittle failure and plastic deformation exhibited in fractured PCUs. The behavior of the composites in compression varied with respect to the PCU matrix materials. The mechanical characteristics demonstrated by the developed PCU composites suggest that fiber reinforcements have a considerable potential to duplicate the distinct and multifaceted biomechanical roles of the meniscus.

Keywords: composite; fibers; polycarbonate-urethane; meniscal replacement; mechanical properties; meniscus

1. Introduction

The meniscus is a complex and vital biomechanical fibrocartilaginous tissue in the knee joint. The menisci are important structures, as they have a participatory role in shock absorption, joint lubrication and joint congruency [1–4]. As a result of a large amount of force borne by the meniscus, it is frequently subjected to tear, and it wears out with time. Meniscal tears have been extensively reported as one of the most recurrent knee injuries [5–9]. Damage to the meniscus affects its load sharing and distribution roles performed in the knee, which has been linked to the degeneration of the articular cartilage and a high risk of the devastating disease, osteoarthritis [9–11].

Meniscus removal has considerable repercussions for the joint, as it causes abnormal contact pressures, resulting in joint degeneration [12]. As a result of the consequences of meniscectomy, alternative measures have involved the repair or replacement of the meniscus. However, limited success has been achieved with available options owing to various limitations, such as repair only being possible when the tear occurs in the vascularized area, which does not heal well due to shortage of blood supply [13]. Allografts are alternatives for replacing the meniscus of younger patients with a meniscectomized knee joint [14]. Although allografts have acceptable clinical outcomes, the long-term examinations revealed debatable protective effects to the cartilage [15]. Besides, meniscal allografts have difficulty with remodeling and lack adequate strength [12]. Consequently, allografts cannot be an absolute cure for post-meniscectomy pain [14]. On the other hand, meniscal replacements are biomechanically suitable, and have a distinct advantage of acting as substitutes in the cases of multifaceted tears, as well as alleviating the excruciating pain related to meniscus damage [16].

Different replacements have been sought for the meniscus using either synthetic scaffolds [17,18], natural meniscal tissues or composite materials [19–24]. Amongst the first meniscus replacements developed were permanent meniscal substitutes made from teflon and dacron [11,25,26]. These alternatives were found incompatible after in-vivo testing in rabbits as particles resulting from wear were deposited on the implant. Besides, mechanical integrity was compromised. Similarly, researchers have worked on scaffolds of poly(lactic-co-glycolic acid) embedded with polyglycolic acid fiber [27]. Although the in-vivo studies showed that the scaffolds produced meniscus-resembling tissue after ten weeks' implantation, mechanically the modulus of the tissue was inadequate, as it was not comparable with that of the native meniscus. Another development was a composite of hyaluronic acid and the polycaprolactone matrix with poly-lactic acid reinforcement fibers. The in-vivo studies established that the composite materials support meniscal tissue growth devoid of any adverse effect on the cartilage after implantation, but there was failure resulting from implant extrusion [22,28].

Balint et al. [29] used a different approach in developing a total meniscal substitute where a porous scaffold of collagen-hyaluronan matrix with degradable poly (desaminotyrosyl-tyrosine dodecyl ester dodecanoate) reinforcement fibers was studied. These scaffolds proved to be successful, with considerable mechanical properties suitable as meniscal substitutes; however, implant extrusion remains a challenge [30,31]. The use of polyvinyl alcohol (PVA) hydrogel as a choice material for permanent meniscus replacement has been investigated [32,33]. The study showed the inability of the implant to protect the articular cartilage, and the authors concluded that the failure could be due to size incongruity and inefficient fixation [34].

Studies have explored the possibility of incorporating Ultra High Molecular Weight Polyethylene (UHMWPE) fibers into different matrices, such as poly (vinyl alcohol) hydrogels and polycarbonate-urethane [35,36]. The former group fabricated meniscal substitutes that showed promising mechanical properties and manufacturability. They reported limitations, including delamination and implant extrusion [37,38]. In the latter study, the mechanical behavior of the meniscal prosthesis was determined mathematically and experimentally using a knee model subjected to compressive loading. However, it is indispensable to fully characterize the mechanical properties of the developed composite material in-vitro. Moreover, the finite element model is an approximation of the numerical model [39].

Although several researchers have worked on the development of meniscal substitutes, most of their attempts have been focused on investigating the biological characteristics of the implants from a tissue engineering perspective. While these factors are critical for body replacement parts, the mechanical properties of the developed implants are not given adequate consideration. This gap is evident from the failure reported in literature [40–42]. As a result, there is a dearth of information describing the mechanical requirements and functional performance of replacement for the meniscal tissue. Due to the load-bearing capabilities of the meniscus and its exposure to millions of cycles on an annual basis [43,44], the importance of evaluating the mechanical characteristics of the implant material intended to replace the meniscus cannot be overemphasized, since this will provide critical information, such as safety before implantation. Furthermore, current failures associated with existing permanent synthetic meniscal implants, such as insufficient strength, durability, dislocation, wear and fracture [40] further buttressed the need for assessing the mechanical behavior of a material proposed to replace the meniscus. Thus, the choice of appropriate materials for design, the geometry and the mechanical attributes, could produce a suitable candidate for replacing the meniscus.

The meniscus, being a bundle of cartilaginous tissue, has a complex make-up of material properties which vary with location as well as direction [45]. Its peculiar and particular roles emanate from its unique chemical, physical and biomechanical composition, as well as its distinctive structural architecture [46].

Therefore, in order to restore the biomechanical tasks of a worn-out meniscus, it is important to circumspectly select a replacement material having biomechanical properties as close as possible to those of a native meniscus. Though the meniscus demonstrates site-dependent properties, both in tension and compression, it is anticipated that a composite material can be precisely fitted to duplicate these properties, and hence replace the meniscus. Thus, an isotropic matrix material may not be able to produce the inhomogeneous and anisotropic characteristics exhibited by the natural meniscus [45], but fiber-reinforced polycarbonate-urethanes (PCU) could be customized and adapted based on a suitable choice of reinforcement fiber and its orientation within the PCU matrix. Additionally, the use of fiber-reinforced composite will enable the reinforcing fibers to act as a channel for attaching the implant to the joint capsule to cater for dislocation during motion. Incorporating fibers with high strength into soft polymeric matrices to engineer synthetic meniscal replacement has not been extensively explored. Consequently, this work aimed at developing a structured, tailor-made, PCU-reinforced UHMWPE (referred to as PE) composite as a meniscal substitute. Several surface engineering approaches have been introduced in order to modify the surface structure of UHMWPE for improved biomedical applications. These methods, which include plasma techniques [47], laser surface modification [48] and incorporating particulate or fibrous reinforcements [49,50], have been used to enhance the mechanical, tribological and biological properties of UHMWPE substrates.

Medical grade polyurethanes have been widely promoted for biomedical applications [51]. In particular, the use of PCU has drawn considerable attention in the orthopedic device industry as a result of their excellent mechanical properties, biostability and biocompatibility [52]. PCUs have been extensively utilized in vascular grafts, stents, catheters, pacemaker leads and artificial heart valves [53]. Specifically, bionate thermoplastic PCU, commercially produced by DSM PTG (Berkeley, CA, USA), has been of great interest in the field of orthopedics, because of its outstanding load-bearing properties and excellent wear resistance, which enables it to overcome the setback of osteolysis. Its superior characteristics have made it an excellent material selected for hip and knee joint prostheses, prosthetic spinal discs and the shoulder joint system [54]. Besides, they offer long term durability and resistance to hydrolytic degradation, making them outstanding for in-vivo orthopedic applications [55]. The use of PCU for meniscus replacement stemmed from its unique weight-bearing capabilities, the ability to withstand intense forces within the knee joint [56], and an ease of lubrication due to its hydrophilic nature. Moreover, it proffers low friction properties [57–59] to promote movement within the meniscal compartment, while withstanding repeated stresses from the femoral condyle during flexion and extension motions.

Therefore, the overall goal of this study was to develop a meniscal substitute with mechanical properties closely matching those of the native meniscus. For that reason, test pieces of PCU-PE composites were designed, fabricated and evaluated to determine their suitability as a replacement for the meniscus capable of replicating the closest possible mechanical behavior of the native meniscal tissue. The mechanical test pieces were made up of longitudinally-arranged fibers, such as to duplicate the orientation of the circumferential collagen fiber existing in the human meniscus. The effects of the reinforcing fibers on the mechanical properties of the PCU were investigated, and the composites examined and appraised as meniscal implants. The fibers were able to provide a substantial increase in the stiffness of the PCU matrix and enhance the fatigue and abrasion resistance for long term implantation in the knee joint capsule.

2. Materials and Methods

2.1. Materials and Processing

Biomedical Bionate 80A and 90A polycarbonate urethane (PCU) from DSM (PTG, Berkeley, CA, USA) were used as the matrices, and the reinforcing fibers were Ultra High Molecular Weight Polyethylene (UHMWPE) continuous strand fibers, Dyneema Purity® UG from DSM (designated as PE). Initial drying of the PCU pellets was done at 100 °C for 14 h in a vacuum oven, as stipulated by the supplier and established by Geary et al. [55] as the optimal drying conditions for PCUs to reduce the moisture content to about 0.01%. The intrinsic properties of the PCUs and the reinforcing fibers are detailed in Table 1, as stated in the supplier's data sheets.

Table 1. Characteristic properties of the polycarbonate-urethanes (PCUs) and PCU-reinforced Ultra-High Molecular Weight Polyethylene (PE) fiber.

Material Property	80A	90A	PE Fiber
Tear Strength (kN/m)	64.90	96.40	—
Ultimate Tensile Strength (MPa)	46.64	55.11	—
Density (kg/m^3)	1190	1200	960
Elastic Modulus (MPa)	12	29	126,000
Melting temperature (°C)	—	—	220
Diameter (mm)	—	—	0.31
Elongation at break (%)	531	406	—

2.2. Composite Preparation

A stainless-steel, tailor-made mold was fabricated to produce the composite mechanical test samples (Figure 1). The mold, which encloses the fiber, consists of different parts that are assembled to provide the facile removal of cured samples. The composite samples were formulated using a 5% fiber volume fraction. The percentage of the various constituents was based on a computation of fiber diameter, fiber length and the number of fibers, as discussed in our previous work [60]. The composite material was prepared using the combinations of the different PCUs and the reinforcement fibers. The specimen types and their constituents are described in Table 2.

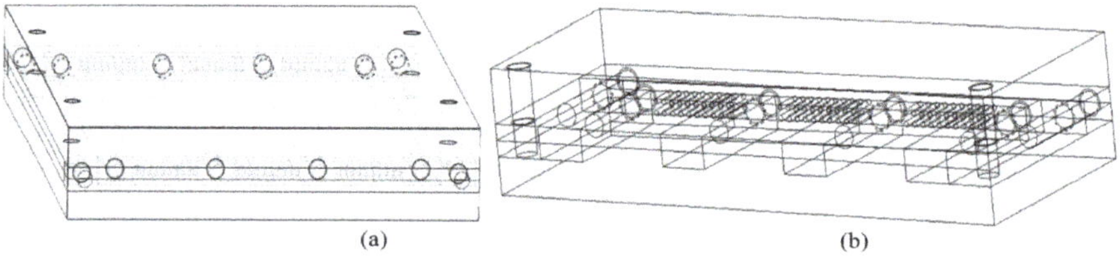

Figure 1. The fabricated mold for the test pieces showing (**a**) The entire mold assemblage (**b**) The fiber holes within the mold.

Table 2. Material constituents for the specimen types.

Matrix	Fiber	Specimen Type
Bionate 80A	—	MX1
	PE	MP1
Bionate 90A	—	MX2
	PE	MP2

The reinforcing fibers were arranged prior to the composite preparation and inflexibly gripped within the mold, such that they were embedded equidistantly within the PCU matrix. After that, the PCU pellets were dispersed to fill the mold. Subsequently, the test pieces were compacted in a pre-heated, custom-built hot press at a pressure of 15 MPa. The PCU and their reinforced samples were cured at 190 °C for MX1 and MP1 samples, and 200 °C for MX2 and MP2 samples for about 10 min. Preliminary experiments were performed to ascertain these molding temperatures, and the amount of pellets required for optimal sample production, since the polymeric matrices and the fibers have varying melting temperatures. The mold was then cooled to room temperature before taking out the samples.

2.3. Mechanical Evaluation

The mechanical test samples were tested for tensile, F_t, and compressive, F_c, properties (Figure 2) using a Zwick/Roell 1484 Material Testing Machine (Zwick GmbH & Co. KG, Ulm, Germany). Each rectangular cuboid tensile specimen of 70 × 19 × 6 mm, was tested at a crosshead speed of 12 mm/min (Figure 2a). Cubic specimens of 6 × 6 × 6 mm were tested for compression, F_c, at a crosshead speed of 5 mm/min (Figure 2b). The moduli for the tensile and compressive tests were calculated from the slope of the linear region of the stress–strain plots. The tensile modulus was taken as the slope of a linear curve fit between 0% to 5% strain, while the compressive modulus was between 2% and 8% strain. These tests were also performed on 100% virgin PCU samples. Three specimens were analyzed in each of the tests. All the results were computed as mean ± standard deviation, and with Excel software (Microsoft, Washington, USA), the unpaired Student's t-test was used to assess the data statistically for significant difference at $p < 0.05$.

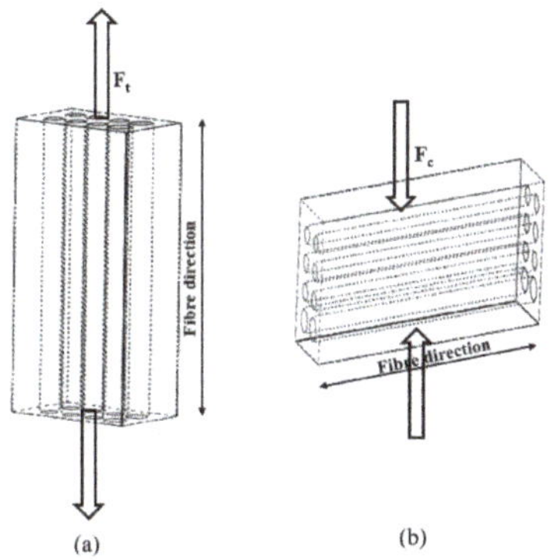

Figure 2. Graphical representation of the mechanical test pieces describing the fiber arrangement and direction for (**a**) tensile testing, F_t and (**b**) compression testing, F_c. The diagrams are not drawn to scale.

2.4. Microstructural Analysis

The arrangements and alignments of the fibers in the matrices were studied using a Wild M400 photomacroscope (Wild Heerbrugg, Gais, Switzerland). In addition, the Nova NanoSem 230 scanning electron microscope (SEM) (FEI, Holland, Netherlands) was used to examine the morphology of the PCUs and their composites, and to investigate the fractured surfaces of the tested samples after failure during the mechanical testing.

3. Results and Discussion

3.1. Microstructural Characterization

In order to study the distribution of fibers in polymeric matrices, a useful approach is to examine their fractured surfaces after mechanical testing through scanning electron microscopy. An accurate and detailed investigation of the fractured surfaces provides information on the nature of the interfacial bonding existing within the fiber–matrix interface. It explains the phenomenon taking place during the deformation process. Also, the knowledge of the failure mechanisms of the PCU and their composites is crucial in evaluating their long-term mechanical reliability for the proposed applications.

Consequently, the surface morphology of the fabricated samples was microscopically viewed to investigate the effects of the reinforcement fibers. The observed micrographs of the PCU matrices and their reinforced composites from the light microscope are presented in Figure 3a–d. The photomicrographs showed that the surface of the reinforced matrices appeared relatively smoother and cleaner with minimal surface imperfections, compared to the as-molded polymeric matrices.

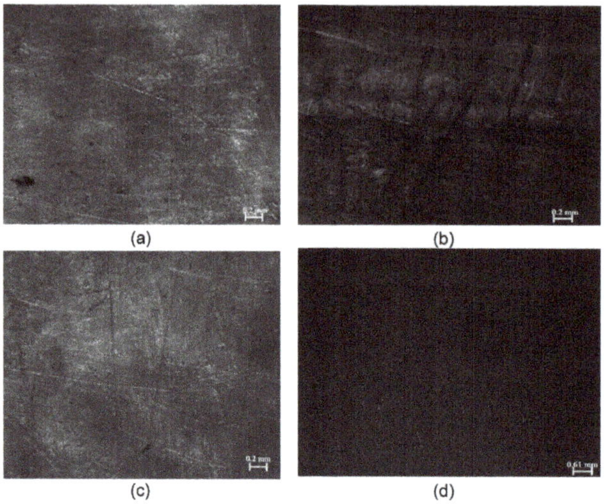

Figure 3. Photomicrographs showing the molded samples of (**a**) MX1 (**b**) MX2 (**c**) MP1 (**d**) MP2.

The fibers were almost evenly distributed within the PCU matrix and at equal distances to one another (Figure 4a). A micrograph of the cross-sectional view of the sample observed from the same microscope corroborated the previous observation (Figure 4b).

Furthermore, the SEM examination, as indicated in Figure 4c, ascertained the uniform distribution of the fiber orientation. The fiber holes, as seen in Figure 4c, were respectively arranged parallel to one another. At higher magnification, the PE fiber was observed to be thoroughly embedded in the polymeric PCU matrix with the surrounding formed ring of PCU-PE fiber interface (Figure 4d).

The fracture behavior of the as-molded PCU samples is shown in Figure 5a,b. The illustrative SEM images showed a sharp, rough and angular fractured surface of the PCU polymer. This surface is characterized by striations suggesting the direction of the crack propagation. The large surface area characterized by smooth regions reveals fast brittle fracture, which is typical of elastomeric PCUs [61,62]. The minimal protrusions disintegrating at the fractured side on the relatively smooth surface (Figure 5b), suggest somewhat the plastic deformation of the PCU polymeric matrix.

On the other hand, PCU-PE composites showed a smooth surface with the fibers firmly held intact in the matrix (Figure 5c,d). A magnified view of the fiber showed an irregular, well-bonded PCU-PE surface. A comparison of the fiber arrangement in the PCU matrix before and after tensile loading gives the information on what ensued when the tensile load was applied (Figures 4b and 5c). SEM images showed the fractured surfaces of the composite material under tensile loading along the fiber direction, with none of the fibers displaced from their average initial positions (Figure 5c).

The SEM images of the PCU-PE composites after the tensile loading showed no evidence of PE fiber pull-out from the PCU matrix, as no void was observed on the microscopically examined surfaces (Figure 5c,d). This observation suggests that there is a strong interfacial bonding that exists between the PCU and the PE fibers. The strong adhesion transmits the applied load from the PCU matrix to the PE fibers, producing a significant rise in the overall mechanical characteristics of the composite material. The existence of the PCU matrix on the surface of the fiber shows there is a strong interaction between the fibers and the PCU, as seen in Figure 5d. The interfacial bonding strength has a corresponding positive effect on the mechanical properties of composite materials [63], hence the increased stiffness observed in the reinforced PCU compared to the virgin PCUs.

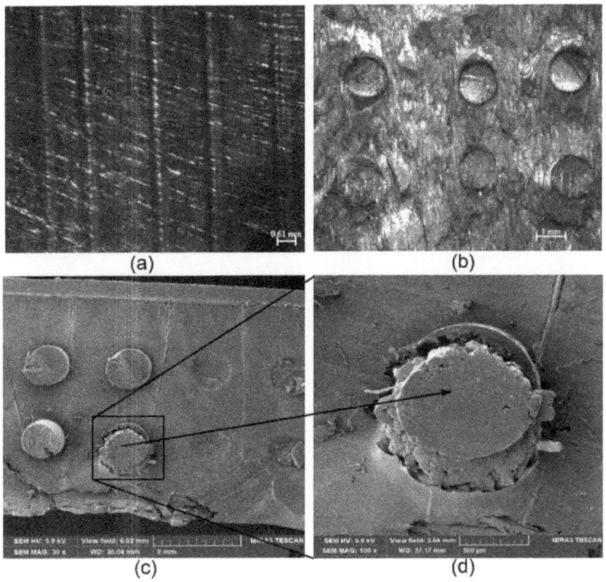

Figure 4. Representative photomicrographs showing (**a**) a transverse sectional view of the arrangement and alignment of the fibers within the composite (**b**) a cross-sectional view of the arrangement and alignment of the fibers within the composite. Scanning electron microscopy showing a cross-section of (**c**) the distribution of the PE fibers within the PCU (**d**) a magnified view of the PE fiber embedded in the PCU matrix.

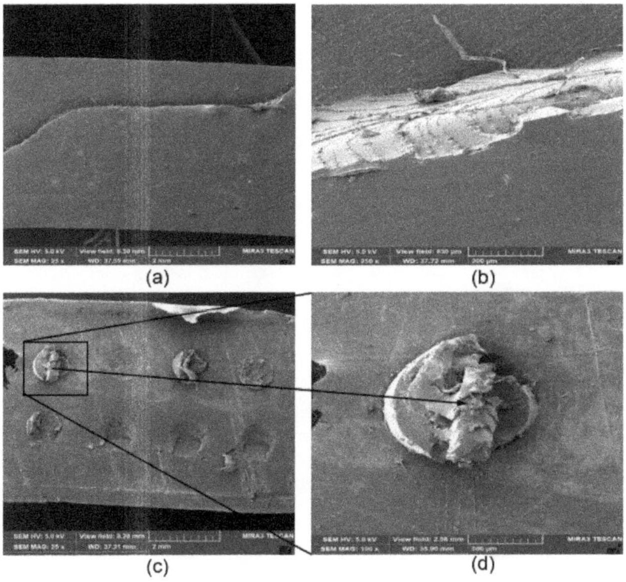

Figure 5. Scanning electron microscopy (SEM) showing fractured surfaces of the samples under tensile loading for (**a**) B8 matrix (**b**) cross-sectional view of the fractured surface of B8 matrix. (**c**) PE-PCU composite along the fiber direction B8 composite (**d**) a closer view of the fractured PE fiber.

3.2. Mechanical Properties

The stress–strain graphs for the average values of the tested samples for both tension and compression tests were plotted in Figures 6 and 7, respectively. These curves exhibited a linear pattern at low strains. Subsequently, a considerable change followed in the slope presenting a nonlinear behavior that continued until the tested samples began to fail. Both MX1 and MX2 demonstrated typical elastomeric stress–strain behavior [55,64]. The PCUs were not as stiff as their fiber-reinforced composites, which indicates that the increase is a function of the stiffness of both the matrix and the interspersed fibers.

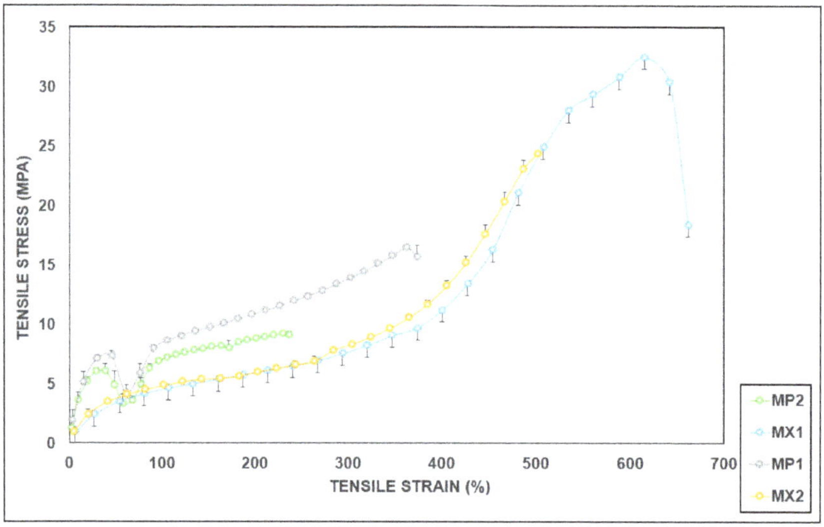

Figure 6. Average tensile stress–strain plots for the PCUs and the PCU-PE composites.

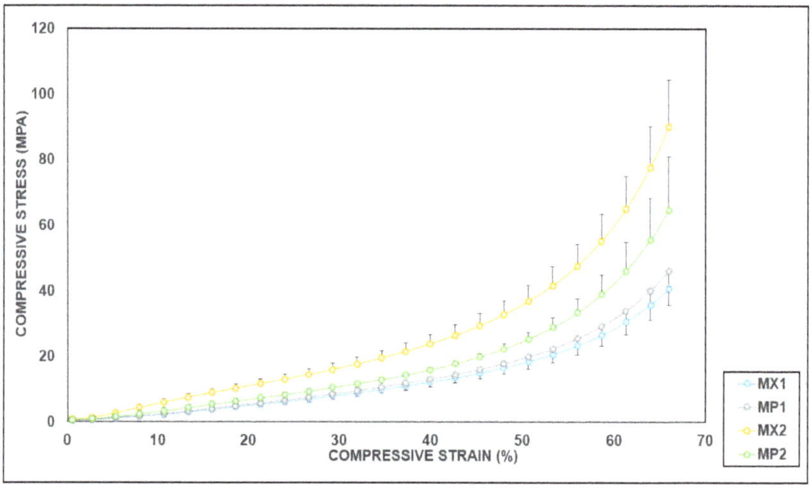

Figure 7. Average compressive stress–strain plots for the PCUs and the PCU-PE composites.

A meniscal substitute must be able to perform similar functions as the natural weight-bearing meniscal structure. To this end, the tensile and compressive properties of the PCUs and their fiber-reinforced composites have been evaluated relative to the meniscus tissue to optimize them as potential meniscal replacements. The empirically calculated tensile moduli for the PCU matrices were rather disparate to those of the supplier (Table 1). These discrepancies could be as a result of the manufacturer's test conducted at 37 °C conditioned in water, since PCU properties are temperature-dependent (DSM-PTG). Besides, their samples were annealed for 24 h at 70 °C before testing. Generally, the moduli of all reinforced specimens were higher than their unreinforced counterparts. The tensile modulus of the meniscus is site-dependent, and hence varies relative to the area and direction. The circumferential tensile modulus of the human meniscus varies between approximately 58 MPa and 295 MPa, while the radial tensile modulus varies between approximately 3 MPa and 60 MPa [65–68]. Therefore, it is expedient that a meniscal device should possess a circumferential tensile modulus of at least 58 MPa in order to prevent deformation, as well as implant extrusion maximally. Both of the PCUs studied offer much lower stiffness (Figure 8). Thus, they will not appropriately perform the rigorous tasks that the meniscal tissue is subjected to on a routine basis. Reinforcing the soft polymeric matrices with durable, high-performance fibers such as PE could therefore possibly construct a composite material which is biomechanically acceptable to replace the worn-out meniscus.

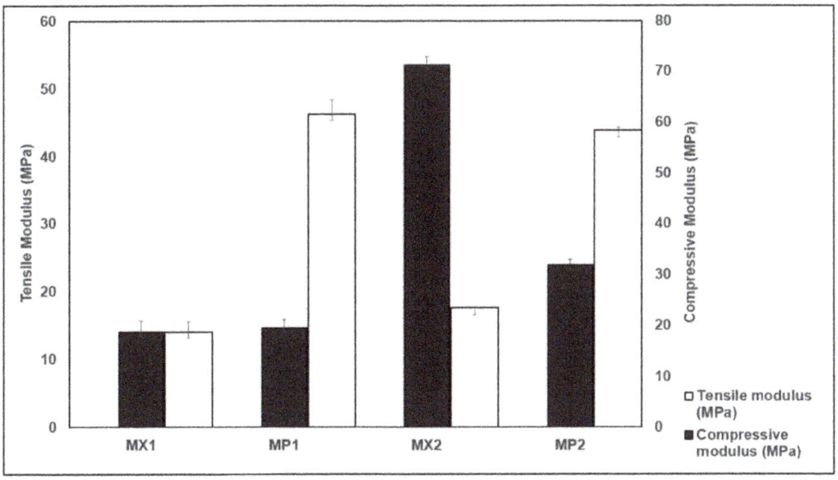

Figure 8. Tensile and compressive properties of the PCU-PE composites compared with their unreinforced matrices.

The tensile moduli of composites of MX1 and MX2 increased appreciably with the incorporation of the PE fibers. MP1 and MP2 exhibited higher stiffness than the PCU matrices, as the fibers significantly ($p < 0.05$) enhanced the tensile properties of the PCUs. Percentage increases of 227% and 148% were obtained for MX1 and MX2, respectively. Interestingly, a similar trend was observed in the tensile characteristics of the curves of the following pairs: MX1 and MX2, and also MP1 and MP2, which revealed similar patterns (Figures 6 and 7). This suggests that the behavior of the fiber is similar, irrespective of the matrix material, further establishing the role of the fibers in influencing the overall performance of a composite material. The function of the fibers could be further understood by the details seen in the micrographs (Figure 5c,d), where the fibers were oriented, and remained in their original positions as they "fought" to withstand the applied tensile load.

The compression behavior of the implant device is pivotal to its overall performance, since the meniscus provides support for knee joint stability [69]. In addition, the compressive modulus is of great significance, as it resists the high stresses and transmits the compression loads exerted by the femur on the tibia.

Unlike the tensile moduli, there was a variation in the changes observed in the compressive moduli with the inclusion of fibers in the PCUs compared to their unreinforced counterparts. All the specimen types behave similarly in compression irrespective of their constituent's composition. The menisci are reported to transmit approximately 50%, and about 85%, of the compressive forces exerted in the knee in extension and 90° in flexion, respectively [70,71]. This role is made possible by the distinctive arrangement of the collagen fibers, which withstand the high stresses produced during the load-bearing. Consequently, a meniscal replacement must be able to reproduce the aforementioned characteristics and peculiarities associated with the native meniscus. An extensive range of compressive moduli values has been published, in which the variation was controlled by strain and loading rates, as well as the type of test conducted. The aggregate compressive modulus varying between 0.10 and 1.13 MPa has been reported for the native meniscus [65,72–74]. In this study, the minimum compressive modulus was recorded for MP1 composite with a 4% rise with the addition of fibers, while MP2 produced a considerable 55% reduction with the inclusion of fibers. The addition of the PE fibers to MX2 were found to be statistically significant in compression. The difference observed in the compressive moduli of MX1 and MX2, and their composites, showed the influence of fiber reinforcements in the compressive properties of the PCU matrices. Although the values obtained are not comparable to those reported for the human meniscus, a higher compressive modulus will be tolerable and acceptable for a meniscal substitute, since some polymers and metals whose compressive moduli are much higher have been utilized as spacers in knee replacement devices [75]. The mechanical characteristics of composites are ultimately determined by the interfacial bond strength between the fiber and the matrix, which is dependent upon the type, shape, orientation and texture of the fiber surface. Consequently, the PCU-reinforced composites can be customized to mimic the desired properties of the native meniscus.

While the PCU composites exhibited excellent mechanical properties in tension and compression, MP1 produced a relatively high tensile modulus close to the natural meniscal range of values and a lower desirable compressive modulus. This comparatively low stiffness could be attributed to a drawback of the fibers encountered during the processing of the composite samples. The PE molecules tend to exhibit some form of relaxation and reorientation, even below the melting point. Besides, under rigorous loading conditions, the fiber molecules can slide, forming new arrangements, which phenomenon in the long run elongates the fiber, thus causing a reduction in tension leading to failure. These molecular changes may trigger a loss in the tensile properties depending on factors like temperature, time and loading conditions (Dyneema Purity® UG, DSM). Therefore, it is anticipated that PE fibers with a higher melting point than the MX1 will produce exceptional results, both in tension and compression. During repetitive tensile stress, failure of the PCU composite can result from fiber fracture or fiber-matrix interfacial debonding. In such a case, the fibers can be detached from the matrix. When the applied tensile load extends the matrix beyond the fibers, the PCU composite will withstand shear at the fiber–matrix interface, which could cause it to fracture [76].

When the meniscus is subjected to an axially compressive force, the load is distributed over its surface area. Due to the meniscal structure, the transmitted force tends to cause the tissue to extrude radially. This structural malalignment is opposed by the hoop stresses generated in the circumferential collagen fibers [2,3]. These tensile stresses, developed within the meniscus during loading, control their function, and are responsible for failure [67]. The ultimate tensile stress of the native meniscus varies with respect to region. The average maximum stress has been reported to be 18.8 MPa, 17.6 MPa and approximately 4 MPa for lateral, medial and radial meniscal tissue, respectively [66]. Consequently, an average ultimate tensile stress of at least 18.8 MPa would be ideal for a meniscal substitute material. Of the composites, the MP1 performs extremely well within this limit.

The ultimate tensile strengths of the PCUs were lower than those from the manufacturer's data sheets (Table 1 and Figure 9). This could be as a result of the disparities in the shape and size of the tested samples, and also the testing technique employed. The average ultimate tensile strength of both composites decreased with fiber reinforcement, and reductions of 36% and 70% were calculated for both MP1 and MP2, respectively. These decreases could be because of the integral sample property, following that tensile strength is a characteristic of both the component materials and the composite samples under examination. Moreover, the tensile properties of polymeric matrix composites are considerably dependent on several factors like the matrix–fiber interface, geometry, distribution and micromechanical deformation [77].

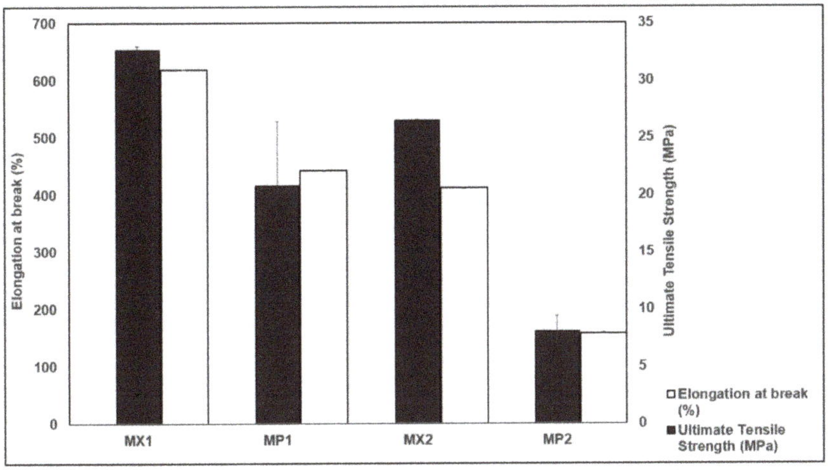

Figure 9. Elongation at break and ultimate tensile strength of the PCU-PE composites compared with their unreinforced matrices.

The elongation at break decreased when the PE fibers were incorporated into the PCU matrices (Figure 9). Elongation at break is a measure of the ductility of a polymeric material, and indicates its ability to resist changes in shape without failing. With PE fibers-incorporated PCU composites, the elongation at break for MX1 and MX2 were reduced by about 28.5% and 61.8%, respectively. Fiber reinforcement of polymeric matrices increases the stiffness and toughness of composites. As the stiffness increases (Figure 6), there is a decrease in the ductility of the composite material, and hence the calculated elongation at break. Previous work has reported similar occurrences [78–80]. The value reported in the manufacturer's datasheet for the PCU suggests that it can attain percentage elongation as high as 531%. At the same time, results from this study showed a higher maximum strain to failure of 619%. The changes in the percentage fracture strains could be due to factors like better surface finish and the firmness of the tensile grips. The wide-ranging tensile and compressive characteristics of the native meniscus complicates the development of a comparable meniscal implant.

Since the functioning of the meniscus is critically reliant on its multiplex shape and structure, it is anticipated that a "true" reflection of the mechanical properties of the fabricated composites could be appropriately exhibited when the prosthesis is designed and manufactured to reproduce the structural configuration of the normal meniscus.

4. Conclusions

The mechanical performance of the PCU matrices determined from this study for the PCU matrices showed they are inadequate, and cannot replace nor sufficiently perform the load-bearing functions of the meniscus. In general, the effect of the fiber reinforcement was favorable, as the tensile modulus was significantly raised to a value within an acceptable tensile modulus of the human meniscus. The results from this study demonstrate that the PCUs can be customized to fit that of the meniscal tissue, by methodically implanting circumferential fibers into the PCU matrix to obtain a meniscal device with desirable mechanical properties. These results visibly revealed the positive effects of the reinforcing fibers.

Furthermore, the microstructural analysis revealed the failure mechanisms during mechanical testing. The embedded fibers in the PCU-PE composites prevented the brittle failure and plastic deformation exhibited in the fractured PCUs. The excellent interfacial bonding strength within the PCU-PE composites produced a corresponding positive effect on the mechanical properties of composite materials. Hence the increased stiffness observed in the reinforced PCUs compared to the virgin PCUs.

This work provides an insight into the mechanical and microstructural performance exhibited by the PCUs and their composites, hence their suitability for artificial bearing surfaces. Further characterization of the composite materials is required to determine their tribological behavior as a meniscal replacement.

Author Contributions: Conceptualization, A.O.I. and C.L.V.; Methodology, A.O.I.; Investigation, A.O.I.; Writing—Original Draft Preparation, A.O.I.; Writing—Review and Editing, A.O.I. and C.L.V.; Supervision, C.L.V. All authors have read and agreed to the published version of the manuscript.

Funding: This research received no funding from any funding agency in the public, commercial, or not-for-profit sectors.

Acknowledgments: The authors will like to thank Mrs. Miranda Waldron for her assistance with the SEM at the Aaron Klug Centre for Imaging and Analysis, Electron Microscopy Unit, University of Cape Town.

Conflicts of Interest: The authors declare no conflict of interest.

References

1. Majd, S.E.; Rizqy, A.I.; Kaper, H.J.; Schmidt, T.A.; Kuijer, R.; Sharma, P.K. An in vitro study of cartilage–meniscus tribology to understand the changes caused by a meniscus implant. *Colloids Surf. B Biointerfaces* **2017**, *155*, 294–303. [CrossRef]
2. McDermott, I.D.; Masouros, S.D.; Amis, A.A. Biomechanics of the menisci of the knee. *Curr. Orthop.* **2008**, *22*, 193–201. [CrossRef]
3. Makris, E.A.; Hadidi, P.; Athanasiou, K.A. The knee meniscus: Structure–function, pathophysiology, current repair techniques, and prospects for regeneration. *Biomaterials* **2011**, *32*, 7411–7431. [CrossRef] [PubMed]
4. Voloshin, A.S.; Wosk, J. Shock absorption of meniscectomized and painful knees: A comparative in vivo study. *J. Biomed. Eng.* **1983**, *5*, 157–161. [CrossRef]
5. Vrancken, A.C.T.; Buma, P.; van Tienen, T.G. Synthetic meniscus replacement: A review. *Int. Orthop.* **2013**, *37*, 291–299. [CrossRef] [PubMed]
6. Garrett, W.E.J.; Swiontkowski, M.F.; Weinstein, J.N.; Callaghan, J.; Rosier, R.N.; Berry, D.J.; Harrast, J.; Derosa, G.P. American Board of Orthopedic Surgery Practice of the Orthopedic Surgeon: Part-II, Certification Examination Case Mix. *J. Bone Joint Surg. Am.* **2006**, *88*, 660–667. [CrossRef] [PubMed]
7. Majewski, M.; Susanne, H.; Klaus, S. Epidemiology of athletic knee injuries: A 10-year study. *Knee* **2006**, *13*, 184–188. [CrossRef] [PubMed]
8. Baynat, C.; Andro, C.; Vincent, J.P.; Schiele, P.; Buisson, P.; Dubrana, F.; Gunepin, F.X. Actifit® synthetic meniscal substitute: Experience with 18 patients in Brest, France. *Orthop. Traumatol. Surg. Res.* **2014**, *100*, S385–S389. [CrossRef]
9. McDermott, I.D.; Amis, A.A. The consequences of meniscectomy. *J. Bone Jt. Surg. Br.* **2006**, *88*, 1549–1556. [CrossRef]
10. Lohmander, L.S.; Englund, P.M.; Dahl, L.L.; Roos, E.M. The Long-term Consequence of Anterior Cruciate Ligament and Meniscus Injuries: Osteoarthritis. *Am. J. Sports Med.* **2007**, *35*, 1756–1769. [CrossRef]
11. Messner, K.; Gillquist, J. Prosthetic replacement of the rabbit medial meniscus. *J. Biomed. Mater. Res.* **1993**, *27*, 1165–1173. [CrossRef] [PubMed]
12. McCann, L.; Ingham, E.; Jin, Z.; Fisher, J. Influence of the meniscus on friction and degradation of cartilage in the natural knee joint. *Osteoarthr. Cartil.* **2009**, *17*, 995–1000. [CrossRef] [PubMed]
13. Sanchez-Adams, J.; Athanasiou, K.A. The Knee Meniscus: A Complex Tissue of Diverse Cells. *Cell. Mol. Bioeng.* **2009**, *2*, 332. [CrossRef]
14. De Bruycker, M.; Verdonk, P.C.; Verdonk, R.C. Meniscal allograft transplantation: A meta-analysis. *SICOT-J.* **2017**, *3*, 33. [CrossRef] [PubMed]

15. Pereira, H.; Fatih Cengiz, I.; Gomes, S.; Espregueira-Mendes, J.; Ripoll, P.L.; Monllau, J.C.; Reis, R.L.; Oliveira, J.M. Meniscal allograft transplants and new scaffolding techniques. *EFORT Open Rev.* **2019**, *4*, 279–295. [CrossRef] [PubMed]
16. Sweigart, M.A.; Athanasiou, K.A. Toward Tissue Engineering of the Knee Meniscus. *Tissue Eng.* **2001**, *7*, 111–129. [CrossRef]
17. Zaffagnini, S.; Giordano, G.; Vascellari, A.; Bruni, D.; Neri, M.P.; Iacono, F.; Kon, E.; Presti, M.L.; Marcacci, M. Arthroscopic collagen meniscus implant results at 6 to 8 years follow up. *Knee Surg. Sports Traumatol. Arthrosc.* **2007**, *15*, 175–183. [CrossRef]
18. Bulgheroni, P.; Murena, L.; Ratti, C.; Bulgheroni, E.; Ronga, M.; Cherubino, P. Follow-up of collagen meniscus implant patients: Clinical, radiological, and magnetic resonance imaging results at 5years. *Knee* **2010**, *17*, 224–229. [CrossRef]
19. Mandal, B.B.; Park, S.-H.; Gil, E.S.; Kaplan, D.L. Multilayered silk scaffolds for meniscus tissue engineering. *Biomaterials* **2011**, *32*, 639–651. [CrossRef]
20. Scotti, C.; Hirschmann, M.T.; Antinolfi, P.; Martin, I.; Peretti, G.M. Meniscus repair and regeneration: Review on current methods and research potential. *Eur. Cells Mater.* **2013**, *26*, 150–170. [CrossRef]
21. Sgaglione, N.A.; Steadman, J.R.; Shaffer, B.; Miller, M.D.; Fu, F.H. Current Concepts in Meniscus Surgery: Resection to Replacement. *Arthrosc. - J. Arthrosc. Relat. Surg.* **2003**, *19*, 161–188. [CrossRef] [PubMed]
22. Chiari, C.; Koller, U.; Dorotka, R.; Eder, C.; Plasenzotti, R.; Lang, S.; Ambrosio, L.; Tognana, E.; Kon, E.; Salter, D.; et al. A tissue engineering approach to meniscus regeneration in a sheep model. *Osteoarthr. Cartil.* **2006**, *14*, 1056–1065. [CrossRef] [PubMed]
23. Kun, M.; Chan, C.; Ramakrishna, S.; Kulkarni, A.; Vadodaria, K. 12—Textile-based scaffolds for tissue engineering. In *The Textile Institute Book Series*; Woodhead Publishing: Sawston, Cambridge, UK, 2019; pp. 329–362. ISBN 978-0-08-102192-7.
24. Hutmacher, D.W. Scaffold design and fabrication technologies for engineering tissues—State of the art and future perspectives. *J. Biomater. Sci. Polym. Ed.* **2001**, *12*, 107–124. [CrossRef] [PubMed]
25. Messner, K. Meniscal substitution with a Teflon-periosteal composite graft: A rabbit experiment. *Biomaterials* **1994**, *15*, 223–230. [CrossRef]
26. Sommerlath, K.; Gallino, M.; Gillquist, J. Biomechanical characteristics of different artificial substitutes for rabbit medial meniscus and effect of prosthesis size on knee cartilage. *Clin. Biomech.* **1992**, *7*, 97–103. [CrossRef]
27. Kang, S.-W.; Sun-Mi, S.; Jae-Sun, L.; Eung-Seok, L.; Kwon-Yong, L.; Sang-Guk, P.; Jung-Ho, P.; Byung-Soo, K. Regeneration of whole meniscus using meniscal cells and polymer scaffolds in a rabbit total meniscectomy model. *J. Biomed. Mater. Res. Part A* **2006**, *77A*, 659–671. [CrossRef]
28. Kon, E.; Filardo, G.; Tschon, M.; Fini, M.; Giavaresi, G.; Reggiani, L.M.; Chiari, C.; Nehrer, S.; Martin, I.; Salter, D.M.; et al. Tissue Engineering for Total Meniscal Substitution: Animal Study in Sheep Model—Results at 12 Months. *Tissue Eng. Part A* **2012**, *18*, 1573–1582. [CrossRef]
29. Balint, E.; Gatt, C.J., Jr.; Dunn, M.G. Design and mechanical evaluation of a novel fiber-reinforced scaffold for meniscus replacement. *J. Biomed. Mater. Res. Part A* **2012**, *100A*, 195–202. [CrossRef] [PubMed]
30. Merriam, A.R.; Patel, J.M.; Culp, B.M.; Gatt, C.J.; Dunn, M.G. Successful Total Meniscus Reconstruction Using a Novel Fiber-Reinforced Scaffold: A 16- and 32-Week Study in an Ovine Model. *Am. J. Sports Med.* **2015**, *43*, 2528–2537. [CrossRef]
31. Patel, J.M.; Merriam, A.R.; Culp, B.M.; Gatt, C.J.; Dunn, M.G. One-Year Outcomes of Total Meniscus Reconstruction Using a Novel Fiber-Reinforced Scaffold in an Ovine Model. *Am. J. Sports Med.* **2016**, *44*, 898–907. [CrossRef] [PubMed]
32. Kobayashi, M.; Toguchida, J.; Oka, M. Development of an artificial meniscus using polyvinyl alcohol-hydrogel for early return to, and continuance of, athletic life in sportspersons with severe meniscus injury. I: Mechanical evaluation. *Knee* **2003**, *10*, 47–51. [CrossRef]
33. Kobayashi, M.; Chang, Y.-S.; Oka, M. A two year in vivo study of polyvinyl alcohol-hydrogel (PVA-H) artificial meniscus. *Biomaterials* **2005**, *26*, 3243–3248. [CrossRef] [PubMed]
34. Kelly, B.T.; Robertson, W.; Potter, H.G.; Deng, X.-H.; Turner, A.S.; Lyman, S.; Warren, R.F.; Rodeo, S.A. Hydrogel Meniscal Replacement in the Sheep Knee: Preliminary Evaluation of Chondroprotective Effects. *Am. J. Sports Med.* **2007**, *35*, 43–52. [CrossRef] [PubMed]

35. Holloway, J.L.; Lowman, A.M.; Palmese, G.R. Mechanical evaluation of poly(vinyl alcohol)-based fibrous composites as biomaterials for meniscal tissue replacement. *Acta Biomater.* **2010**, *6*, 4716–4724. [CrossRef] [PubMed]
36. Elsner, J.J.; Portnoy, S.; Zur, G.; Guilak, F.; Shterling, A.; Linder-Ganz, E. Design of a Free-Floating Polycarbonate-Urethane Meniscal Implant Using Finite Element Modeling and Experimental Validation. *J. Biomech. Eng.* **2010**, *132*, 095001. [CrossRef] [PubMed]
37. Holloway, J.L. Development and Characterization of UHMWPE Fiber-Reinforced Hydrogels for Meniscal Replacement. Ph.D. Thesis, Drexel University, Philadelphia, PA, USA, 2012.
38. Holloway, J.L.; Lowman, A.M.; VanLandingham, M.R.; Palmese, G.R. Interfacial optimization of fiber-reinforced hydrogel composites for soft fibrous tissue applications. *Acta Biomater.* **2014**, *10*, 3581–3589. [CrossRef] [PubMed]
39. Geng, J.; Yan, W.; Xu, W. *Application of the Finite Element Method in Implant Dentistry*; Advanced Topics in Science and Technology in China; Springer: Berlin/Heidelberg, Germany, 2008; ISBN 9783540737643.
40. Shriram, D.; Praveen Kumar, G.; Cui, F.; Lee, Y.H.D.; Subburaj, K. Evaluating the effects of material properties of artificial meniscal implant in the human knee joint using finite element analysis. *Sci. Rep.* **2017**, *7*, 6011. [CrossRef]
41. Trebše, R.; Mihelič, A. *Joint Replacement: Historical Overview BT—Infected Total Joint Arthroplasty: The Algorithmic Approach*; Trebše, R., Ed.; Springer: London, UK, 2012; pp. 7–11. ISBN 978-1-4471-2482-5.
42. Oonishi, H.; Kadoya, Y.; Masuda, S. Gamma-irradiated cross-linked polyethylene in total hip replacements—Analysis of retrieved sockets after long-term implantation. *J. Biomed. Mater. Res.* **2001**, *58*, 167–171. [CrossRef]
43. Kutzner, I.; Heinlein, B.; Graichen, F.; Bender, A.; Rohlmann, A.; Halder, A.; Beier, A.; Bergmann, G. Loading of the knee joint during activities of daily living measured in vivo in five subjects. *J. Biomech.* **2010**, *43*, 2164–2173. [CrossRef]
44. Tudor-Locke, C.; Craig, C.L.; Brown, W.J.; Clemes, S.A.; De Cocker, K.; Giles-Corti, B.; Hatano, Y.; Inoue, S.; Matsudo, S.M.; Mutrie, N.; et al. How many steps/day are enough? for adults. *Int. J. Behav. Nutr. Phys. Act.* **2011**, *8*, 79. [CrossRef]
45. Pangborn, C.A.; Athanasiou, K.A. Knee Meniscus, Biomechanics of. In *Wiley Encyclopedia of Biomedical Engineering*, 1st ed.; Akay, M., Ed.; Wiley-Interscience: Hoboken, NJ, USA, 28 April 2006.
46. Abbot, A.E.; Levine, W.N.; Mow, V.C. Biomechanics of articular cartilage and menisci of the adult knee. In *The Adult Knee*; Callaghan, J.J., Rosenberg, A.G., Rubash, H.E., Simonian, P.T., Wickiewicz, T.L., Eds.; Lippincott Williams & Wilkins: Philadelphia, PA, USA, 2003; pp. 81–104.
47. Zhang, J. Surface modification of ultra-high-molecular-weight polyethylene by argon plasma. J. Thermoplast. *Compos. Mater.* **2014**, *27*, 758–764.
48. Riveiro, A.; Soto, R.; Del Val, J.; Comesaña, R.; Boutinguiza, M.; Quintero, F.; Lusquiños, F.; Pou, J. Laser surface modification of ultra-high-molecular-weight polyethylene (UHMWPE) for biomedical applications. *Appl. Surf. Sci.* **2014**, *302*, 236–242. [CrossRef]
49. Ge, S.; Wang, S.; Huang, X. Increasing the wear resistance of UHMWPE acetabular cups by adding natural biocompatible particles. *Wear* **2009**, *267*, 770–776. [CrossRef]
50. Plumlee, K.; Schwartz, C.J. Improved wear resistance of orthopedic UHMWPE by reinforcement with zirconium particles. *Wear* **2009**, *267*, 710–717. [CrossRef]
51. Wright, J.I. Using Polyurethanes in Medical Applications. *Med. Device Diagn. Ind.* **2006**, *28*, 98–109.
52. Christenson, E.M.; Dadsetan, M.; Wiggins, M.; Anderson, J.M.; Hiltner, A. Poly(carbonate urethane) and poly(ether urethane) biodegradation: In vivo studies. *J. Biomed. Mater. Res. Part A* **2004**, *69A*, 407–416. [CrossRef]
53. Khan, I.; Smith, N.; Jones, E.; Finch, D.S.; Cameron, R.E. Analysis and evaluation of a biomedical polycarbonate urethane tested in an in vitro study and an ovine arthroplasty model. Part I: Materials selection and evaluation. *Biomaterials* **2005**, *26*, 621–631. [CrossRef]
54. Khan, I.; Smith, N.; Jones, E.; Finch, D.S.; Cameron, R.E. Analysis and evaluation of a biomedical polycarbonate urethane tested in an in vitro study and an ovine arthroplasty model. Part II: In vivo investigation. *Biomaterials* **2005**, *26*, 633–643. [CrossRef]
55. Geary, C.; Birkinshaw, C.; Jones, E. Characterisation of Bionate polycarbonate polyurethanes for orthopedic applications. *J. Mater. Sci. Mater. Med.* **2008**, *19*, 3355–3363. [CrossRef]

56. van Tienen, T.G.; Hannink, G.; Buma, P. Meniscus Replacement Using Synthetic Materials. *Clin. Sports Med.* **2009**, *28*, 143–156. [CrossRef]
57. Scholes, S.C.; Burgess, I.C.; Marsden, H.R.; Unsworth, A.; Jones, E.; Smith, N. Compliant Layer Acetabular Cups: Friction Testing of a Range of Materials and Designs for a New Generation of Prosthesis that Mimics the Natural Joint. *Proc. Inst. Mech. Eng. Part H J. Eng. Med.* **2006**, *220*, 583–596. [CrossRef] [PubMed]
58. Smith, S.L.; Ash, H.E.; Unsworth, A. A tribological study of UHMWPE acetabular cups and polyurethane compliant layer acetabular cups. *J. Biomed. Mater. Res.* **2000**, *53*, 710–716. [CrossRef]
59. Caravia, L.; Dowson, D.; Fisher, J. Start up and steady state friction of thin polyurethane layers. *Wear* **1993**, *160*, 191–197. [CrossRef]
60. Inyang, A.O.; Abdalrahman, T.; Bezuidenhout, D.; Bowen, J.; Vaughan, C.L. Suitability of developed composite materials for meniscal replacement: Mechanical, friction and wear evaluation. *J. Mech. Behav. Biomed. Mater.* **2019**, *89*, 217–226. [CrossRef] [PubMed]
61. Hertzberg, R.W.; Manson, J.A. *Fatigue of Engineering Plastics*; Academic Press: Cambridge, MA, USA, 1980.
62. Ford, A.C.; Gramling, H.; Li, S.C.; Sov, J.V.; Srinivasan, A.; Pruitt, L.A. Micromechanisms of fatigue crack growth in polycarbonate polyurethane: Time dependent and hydration effects. *J. Mech. Behav. Biomed. Mater.* **2018**, *79*, 324–331. [CrossRef] [PubMed]
63. Jian, K.; Chen, Z.-H.; Ma, Q.-S.; Zheng, W.-W. Effects of pyrolysis processes on the microstructures and mechanical properties of Cf/SiC composites using polycarbosilane. *Mater. Sci. Eng. A* **2005**, *390*, 154–158. [CrossRef]
64. Inyang, A.O.; Abdalrahman, T.; Vaughan, C.L. Novel composites for human meniscus replacement. In *Orthop Biomaterials—Advances and Applications*; Li, B., Webster, T., Eds.; Springer Pub Company, Inc.: Berlin/Heidelberg, Germany, 2018; pp. 547–568.
65. Almarza, A.J.; Athanasiou, K.A. Design Characteristics for the Tissue Engineering of Cartilaginous Tissues. *Ann. Biomed. Eng.* **2004**, *32*, 2–17. [CrossRef]
66. Tissakht, M.; Ahmed, A.M. Tensile stress-strain characteristics of the human meniscal material. *J. Biomech.* **1995**, *28*, 411–422. [CrossRef]
67. Fithian, D.C.; Kelly, M.A.; Mow, V.C. Material properties and structure-function relationships in the menisci. *Clin. Orthop. Relat. Res.* **1990**, 19–31. [CrossRef]
68. Proctor, C.S.; Schmidt, M.B.; Whipple, R.R.; Kelly, M.A.; Mow, V.C. Material properties of the normal medial bovine meniscus. *J. Orthop. Res.* **1989**, *7*, 771–782. [CrossRef]
69. Heijkants, R.G.J.C.; van Calck, R.V.; de Groot, J.H.; Pennings, A.J.; Schouten, A.J.; van Tienen, T.G.; Ramrattan, N.; Buma, P.; Veth, R.P.H. Design, synthesis and properties of a degradable polyurethane scaffold for meniscus regeneration. *J. Mater. Sci. Mater. Med.* **2004**, *15*, 423–427. [CrossRef] [PubMed]
70. Whipple, R.R.; Wirth, C.R.; Mow, V.C. Mechanical properties of the meniscus. In *Advances in Bioengineering*; Spilker, R.L., Ed.; ASME Press: New York, NY, USA, 1984; pp. 32–33.
71. Rennie, W.J.; Finlay, D.B.L. Meniscal Extrusion in Young Athletes: Associated Knee Joint Abnormalities. *Am. J. Roentgenol.* **2006**, *186*, 791–794. [CrossRef]
72. Sweigart, M.A.; Zhu, C.F.; Burt, D.M.; deHoll, P.D.; Agrawal, C.M.; Clanton, T.O.; Athanasiou, K.A. Intraspecies and Interspecies Comparison of the Compressive Properties of the Medial Meniscus. *Ann. Biomed. Eng.* **2004**, *32*, 1569–1579. [CrossRef]
73. Chia, H.N.; Hull, M.L. Compressive moduli of the human medial meniscus in the axial and radial directions at equilibrium and at a physiological strain rate. *J. Orthop. Res.* **2008**, *26*, 951–956. [CrossRef] [PubMed]
74. Joshi, M.D.; Suh, J.-K.; Marui, T.; Woo, S.L.-Y. Interspecies variation of compressive biomechanical properties of the meniscus. *J. Biomed. Mater. Res.* **1995**, *29*, 823–828. [CrossRef]
75. Bruni, D.; Iacono, F.; Akkawi, I.; Gagliardi, M.; Zaffagnini, S.; Marcacci, M. Unicompartmental knee replacement: A historical overview. *Joints* **2013**, *1*, 45–47. [PubMed]
76. Harris, B. (Ed.) *Fatigue in Composites: Science and Technology of the Fatigue Response of Fiber-Reinforced Plastics*; Woodhead Publishing: Sawston, Cambridge, UK, 2003.
77. Neppalli, R.; Marega, C.; Marigo, A.; Bajgai, M.P.; Kim, H.Y.; Causin, V. Poly(ε-caprolactone) filled with electrospun nylon fibers: A model for a facile composite fabrication. *Eur. Polym. J.* **2010**, *46*, 968–976. [CrossRef]
78. Atuanya, C.U.; Edokpia, R.O.; Aigbodion, V.S. The physio-mechanical properties of recycled low density polyethylene (RLDPE)/bean pod ash particulate composites. *Results Phys.* **2014**, *4*, 88–95. [CrossRef]

79. Sanadi, A.R.; Caulfield, D.F.; Jacobson, R.E.; Rowell, R.M. Renewable Agricultural Fibers as Reinforcing Fillers in Plastics: Mechanical Properties of Kenaf Fiber-Polypropylene Composites. *Ind. Eng. Chem. Res.* **1995**, *34*, 1889–1896. [CrossRef]
80. Shuhadah, S.; Supri, A.G. LDPE-Isophthalic Acid-Modified Egg Shell Powder Composites (LDPE/ESP I). *J. Phys. Sci.* **2009**, *20*, 87–98.

materials

MDPI

Article

Quantitative Measurements of Backside Wear in Acetabular Hip Joint Replacement: Conventional Polyethylene Versus Cross-Linked Polyethylene

Steffen Braun, Sebastian Jaeger, Robert Sonntag, Stefan Schroeder and J. Philippe Kretzer *

Laboratory of Biomechanics and Implant Research, Heidelberg University Hospital, Schlierbacher Landstraße 200a, 69118 Heidelberg, Germany; Steffen.Braun@med.uni-heidelberg.de (S.B.); Sebastian.Jaeger@med.uni-heidelberg.de (S.J.); Robert.Sonntag@med.uni-heidelberg.de (R.S.); Stefan.Schroeder@med.uni-heidelberg.de (S.S.)

* Correspondence: Philippe.Kretzer@med.uni-heidelberg.de; Tel.: +49-6221-56-29209

Received: 25 February 2020; Accepted: 7 April 2020; Published: 15 April 2020

check for updates

Abstract: As shown in previous studies, the modification of conventional polyethylene (CPE) to cross-linked polyethylene (XLPE) and the contribution of antioxidants result in a reduction in total wear. The aim of this study was to evaluate XLPE inserts with vitamin E and CPE regarding their resistance to the backside wear mechanism. A cementless hip cup system (Plasmafit® Plus 7, Aesculap) was dynamically loaded using CPE and XLPE inserts. The backside wear was isolated, generated and collected using the two-chamber principle. The chambers were filled with ultrapure water. After 2×10^6 cycles, the fluids were examined for wear particles according to a particle analysis. Using XLPE inserts, the backside wear was significantly reduced by 35%. While XLPE backside wear particles are significantly larger than CPE particles, they do not differ in their morphology. This study confirms the greater resistance to backside wear of XLPE compared to CPE. It can be assumed that the improved fatigue resistance of the vitamin E-stabilized XLPE inserts demonstrates XLPE's effectiveness against micro-motion and the resulting changing tensions in interface areas like surface breakdown, pitting and the release of very small particles.

Keywords: backside wear; cross-linked; total hip replacement; hip cup system

---

1. Introduction

The modification of conventional polyethylene (CPE) to cross-linked polyethylene (XLPE) and the introduction of antioxidants results in a reduction in joint articular wear [1–3].

While CPE has been the gold standard in hip joint replacement for many years, it is being increasingly replaced by XLPE at a rate of over 95% in hip arthroplasty [4,5]. There is a continuing trend towards bearings with ceramic heads and XLPE inserts [4,5]. Polymer chains that are cross-linked as a result of defined gamma or electron radiation in oxygen-free settings and subsequent thermal treatment show a higher articular wear resistance compared to CPE [6]. In addition, antioxidants such as vitamin E are used to bind free radicals in order to improve the mechanical stability, fatigue strength and oxidation resistance of XLPE [7], which results in a further reduction in articular wear [1–3]. Furthermore, XLPE with vitamin E shows an effective prevention of oxidation even with long aging and leads to more consistent wear behavior compared to CPE and XLPE [8].

Established and internationally standardized examination methods are frequently used to evaluate the articular wear resistance of joint replacements. In addition, a new investigation method enables the isolated quantitative measurement of polyethylene (PE) backside wear in modular acetabular cups [9]. While the resistance to articular wear processes of XLPE has already been demonstrated in many experimental [1,10–13] and clinical studies [3,14,15], no quantitative evidence exists about the supposed wear advantage of XLPE in terms of backside wear.

Therefore, the aim of this study was to compare the backside wear behavior of CPE and XLPE inserts. The questions were defined as follows: Is less PE backside wear generated by using XLPE than using CPE? Is there a difference in the size and morphology of the backside wear particles of XLPE compared to those of CPE?

2. Materials and Methods

2.1. Experimental Groups

Two test groups were compared to each other (CPE versus XLPE). Each group included three cups and three inserts, which were tested independently of each other under the same test conditions.

2.2. Analyzed Components

The cementless cup system Plasmafit® Plus 7 (Aesculap, Tuttlingen, Germany, Ref: NV352T) with a cup size of 52 mm was used. The locking mechanism for fixing the PE inserts was based on a conical width and rough striking surface (R_a = 3.7 μm, R_z = 24.7 μm). PE inserts with an inner diameter of 32 mm were used for all investigations. For the first group (CPE), conventional PE (Ref: NV203) in accordance with ISO 5834-2 was used. Cross-linked PE Vitelene® (Aesculap, Tuttlingen, Germany, Ref: NV203E) with vitamin E as the antioxidant was used for the second group (XLPE).

2.3. Test Setup

The method used for the quantitative measurement of PE backside wear has been validated and described in detail [9].

The cup system (Figure 1(5)) was fixed in polyurethane and dynamically loaded with a frequency of 3 Hz and a force of up to 2.5 kN. Simultaneously, a moment of about 5 Nm (Figure 1) was induced in the cup system. The inclination of the cup was 30° to the load axis (Figure 1(1)). The test duration was 2×10^6 cycles.

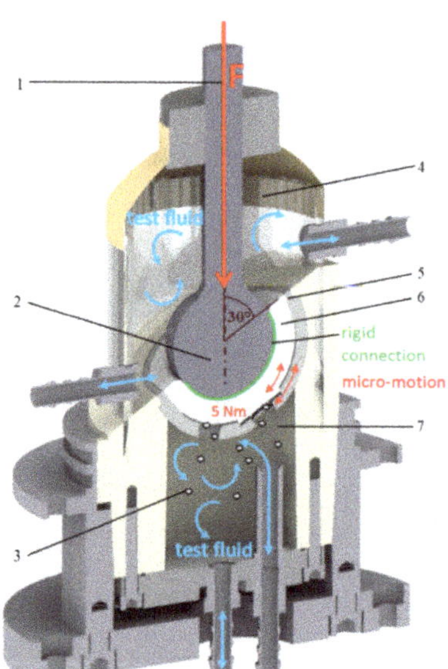

Figure 1. Test setup according to the two-chamber principle: (**1**) load axis, (**2**) artificial femoral head, (**3**) backside wear particle, (**4**) articulation area, (**5**) cup system, (**6**) PE insert, (**7**) backside area [9].

According to the two-chamber principle, the backside area (Figure 1(7)) was separated from the articulation area (Figure 1(4)) and each area represented a chamber system. Both chambers were filled with ultrapure water as the test fluid. However, the interface between the PE insert and the cup could not be considered a reliable seal between the articulation area and the backside cup area. To ensure an isolated generation of backside wear particles, the articulation of the sliding partners (head and PE insert) and thus the articular wear must be prevented. Due to the rigid and resistant cohesive connection (Loctite® 406/SF770, Henkel AG & Co. KGaA, Duesseldorf, Germany) (Figure 1) between the articulation surfaces (femoral head and insert (Figure 1(2,6)), an isolated generation of insert backside wear (Figure 1(3)) was provided. Thereby, backside-generated PE particles were collected in the test fluid. The integrity of the rigid connection was checked and confirmed before and after each test.

2.4. Wear Analysis

After the test was completed, the test fluid was removed from the chambers and analyzed for wear particles. For this analysis, the test fluid was vacuum filtered (pore size: 0.02 μm) and the filters were examined by a scanning electron microscope (SEM Leo 1530, Carl Zeiss AG, Oberkochen, Germany) at a 20,000 times magnification (3 filters/3 SEM images per chamber). The SEM images created were then analyzed using image processing software (ImageJ, version 1.48, public domain) and a particle analysis was carried out in accordance with ASTM F1877-16 [16]. The wear particles were characterized in terms of their size (ECD) and morphology (aspect ratio (AR) and roundness (R)). The particle shapes were characterized as round, oval and fibril-like according to their AR [17]. In addition, the number of analyzed particles was extrapolated to obtain a total number of particles (ETN) as a measure of PE wear [9].

2.5. Statistics

Descriptive statistics with the mean and standard deviations of nine individual values were given for all results. An independent t-test was used to compare all the mean parameters (ETN, ECD, AR and R) of the particle analysis of the two groups. All the requirements for the implementation of statistical procedures were confirmed. The software SPSS (Version 22, IBM, Amonk, NY, USA) was used for the statistical evaluation. The level of significance was set at 5% ($p < 0.05$).

3. Results

Figure 2 shows an example of isolated backside PE wear particles. Figure 3 shows the mean and standard deviations of the PE backside wear in a direct comparison between the use of CPE and XLPE in the investigated cup system. The results of the CPE backside wear with the parameters ETN, ECD, AR and R were partially published in a previous study [9]. In addition, the amounts of different particle shapes and the median (max and min) are shown in Figure 3.

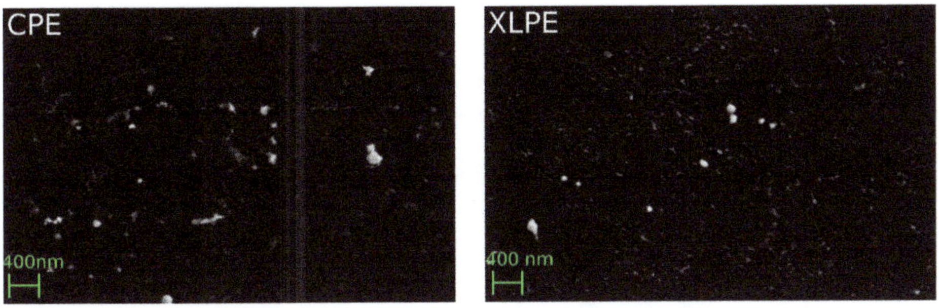

Figure 2. Filter images of the analyzed backside PE wear particles.

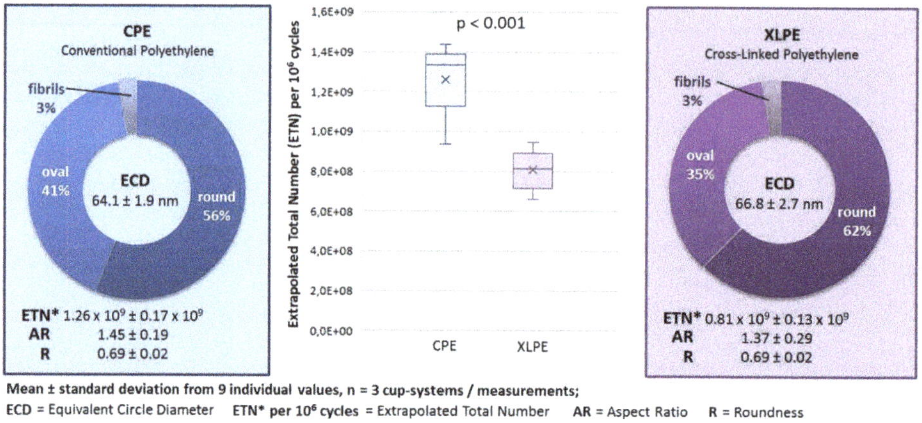

Figure 3. Results of the particle analysis and comparison between the conventional polyethylene (CPE) and cross-linked polyethylene (XLPE) backside wear.

Using the XLPE, about 35% less PE backside wear was generated compared to when the CPE was used. This difference was statistically significant (ETN: $p < 0.001$).

The morphology parameters AR and R did not differ significantly between the CPE and XLPE (AR: $p = 0.465$ and R: $p = 0.126$). The particles were predominantly characterized by a round morphology. The XLPE tended to have a larger amount of round particles than the CPE. In addition, the XLPE particles were slightly but significantly larger ($p = 0.028$). The PE backside wear particles were generally nanoparticles in a size range of between 40 and 100 nm (Figure 4).

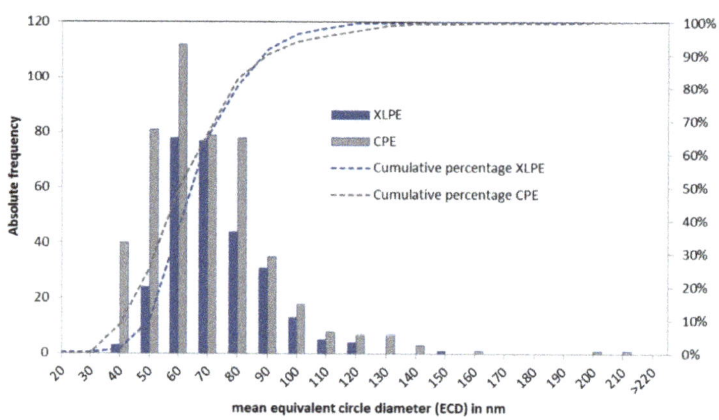

Figure 4. The size distribution of the analyzed particles from the CPE and XLPE inserts.

Figure 5 shows an example of the backside of the CPE and XLPE inserts. It seems that micro-scratches are much more pronounced on the CPE inserts than on the XLPE inserts, whereas the XLPE inserts show a higher proportion of pitting. However, the damage to the XLPE inserts is generally less pronounced.

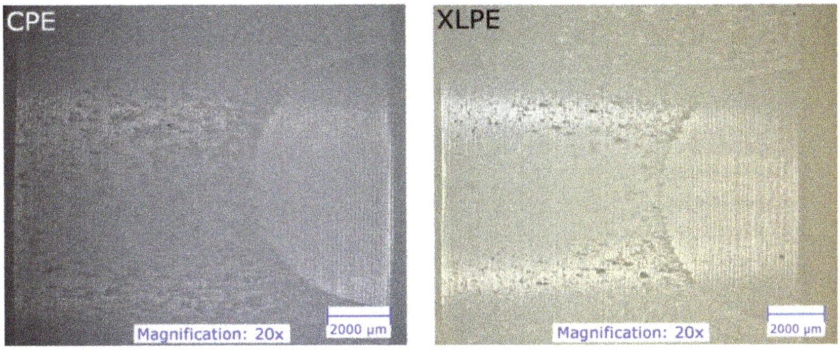

Figure 5. Photographs (Digital Microscope VHX-5000, Keyence, Japan) of the backside of the tested PE inserts.

4. Discussion

The use of XLPE generated significantly less PE backside wear than the use of CPE for the investigated cup system. With the CPE and XLPE inserts, predominantly round particles (56–62%) were generated on the backside and the XLPE particles were significant larger.

The particle analysis of the reported articular wear showed comparable results regarding the morphology [10,12,18]. Illgen et al. described predominantly round particles from CPE (amount: 78%) and from XLPE (amount: 87%) [10]. The amount of elongated fibrils was significantly higher for CPE (14%) than for XLPE (6%). In addition, the CPE particles in published articular wear studies (approx. 196–710 nm) are significantly larger than the XLPE particles (approx. 110–260 nm) [10,12,13]. There was a significant difference in the size of backside-generated particles in this study. The difference in size between the CPE and XLPE particles was only around 3 nm. Considering the size distribution (about 40–100 nm) of the analyzed PE particles, the difference seems negligibly small. However, this 3 nm was a difference of 4%. Fewer wear particles that are of a larger size could result in a higher volume of wear. However, the significantly higher amount of wear particles generated by the CPE inserts (by 35%) seems to have been more relevant than the 4% larger size of the XLPE particles. In addition, we only have two-dimensional images of the particles, which makes it impossible to determine the volume of wear.

The reason for the comparable particle shape was probably the rough locking surface of the cup. The rough peaks of the titanium locking surface engaged the soft PE insert. Due to the micro-motion at this interface, the rough peaks plowed grooves into the soft surface layer of the PE. Therefore, the resulting wear mechanism was dominated by scratching and micro-machining and thus produced comparable wear particles in terms of size and shape. This could be confirmed by the similar type of damage on the backside of the CPE and XLPE inserts. In addition, the permanent alternating stress could cause material fatigue in the PE interface areas and could lead to the release of very small particles.

The higher articular wear resistance of XLPE has not only already been demonstrated in some experimental studies [1,10–13], but also in clinical trials [3,14,15,19,20], which examined the primary wear process. In vitro wear simulating showed a great decrease in wear rate, significantly higher fatigue resistance and improved mechanical properties for XLPE compared to CPE inserts [1,8,11]. In this study, the XLPE also showed greater resistance to backside wear. An explanation for this might be that the improved fatigue resistance of the vitamin E-stabilized inserts was effective against the occurring micro-motions and resulting variations in stress in the PE interface areas (surface disruption, pitting and release of very small particles).

In addition to lower total wear, previous studies have documented a significantly lower occurrence of osteolysis and lower revision rate in patients with XLPE components [3,15,21]. Hanna et al. described a survival rate of 86% for CPE and 100% for XLPE after 13 years with revisions due to excessive wear or osteolysis [15]. While osteolysis was identified in patients with CPE in up to 25% of cases, osteolysis could only be detected in about 2% of cases in patients with XLPE [3,15,21]. According to Fukui et al., patients with XLPE had no wear rate above the osteolysis threshold of 0.1 mm/year, in contrast to 35% of cases in patients with CPE [3]. Cellular responses to the wear particles of XLPE are controversial. While no cell responses [22] or significantly reduced functional biological activity [23] were detected in patients with XLPE in some studies, other studies have shown a significantly larger inflammatory response to XLPE particles [10,13,24,25] than to CPE particles. Illgen et al. showed a concentration-dependent inflammatory response of macrophages against CPE and XLPE particles [10]. At low concentrations, no significant differences in the biological inflammatory response could be observed between CPE and XLPE. However, a significantly higher response was shown to XLPE particles at higher particle concentrations than to CPE wear particles [10]. This higher inflammatory response with a high particle concentration could certainly play an important role with regards to pelvic lysis behind the cups. A particle migration within the cup system [26] could lead to a local accumulation of wear products in the area of screw holes on the bony acetabulum and thereby steadily increase the concentration of PE particles. Therefore, the migration of articulating and backside-generated PE particles should be avoided as far as possible [26].

5. Limitations

This study was an experimental investigation. Deviations from the clinical situation were unavoidable. Therefore, the following limitations must be taken into account.

Instead of bovine serum or a similar lubricant used for in vitro wear studies in accordance with ISO 14242, ultrapure water was used as the test fluid. However, the backside wear mechanism was completely different to the articulation wear mechanism. On the backside, the rough peaks of the harder inner surface of the cup plowed grooves into the surface of the softer PE. Therefore, the influence of the lubricant on the backside wear mechanism was assumed to be negligible. In addition, ultrapure water proved to be advantageous due to its simpler handling and lower risk of contamination by proteins or other residues of biological substances. On the basis of the structure of the artificial cup system, a reliable hermetic sealing of the articulation area from the backside of the PE insert could hardly be achieved. Therefore, articulation between the head and PE insert was prevented (due to the rigid connection). Subsequently, the usual articulation between the sliding bearings was not possible and the applied force represented a simplified load condition compared to physiological hip loads. However, together with the introduction of moments, the load situation approximated the clinical situation in the cranial–caudal axis [27].

6. Conclusions

XLPE showed a significantly higher backside wear resistance compared to CPE. It can be assumed that the improved fatigue resistance of the vitamin E-stabilized XLPE inserts demonstrated XLPE's effectiveness against the occurring micro-motion and resulting changing tensions in and below the interface areas like surface breakdown, pitting and the release of very small particles. The detected backside wear particles were smaller than the reported particles of articular wear, while the morphology of the backside-generated CPE particles did not differ from those of the XLPE.

Author Contributions: Conceptualization, S.B. and J.P.K.; methodology, S.B.; formal analysis, S.B. and S.J.; investigation, S.B., R.S. and S.S.; resources, J.P.K.; data curation, S.B. and S.J.; writing—Original draft preparation, S.B.; review and editing, S.J., R.S., S.S. and J.P.K.; visualization, S.B.; supervision, S.B. and J.P.K.; project administration, J.P.K. All authors have read and agreed to the published version of the manuscript.

Funding: This research was funded by Aesculap AG, Tuttlingen, Germany who provided the implant components.

Acknowledgments: The authors would like to thank Aesculap, Tuttlingen, Germany for the study support.

Conflicts of Interest: The authors declare no conflict of interest in relation to this study. The funders had no role in the design of the study; in the collection, analyses, or interpretation of data; in the writing of the manuscript, or in the decision to publish the results.

References

1. Affatato, S.; De Mattia, J.S.; Bracco, P.; Pavoni, E.; Taddei, P. Wear performance of neat and vitamin E blended highly cross-linked PE under severe conditions: The combined effect of accelerated ageing and third body particles during wear test. *J. Mech. Behav. Biomed. Mater.* **2016**, *64*, 240–252. [CrossRef]
2. Garvin, K.L.; White, T.C.; Dusad, A.; Hartman, C.W.; Martell, J. Low Wear Rates Seen in THAs With Highly Crosslinked Polyethylene at 9 to 14 Years in Patients Younger Than Age 50 Years. *Clin. Orthop. Relat. Res.* **2015**, *473*, 3829–3835. [CrossRef] [PubMed]
3. Fukui, K.; Kaneuji, A.; Sugimori, T.; Ichiseki, T.; Matsumoto, T. Wear Comparison between Conventional and Highly Cross-Linked Polyethylene Against a Zirconia Head. *J. Arthroplast.* **2013**, *28*, 1654–1658. [CrossRef] [PubMed]
4. Deutschland, E. *Jahresbericht 2017—Mit Sicherheit mehr Qualität*; EPRD Deutsches Endoprothesenregister gGmbH: Berlin, Germany, 2017.
5. Graves, S.; Davidson, D.; Ingerson, L.; Ryan, P.; Griffith, E.C.; McDermott, B.F.J.; McElroy, H.J.; Pratt, N. The Australian Orthopaedic Association National Joint Replacement Registry. *Med. J. Aust.* **2004**, *180*, S31–S34. [CrossRef] [PubMed]
6. Bistolfi, A.; Bellare, A. The relative effects of radiation crosslinking and type of counterface on the wear resistance of ultrahigh-molecular-weight polyethylene. *Acta Biomater.* **2011**, *7*, 3398–3403. [CrossRef]
7. Oral, E.; Muratoglu, O.K. Vitamin E diffused, highly crosslinked UHMWPE: A review. *Int. Orthop.* **2010**, *35*, 215–223. [CrossRef]
8. Grupp, T.M.; Holderied, M.; Mulliez, M.A.; Streller, R.; Jäger, M.; Blömer, W.; Utzschneider, S. Biotribology of a vitamin E-stabilized polyethylene for hip arthroplasty—Influence of artificial ageing and third-body particles on wear. *Acta Biomater.* **2014**, *10*, 3068–3078. [CrossRef]
9. Braun, S.; Sonntag, R.; Schroeder, S.; Mueller, U.; Jäger, S.; Gotterbarm, T.; Kretzer, J.P. Backside wear in acetabular hip joint replacement. *Acta Biomater.* **2019**, *83*, 467–476. [CrossRef]
10. Illgen, R.L.; Forsythe, T.M.; Pike, J.W.; Laurent, M.P.; Blanchard, C.R. Highly Crosslinked vs Conventional Polyethylene Particles—An In Vitro Comparison of Biologic Activities. *J. Arthroplast.* **2008**, *23*, 721–731. [CrossRef]
11. Oral, E.; Christensen, S.D.; Malhi, A.S.; Wannomae, K.K.; Muratoglu, O.K. Wear Resistance and Mechanical Properties of Highly Cross-linked, Ultrahigh–Molecular Weight Polyethylene Doped With Vitamin E. *J. Arthroplast.* **2006**, *21*, 580–591. [CrossRef]
12. Affatato, S.; Bersaglia, G.; Rocchi, M.; Taddei, P.; Fagnano, C.; Toni, A. Wear behaviour of cross-linked polyethylene assessed in vitro under severe conditions. *Biomaterials* **2005**, *26*, 3259–3267. [CrossRef] [PubMed]
13. Ingram, J.H.; Stone, M.; Fisher, J.; Ingham, E. The influence of molecular weight, crosslinking and counterface roughness on TNF-alpha production by macrophages in response to ultra high molecular weight polyethylene particles. *Biomaterials* **2004**, *25*, 3511–3522. [CrossRef] [PubMed]
14. Scemama, C.; Babinet, A.; Dumaine, V.; Courpied, J.P.; Hamadouche, M.; Anract, P. Does vitamin E-blended polyethylene reduce wear in primary total hip arthroplasty: A blinded randomised clinical trial. *Int. Orthop.* **2016**, *41*, 1113–1118. [CrossRef] [PubMed]
15. Hanna, S.; Somerville, L.; McCalden, R.W.; Naudie, D.D.; Macdonald, S.J. Highly cross-linked polyethylene decreases the rate of revision of total hip arthroplasty compared with conventional polyethylene at 13 years' follow-up. *Bone Jt. J.* **2016**, *98*, 28–32. [CrossRef]
16. *Standard Practice for Characterization of Particles*; ASTM F1877-16; American Society for Testing and Materials: West Conshohocken, PA, USA, 2010.
17. Catelas, I.; Bobyn, J.D.; Medley, J.B.; Krygier, J.J.; Zukor, D.J.; Huk, O.L. Size, shape, and composition of wear particles from metal-metal hip simulator testing: Effects of alloy and number of loading cycles. *J. Biomed. Mater. Res.* **2003**, *67*, 312–327. [CrossRef]

18. Oral, E.; Wannomae, K.K.; Bichara, D.A.; Micheli, B.; Doshi, B.N.; O'Brien, C.; Nielsen, G.P.; Muratoglu, O.K. An antioxidant stabilized, chemically cross-linked UHMWPE with superior toughness. *J. Biomed. Mater. Res. Part B Appl. Biomater.* **2018**, *107*, 1945–1952. [CrossRef] [PubMed]
19. Nakashima, Y.; Sato, T.; Yamamoto, T.; Motomura, G.; Ohishi, M.; Hamai, S.; Akiyama, M.; Hirata, M.; Hara, D.; Iwamoto, Y. Results at a minimum of 10years of follow-up for AMS and PerFix HA-coated cementless total hip arthroplasty: Impact of cross-linked polyethylene on implant longevity. *J. Orthop. Sci.* **2013**, *18*, 962–968. [CrossRef]
20. Beksaç, B.; Salas, A.; Della Valle, A.G.; Salvati, E.A. Wear is Reduced in THA Performed with Highly Cross-linked Polyethylene. *Clin. Orthop. Relat. Res.* **2008**, *467*, 1765–1772. [CrossRef]
21. Mall, N.A.; Nunley, R.M.; Zhu, J.J.; Maloney, W.J.; Barrack, R.L.; Clohisy, J.C. The Incidence of Acetabular Osteolysis in Young Patients with Conventional versus Highly Crosslinked Polyethylene. *Clin. Orthop. Relat. Res.* **2010**, *469*, 372–381. [CrossRef]
22. Sethi, R. Macrophage response to cross-linked and conventional UHMWPE. *Biomaterials* **2003**, *24*, 2561–2573. [CrossRef]
23. Baxter, R.M.; Macdonald, D.W.; Kurtz, S.M.; Steinbeck, M.J. Characteristics of highly cross-linked polyethylene wear debris in vivo. *J. Biomed. Mater. Res. Part B Appl. Biomater.* **2013**, *101*, 467–475. [PubMed]
24. Ormsby, R.T.; Solomon, L.B.; Yang, D.; Crotti, T.N.; Haynes, D.R.; Findlay, D.M.; Atkins, G.J. Osteocytes respond to particles of clinically-relevant conventional and cross-linked polyethylene and metal alloys by up-regulation of resorptive and inflammatory pathways. *Acta Biomater.* **2019**, *87*, 296–306. [CrossRef] [PubMed]
25. Fisher, J.; McEwen, H.M.J.; Tipper, J.L.; Galvin, A.; Ingram, J.; Kamali, A.; Stone, M.H.; Ingham, E. Wear, debris, and biologic activity of cross-linked polyethylene in the knee: Benefits and potential concerns. *Clin. Orthop. Relat. Res.* **2004**, 114–119. [CrossRef] [PubMed]
26. Braun, S.; Vardag, S.; Mueller, U.; Schroeder, S.; Sonntag, R.; Bormann, T.; Gotterbarm, T.; Kretzer, J.; Gotterabarm, T. Backside wear, particle migration and effectiveness of screw hole plugs in acetabular hip joint replacement with cross-linked polyethylene. *Acta Biomater.* **2019**, *97*, 239–246. [CrossRef]
27. Sonntag, R.; Braun, S.; Al-Salehi, L.; Reinders, J.; Mueller, U.; Kretzer, J.P. Three-dimensional friction measurement during hip simulation. *PLoS ONE* **2017**, *12*, e0184043. [CrossRef]

materials

MDPI

Article

Comparison of the Primary Stability of Porous Tantalum and Titanium Acetabular Revision Constructs

Nicholas A. Beckmann [1,2,*], Rudi G. Bitsch [3], Mareike Schonhoff [4], Klaus-Arno Siebenrock [2], Martin Schwarze [1] and Sebastian Jaeger [4]

1 Clinic for Orthopedics and Trauma Surgery, Heidelberg University Hospital, 69118 Heidelberg, Germany; martin.schwarze@med.uni-heidelberg.de
2 Department of Orthopaedic Surgery and Traumatology, Inselspital, Bern University Hospital, 3010 Bern, Switzerland; klaus.siebenrock@insel.ch
3 National Joint Center, ATOS Clinics, 69115 Heidelberg, Germany; rudi.bitsch@atos.de
4 Laboratory of Biomechanics and Implant Research, Clinic for Orthopedics and Trauma Surgery, Heidelberg University Hospital, 69118 Heidelberg, Germany; mareike.schonhoff@med.uni-heidelberg.de (M.S.); sebastian.jaeger@med.uni-heidelberg.de (S.J.)
* Correspondence: nicholasandreas.beckmann@insel.ch or nicholas.beckmann@med.uni-heidelberg.de

Received: 29 February 2020; Accepted: 7 April 2020; Published: 10 April 2020

Abstract: Adequate primary stability of the acetabular revision construct is necessary for long-term implant survival. The difference in primary stability between tantalum and titanium components is unclear. Six composite hemipelvises with an acetabular defect were implanted with a tantalum augment and cup, using cement fixation between cup and augment. Relative motion was measured at cup/bone, cup/augment and bone/augment interfaces at three load levels; the results were compared to the relative motion measured at the same interfaces of a titanium cup/augment construct of identical dimensions, also implanted into composite bone. The implants showed little relative motion at all load levels between the augment and cup. At the bone/augment and bone/cup interfaces the titanium implants showed less relative motion than tantalum at 30% load ($p < 0.001$), but more relative motion at 50% (p = n.s.) and 100% ($p < 0001$) load. The load did not have a significant effect at the augment/cup interface ($p = 0.086$); it did have a significant effect on relative motion of both implant materials at bone/cup and bone/augment interfaces ($p < 0.001$). All interfaces of both constructs displayed relative motion that should permit osseointegration. Tantalum, however, may provide a greater degree of primary stability at higher loads than titanium. The clinical implication is yet to be seen

Keywords: porous implants; tantalum; titanium; acetabulum; hip arthroplasty; hip replacement; revision hip arthroplasty; acetabular revision; primary stability

1. Introduction

Total hip arthroplasty (THA) is a highly successful surgical intervention that is being performed with increasing frequency in cases of advanced osteoarthritis, and in patients of decreasing age [1]. The increased incidence of primary THA is accompanied by a corresponding increase in revision THA with the associated concerns of diminished bone quality, bone loss and compromised soft tissue [2]. Earlier interventions that addressed these concerns included the use of large structural allografts that had mixed results with loosening and migration rates of up to 70% [3]. Utilization of metal cages for large defects reduced the loosening rate to 14% at 6 year follow up [3,4]. In addition, the recognition of cement disease as a major cause of loosening and later failure in cemented constructs [5] led to the increasing use of cementless porous metal components that allowed for bone ingrowth that facilitated stable fixation. The introduction of porous metal implants with a range of accessory porous metal augments, buttresses and shims has led to a further improvement in revision THA outcome.

Currently, the most frequently used porous metal implants have either a tantalum or titanium porous metal surface and with press-fit implantation they provide a stable mechanical surface between implant and host bone in the short term (primary stability), and facilitate osseointegration in the mid and long term [6,7]. Tantalum in the form of Trabecular Metal™ (TM) (Zimmer Biomet, Warsaw, Indiana) is currently one of the more frequently used porous implants [8], and has been used to treat very extensive acetabular defects [9] as well as for neoplastic periacetabular lesions [10].

Optimal primary stability and ultimately adequate osseointegration and successful outcome is dependent on minimal relative motion at the component/bone interface. Prior experimental studies have shown that successful osseointegration occurs with relative motion between surfaces of up to 40μm, and that fibrous attachment occurs at 150 μm [11]. In addition, increased relative motion between components and bone can lead to particle generation and shedding that promotes later loosening and failure [12]. The use of additional components such as augments and buttresses increases the number of opposed surfaces and also the potential for increased relative motion in the construct as a whole, with possible consequences for the stability of the construct [13].

The aim of this study was to utilize an experimental biomechanical set up to evaluate the primary stability of a tantalum acetabular cup and augment construct as used in the treatment of larger acetabular defects [14], and compare the results with those of a similar porous titanium construct and augment that were published previously in an identical experimental set up [13].

2. Materials and Methods

Six Trabecular Metal™ acetabular cup components of 56 mm diameter and corresponding Trabecular Metal™ augments of 54/56 × 1 cm size were utilized in our biomechanical set up (see Figure 1). We also utilized six large fourth generation composite left hemipelvises (#3405 Sawbones; Sawbones Europe AB, Malmo, Sweden) each with a created Paprosky 2b defect of 1 cm thickness that was segmental and constituted less than one third of the acetabular circumference. Each defect was created in an identical standardized manner at the postero-cranial aspect of each acetabulum, with the edge of the defect adjacent to the antero-inferior iliac spine. To accomplish this, the periphery of the defect was first marked on each hemipelvis and the central synthetic bone was reamed and burred to the peripheral mark and to 1cm depth. This created defect was then completely covered with a TM augment according to manufacturer's instructions, and fixed to host bone with two 5.5 × 30 mm screws. Prior to acetabular cup implantation, premixed cement was then applied to the aspect of the augment that apposed the acetabular cup. A medium viscosity bone cement (Palacos R + G pro; Heraeus Medical Gmbh, Wehrheim, Germany) was used at this interface. The cement was vacuum mixed (Optivac Cement Mixing System; Zimmer Biomet, Warsaw, Indiana) and applied 120 s after the start of mixing. The cement was dispensed with a cement gun and 1.5 cm^3 was hand-modeled on the augment surface. Excess cement was carefully removed from the multi-hole acetabular component. All cementing was done under standardized conditions with the same mean room temperature and humidity as in the prior experiment [13]. The acetabular cup component was then implanted according to manufacturer's directions, and attached to host composite bone with one each of 6.5 × 40 mm and 6.5 × 30 mm screws, with the screws directed towards the sacro-iliac joint. The acetabular cup was press-fit as well as the rim defect would allow. All augments and cups were implanted by a single experienced surgeon (R.G.B.).

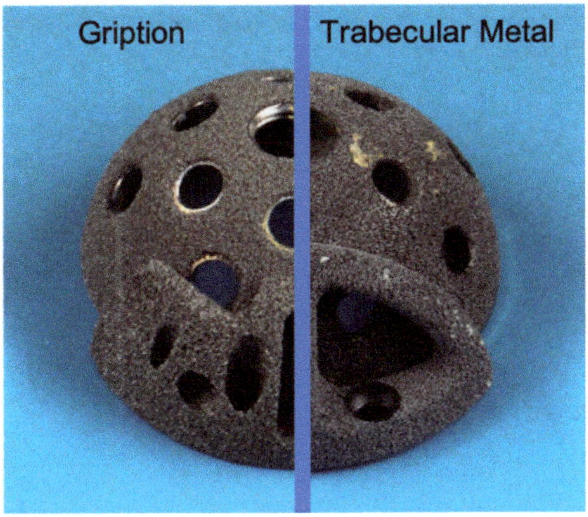

Figure 1. Photograph of titanium Gription cup and augment (**left**) and tantalum Trabecular Metal augment and cup (**right**) after implant explantation, demontrating the differences in their hole positions and augment geometry.

Following implantation, the hemipelvises were secured along the sacral side of the ilium using polyurethane foam (RenCast FC 53 A/B; Goessl + Pfaff Gmbh, Karlskron, Germany) in a containment device [13]. The symphysis was also secured to a two-component casting resin block that had an attached stainless-steel ball on the under-side that was placed on a metal plate. This constituted a two-point pelvic fixation, with the pelvis fixed in only one degree of freedom to allow for multi-planar movement and rotation of the symphysis that mimics a physiologic fixation, as described in prior studies [13].

Optical markers of 0.8 mm diameter (uncoded passive white markers, GOM Item Number 21874; GOM Gmbh, Braunschweig, Germany) were placed in adjacent rows along the rims of the acetabular cup component, adjacent augment and the host bone [13]. These adjacent rows of markers were detected in grey-scale by a stereo camera system that provided 3D discrimination and recording of relative motion between components and bone during loading. This was achieved by using 3D point triangulation to calculate the 3D marker position in the x, y and z axes of the defined coordinate system [15]. The 3D relative motion in the x, y and z axes were measured simultaneously between the acetabular component and bone, acetabular component and augment and augment and bone using an optical measuring system (PONTOS, GOM Gmbh, Braunschweig, Germany).

We pre-tested the hemipelvis set-up using a materials testing machine (MTS Mini Bionix 359; MTS Systems Corporation, Eden Prairie, Minnesota), with the load applied in the direction of the greatest load that occurs during the normal gait cycle, as defined by Bergmann et al. [16]. The maximum load during normal walking was found to be 233% of the individual's body weight at 31 degrees of rotation around the x axis and 5 degrees around the z axis relative to the acetabular component system described by Bergmann et al. [17]. We arbitrarily chose a body weight of 80 kg for each specimen, as in our prior study [13], that was equivalent to 1.8 kN at 100% load at the hip during normal gait.

Three load levels were chosen; 3–30% load (equal to 0.5 kN), 5–50% load (equal to 0.9 kN) and 10–100% load (equal to 1.8 kN) (see Figure 2). A total of 1000 cycles were applied sequentially in a sinusoidal wave-form at 1 Hz at each of the three load levels. To ensure good force closure between force plate and testing sample, 0.2 kN was applied prior to testing. The dependent variable (measured in μm, with average and variance) was the relative motion between components and bone, measured at the following groups of cycles; 1 to 50, 51 to 200, 201–500, 501–800, 801–995.

The results of measurements obtained as described above were compared to results obtained in an identical manner during a prior experimental set-up using Gription® titanium components instead of tantalum components of the same diameter/size [13].

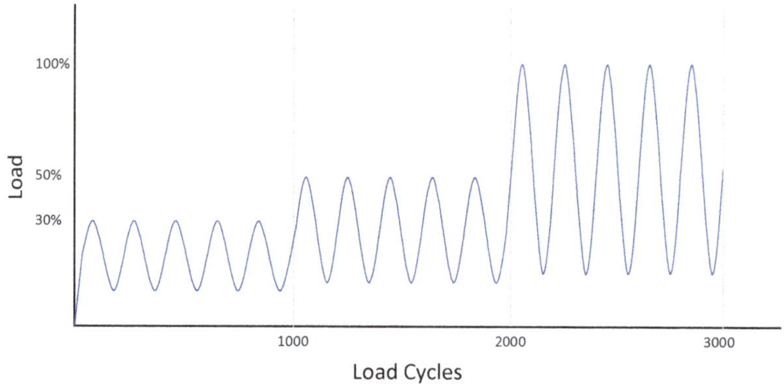

Figure 2. Schematic graph displaying the load applied for each sample over the 3000 test cycles.

3. Statistical Analysis

Statistical evaluation was carried out descriptively (arithmetic mean, standard deviation, minimum and maximum). After confirmation of normal distribution using a Shapiro–Wilk test, a t-test of independent variables was performed. To evaluate differences in both groups during the cyclic loading, we performed an analysis of variance with repeated measures (ANOVA). The effects with regard to implant type and time points were evaluated. A p-value of ≤ 0.05 was considered significant. Results were presented using statistical graphics when necessary. Statistical evaluation was performed using Microsoft Excel (Microsoft Corporation, Santa Rosa, CA, USA), and the analytical software SPSS 25 (IBM Inc., Armonk, New York, USA).

4. Results

One of the six samples was excluded, since the fixation of the hemipelvis in the containment device failed. Table 1 shows the average relative motion between the tantalum augment/cup, tantalum augment/bone and tantalum cup/bone interfaces at 30%, 50% and 100% load for the remaining samples and compares it to the average relative motion between titanium cups/augment, titanium augment/bone and titanium cup/bone interfaces.

The t-test revealed a statistically significant difference in the relative motion between titanium augment/cup and tantalum augment/cup at all load levels (30% load: $t(8) = -20.34$, $p < 0.001$; 50% load: $t(8) = -30.06$, $p < 0.001$; 100% load: $t(8) = -14.32$, $p < 0.001$) (see Figure 3).

The titanium augment/sawbone interface displayed less relative motion at 30% load than the tantalum augment/sawbone interface (30% load: $t(8) = -8.81$, $p < 0.001$). At 50% ($t(8) = 1.59$, $p = 0.151$) and 100% ($t(8) = 15.47$, $p < 0.001$) load there was an increased average relative motion of the titanium augment/Sawbone interface when compared to the relative motion at the tantalum/Sawbone interface (see Figure 4).

At 30% load, the titanium displayed significantly lower relative motion ($t(8) = -13.00$, $p < 0.001$) at the bone/cup interface, while at 50% load ($t(8) = -0.20$, $p = 0.843$) and at 100% load ($t(8) = 11.76$, $p < 0.001$) the tantalum displayed lower relative motion (see Figure 5).

No significant difference was noted at the augment/cup interface with regard to the load level ($F(2, 16) = 2.87$, $p = 0.086$). The load level did, however, have a significant effect on the relative motion at the bone/augment ($F(2, 16) = 352.66$, $p < 0.001$) and bone/cup ($F(2, 16) = 331.96$, $p < 0.001$) interfaces.

Table 1. Table showing the mean and standard deviation (SD) of the relative motion (μm) of tantalum (Trabecular Metal) and titanium (Gription) implants at the respective implant/bone interfaces and load levels.

Interface	**Augment/Cup**		**Bone/Augment**		**Bone/Cup**	
Implant Material	**Titanium**	**Tantalum**	**Titanium**	**Tantalum**	**Titanium**	**Tantalum**
Load	Mean (SD)	Mean (SD)	Mean (SD)	Mean (SD)	Mean (SD)	Mean (SD)
30%	11.0 (1.9)	22.5 (6.1)	20.0 (7.3)	29.7 (8.1)	27.9 (8.0)	46.3 (18.6)
50%	10.9 (2.1)	24.7 (5.7)	38.7 (17.8)	31.7 (9.7)	50.2 (18.6)	51.4 (19.8)
100%	11.3 (4.2)	23.7 (6.6)	84.3 (40.2)	39.4 (15.0)	107.2 (44.0)	61.6 (20.5)

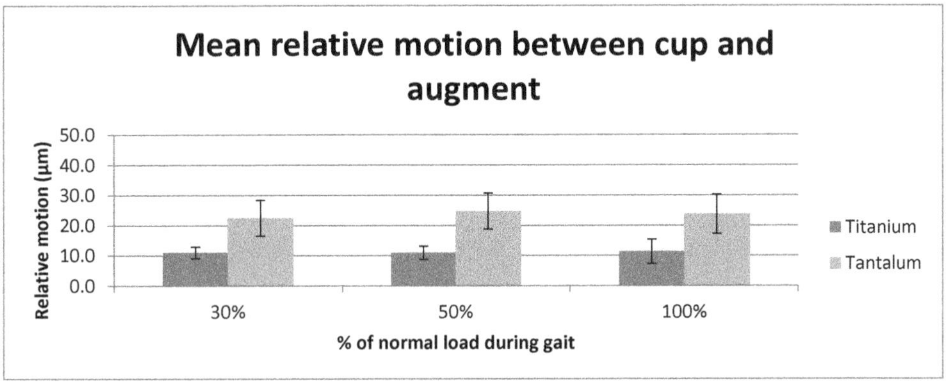

Figure 3. Graph displaying the average relative motion (μm) at the tantalum and titanium augment and cup interfaces at the three tested load levels (30%, 50% and 100% load).

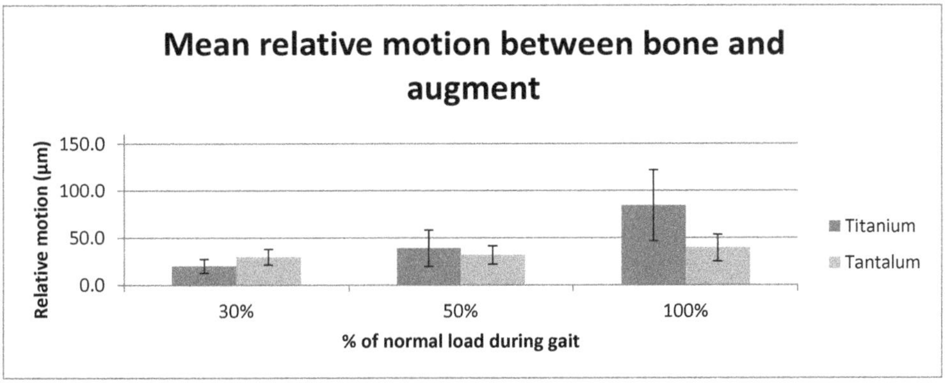

Figure 4. Graph shows the average relative motion (μm) between the tantalum and titanium augment and adjacent composite bone at the three tested load levels (30%, 50% and 100% load).

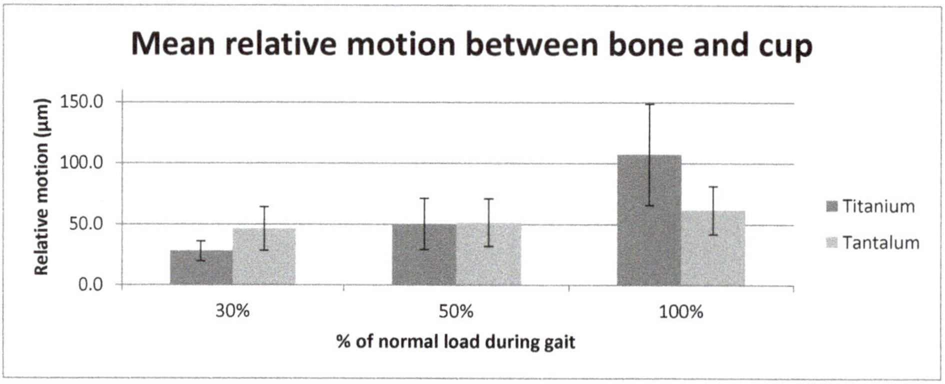

Figure 5. Graph showing the average relative motion (µm) between the tantalum and titanium cup and composite bone at the three tested load levels (30%, 50% and 100% load).

5. Discussion

The incidence of revision total hip arthroplasty (RTHA) is continuing to increase, particularly in younger patients [1,18] and is predicted to increase to 14.5% of all THAs and to increase by 174% from 2005 to 2030 [1].

Aseptic loosening has been reported to be the major reason for THA revision [2,18], and the frequently associated osteolytic defects that result from particulate debris and component wear can present a significant surgical challenge [2,19,20]. Revision THA consequently has a greater incidence of failure than primary THA because of the compromised soft tissue, bone loss and increased complexity of the procedure. This has prompted the ongoing search for improved components. Cemented acetabular cups allow only bone ongrowth rather than bone ingrowth [21,22] and have been associated with poor integration into the sclerotic host bone, increased rates of bone resorption and increased difficulty with later revision procedures [23]. Porous coated uncemented acetabular implants depend upon press-fit implantation to provide adequate primary stability during the surgical and early postoperative phase and secondary stability from later adequate osseointegration [22].

Currently tantalum and titanium are the most frequently used metals in uncemented porous components due to their biologically inert nature and their physical properties that are close to those of cancellous bone [24–27]. In addition, a recent study by Brüggemann et al. has shown little systemic response to tantalum implants, underscoring their safety in joint replacement procedures [28]. A large body of literature documents the success of porous trabecular tantalum constructs in RTHA [4,29,30]. In contrast, there is a relatively small body of literature documenting outcome with titanium constructs that vary in type and physical properties as a result of differing manufacturing processes [6,30].

The goal of our study was the evaluation of relative motion occurring at all interfaces of an implanted TM acetabular/augment construct and comparison of the results with the previously recorded relative motion occurring at the same interfaces of a porous titanium (Gription®) acetabular cup/augment construct, implanted under identical technical and environmental conditions. The inclusion of an augment in the construct added an additional interface (cup/augment) with the potential for additional relative motion. In all our tantalum and titanium constructs we found minimal relative motion at the cup/augment interface at all load levels, and we therefore interpreted this interface as having no significant negative impact on the stability of the construct as a whole. The tantalum and titanium constructs also displayed minimal relative motion of 30–50 µm at the bone/cup and bone/augment interfaces at the 30% and 50% load level. At the 100% load only, the Gription constructs displayed increased relative motion of the bone/cup and bone/augment interfaces of 107 and 84 µm, respectively. This may be due to the properties of the materials and implants, and their respective elastic modulus. Differences in the coefficient of friction alone have been shown in a prior study to have little impact on the primary stability of the acetabular component [31]. In all instances these values are below the previously recorded levels of relative motion that are thought to result in fibrous attachment. Prior in-vivo animal studies and studies on human autopsy bones have shown that successful osseointegration occurs with up to 40 μm relative motion between implant and bone, and fibrous attachment occurs with 150 μm relative motion [11,32]. It has also been shown that successful osseointegration can occur with bony attachment that involves substantially less than 100% of the bone/implant interface [33,34] and most of the osseointegration occurs around the acetabular rim, and decreases towards the pole [34,35]. One study by Bondarenko et al. showed that osteoporotic bone has worse osseointegration than healthy bone, and also that the implant can have a significant effect on the osseointegration, or bone-implant-contact [36]. In their study the tantalum implant Trabecular Metal® and the Trabecular Titanium® showed better osseointegration than the titanium implants Stiktite®, titanium with Gription® coating or Tritanium® [36].

The minimal levels of relative motion between porous implant and bone promote successful osseointegration, secondary stability and good surgical outcome, as documented in clinical reports. In RTHA in particular, tantalum components have been reported to have excellent results in complex cases, even with large bone deficiencies [21,37,38]. The ancillary use of porous tantalum augments as buttresses in cases of insufficient acetabular rim support has also been reported to have superior results [2,39,40]. Konan et al. reported a 96% survivorship of the TM acetabular component and good functional outcome at a mean 11 year follow up in patients with Paprosky 11 and 111 defects [41]. Morselized allograft was used in most cases, and no augments. Survivorships of 10 years for tantalum cup/augment constructs have been reported by a number of authors as 91–97% [20,27,37]. There are very few studies of porous titanium components in RTHA, and also little research of the titanium Gription cup/augment constructs in RTHA. One study evaluating 146 Pinnacle Gription cups, 1 of which was used in combination with an augment, showed good short-term results after RTHA [42]. In addition, studies with Gription augments used with various other cup types have shown good functional results [2]. However, there have been recent reports of studies that used other titanium components of different composition and manufacture, such as Trabecular Titanium and Tritanium® [43]. Hosny in 2018 reported a 98.4% aseptic acetabular cup survivorship at mean follow up of 87.6 months using Tritanium® revision cups in 62 patients with Paprosky 1–3 defects [6]. No augments were used. Delanois reported a 97% aseptic acetabular cup survivorship in 35 patients with a mean 6 year follow up, also using Tritanium® cups [23].

Our study has several limitations. Although the tantalum and titanium set-ups were done under identical technical and environmental conditions with implantations performed by a single surgeon in all cases (RGB), small differences in implantation technique cannot be ruled out.

Results for the Gription® samples cannot be extrapolated to other types of titanium implants in biomechanical experiments or in the clinical scenario. There are currently several different titanium product lines that differ in composition, architecture and manufacture, with differing biophysical properties.

We chose to use synthetic composite bone (Sawbone®) rather than cadaveric bone because of the uniformity in composition that is particularly important when working with a small sample, although the biomechanical properties are not identical to bone. Our results therefore may not reflect the clinical scenario.

Increasing loads were applied up to 100% of average normal body weight (80 kg), that was our best estimate of the limited weight bearing experienced during the postoperative period. The maximum load experienced during normal walking conditions is 233% of body weight [16,17]. Joint loading was applied only in the direction of maximal load as defined by Bergmann, and did not reflect the cyclical pattern of loading during normal walking conditions.

6. Conclusions

The samples in our study showed minimal relative motion that should promote successful osseointegration. The Gription construct showed more relative motion than TM at the cup/bone and augment/bone interfaces at 100% load only, that was below the value thought to promote fibrous attachment. Relative motion at the cup/augment interface of both TM and Gription constructs was of a degree that should not negatively impact the stability of the construct as a whole. Our biomechanical results are consistent with the positive clinical experience with TM components. There are too few reports on Gription constructs to make any clinical correlation, but our test results suggest that they should function satisfactorily.

Author Contributions: Conceptualization, N.A.B., M.S. (Martin Schwarze), M.S. (Mareike Schonhoff) and S.J.; methodology, N.A.B. and S.J.; validation, M.S. (Mareike Schonhoff), N.A.B. and S.J.; formal analysis, data curation, and investigation S.J., N.A.B., R.G.B. and M.S. (Mareike Schonhoff); writing—original draft preparation, N.A.B.; writing—review and editing, N.A.B., R.G.B., M.S. (Martin Schwarze), M.S. (Mareike Schonhoff), K.-A.S., S.J.; visualization, N.A.B. and S.J.; supervision, R.G.B., S.J., K.-A.S.; project administration, S.J.; funding acquisition, S.J.. All authors have read and agreed to the published version of the manuscript.

Funding: We acknowledge financial support by the Baden-Württemberg Ministry of Science, Research and the Arts and by Ruprecht-Karls-Universität Heidelberg.

Acknowledgments: We have no further acknowledgements.

Conflicts of Interest: The authors declare no pertinent conflicts of interest.

References

1. Kurtz, S.; Ong, K.; Lau, E.; Mowat, F.; Halpern, M. Projections of primary and revision hip and knee arthroplasty in the United States from 2005 to 2030. *J. Bone Jt. Surg. Am.* **2007**, *89*, 780–785. [CrossRef]
2. Mahmoud, A.N.; Sundberg, M.; Flivik, G. Comparable Results With Porous Metal Augments in Combination With Either Cemented or Uncemented Cups in Revision Hip Arthroplasty: An Analysis of One Hundred Forty-Seven Revisions at a Mean of Five Years. *J. Arthroplast.* **2017**, *32*, 1612–1617. [CrossRef] [PubMed]
3. Pollock, F.H.; Whiteside, L.A. The fate of massive allografts in total hip acetabular revision surgery. *J. Arthroplast.* **1992**, *7*, 271–276. [CrossRef]
4. Beckmann, N.A.; Weiss, S.; Klotz, M.C.; Gondan, M.; Jaeger, S.; Bitsch, R.G. Loosening after acetabular revision: Comparison of trabecular metal and reinforcement rings. A systematic review. *J. Arthroplast.* **2014**, *29*, 229–235. [CrossRef] [PubMed]
5. Jones, L.C.; Hungerford, D.S. Cement disease. *Clin. Orthop. Relat. Res.* **1987**, *225*, 192–206. [CrossRef]
6. Hosny, H.A.H.; El-Bakoury, A.; Srinivasan, S.C.M.; Yarlagadda, R.; Keenan, J. Tritanium Acetabular Cup in Revision Hip Replacement: A Six to Ten Years of Follow-Up Study. *J. Arthroplast.* **2018**, *33*, 2566–2570. [CrossRef]
7. Meneghini, R.M.; Meyer, C.; Buckley, C.A.; Hanssen, A.D.; Lewallen, D.G. Mechanical stability of novel highly porous metal acetabular components in revision total hip arthroplasty. *J. Arthroplast.* **2010**, *25*, 337–341. [CrossRef]
8. Kärrholm, J.; Mohaddes, M.; Odin, D.; Vinblad, J.; Rogmark, C.; Rolfson, O. *Swedish Hip Arthroplasty Register Annual Report 2017*; Svenska Höftprotesregistret Registercentrum Västra Götaland SE-413 45: Göteborg, Sweden, 2018.
9. Rossmann, M.; Ansorge, C.; Lausmann, C.; Suero, E.M.; Gehrke, T.; Citak, M. An alternative treatment option for Paprosky Type IIIb acetabular defect using multiple tantalum wedges—A case report. *J. Clin. Orthop. Trauma* **2020**, *11*, 70–72. [CrossRef]
10. Houdek, M.T.; Abdel, M.P.; Perry, K.I.; Salduz, A.; Rose, P.S.; Sim, F.H.; Lewallen, D.G. Outcome of Patients Treated With Porous Tantalum Acetabular Implants for Neoplastic Periacetabular Lesions. *J. Am. Acad. Orthop. Surg.* **2020**, *28*, 256–262. [CrossRef]
11. Pilliar, R.M.; Lee, J.M.; Maniatopoulos, C. Observations on the effect of movement on bone ingrowth into porous-surfaced implants. *Clin. Orthop. Relat. Res.* **1986**, *208*, 108–113. [CrossRef]
12. Aspenberg, P.; Goodman, S.; Toksvig-Larsen, S.; Ryd, L.; Albrektsson, T. Intermittent micromotion inhibits bone ingrowth. Titanium implants in rabbits. *Acta Orthop. Scand.* **1992**, *63*, 141–145. [CrossRef] [PubMed]

13. Beckmann, N.A.; Bitsch, R.G.; Gondan, M.; Schonhoff, M.; Jaeger, S. Comparison of the stability of three fixation techniques between porous metal acetabular components and augments. *Bone Jt. Res.* **2018**, *7*, 282–288. [CrossRef] [PubMed]
14. Paprosky, W.G.; Perona, P.G.; Lawrence, J.M. Acetabular defect classification and surgical reconstruction in revision arthroplasty. A 6-year follow-up evaluation. *J. Arthroplast.* **1994**, *9*, 33–44. [CrossRef]
15. Bergmann, G.; Graichen, F.; Rohlmann, A.; Bender, A.; Heinlein, B.; Duda, G.N.; Heller, M.O.; Morlock, M.M. Realistic loads for testing hip implants. *Biomed. Mater. Eng.* **2010**, *20*, 65–75. [CrossRef] [PubMed]
16. Bergmann, G.; Deuretzbacher, G.; Heller, M.; Graichen, F.; Rohlmann, A.; Strauss, J.; Duda, G.N. Hip contact forces and gait patterns from routine activities. *J. Biomech.* **2001**, *34*, 859–871. [CrossRef]
17. Bergmann, G.; Bender, A.; Dymke, J.; Duda, G.; Damm, P. Standardized Loads Acting in Hip Implants. *PLoS ONE* **2016**, *11*, e0155612. [CrossRef]
18. Ulrich, S.D.; Seyler, T.M.; Bennett, D.; Delanois, R.E.; Saleh, K.J.; Thongtrangan, I.; Kuskowski, M.; Cheng, E.Y.; Sharkey, P.F.; Parvizi, J.; et al. Total hip arthroplasties: What are the reasons for revision? *Int. Orthop.* **2008**, *32*, 597–604. [CrossRef]
19. Davies, J.H.; Laflamme, G.Y.; Delisle, J.; Fernandes, J. Trabecular metal used for major bone loss in acetabular hip revision. *J. Arthroplast.* **2011**, *26*, 1245–1250. [CrossRef]
20. Abolghasemian, M.; Tangsataporn, S.; Sternheim, A.; Backstein, D.; Safir, O.; Gross, A.E. Combined trabecular metal acetabular shell and augment for acetabular revision with substantial bone loss: A mid-term review. *Bone Jt. J.* **2013**, *95*, 166–172. [CrossRef]
21. Jafari, S.M.; Bender, B.; Coyle, C.; Parvizi, J.; Sharkey, P.F.; Hozack, W.J. Do tantalum and titanium cups show similar results in revision hip arthroplasty? *Clin. Orthop. Relat. Res.* **2010**, *468*, 459–465. [CrossRef]
22. Steno, B.; Kokavec, M.; Necas, L. Acetabular revision arthroplasty using trabecular titanium implants. *Int. Orthop.* **2015**, *39*, 389–395. [CrossRef] [PubMed]
23. Delanois, R.E.; Gwam, C.U.; Mohamed, N.; Khlopas, A.; Chughtai, M.; Malkani, A.L.; Mont, M.A. Midterm Outcomes of Revision Total Hip Arthroplasty With the Use of a Multihole Highly-Porous Titanium Shell. *J. Arthroplast.* **2017**, *32*, 2806–2809. [CrossRef] [PubMed]
24. Gallart, X.; Fernandez-Valencia, J.A.; Riba, J.; Bori, G.; Garcia, S.; Tornero, E.; Combalia, A. Trabecular TitaniumTM cups and augments in revision total hip arthroplasty: Clinical results, radiology and survival outcomes. *HIP Int.* **2016**, *26*, 486–491. [CrossRef] [PubMed]
25. De Meo, F.; Cacciola, G.; Bellotti, V.; Bruschetta, A.; Cavaliere, P. Trabecular Titanium acetabular cups in hip revision surgery: Mid-term clinical and radiological outcomes. *HIP Int.* **2018**, *28*, 61–65. [CrossRef]
26. Banerjee, S.; Issa, K.; Kapadia, B.H.; Pivec, R.; Khanuja, H.S.; Mont, M.A. Systematic review on outcomes of acetabular revisions with highly-porous metals. *Int. Orthop.* **2014**, *38*, 689–702. [CrossRef]
27. Levine, B.; Della Valle, C.J.; Jacobs, J.J. Applications of porous tantalum in total hip arthroplasty. *J. Am. Acad. Orthop. Surg.* **2006**, *14*, 646–655. [CrossRef]
28. Bruggemann, A.; Mallmin, H.; Bengtsson, M.; Hailer, N.P. Safety of Use of Tantalum in Total Hip Arthroplasty. *J. Bone Jt. Surg. Am.* **2020**, *102*, 368–374. [CrossRef]
29. Bruggemann, A.; Fredlund, E.; Mallmin, H.; Hailer, N.P. Are porous tantalum cups superior to conventional reinforcement rings? *Acta Orthop.* **2017**, *88*, 35–40. [CrossRef]
30. Pulido, L.; Rachala, S.R.; Cabanela, M.E. Cementless acetabular revision: Past, present, and future. Revision total hip arthroplasty: The acetabular side using cementless implants. *Int. Orthop.* **2011**, *35*, 289–298. [CrossRef]
31. Goldman, A.H.; Armstrong, L.C.; Owen, J.R.; Wayne, J.S.; Jiranek, W.A. Does Increased Coefficient of Friction of Highly Porous Metal Increase Initial Stability at the Acetabular Interface? *J. Arthroplast.* **2016**, *31*, 721–726. [CrossRef]
32. Engh, C.A.; O'Connor, D.; Jasty, M.; McGovern, T.F.; Bobyn, J.D.; Harris, W.H. Quantification of implant micromotion, strain shielding, and bone resorption with porous-coated anatomic medullary locking femoral prostheses. *Clin. Orthop. Relat. Res.* **1992**, *285*, 13–29. [CrossRef]
33. MacKenzie, J.R.; Callaghan, J.J.; Pedersen, D.R.; Brown, T.D. Areas of contact and extent of gaps with implantation of oversized acetabular components in total hip arthroplasty. *Clin. Orthop. Relat. Res.* **1994**, *298*, 127–136. [CrossRef]
34. Schwartz, J.T., Jr.; Engh, C.A.; Forte, M.R.; Kukita, Y.; Grandia, S.K. Evaluation of initial surface apposition in porous-coated acetabular components. *Clin. Orthop. Relat. Res.* **1993**, *293*, 174–187. [CrossRef]

35. Ong, K.L.; Lehman, J.; Notz, W.I.; Santner, T.J.; Bartel, D.L. Acetabular cup geometry and bone-implant interference have more influence on initial periprosthetic joint space than joint loading and surgical cup insertion. *J. Biomech. Eng.* **2006**, *128*, 169–175. [CrossRef] [PubMed]
36. Bondarenko, S.; Dedukh, N.; Filipenko, V.; Akonjom, M.; Badnaoui, A.A.; Schwarzkopf, R. Comparative analysis of osseointegration in various types of acetabular implant materials. *HIP Int.* **2018**, *28*, 622–628. [CrossRef]
37. Lochel, J.; Janz, V.; Hipfl, C.; Perka, C.; Wassilew, G.I. Reconstruction of acetabular defects with porous tantalum shells and augments in revision total hip arthroplasty at ten-year follow-up. *Bone Jt. J.* **2019**, *101*, 311–316. [CrossRef]
38. Fraile Suari, A.; Marques Lopez, F.; Cuenca Llavall, M.; Tey Pons, M.; Leon Garcia, A. Reconstruction for pelvic discontinuity and massive acetabular defects. *Rev. Esp. Cir. Ortop. Traumatol.* **2020**, *64*, 64–73. [CrossRef]
39. Whitehouse, M.R.; Masri, B.A.; Duncan, C.P.; Garbuz, D.S. Continued good results with modular trabecular metal augments for acetabular defects in hip arthroplasty at 7 to 11 years. *Clin. Orthop. Relat. Res.* **2015**, *473*, 521–527. [CrossRef]
40. Meneghini, R.M.; Hull, J.R.; Russo, G.S.; Lieberman, J.R.; Jiranek, W.A. Porous Tantalum Buttress Augments for Severe Acetabular Posterior Column Deficiency. *Surg. Technol. Int.* **2015**, *27*, 240–244.
41. Konan, S.; Duncan, C.P.; Masri, B.A.; Garbuz, D.S. Porous tantalum uncemented acetabular components in revision total hip arthroplasty: A minimum ten-year clinical, radiological and quality of life outcome study. *Bone Jt. J.* **2016**, *98*, 767–771. [CrossRef]
42. Chacko, V.; Agrawal, P.; Porter, M.L.; Board, T.N. Early results of a high friction surface coated uncemented socket in revision hip arthroplasty. *HIP Int.* **2019**. [CrossRef] [PubMed]
43. Naziri, Q.; Issa, K.; Pivec, R.; Harwin, S.F.; Delanois, R.E.; Mont, M.A. Excellent results of primary THA using a highly porous titanium cup. *Orthopedics* **2013**, *36*, e390–e394. [CrossRef] [PubMed]

materials

MDPI

Article

Optimizing Manufacturing and Osseointegration of Ti6Al4V Implants through Precision Casting and Calcium and Phosphorus Ion Implantation? In Vivo Results of a Large-Scale Animal Trial

Wölfle-Roos JV [1,*], Katmer Amet B [2], Fiedler J [2], Michels H [3], Kappelt G [4], Ignatius A [5], Dürselen L [5], Reichel H [1] and Brenner RE [2]

1 Department of Orthopaedic Surgery, Ulm University, 89081 Ulm, Germany; heiko.reichel@rku.de
2 Department of Orthopaedic Surgery, Division for Biochemistry of Joint and Connective Tissue Diseases, Ulm University, 89081 Ulm, Germany; betuel.katmer@uni-ulm.de (K.A.B.); joerg.fiedler@uni-ulm.de (F.J.); rolf.brenner@uni-ulm.de (B.R.E.)
3 Access e.V., 52072 Aachen, Germany; h.michels@access-technology.de
4 Peter Brehm GmbH, 91085 Weisendorf, Germany; gerhard.kappelt@peter-brehm.de
5 Institute of Orthopaedic Research and Biomechanics, Ulm University, 89081 Ulm, Germany; anita.ignatius@uni-ulm.de (I.A.); lutz.duerselen@uni-ulm.de (D.L.)
* Correspondence: julia.woelfle@uni-ulm.de

Received: 25 February 2020; Accepted: 26 March 2020; Published: 3 April 2020

Abstract: Background: Uncemented implants are still associated with several major challenges, especially with regard to their manufacturing and their osseointegration. In this study, a novel manufacturing technique—an optimized form of precision casting—and a novel surface modification to promote osseointegration—calcium and phosphorus ion implantation into the implant surface—were tested in vivo. Methods: Cylindrical Ti6Al4V implants were inserted bilaterally into the tibia of 110 rats. We compared two generations of cast Ti6Al4V implants (CAST 1st GEN, n = 22, and CAST 2nd GEN, n = 22) as well as cast 2nd GEN Ti6Al4V implants with calcium (CAST + CA, n = 22) and phosphorus (CAST + P, n = 22) ion implantation to standard machined Ti6Al4V implants (control, n = 22). After 4 and 12 weeks, maximal pull-out force and bone-to-implant contact rate (BIC) were measured and compared between all five groups. Results: There was no significant difference between all five groups after 4 weeks or 12 weeks with regard to pull-out force ($p > 0.05$, Kruskal Wallis test). Histomorphometric analysis showed no significant difference of BIC after 4 weeks ($p > 0.05$, Kruskal–Wallis test), whereas there was a trend towards a higher BIC in the CAST + P group (54.8% ± 15.2%), especially compared to the control group (38.6% ± 12.8%) after 12 weeks ($p = 0.053$, Kruskal–Wallis test). Conclusion: In this study, we found no indication of inferiority of Ti6Al4V implants cast with the optimized centrifugal precision casting technique of the second generation compared to standard Ti6Al4V implants. As the employed manufacturing process holds considerable economic potential, mainly due to a significantly decreased material demand per implant by casting near net-shape instead of milling away most of the starting ingot, its application in manufacturing uncemented implants seems promising. However, no significant advantages of calcium or phosphorus ion implantation could be observed in this study. Due to the promising results of ion implantation in previous in vitro and in vivo studies, further in vivo studies with different ion implantation conditions should be considered.

Keywords: ion implantation; precision casting; Ti6Al4V; calcium; phosphorus; centrifugal casting

---

1. Introduction

Uncemented implants are widely used in arthroplasty, in primary total hip arthroplasty (THA), up to 80% of implants are uncemented [1]. As the long-term stability of uncemented implants can only be achieved by osseointegration—i.e., a stable connection between the implant surface and the adjacent bone—titanium-based implants are preferred for uncemented implantation due to their high biocompatibility [2,3].

However, the manufacturing process of titanium and its alloys is much more complex than that of cobalt–chromium-based implants [4]. Due to the high melting point of titanium of >1700 °C, its high reactivity to oxygen [5] and its unfavorable fluidity properties [6], precision casting of titanium implants is challenging and currently not routinely used [7]. At present, standard manufacturing of titanium implants involves machining blanks on a milling machine, which results in a high waste of Ti6Al4V material and fabrication costs, as well as a limitation of titanium implants to comparatively simple implant designs [8]. Recent innovations have led to the optimization of centrifugal casting units, see Figure 1. By introducing a cold wall induction crucible and by improving casting conditions, mould material and subsequent heat treatment, centrifugal precision casting has become feasible for the manufacturing of titanium-based implants [9].

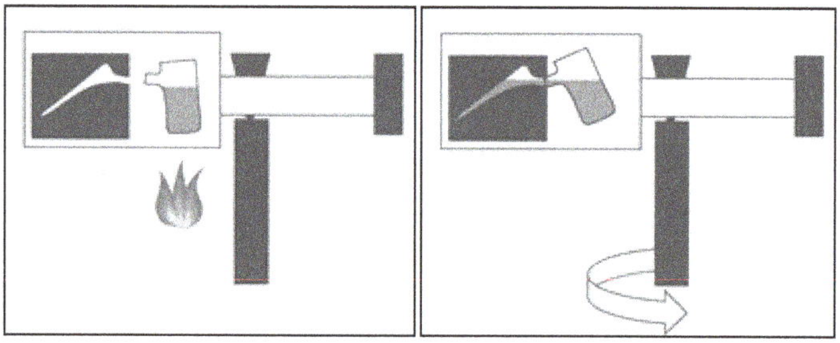

(**A**) (**B**)

Figure 1. Schematic illustration of a centrifugal precision casting unit: The ingot-casting material is melted (**A**) and the liquid melt is then pressed into the casting mould by rotating the casting arm around a vertical shaft (**B**).

The technically demanding manufacturing process of titanium-based implants is not the only challenge of uncemented implantation. Osseointegration of the uncemented implant still remains a crucial issue, as the stability of the implantation depends on a strong connection between implant surface and adjacent bone [10]. To enhance osseointegration, countless different methods of surface modification have been developed. One of the most prominent surface modifications is hydroxyapatite coating (HA-coating) [11], which has been widely used in uncemented THA. As, however, delamination of the HA-coating and three-body-wear due to HA-molecules have been increasingly reported in the literature [11–13], attempts have been made to implant calcium and phosphorus ions—the chemical components of HA—directly into the implant surface [14], encouraging the formation of calcium phosphate precipitates [15,16]. For complex implant shapes, the so-called plasma immersion ion implantation is currently used [17], see Figure 2.

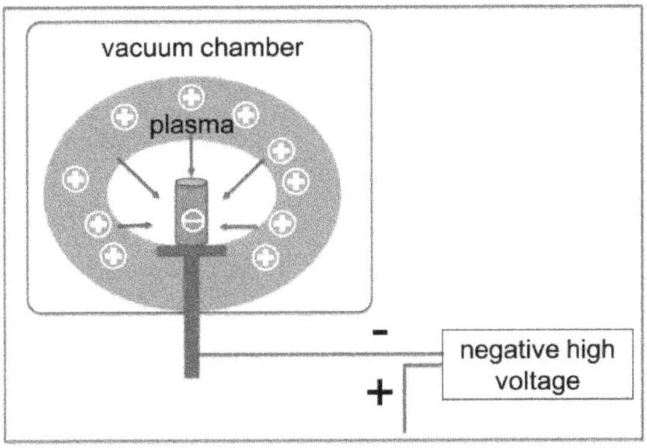

Figure 2. Schematic illustration of plasma immersion ion implantation. In a vacuum chamber, calcium ions are extracted from the plasma—i.e., ionized gas—and accelerated towards the negatively charged implant surface by applying a high voltage direct current.

To improve both the manufacturing and osseointegration of uncemented titanium implants, a novel manufacturing technique—an optimized form of centrifugal precision casting—and a novel surface modification to promote osseointegration—calcium and phosphorus ion implantation into the implant surface—were tested in vivo in this study.

2. Methods

2.1. Animals

110 adult female WISTAR rats were supplied by Charles River Laboratories (Kißlegg, Germany). They were kept under climate-controlled conditions (21 ± 1.5 °C, 47.5% ± 7.5% humidity, light–dark cycle 12/12 h). Access to food and tap water was ad libitum. Age at the time of surgery was ten weeks; mean body weight was 277 ± 15 g. The animal trial was conducted according to relevant national and international guidelines, such as the ARRIVE guidelines; the study was approved by the Regional Administrative Council (registration number 1246).

2.2. Implants

The standard implants consisted of machined, aluminium-oxide-blasted Ti6AlV4 cylindrical rods of 5.2 mm length with a diameter of 1.6 mm (provided by Peter Brehm GmbH, Weisendorf, Germany). One end of the rod was threaded (length 1.2 mm) to enable biomechanical pull-out testing.

All other implants had identical dimensions compared to the standard implants and were manufactured with the optimized centrifugal precision-casting technique, which had been developed in cooperation between the Technical University Aachen (RWTH, Germany), the associated research center Access technology e. V. (Aachen, Germany), and the implant manufacturer Peter Brehm GmbH (Weisendorf, Germany). Implants were cast in a combined vacuum-induction centrifugal casting device Leicomelt 5 TP (ALD Vacuum Technologie GmbH, Hanau, Germany). As the optimized centrifugal precision casting technique is currently undergoing a patent grant procedure, only a general overview of the manufacturing process is given here.

Casting moulds were made of multilayer ceramics with the lost wax technique, with the help of 3D-simulation programs for the casting and solidification process. The ingot-casting material Ti6Al4V was melted in a cold wall induction crucible in a vacuum system with inert gas flushing. The liquid melt was then pressed into the casting mould by rotating the casting arm around a vertical shaft, see Figure 1.

In contrast to other metals' higher centrifugal forces and higher speed are required for titanium implants to completely fill out the mould due to its low specific weight, high melting point and high solidification speed. After solidification of the melt, the casting was revealed under the destruction of the ceramic form, see Figure 3. The cast implants were separated from the feed channel with a cutting disc. The implants were then submitted to hot isostatic pressing with 920 ± 10 °C at a pressure of 1000 ± 50 bar for 120 ± 30 min to reduce porosity, and subsequent heat treatment.

Figure 3. Cast cylindrical implants (→) before separation from the feed channel.

In cast Ti6Al4V implants of the first generation (CAST 1st GEN), the so-called alpha case layer—a hardened brittle layer at the implant surface due to the reaction between the oxides of the mould and the molten titanium, see Figure 4—had to be removed by acid etching at room temperature with Ceramex® (Renfert GmbH, Hilzingen, Germany), i.e., a mixture of 3% hydrofluoric acid and10% sulphuric acid (MediMet Precision Casting and Implants Technology GmbH, Stade, Germany).

Figure 4. Alpha case layer on the implant surface of cast Ti6Al4V implants of the first generation.

By further optimizing the conditions of the casting process, the alpha case layer of cast Ti6Al4V implants of the second generation (CAST 2nd GEN) was negligible, so acid etching could be dispensed with. All cast implants were then submitted to identical aluminium-oxide-blasting to the standard implants. Material analysis showed that the cast material met with the standard grade for Ti6Al4V alloys [4].

Cast Ti6Al4V implants of the second generation were then submitted to plasma immersion ion implantation (PIII) with calcium (CAST + CA group) or phosphorus (CAST + P group) in cooperation with Helmholtz Center Dresden-Rossendorf in a 5-MV-Tandem EGP-10-1 (Efremov-Institut NIIEFA Leningrad, St. Petersburg, Russia). As an ion source, H_3P, and a calcium arc source, respectively, were used. PIII was conducted at room temperature, ion energy was 30 keV, and ion current 2.4 µA (calcium) and 3–500 µA (phosphorus), respectively, at a dose of $1 \times 10^{16}/cm^2$.

2.3. Experimental Design

The animals were randomly assigned 22 animals each to one of the following groups: (1) control group, (2) CAST 1st GEN, (3) CAST 2nd GEN, (4) CAST + CA, or (5) CAST + P group. According to which group they belonged to, one of the following implants was inserted:

1. Standard Ti6Al4V implants;
2. Cast Ti6Al4V implants of the first generation;
3. Cast Ti6Al4V implants of the second generation;
4. Cast Ti6Al4V implants of the second generation with calcium ion implantation;
5. Cast Ti6Al4V implants of the second generation with phosphorus ion implantation.

Eleven animals of each group were sacrificed after 4 weeks, and the remaining animals 12 weeks after implant insertion. The right tibia was prepared for biomechanical testing and the left tibia for histomorphometric analysis.

2.4. Surgical Procedure

The above-mentioned implants were inserted into the proximal tibia on both sides. Anaesthesia was administered by means of an inhalation device (isoflurane 2%) and the subcutaneous injection of analgetics (tramadolor 20 mg/kg) and buprenorphin (0.03 mg/kg). The surgical technique was identical to previous studies conducted in our study group [18]: a 10 mm incision was made at the medial aspect of the proximal tibia, and the periosteum was incised ventrally to the medial collateral ligament. A 1.7 mm drill hole was made level with the insertion of the patella tendon ventrally to the medial collateral ligament using hand-held drills held strictly perpendicular to the longitudinal axis of the tibia. The implant was then inserted into the bone. The threaded part remained outside and was covered by a 2 mm tube cut off from a venous catheter (fluorinated ethylene propylene, Vasofix® Braunüle® 18 G, Braun B., Melsungen, Germany) in order to prevent osseous overgrowth. Postoperative analgesia was ensured by adding tramadolor to the drinking water (25 mg per litre). Antibiotics (clindamycine 45 mg/kg) were administered subcutaneously daily on the first three postoperative days.

2.5. Biomechanical Testing

For biomechanical testing, a specifically designed cylindrical device with a matching internal thread was screwed onto the threaded part of the implant of the right tibia. This device was passed through a perforated plate, which served as a mobile fixation of the implant, and then attached to a 200 N load cell (HBM, Darmstadt) of a standard testing machine (Z010, Zwick, Ulm, Germany; see Figure 1A). The surrounding soft tissue around the implant had been purposely left in place to ensure an even contact between the specimen and the perforated plate. The implant was then aligned straight—i.e., longitudinally to the tensile axis—by a low preload of 0.5 N. A force-displacement diagram (test speed 10 mm/s, preload 0.5 N) was recorded by a testing software (testXpert II, Zwick, Ulm, Germany) and the load occurring before the first sudden drop in the tensile force was defined as maximum pull-out force. To calculate the pull-out force per area, a Micro-CT scanning of the former implant bed of the first 44 animals was conducted with a μCT system (Skyscan 1172, Kontich, Belgium). As the correlation between pull-out force and pull-out force per area was found to be highly significant and very strong (Pearson's correlation coefficient $R = 0.970$, $p < 0.001$), pull-out force was used as the only biomechanical parameter for the remaining animals.

2.6. Histomorphometric Analysis

The left tibia including the inserted implant was embedded in Technovit VLC7200 (Kulzer, Germany) and ground down to sections of 100 μm along the longitudinal axis of the tibia. Masson–Goldner staining of the sections was used to visualize the connective tissue surrounding the implant. The sections were inspected and scanned with a fully automated inverted light microscope (Leica DMI6000B, Wetzlar, Germany). To quantify the amount of bone surrounding the implant, the following parameters were determined semi-automatically with the aid of an imaging analysis software (MetaMorph®, Leica, Wetzlar, Germany). The bone-to-implant contact rate (BIC) was calculated by dividing the total length of bone-to-implant contact by the total length around the implant within the tibia. Histomorphometric analysis was conducted by two independent observers blinded to the implant material used.

2.7. Statistical Analysis

Statistical advice, including an estimation of required sample size, was gained before the planning of the study based on the only comparable study with a similar animal model and an identical location of the implant insertion [19,20]. For statistical analysis, the Statistical Package for Social Sciences (SPSS® Inc., IBM, version 24) was used. Continuous variables were summarized as mean ± standard deviation. To compare the results of the five groups after 4 and 12 weeks, Kruskal–Wallis test was used. Pearson's correlation coefficient was used to calculate interobserver reliability and correlation between pull-out force and pull-out force per area. A probability value of less than 0.05 was considered to indicate statistical significance.

3. Results

With regard to maximum pull-out force, there was no significant difference between all five groups after 4 weeks and after 12 weeks ($p = 0.596$ and $p = 0.127$ respectively, Kruskal–Wallis test). After 12 weeks, cast Ti6Al4V implants of the first generation (CAST 1st GEN) showed a slightly lower maximum pull-out force (78.4 ± 16.9 N) compared to the control group (99.8 ± 25.0 N), though this difference did not reach statistical significance. However, no disadvantage of the cast Ti6Al4V implants of the second generation could be observed (95.4 ± 23.0 N) after 12 weeks, see Figure 5.

Analysis of the histological sections showed a thin layer of osseous tissue covering a large part of the implant surface on the section of all five groups. Examples of histologic sections of both cast and machined Ti6Al4V implants are depicted in Figure 6.

Histomorphometric measurement revealed no significant difference in BIC after 4 weeks between all five groups ($p > 0.05$, Kruskal–Wallis test), whereas there was a trend towards higher BIC in the CAST + P group (54.8% ± 15.2%), especially when compared to the control group (38.6% ± 12.8%) after 12 weeks ($p = 0.053$, Kruskal–Wallis test), see Figure 7.

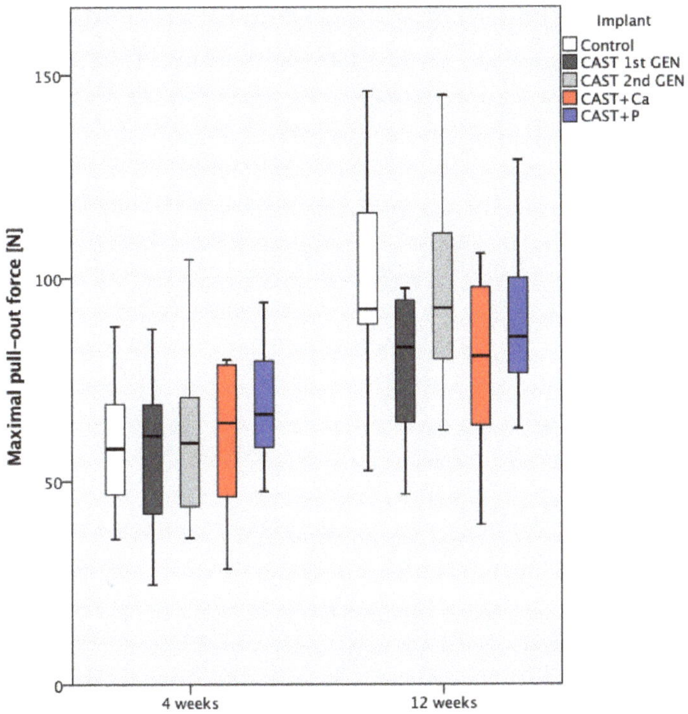

Figure 5. Boxplot of maximum pull-out force after 4 and 12 weeks. The 25th and the 75th percentile as well as the median are indicated by the box, while the whiskers depict maximum and minimum values. No significant difference between all 5 groups was observed ($p > 0.05$, Kruskal–Wallis test).

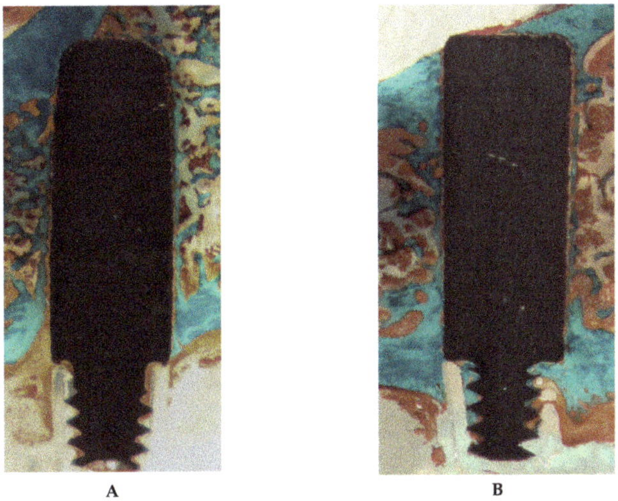

Figure 6. Exemplary Masson–Goldner stains of cast (**A**) and standard (**B**) Ti6l4V implants after 12 weeks. Mature osseous tissue is stained blue, while newly formed bone as well as connective tissue appears red. In both sections, a thin layer of mature bone covers the implant surface.

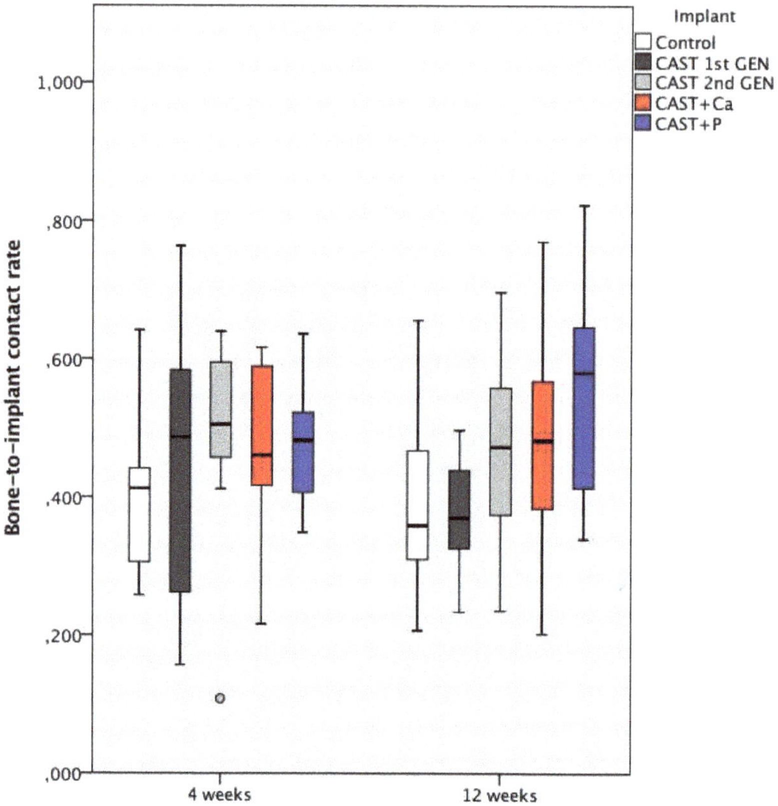

Figure 7. Boxplot of bone-to-implant contact rate (BIC) after 4 and 12 weeks. The 25th and the 75th percentile as well as the median are indicated by the box while the whiskers depict maximum and minimum values. Outlies are plotted as individual dots. There was no significant difference between all five groups after 4 weeks ($p > 0.05$), whereas there was a trend towards higher BIC in the CAST + P group after 12 weeks ($p = 0.053$, Kruskal–Wallis test).

4. Discussion

The most important findings of this study were that no indication of the inferiority of the cast Ti6Al4V implants of the second generation could be observed with regard to osseointegration when compared to standard machined implants, and that no significant advantages after calcium and phosphorus ion implantation could be seen.

4.1. Centrifugal Precision Casting

Precision casting of titanium-based implants is technically challenging due to the high melting point of titanium of >1700 °C, its high reactivity to oxygen and its unfavorable fluidity properties. Only in highly specialized centrifugal precision casting units are sufficient centrifugal forces and speed reached to completely fill out the mould, due to the low specific weight and high solidification speed. However, precision-casting of titanium implants holds vast potential economic advantages over the standard milling process [21]. On the one hand, the process offers a significantly decreased material demand per implant by directly casting near net-shape instead of milling away most of the starting ingot, thus greatly lowering the base material demand and cost. Adding to this lower base demand is the inherent improvement in sustainability. Milling waste must be recycled by complex reclaiming processes before it can be reused, while the direct reclamation of casting scrap is regular practice in casting processes, given that constant cast part quality is proven. On the other hand, the centrifugal casting process offers a great freedom of design, in particular by using custom-printed wax patterns in the future, which will enable us to produce implants of complex geometrical shapes (especially those involving undercuts) in large numbers.

Mechanical properties of cast and wrought titanium—the latter being the raw material from which current standard titanium implants are machined—need not necessarily be identical, as the solidification process and the exposure of molten titanium to oxygen varies between the two manufacturing techniques. However, Nastac et al. conducted a review on investment casting of Ti6Al4V alloys and found that almost all the static and dynamic mechanical properties of cast Ti6Al4V alloys are similar to wrought Ti6Al4V. Only the fatigue strength—especially high-cycle fatigue—of cast Ti6Al4V alloys was found to be inferior, which, however, can be improved by subsequent heat treatment [6]. To achieve optimal mechanical properties of cast Ti6Al4V alloys, countless improvements in centrifugal precision casting have been introduced in recent years, e.g., a vacuum system and inert gas flushing due to the high reactivity of titanium with oxygen, high centrifugal forces to ensure complete filling of the mould, or hot isostatic pressing to reduce porosity [6,21]. The alpha case layer—a hardened brittle layer at the implant surface due to the reaction between the oxides of the mould and the molten titanium—still remains a challenge for precision casting of Ti6Al4V [22]. This layer is most commonly removed by acid etching, a technique which in itself bears the risk of hydrogen embrittlement of the implant surface through hydrogen absorption and hydride formation [23]. In our study, we found slightly inferior maximum pull-out force after 12 weeks in cast Ti6Al4V of the first generation, in which the alpha case layer had been removed by acid etching (CAST 1st GEN). By further optimizing the conditions of the casting process, the thickness of the alpha case layer was further reduced (CAST 2nd GEN), so that acid etching could be dispensed with. In these cast Ti6Al4V implants of the second generation, we did not observe any indication of inferiority in comparison to standard Ti6Al4V either on biomechanical testing or on histomorphometric analysis. Standard material testing carried out by our project partners before delivering the implants showed no deviation from standard implant requirements, but further biomechanical studies will have to follow to prove the equivalence of the main mechanical properties such as modulus of elasticity, hardness, toughness, fatigue strength, or tensile strength.

There are only a few in vivo studies comparing cast and machined implants with regard to biocompatibility and osseointegration. Mohammadi et al. implanted cast and machined titanium implants into the abdominal wall of rats so that part of the implant was located in the abdominal wall and part in the peritoneal cavity. On light and electron microscopy they found no difference in peri-implant tissue in the abdominal wall between the two groups. For intraperitoneal implants, significantly more fibroblasts and macrophages were observed on the implant surface of cast implants (in 6/8 implants) when compared to machined implants (in 1/9 implants) [24]. However, as there was a significant difference between tissue response in muscular and peritoneal tissue, it seems doubtful that these findings can predict tissue response in osseous tissue. Moreover, cast titanium implants in their study—as opposed to implants cast with modern centrifugal precision casting units—received hardly any post-processing treatment and showed a comparatively rough implant surface [24]. As this rough implant surface is probably the main reason for the wash-out of titanium particles and the subsequent heightened response of macrophages, their results cannot be compared with modern cast titanium implants. In a more recent study, Mohammadi et al. implanted machined cast titanium implants and standard machined titanium implants into the tibia of rabbits and found no significant difference in bone–implant contact after 3 months (19% BIC in cast implants vs. 25% in machined implants) and 6 months (45% in cast implants vs. 37% in machined implants) between both groups [7]. As in their study cast titanium implants were machined after casting and 0.25 mm of the implant surface were removed, again, their results cannot be compared to ours. There is only one study reporting the application of cast titanium implants in humans: 15 patients received individualized cast titanium implants for the reconstruction of bony skull defects using data from 3D computer tomography scans.

There was no control group and mean follow-up was 16.6 months. With the exception of one case of early infection which resulted in the removal of the implant, no complications were noted and osseointegration was successful on clinical and radiological examination [25]. These findings agree with our study, which showed no indication of the inferiority of cast Ti6Al4V implants of the second generation compared to standard Ti6Al4V implants. However, given the small number of patients, the short follow-up, the lack of a control group, and the location of the implants in a non-weight-bearing part of the skeleton [23], evidence regarding the long-term equivalence of cast orthopedic titanium implants when compared to today's standard implants has yet to be provided. Thus, further in vivo trials involving a large animal model in a biomechanically loaded position might be considered before establishing centrifugal precision casting as a manufacturing technique of titanium orthopedic implants.

4.2. Calcium and Phosphorus Ion Implantation

The second part of this study was dedicated to calcium and phosphorus ion implantation into the surface of cast Ti6Al4V implants of the second generation and its effects on osseointegration. Ion implantation represents an ultra-clean process in which the concentration and depth distribution of ions can be controlled with high accuracy [17]. Several in vitro studies have shown the advantageous effects of calcium and phosphorus ion implantation on the mechanical properties of titanium. Corrosion resistance, as measured by electrochemical methods in a simulated body fluid, was significantly increased after calcium and phosphorus ion implantation (ion dose $1 \times 10^{17}/cm^2$) which is a very important aspect of biocompatibility [26,27]. X-ray photoelectron spectroscopy of calcium and phosphorus implanted titanium (ion dose $1.8 \times 10^{17}/cm^2$ and $9 \times 10^{16}/cm^2$ respectively) was conducted to analyze the resulting chemical composition of the implant surface. It was shown that under hydrothermal oxidation a three-step chemical reaction takes place with (1) oxidation of phosphorus ions to P_2O_5, (2) hydrolysis resulting in Ca^{2+}, PO_4^{3-} und H^+, followed by (3) the appearance of needle-like crystallites of calcium phosphates such as hydroxyapatite on the implant surface [16].

The influence of calcium and phosphorus ion implantation on osteogenesis has been thoroughly investigated in vitro by Nayab et al. [28–31] and Krupa et al. [14,26,27]. Nayab et al. seeded radioisotopically tagged alveolar bone cells on calcium-implanted titanium and found that, although cell adhesion was reduced on the calcium ion-implanted surface, cell spreading and subsequent cell growth of the alveolar bone cells was significantly enhanced [28]. This effect seemed to be dependent on the ion dose: in implants with a high calcium ion dose ($1 \times 10^{17}/cm^2$), the adhesion of MG-63-cells, though initially reduced, was substantially increased in time, and cell spreading was significantly enhanced. In contrast, no marked differences were observed with regard to the adhesion and spreading of MG-63-cells on titanium implanted with low ($1 \times 10^{15}/cm^2$) or medium ($1 \times 10^{16}/cm^2$) ion doses [29]. Gene expression analysis showed the up-regulation of key proteins of osteogenesis such as bone sialoprotein, bone morphogenetic protein receptor-1B and osteopontin in MG-63 cells [30] as well as increased markers of proliferation [31] on calcium-implanted titanium (ion dose $1 \times 10^{17}/cm^2$). Krupa et al. seeded human mesenchymal stem cells on titanium implanted with calcium ions (ion dose $1 \times 10^{17}/cm^2$) [26], phosphorus ions (ion dose $1 \times 10^{17}/cm^2$) [27] and calcium/phosphorus ions combined (ion dose $1 \times 10^{17}/cm^2$) [14]. In contrast to Nayab et al., they found no positive effects of calcium and/or phosphorus ion implantation on osteogenesis; there was no difference with regard to the vitality of mesenchymal stem cells and the expression of alkaline phosphatase when compared to the control group [14,26,27]. These findings agree with our study, which did not show significant positive effects of either calcium or phosphorus ion implantation on osseointegration.

The effects of ion implantation on osseointegration has only been studied in vivo with calcium-implanted titanium to date. Hanawa et al. investigated bone formation in rats around a titanium implant which had been implanted with calcium ions (ion dose $1 \times 10^{17}/cm^2$) on one side only; tetracycline and calcein were used as hard-tissue labels. They found more bone formation on the calcium-implanted side of the implant compared to the other side after 2 and 8 days; 18 weeks after surgery, no difference between both sides could be seen [32]. Jinno et al. performed THA in dogs with

Ti6Al4V implants with and without calcium ion implantation; doxycycline and fluorescein were used as labels to investigate new bone formation. Histomorphometric analysis showed significantly greater new bone apposition in calcium-implanted stems compared to the control group; 6 and 12 months after implantation no significant difference could be seen between both groups [33]. Cheng et al. inserted cylindrical titanium implants with and without calcium ion implantation into the femur of rats. After 4, 8, and 12 weeks they found significant advantages of calcium-implanted titanium implants on histomorphometric analysis, computer tomographic assessment of the implant bed, and analysis of new bone formation, as well as the biomechanical testing of the push-force when compared to the control group [34]. The findings of these three in vivo studies did not concur with our study, in which no positive effects of calcium ion implantation on osseointegration could be observed. Possibly, the ion density used in our study ($1 \times 10^{16}/cm^2$) was too low, or the ion implantation energy too high (30 keV) to reach enough ion deposition close to the implant surface to allow for any noticeable effects on osseointegration. Furthermore, the properties of the implant surface of the studies mentioned above were not identical when compared to our study, as in some cases polished implants [32] and/or pure titanium implants [32,34] were used. Additionally, the casting process itself may have influenced the metallurgic properties of the implant surface, and thus the depth of ion implantation. As we did not expect a relevant release of implanted ions, serum calcium or phosphorus levels were not determined. This might be included in future studies on the effects of different dosages of ion implantation.

5. Conclusions

In this study, we found no indication of inferiority of Ti6Al4V implants cast with the optimized centrifugal precision-casting technique of the second generation compared to standard Ti6Al4V implants. As the employed manufacturing process holds considerable economic potential, mainly due to a significantly decreased material demand per implant by casting near net-shape instead of milling away most of the starting ingot, its application in manufacturing uncemented implants seems promising.

However, no significant advantages of calcium or phosphorus ion implantation could be observed in this study with the doses applied. Due to the promising results of ion implantation in previous in vivo studies and its positive effect on the mechanical properties of titanium implants, further in vivo studies with different ion implantation conditions (regarding ion density and implantation energy) should be considered.

Author Contributions: Conceptualization, W.-R.J.V., F.J., M.H., K.G., I.A., D.L., R.H. and B.R.E.; Data curation, W.-R.J.V.; Funding acquisition, B.R.E.; Investigation, W.-R.J.V. and K.A.B.; Methodology, W.-R.J.V.; Project administration, B.R.E.; Supervision, B.R.E.; Writing—original draft, W.-R.J.V.; Writing—review and editing, K.A.B., F.J., M.H., K.G., I.A., D.L., R.H. and B.R.E. All authors have read and agreed to the published version of the manuscript.

Funding: This study has been funded by a grant from the German Federal Ministry of Education and Research (BMBF, No. 13 GW0020E).

Acknowledgments: We wish to acknowledge the support of Andreas Seitz, Annette Palmer, Patrizia Horny, Giovanni Ravalli, and Christiane Schulz in conducting the experimental work.

Conflicts of Interest: The funders had no role in the design of the study; in the collection, analyses, or interpretation of data; in the writing of the manuscript, or in the decision to publish the results. The authors declare no conflict of interest.

References

1. Grimberg, A.; Jansson, V.; Liebs, T.; Melsheimer, O.; Steinbrück, A. *Endoprothesenregister Deutschland: Jahresbericht 2015*; EPRD Deutsche Endoprothesenregister gGmbH: Berlin, Germany, 2016.
2. Garellick, G.; Karrholm, J.; Rogmark, C.; Rolfson, O.; Herberts, P. *Swedish Hip Arthroplasty Register. Annual Report 2011*; Swedish Hip Arthroplasty Register: Goteborg, Sweden, 2011.
3. Marco, F.; Milena, F.; Gianluca, G.; Vittoria, O. Peri-implant osteogenesis in health and osteoporosis. *Micron* **2005**, *36*, 630–644. [CrossRef]

4. Ha, S.W.; Wintermantel, E. Part IV Werkstoffe in der Medizintechnik, Kapitel 11 Metalle. In *Medizinitechnik Life Science Engineering*, 4th ed.; Wintermantel, E., Ha, S.W., Eds.; Springer: Berlin/Heidelberg, Germany, 2008; pp. 183–210. [CrossRef]
5. Sung, S.-Y.; Kim, Y.-J. Alpha-case formation mechanism on titanium investment castings. *Mat. Sci. Eng. A Struct.* **2005**, *405*, 173–177. [CrossRef]
6. Nastac, L.; Gungor, M.N.; Ucok, I.; Klug, K.L.; Tack, W.T. Advances in investment casting of Ti–6Al–4V alloy: A review. *Int. J. Cast Metal. Res.* **2006**, *19*, 73–93. [CrossRef]
7. Mohammadi, S.; Esposito, M.; Wictorin, L.; Aronsson, B.-O.; Thomsen, P. Bone response to machined cast titanium implants. *J. Mat. Sci.* **2001**, *36*, 1987–1993. [CrossRef]
8. Hagemann, K. Gießtechnische Herstellung offenporiger Implantate aus Ti6Al7Nb. Ph.D. Thesis, Rheinisch-Westfälische Technische Hochschule Aachen, Aachen, Germany, 2012.
9. Billhofer, H.; Hauptmann, T. Fine casting system for titanium and titanium alloys. *Lightweight Des.* **2010**, *3*, 47–51. [CrossRef]
10. Kienapfel, H.; Sprey, C.; Wilke, A.; Griss, P. Implant fixation by bone ingrowth. *J. Arthroplast.* **1999**, *14*, 355–368. [CrossRef]
11. Surmenev, R.A.; Surmeneva, M.A.; Ivanova, A.A. Significance of calcium phosphate coatings for the enhancement of new bone osteogenesis—A review. *Acta Biomater.* **2014**, *10*, 557–579. [CrossRef]
12. Lazarinis, S.; Makela, K.T.; Eskelinen, A.; Havelin, L.; Hallan, G.; Overgaard, S.; Pedersen, A.B.; Karrholm, J.; Hailer, N.P. Does hydroxyapatite coating of uncemented cups improve long-term survival? An analysis of 28,605 primary total hip arthroplasty procedures from the Nordic Arthroplasty Register Association (NARA). *Osteoarthr. Cartil.* **2017**, *25*, 1980–1987. [CrossRef]
13. Morscher, E.W.; Hefti, A.; Aebi, U. Severe osteolysis after third-body wear due to hydroxyapatite particles from acetabular cup coating. *J. Bone Joint. Surg. Br.* **1998**, *80*, 267–272. [CrossRef]
14. Krupa, D.; Baszkiewicz, J.; Kozubowski, J.A.; Barcz, A.; Sobczak, J.W.; Bilinski, A.; Lewandowska-Szumiel, M.; Rajchel, B. Effect of dual ion implantation of calcium and phosphorus on the properties of titanium. *Biomaterials* **2005**, *26*, 2847–2856. [CrossRef]
15. Hanawa, T.; Ukai, H.; Murakami, K. X-ray photoelectron spectroscopy of calcium-ion-implanted titanium. *J. Electron. Spectrosc. Relat. Phenom.* **1993**, *63*, 347–354. [CrossRef]
16. Pham, M.T.; Reuther, H.; Matz, W.; Mueller, R.; Steiner, G.; Oswald, S.; Zyganov, I. Surface induced reactivity for titanium by ion implantation. *J. Mater. Sci. Mater. Med.* **2000**, *11*, 383–391. [CrossRef] [PubMed]
17. Rautray, T.R.; Narayanan, R.; Kwon, T.Y.; Kim, K.H. Surface modification of titanium and titanium alloys by ion implantation. *J. Biomed. Mater. Res. B Appl. Biomater.* **2010**, *93*, 581–591. [CrossRef] [PubMed]
18. Wolfle, J.V.; Fiedler, J.; Durselen, L.; Reichert, J.; Scharnweber, D.; Forster, A.; Schwenzer, B.; Reichel, H.; Ignatius, A.; Brenner, R.E. Improved anchorage of Ti6Al4V orthopaedic bone implants through oligonucleotide mediated immobilization of BMP-2 in osteoporotic rats. *PLoS ONE* **2014**, *9*, e86151. [CrossRef] [PubMed]
19. Dayer, R.; Badoud, I.; Rizzoli, R.; Ammann, P. Defective implant osseointegration under protein undernutrition: Prevention by PTH or pamidronate. *J. Bone Miner. Res.* **2007**, *22*, 1526–1533. [CrossRef]
20. Dayer, R.; Rizzoli, R.; Kaelin, A.; Ammann, P. Low protein intake is associated with impaired titanium implant osseointegration. *J. Bone Miner. Res.* **2006**, *21*, 258–264. [CrossRef]
21. Nicolai, H.P.; Liesner, C. Investment Casting of Titanium. In *Titanium and Titanium Alloys: Fundamentals and Applications*, 1st ed.; Leyens, C., Peters, M., Eds.; Wiley-Vch: Weinheim, Germany, 2003; pp. 263–272.
22. Guilin, Y.; Nan, L.; Yousheng, L.; Yining, W. The effects of different types of investments on the alpha-case layer of titanium castings. *J. Prosth. Dent.* **2007**, *97*, 157–164. [CrossRef]
23. Nagaoka, A.; Yokoyama, K.i.; Sakai, J.i. Evaluation of hydrogen absorption behaviour during acid etching for surface modification of commercial pure Ti, Ti–6Al–4V and Ni–Ti superelastic alloys. *Corros. Sci.* **2010**, *52*, 1130–1138. [CrossRef]
24. Mohammadi, S.; Wictorin, L.; Ericson, L.E.; Thomsen, P. Cast titanium as implant material. *J. Mater. Sci. Mater. Med.* **1995**, *6*, 435–444. [CrossRef]
25. Heissler, E.; Fischer, F.S.; Boiouri, S.; Lehrnann, T.; Mathar, W.; Gebhardt, A.; Lanksch, W.; Bler, J. Custom-made cast titanium implants produced with CAD/CAM for the reconstruction of cranium defects. *Int. J. Oral. Maxillofac. Surg.* **1998**, *27*, 334–338. [CrossRef]

26. Krupa, D.; Baszkiewicz, J.; Kozubowski, J.A.; Barcz, A.; Sobczak, J.W.; Biliniski, A.; Lewandowska-Szumiel, M.D.; Rajchel, B. Effect of calcium-ion implantation on the corrosion resistance and biocompatibility of titanium. *Biomaterials* **2001**, *22*, 2139–2151. [CrossRef]
27. Krupa, D.; Baszkiewicz, J.; Kozubowski, J.A.; Barcz, A.; Sobczak, J.W.; Bilinski, A.; Lewandowska-Szumiel, M.; Rajchel, B. Effect of phosphorus-ion implantation on the corrosion resistance and biocompatibility of titanium. *Biomaterials* **2002**, *23*, 3329–3340. [CrossRef]
28. Nayab, S.N.; Jones, F.H.; Olsen, I. Human alveolar bone cell adhesion and growth on ion-implanted titanium. *J. Biomed. Mater. Res. A* **2004**, *69*, 651–657. [CrossRef] [PubMed]
29. Nayab, S.N.; Jones, F.H.; Olsen, I. Effects of calcium ion implantation on human bone cell interaction with titanium. *Biomaterials* **2005**, *26*, 4717–4727. [CrossRef]
30. Nayab, S.N.; Jones, F.H.; Olsen, I. Effects of calcium ion-implantation of titanium on bone cell function in vitro. *J. Biomed. Mater. Res. A* **2007**, *83*, 296–302. [CrossRef]
31. Nayab, S.N.; Jones, F.H.; Olsen, I. Modulation of the human bone cell cycle by calcium ion-implantation of titanium. *Biomaterials* **2007**, *28*, 38–44. [CrossRef]
32. Hanawa, T.; Kamiura, Y.; Yamamoto, S.; Kohgo, T.; Amemiya, A.; Ukai, H.; Murakami, K.; Asaoka, K. Early bone formation around calcium-ion-implanted titanium inserted into rat tibia. *J. Biomed. Mater. Res.* **1997**, *36*, 131–136. [CrossRef]
33. Jinno, T.; Kirk, S.K.; Morita, S.; Goldberg, V.M. Effects of calcium ion implantation on osseointegration of surface-blasted titanium alloy femoral implants in a canine total hip arthroplasty model. *J. Arthroplast.* **2004**, *19*, 102–109. [CrossRef]
34. Cheng, M.; Qiao, Y.; Wang, Q.; Jin, G.; Qin, H.; Zhao, Y.; Peng, X.; Zhang, X.; Liu, X. Calcium Plasma Implanted Titanium Surface with Hierarchical Microstructure for Improving the Bone Formation. *ACS Appl. Mater. Interfaces* **2015**, *7*, 13053–13061. [CrossRef]

materials

MDPI

Article

Model-Based Roentgen Stereophotogrammetric Analysis to Monitor the Head–Taper Junction in Total Hip Arthroplasty in Vivo—And They Do Move

Jing Xu [1], Robert Sonntag [2], J. Philippe Kretzer [2], Dominic Taylor [1], Raimund Forst [1] and Frank Seehaus [1,*]

1 Department of Orthopaedic Surgery, Faculty of Medicine, University of Erlangen-Nürnberg, 91054 Erlangen, Germany; jing.xu@fau.de (J.X.); dominictaylor2000@yahoo.com (D.T.); raimund.forst@fau.de (R.F.)

2 Laboratory of Biomechanics and Implant Research, Clinic for Orthopedics and Trauma Surgery, Center for Orthopedics, Trauma Surgery and Spinal Cord Injury, Heidelberg University Hospital, 69118 Heidelberg, Germany; Robert.Sonntag@med.uni-heidelberg.de (R.S.); Philippe.Kretzer@med.uni-heidelberg.de (J.P.K.)

* Correspondence: frank.seehaus@fau.de

Received: 28 February 2020; Accepted: 24 March 2020; Published: 27 March 2020

check for updates

Abstract: Model-based Roentgen stereophotogrammetric analysis (RSA) using elementary geometrical shape (EGS) models allows migration measurement of implants without the necessity of additional attached implant markers. The aims of this study were: (*i*) to assess the possibility of measuring potential head–taper movement in THA in vivo using model-based RSA and (*ii*) to prove the validity of measured head–taper migration data in vitro and in vivo. From a previous RSA study with a 10 years follow-up, retrospectively for $n = 45$ patients head–taper migration was calculated as the relative migration between femoral ball head and taper of the femoral stem using model-based RSA. A head–taper migration of 0.026 mm/year can be detected with available RSA technology. In vitro validation showed a total migration of 268 ± 11 µm along the taper axis in a similar range to what has been reported using the RSA method. In vivo, a proof for interchangeable applicability of model-based RSA (EGS) and standard marker-based RSA methods was indicated by a significant deviation within the migration result after 12-month follow-up for all translation measurements, which was significantly correlated to the measured head–taper migration (r from 0.40 to 0.67; $p < 0.05$). The results identified that model-based RSA (EGS) could be used to detect head–taper migration in vivo and the measured movement could be validated in vitro and in vivo as well. Those findings supported the possibility of applying RSA for helping evaluate the head–taper corrosion related failure (trunnionosis).

Keywords: Roentgen stereophotogrammetric analysis; hip arthroplasty; elementary geometrical shape model; interchangeability; head–taper junction; migration

1. Introduction

Modern total hip arthroplasty (THA), referred as the "surgery of the century", shows survival rates of >95% after 10 years and of >80% after 25 years [1]. One of the top five clinical challenges within THA is aseptic loosening [2], which presents not only a question of wear [2]. However, implant-to-bone movements, termed as migration, are considered as one the most important mechanical factors for aseptic loosening [3]. Marker-based Roentgen stereophotogrammetric analysis (RSA) presents the common gold standard to detect in vivo implant migration [4] (Figure 1A). However, additional implant marking could be associated with increased production/recertification costs for implant manufacturer as well as with an increased risk of cement cracking [5]. To overcome these disadvantages of standard marker-based RSA, a model-based RSA method was introduced, using 3D surface models of the implant to match models projection contour on the implant contour with radiographic RSA image pairs to calculate the 3D position of the implant [5]. Sourcing of the required surface models can be achieved by reverse engineering (RE) or by computer aided design (CAD) method. Interchangeability of model-based RSA approach using CAD/RE models to standard marker-based RSA could be verified [5–7]. Model-based RSA using elementary geometrical shape (EGS) models is a further RSA approach to measure migration of THA components [8]. The EGS approach using spheres and cylinders or hemispheres, to represent femoral head or stem (Figure 1B) or acetabular cup component, respectively. Bone markers are still required.

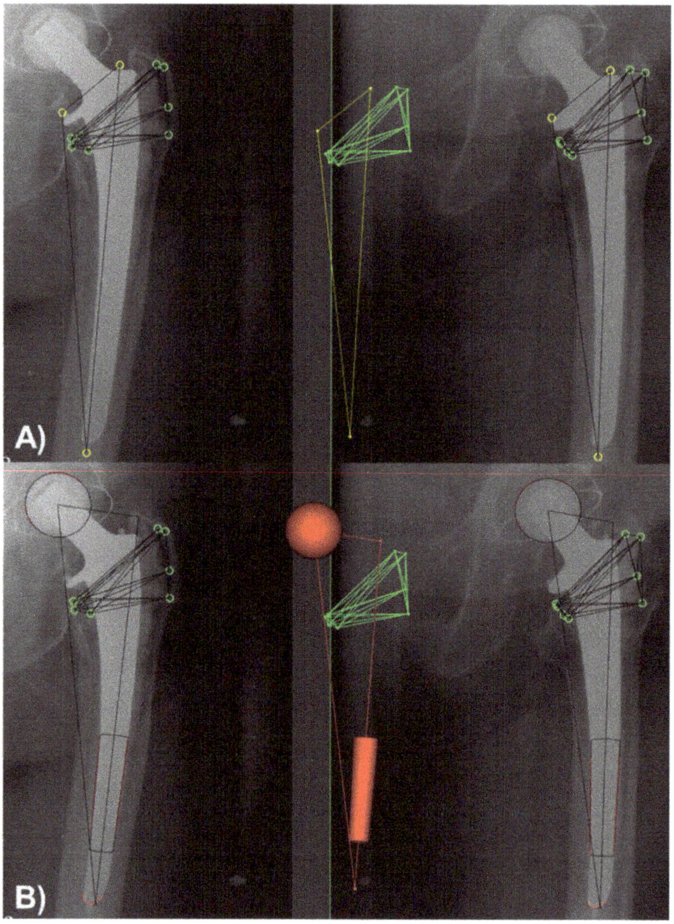

Figure 1. Same pair of Roentgen stereophotogrammetric analysis (RSA) radiographs analyzed by the both RSA approaches (**A**) marker-based and (**B**) model-based RSA (EGS). Bone markers are represent as green circles within the left and right X-ray of RSA image pair as well as green dots in their resulting 3D position (x, y, z coordinate) within the RSA analysis scene. Additional attached implant markers for marker-based RSA approach are illustrated as yellow circles and dots, respectively. The EGS model for model-based RSA approach is represented as red contours at the ball head, with the both X-rays and as red cylinder, sphere with the RSA analysis scene.

In addition to articular wear, many clinical studies have indicated corrosion between the femoral ball head component and the stem-taper as one of the sources of THA failure [9–12]. This phenomenon is also called trunnionosis, which is defined as the wear and corrosion that occurs at the junction of the modular head–taper or neck-stem initialized by micromotion [13]. There are local corrosive processes, but also mechanically induced processes taking place at that interface over the lifetime of the implant. A subsequent head–taper migration respective to the seating of the ball head on the male stem-taper is multifactorial, including the implant design (e.g., technical specification of male/female tapers), the surgical technique (e.g., impaction force) or patient-related factors (e.g.,/ body mass index, activity) [14]. In addition, the geometry and topography of both taper parts, the female ball head and the male stem-taper, determine the local effects. In that context, it is important to know that even though many tapers are called the same, e.g., a 12/14 standard taper—they are far from being standardized and vary between manufacturers [15]. The same is true for the topography, which is mainly characterized for the male stem-taper.

According to previous research, trunnionosis can cause clinical symptoms such as adverse local tissue reactions (ALTRs), thigh pain and local swelling [16]. Instability and special gait patterns was also observed in some cases [17]. The overall incidence of trunnionosis of primary THA cases was reported at approximately 1–3% [18,19]. At present, laboratory investigations of trunnionosis are limited to the detection of the metal ion in serum and joint capsule. Magnetic Resonance Imaging (MRI) and ultrasound present good specificity and sensitivity on detecting pseudotumor and ALTRs [20,21]. Severe osteolysis caused by ALTRs can be viewed on plain radiographs. However the plain radiograph is insufficient for the detection of head–taper micromotions. Sultan et al. [14] stated that the true incidence rate of trunnionosis was underestimated as the release of metal debris can also cause osteolysis and aseptic loosening.

To the authors' knowledge, no clinical measurement tools commonly exists to measure or monitor the possibility of head–taper migration in vivo. Model-based RSA or a combination with standard marker-based RSA offers the opportunity to measure a possible motion within the head–taper junction. However, this option is limited in application to hard-soft bearings only, while visibility of the ball head's contour is necessary within the RSA radiographic pairs.

Therefore, the purpose of this study is (*i*) to assess the feasibility to detect a possible head–taper movement in THA in vivo using model-based RSA and (*ii*) to prove the validity of measured migration data. The authors hypothesize thus a measurement of head–taper migration (if available within the given clinical records) for hard-soft bearings is possible and thus the measured migration is valid.

2. Materials and Methods

Data presented was taken out of a previous RSA study [22] initiated in 1999 (ethical registration number: 1.077). It offers the opportunity to study long-term implant migration as well as to measure it by the both marker-based and model-based RSA (EGS) method, respectively.

2.1. Patient Cohort—Inclusion Criteria

All patients who were included had received a cemented femoral stem (Lubinus SP II, Waldemar Link GmbH, Hamburg, Germany) with the three visible additional attached tantalum markers and a cemented polyethylene acetabular cup (LINK® IP Acetabular Cup, Waldemar Link GmbH, Hamburg, Germany) component. Each femoral stem has a caput–collum–diaphyseal angle of 126 degrees and is combined with a ceramic ball head (BIOLOX®*forte*, Ceram Tec GmbH, Plochingen, Germany) diameter of 28 mm. Three different sizes of head–neck length were used (S, M and L size, Tables A1 and A2). In summary, $n = 45$ cases could be analyzed. To the authors' knowledge, the full weight bearing of the THA treated hip was first allowed within the rehabilitation period.

2.2. Image Acqusition and Analysis

All patients underwent the first reference RSA examination during the first postoperative week and received follow-up RSA examinations at 1.5 months, 3 months, 6 months, 1 year, 2 years, 5 years, and 10 years. In all RSA follow-ups, the same equipment and parameters were used. Patients were positioned in the supine position on the X-ray table. The X-ray table is placed within the uniplanar RSA setup. The RSA setup consists of two X-ray sources, focused at the patient hip with an intersection angle of 40 degrees directly above the patient. Underneath the X-ray table, a calibration box (Umeå 43, RSA BioMedical Innovations AB, Umeå, Sweden) was placed, with both of the X-ray films in the lower level of the calibration box. The X-ray film-focus distance was 140 cm.

All resulting RSA image pairs were analysed using model-based RSA method (MBRSA 4.1, RSA Core, Leiden, The Netherlands). Migration was calculated based on a reference point, which is usually represented by the center of gravity of the rigid body. For image analysis, the RSA standard thresholds (calibration errors translation ≤ 0.05 mm, focus error ≤ 0.5 mm; condition number ≤ 100; rigid-body error ≤ 0.35 mm) were applied to verify the quality of the image calibration procedure [23,24].

Migration results were represented with respect to a global coordinate system defined by the calibration box: translation along the medial–lateral (x) and cranial–caudal (y) axes constitute in-plane motion, and translation along the anterior-posterior-axis (z) out-of-plane motion; rotation around the anterior-posterior axis (Rz) further described in-plane motion, and along the medial–lateral (Rx) and cranial–caudal axes (Ry), and out-of-plane motion, respectively.

2.3. *Head–Taper Migration Measurements*

Using a model-based RSA approach (Figure 2) the head–taper migration was calculated between the rigid body of the femoral stem component (defined by the three additional attached implant markers) and the ball head (represented as a spherical EGS model). Migration results were calculated according to the above-mentioned protocol and displayed by using the allocated coordinate system.

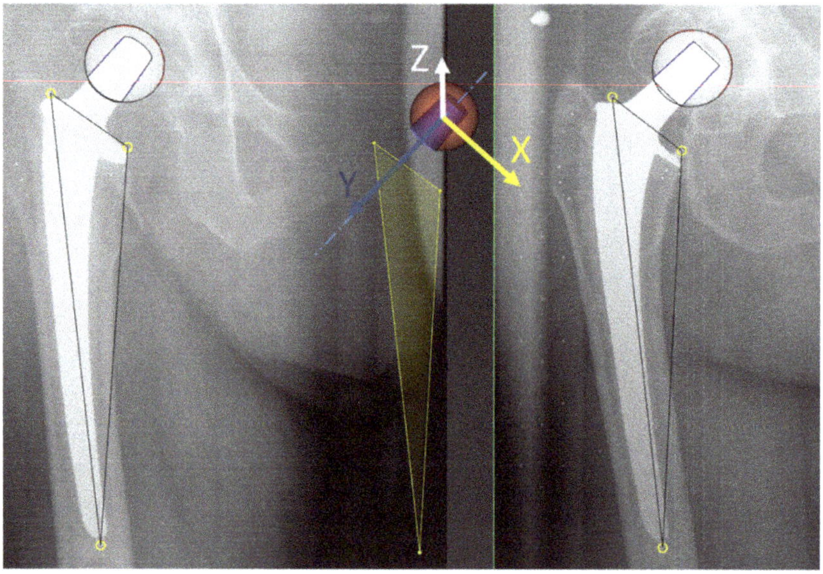

Figure 2. Coordinate system based on the long axis of the taper (by fitting a cone model to the contour of taper) and the three implant markers (yellow circles) to present head–taper migrations. The Y-axis was set on the long axis of the taper. The X-axis was orthogonal to Y-axis and parallel to a plane across three implant markers (yellow plane). The Z-axis of the new coordinate system was orthogonal to the X- and Y-axes.

To take into consideration that the femoral ball head can move mechanically along the long axis of the taper, a new coordinate system based on the direction of the long axis of the taper and three implant markers was set up to measure head–taper migration (Figure 2).

The taper contour was matched with an EGS cone model to find the orientation of its central axis, which became the Y-axis of this new coordinate system. The X-axis of the new coordinate system was set orthogonal to Y-axis and parallel to a plane across three implant markers and pointing in a medial direction. The Z-axis of the new coordinate system was orthogonal to the X- and Y-axes and pointing in an anterior direction. Then the original head–taper migration results were transformed according this new coordinate system.

2.4. Experimental Verification of Head–Taper Migration

Based on a previous study analyzing the taper geometry and topography of different manufacturers [15], $n = 5$ CoCr sample tapers matched with the taper profile of the manufacturer of the cemented femoral stems used in the RSA study (Link GmbH, Hamburg, Germany). The geometry of the sample tapers has been verified by a coordinate measuring machine (Mahr Multisensor/MarVision MS 222; Mahr, Göttingen, Germany; accuracy: ±2.3 μm) and the topography by a tactile roughness measurement instrument (Perthometer M2; Mahr, Göttingen, Germany; accuracy: 12 nm). For the experimental investigation of head–taper migration, ceramic ball heads (BIOLOX®*delta*, Ceram Tec GmbH, Plochingen, Germany) of 28 cm diameter and head–neck length M have been used. The length of the head sample taper assembly has been measured after assembly by hand (0 kN) and after impaction at two load levels (2 kN and 4 kN) using a servo-hydraulic testing machine (MTS 858 Mini Bionix II; MTS, Eden Prairie, MN, USA).

2.5. Clinical Verification of Potential in Vivo Head–Taper Migrations—Proving Interchangeable Applicability of Marker and Model-Based RSA EGS Method

To prove data validity of the in vivo obtained migration results, a test of interchangeable applicability of measured marker-based and model-based RSA (EGS) is necessary. In summary $n = 18$ cases from the above mentioned study cohort were identified, for which a complete set of the RSA follow up examinations were available.

Each pair of RSA radiograph was analyzed three times (one week break for the observer is predefined and respected) by both marker-based and model-based RSA EGS approach according to a written protocol. Since both RSA methods calculate their implant-to-bone migration from the same image data sets, the results need to be identical. However, according applied reference points for migration detection, special attention has to be paid on (1) reference point correction and (2) EGS model generation.

If the implant-to-bone migration including rotational motion, which is present within most of clinical situations, the resulting translational migration results can be variant if different reference points were used (Figure 3) [6]. To enable the comparison of migration results from marker- and model-based RSA (EGS) approaches, a reference point correction described by Hurschler et al. [6] was performed (Mathematica, Wolfram Research, Champaign, IL, USA). The reference point of model-based RSA (EGS) was corrected to the same position as in marker-based RSA with a subsequent recalculation of the translational results in all three axes (Figure 3).

In the model-based RSA (EGS) approach, the femoral stem component of a THA design was displayed by a cylinder and a sphere for the migration measurement (Figure 4A). Three virtual markers resulting out of the arrangement of both the EGS models. In comparing the marker-based RSA (where the reference point is calculated from three rigidly fixed tantalum markers on the femoral stem) with the model-based RSA (EGS) whose reference point could be variable, thus creating head–taper migration (Figure 4B). Due to the reference point variability (in the model-based RSA (EGS) approach) and its effect on the migration results: a head–taper migration can induce additional movement to the migration results.

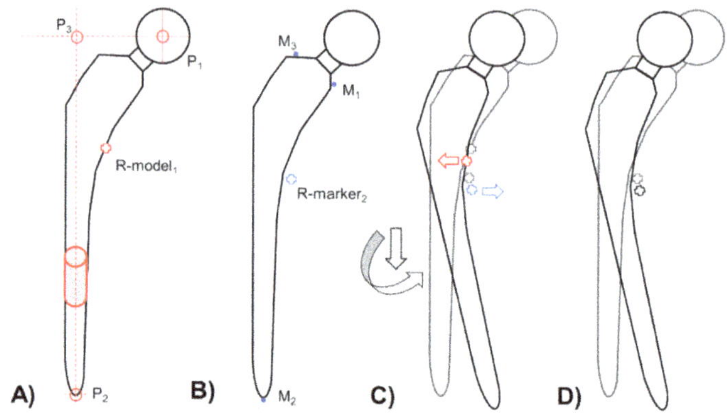

Figure 3. Resulting reference points for migration measurement of the same femoral stem design using (**A**) model-based RSA (red cross; R-model$_1$) and (**B**) marker-based RSA (blue cross, R-marker$_2$) approach. (**C**) If a stem subsidence with a combined rotation (black arrows) around from the anterior–posterior direction occurs, each reference point of the marker (blue cross and arrow) and model-based RSA (red cross and arrow) indicate a different directions of migration (relative to grey crosses). (**D**) To compare in vivo migration of the same RSA image pairs, a reference point correction is necessary. To enable this mentioned comparison of migration results the reference point of model-based RSA (EGS) was corrected to the reference point of marker-based RSA.

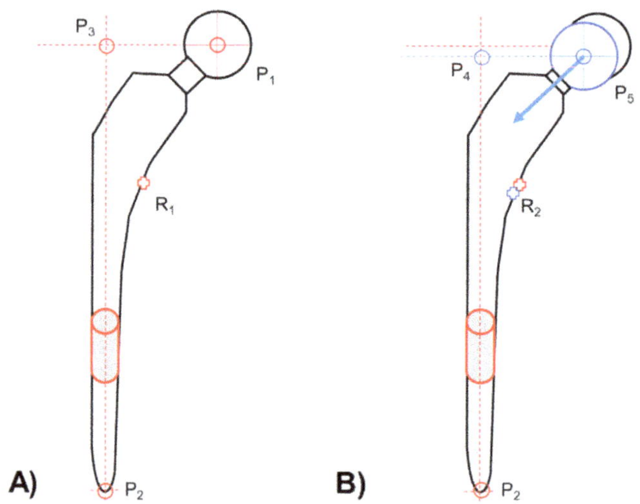

Figure 4. (**A**) An EGS model of a femoral stem component, consist out of three virtual markers (red circles). First marker indicates the center of the femoral ball head (P_1). Second marker indicates the lowest point of the tip (P_2). Last virtual marker (P_3) is defined by the intersection of the horizontal extension line of point P_1 and the perpendicular (vertical) line, resulting out of the point P_2 and longitudinal axis of applied EGS cylinder model (gray cylinder with red contour). This determination of the femoral stem as rigid body with three virtual markers requires a stable head–taper connection. (**B**) Therefore, the three virtual markers are no longer a rigid body if the head center marker (P_1 to P_5) and its projection marker (P_3 to P_4) may migrate relative to the tip marker (P_2) during the follow-up time period, and afterwards introduce a bias in the applied reference point (changing from R_1 to R_2) measurement with EGS stem model.

The two above-mentioned details about reference point behavior, enable the clinical verification of measured head–taper migration. If a head–taper movement exists, the virtual markers and the reference point are changed respectively (from P_1 to P_5 and P_3 to P_4 as well as R_1 to R_2; Figure 4) whilst using model-based RSA approach for migration detection. This indicated a deviating rigid body representation in comparison to standard marker-based RSA, which use three additional fixed tantalum markers to represent a monobloc of the femoral stem (Figure 3B). A marker movement cannot occur, thus a change in the reference point is not possible.

2.6. Statistics

All statistical analyses were carried out by using R (R Foundation, Vienna, Austria). Head–taper migration was presented by using descriptive statistics. Student's *t*-test was performed to evaluate the significance of head–taper migration over the 10 years follow up period.

All experimental data is given as a mean and a standard deviation for the $n = 5$ head-stem samples tested.

To verify measured head–taper migration, interchangeable applicability of marker- and model-based RSA EGS approaches were analysed in six degrees of freedom using the Bland–Altman plots [25]. Bland–Altman plots were used as an unscaled method to calculate the mean inter method difference and limit of agreement (LoA) between the both of the RSA approaches. According to the literatures [26,27], the upper limit of RSA accuracy (0.5 mm for translation, 1.15 degrees for rotation) was used as the upper threshold of LoA to determine whether the two RSA methods are interchangeable or not. A one-sample student's *t*-test (significance level = 0.05) was performed to evaluate the significance of mean difference between two RSA methods. The Pearson correlation coefficient was used to assess a linear correlation between intermethod difference and head–taper migration [28]. To test the working hypothesis, a test based on Fisher's Z transformation [29] was used to establish that the true correlation coefficient is not equal to 0 with the significance level of 0.05.

3. Results

3.1. Head–Taper Migration

Migration of the femoral ball head is caused by a creeping embedment in the direction of the taper throughout the follow-up period. The majority of the migration was discovered to occur in the 6- to 4-month follow-up period (Figure 5B). A translation towards the taper along the longitudinal axis (y-axis) from 0.028 to 0.115 mm ($p < 0.01$) was measured. The migration speed slowed down in 24 to 120 month follow-up period.

Additionally a continuous migration perpendicular to the long axis of the taper and in a medial-posterior direction (x-axis translation: from 0.044 to 0.139 mm, $p < 0.01$; z-axis translation: from −0.109 to −0.279 mm, $p < 0.05$) was also observed in the 6 to 24 month period, and then tended to stabilize (Figure 5A,C). A mean head–taper migration of 0.026 mm per year, over the 10-year follow-up period was identified.

A subgroup analysis using head–neck length, indicated a similar migration pattern for all three groups. The L size group showed a trend of greater migration along medial-lateral (x-axis) direction and in the anterior-posterior direction (z-axis) than the other head–neck length groups (Figure 6). However, results of the student's *t*-test showed that it was insufficient to demonstrate a significant difference between any two groups.

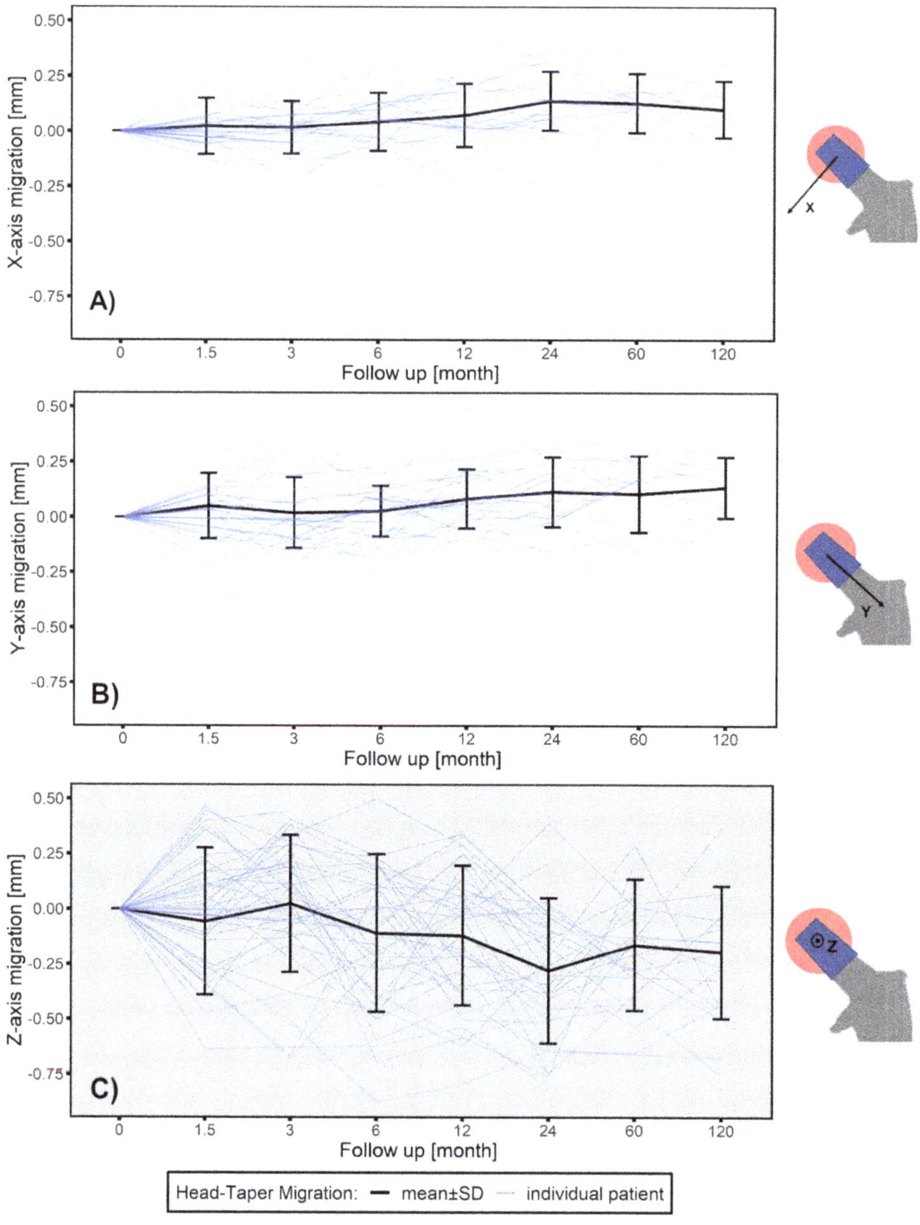

Figure 5. Head–taper migration depicted as mean ± SD (black line) for a ten-year follow-up time period along (**A**) medial-lateral direction perpendicular to the taper (x-axis translation), (**B**) cranial–caudal direction along the taper longitudinal axis (y-axis translation) and (**C**) anterior–posterior direction perpendicular to the taper (z-axis translation). Blue lines indicate each individual data set. Gray background indicate out-of-plane measurement.

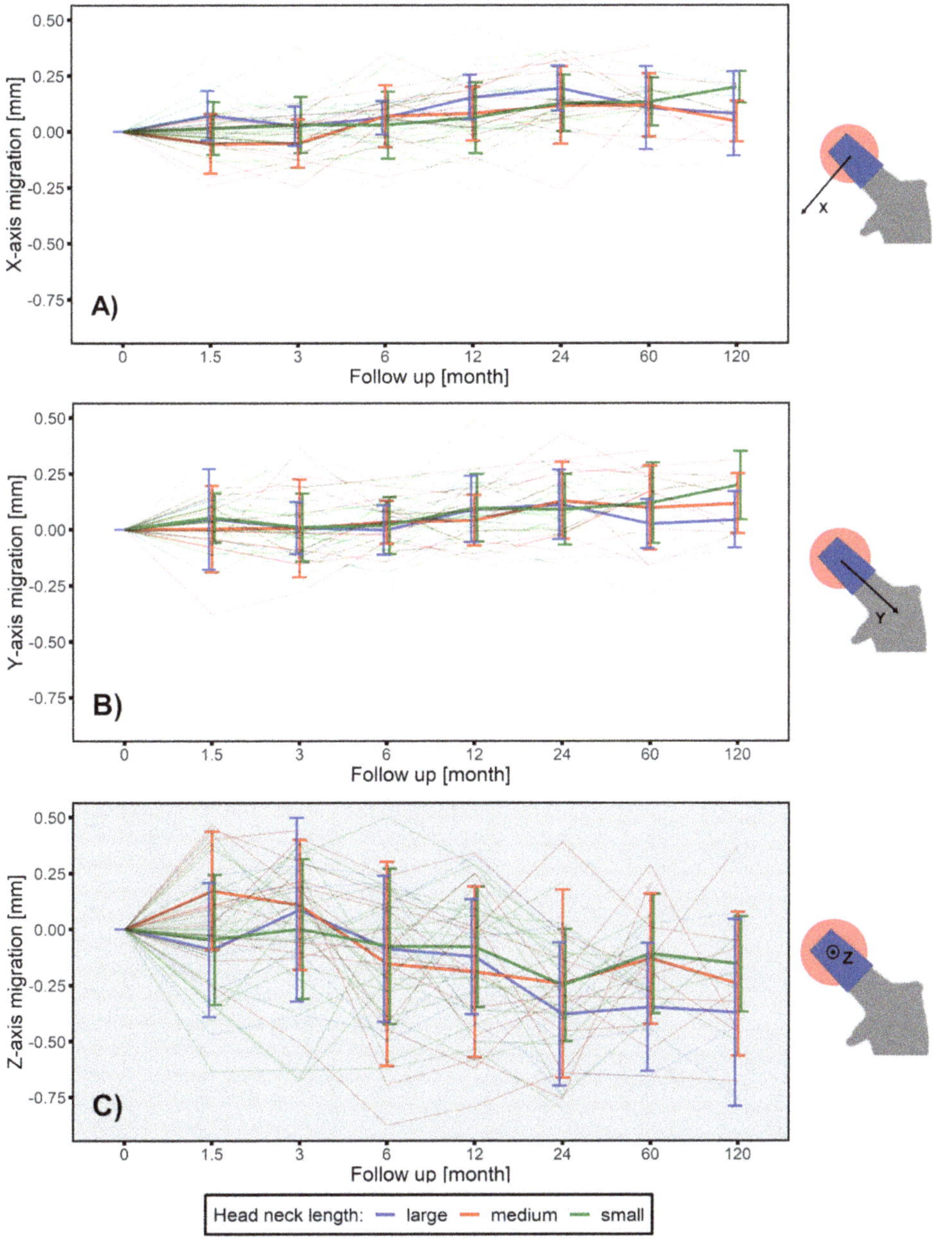

Figure 6. Head–taper migration depicted as mean ± SD for a ten-year follow-up time period along (**A**) medio-lateral direction perpendicular to the taper (x-axis translation), (**B**) cranio–caudal direction along the taper longitudinal axis (y-axis translation) and (**C**) anterior–posterior direction perpendicular to the taper (z-axis translation), grouped by head–neck length: small (green line), medium (red line) and large (blue line). Gray background indicate out-of-plane measurement.

3.2. Experimental Verification of the Measured in Vivo Head–taper Movement

Male taper geometry and topography have been shown to be comparable to what has been reported in the literature for the femoral stem investigated in this study (Table 1) [15]. As expected, migration has been shown to be largest after initial impaction at 2 kN (162 ± 6 μm) and smaller after subsequent impaction at 4 kN (106 ± 10 μm) (Table 2). In total, a head–taper migration of 268 ± 11 μm was measured during the experimental investigation.

Table 1. Stem-taper geometry and topography.

Parameter	Value
Taper angle	5.611 ± 0.004 deg
Ra	3.58 ± 0.06 μm
Rz	14.26 ± 0.29 μm
RSm	122.93 ± 0.20 μm
Rp	7.50 ± 0.19 μm
Rk	13.90 ± 0.34 μm

Table 2. Migration along the taper axis.

Head–Taper Junction	0 > 2 kN	2 > 4 kN	Total
#1	160 μm	105 μm	265 μm
#2	155 μm	95 μm	250 μm
#3	160 μm	125 μm	285 μm
#4	173 μm	100 μm	273 μm
#5	160 μm	105 μm	265 μm

3.3. Clincial Verification—Proof Interchangeability between Marker-Based RSA and Model-based RSA (EGS) Migration Results

A discrepancy between the marker and model-based RSA (EGS) approach, especially in cranial-caudal axis from 12 to 120 months is observable (Figure 7). Mean translation along cranial–caudal axis measured over all follow-ups (mean value from 3 to 120 months) by model-based RSA (EGS) was −0.083 mm, which was much greater than that measured by marker-based RSA (−0.021 mm).

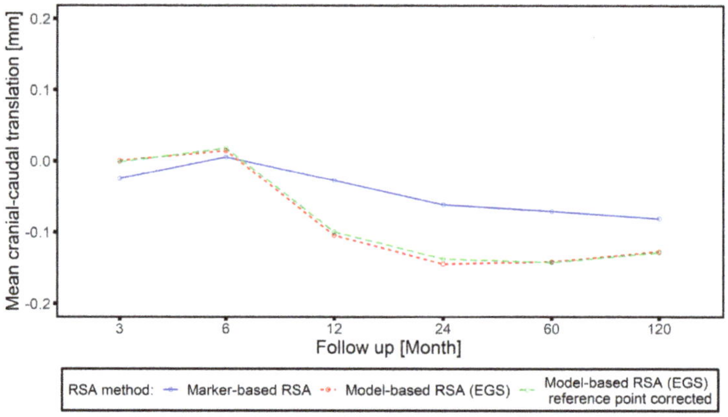

Figure 7. Cranial–caudal translation along follow-up time period analyzed by marker-based RSA (blue solid line), model-based RSA (EGS) (red dotted line) and model-based RSA (EGS) after reference point correction (green dashed line).

The LoA between the both RSA methods were acceptable for in-plane migration (Figure A1), less than 0.04 ± 0.34 mm on translation measurement, and less than 0.02 ± 0.39 degrees on rotation measurement. For out-of-plane migration (Figure A2) LoA was greater than in-plane measurement. Especially on rotational measurement around cranial–caudal axis, the LoA between two methods was −0.47 ± 3.78 degrees, which was much greater than that of other directions (medial-lateral rotation: −0.01 ± 0.88 degrees, anterior–posterior translation: 0.02 ± 0.53 mm).

After dividing all cases into two groups (3 to 6 months and 12 to 120 months), most of the measurements showed a similar range of the LoA between the two groups (Figure 8, Figure A3). Whereas, in all of translational measurements from 12 to 120 months, the mean difference between two RSA methods was found to have a significant bias from 0 ($p < 0.05$) (Figure 8B,D,F).

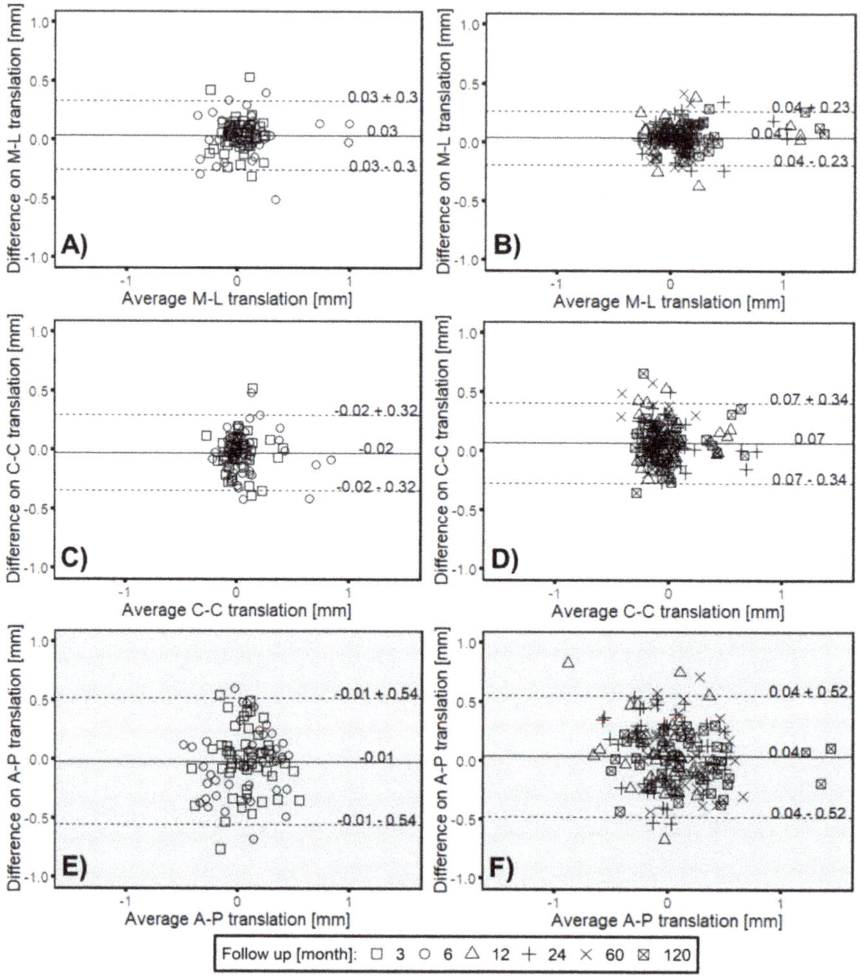

Figure 8. Bland–Altman plots presenting of the calculated average and difference between marker-based RSA and model-based RSA (EGS) for translational migration measurements along medial-lateral (ML), cranial–caudal (C-C) and anterior–posterior direction (AP). Solid line indicates the mean difference with LoA (dashed line) for (**A**,**C**,**E**) 3 (square) and 6 month follow up (circle) as well as for (**B**,**D**,**F**) 12 (triangle), 24 (plus), 60 (cross) and 120 (square with cross) month follow up respectively. Gray background indicate out-of-plane measurement. Mean difference equal to 0 mm: (**A**): $p < 0.05$; (**B**): $p < 0.05$; (**C**): $p = 0.20$; (**D**): $p < 0.05$; (**E**): $p = 0.60$; (**F**): $p < 0.05$.

3.4. Correlation between Intermethod Difference and Head–Taper Migration

The intermethod difference and head–taper migration showed a strong linear correlation in all three axes with r ranging from 0.40 to 0.67 ($p \leq 0.05$) and correlated most in the cranial–caudal axis (r = 0.67, $p \leq 0.01$) (Figure 9). Analyses with each follow up points exhibited a strong correlation at most of the follow-up time points (Table A3).

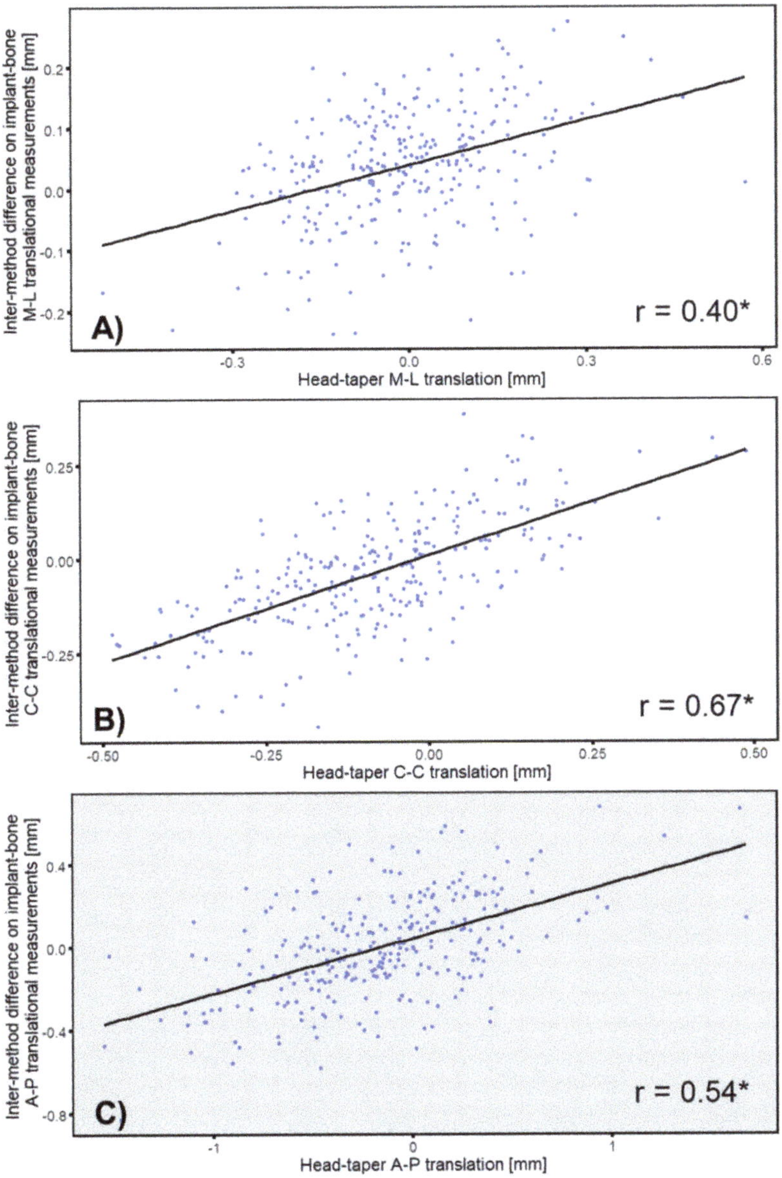

Figure 9. Correlation of intermethod difference and head–taper translation along (**A**) medial-lateral (**B**) cranial–caudal and (**C**): anterior–posterior direction. * $p \leq 0.05$. Gray background indicate out-of-plane measurement.

4. Discussion

Series of clinical complications can be caused by head–taper corrosion, including ALTRs, pseudotumors and even the unacceptable gross taper failures [30]. Recently, MRI and ultrasound are regards as best methods for detecting ALTRs and pseudotumors caused by metal debris [20,21]. To the authors' knowledge, there is no clinical measurement tool available to monitor the head–taper migration in vivo. For hard-soft bearings (MoP and CoP), model-based RSA method or a combination with the standard marker-based RSA represents an approach that offers this option, and could be a potential candidate for the evaluation of head–taper migration. Since this method uses rigid-body kinematics to determine the relative motion between two rigid bodies, an in vivo assessment of the head (using a spherical model of the ball head) and taper (using an EGS model of the stem component or additionally attached implant markers) junction would be enabled. However, this application requires the visibility of the femoral ball head contour within the resulting RSA radiographic pairs.

Head–taper migration could be detected using model-based RSA (EGS) method retrospectively from the available previous clinical RSA images. Measured migration rates within the first 6 months were low, while they increased continuously from 6 to 24 months post-op (along the taper: from 0.028 to 0.115 mm; perpendicular to the taper: from 0.118 to 0.312 mm). Migration slowed from 24 to 120 months. Therefore, the hypothesis (*i*), related to the in vivo assessment of head–taper movement in THA using model-based RSA approach could be accepted. Benefited from the good accuracy of RSA technology, it is possible to measure the submillimeter head–taper migration. Even for cases with stable implants (no case included in this study failed during the 10 years of follow-up period), model-based RSA (EGS) is able to detect the migration of head–taper junction within the necessity of additional implant or bone marking. Therefore, the presented results validated the possibility of RSA technology for head–taper migration measurement and enable a wider applicability. It is hoped that this will improve the ability to evaluate further couplings within THA next to investigated metal-ceramic coupling of head–taper stem.

The in vitro as well as an in vivo validation of measured head–taper migration supports the findings of retrospective analyses. The experimental data shows a migration along the taper axis in a similar range (up to approximately 250 μm) to what has been measured using the RSA approach. The discrepancy between the in vitro and in vivo data may be related to different starting references. While in the RSA group, the head had already been intra-operatively impacted and migration was measured over time with more and more loading of the patient, the experiments started with a nonimpacted status (manual assembly) and increased the impaction force without the investigation of any dynamic loading. However, the findings indicate that the migration measured by the RSA method is in a very reasonable range.

The clinical validation (in vivo) of measured head–taper migration could be confirmed by proving interchangeable applicability of model-based RSA (EGS) approach next to standard marker-based method. In the first 6 months follow up period, the results of individual measured migration by both RSA methods, supported and demonstrated the interchangeable applicability of two methods (Figure 8, Figure A3). The results of this study showed that the migration between investigated femoral stem and applied femoral ball head occurs mostly take part between 6 to 24 months post-operatively, which lead to the significant difference ($p < 0.05$) between the two RSA methods on all translational measurements after 12 months follow up (Figure 8). Before head–taper migration accelerated (between 3 and 6 months post-operation), the migration measurements with EGS model did not have significant difference compared to the gold standard marker-based RSA on cranial–caudal and anterior–posterior translation ($p > 0.05$). Migration and LoA from 12 to 120 months indicate a different behavior. The results of the correlation analysis showed that there was a strong correlation between head–taper migration and intermethod difference (r from 0.40 to 0.67; $p < 0.05$). As an additional factor, head–taper migration could introduce a systematic error to the migration measurement of hip stem prosthesis when using EGS model, or the continuous migration between head and taper can be regarded mistakenly as a continuous instability of the hip stem. If the head–taper migration was extremely large and leads to an obvious change of the relative marker position, the rigid body error would exceed the threshold of 0.35 mm and make the migration analysis meaningless in this situation. However, if the head–taper migration was relatively small, its contribution to the measurement error could be negligible. Hence, the results of EGS stem model need to be treated cautiously only when the head–taper migration was relatively large but the rigid body error was still within the threshold.

Based on this hypothesis, (*ii*) could be accepted. A measurement of head–taper migration for hard-soft bearings is possible and thus the measured migration is valid. However, the measured head–taper migration pattern can be different in other implant designs.

The presented results showed a migration along the taper axis with higher values after implantation and a reduced migration speed over time. This phenomenon can be explained as follows: Insufficient force applied during the impaction of the femoral head is one of the pathogenesis of head–taper migration [31]. According to the presented in vitro experiment, after the femoral head has been impacted with lower force (2 kN), increased force (4 kN) can cause further plastic deformation (seating) on the taper even under noncorrosive conditions. From this it is speculated that in the early post-operative period, the head–taper migration was dominated by reasons induced by mechanical/material deformation, and showed a relatively high speed. In the later period, the corrosion dominated and resulted in a relatively lower migration. However, exploring the causes of head–taper migration in each specific period is beyond the scope of this study, and need to be further confirmed by clinical retrieval studies. In general, high assembly forces generate the safest situation for the head–taper junction, resulting in the greatest resistance to relative motion over time [32].

In addition to the migration along the longitudinal axis of the taper, the femoral head prosthesis revealed also migration in other directions perpendicular to the taper. According to previous literature, the authors believe that this type of migration was caused by the fretting corrosion in-between the head–taper interface [33]. The periodic pressure during walking is transmitted from femoral head prosthesis to the stem prosthesis through the taper, resulting in a long-term oscillatory slip at the head–taper interface. The passive layer of material at the contact surface can be gradually destroyed by this mechanical action and then corrosion is accelerated [34,35]. This phenomenon can be much worse in metal-to-metal contact surfaces (e.g., titanium head vs CoCr taper). The released metal debris could induce ALTRs and aseptic lymphocytic vasculitis associated lesion [36]. The inflammatory response to the metal debris can also be related to the following bone resorption and aseptic loosening. While ceramic head prosthesis can slow down the taper corrosion speed and also reduce the released metal debris [37–40]. Besides the adverse reactions of metal debris, the material loss caused by corrosive and mechanical abrasion can destroy the force-locking of the head–taper junction, further exacerbating the wearing-out of taper material. The end phase of this process was gross stem-taper failure [30]. Monitoring of head–taper migration in context of corrosive processes is essential. It may be that corrosive processes change the location of head-stem-taper, which is generally proximal for a ceramic head. For a more distal contact over time, the risk of a ceramic head fracture may be increased [41].

Head–neck length is another factor that influences the corrosion speed [42]. A head with larger head neck length correspondingly has a smaller head–taper contact area, resulting in a stress concentration and more corrosion at the interface [32,43]. The results of this study showed that the large head neck length group had slightly more corrosion than the other two groups. However, concerning the aspect of head–neck length and interface corrosion, there was insufficient evidence to support the statistical difference. The limiting factor being the small sample size in each group.

5. Conclusions

Overall, RSA is a reliable method for measuring in vivo implant migration. The results of this study demonstrate that model-based RSA can be used to detect head–taper migration for hard-soft bearings in THA and could be a potential candidate for the clinical evaluation of trunnionosis. In addition, future diagnostic studies are essential for further understanding the effect measured by the RSA method. For patients with high risk factors, more detailed tests should be performed during the follow-ups to reduce the possibility of misdiagnosis in clinical.

Furthermore, clinical validation of measured head–taper migration could be confirmed, thus demonstrating the interchangeable applicability of model-based RSA (EGS) to standard marker-based RSA method. Observed deviation of the measurements between the two RSA methods from 12 to 120 months postoperatively may be explained by head–taper migration, which may confirm the validity of the findings. It has been statistically shown that head–taper migration was correlated to the deviation between methods. Thus, the results indicate that it is necessary to prove the head–taper migration during the entire study period using model-based RSA EGS approach, so that no misinterpretations are drawn from the resulting migration values happens.

Author Contributions: Conceptualization, J.X., R.F. and F.S.; Data curation, J.X., R.S. and J.P.K.; Formal analysis, J.X. and R.S.; Methodology, J.X., R.S., J.P.K. and F.S.; Project administration, F.S.; Resources, J.P.K. and R.F.; Supervision, R.F.; Visualization, J.X. and F.S.; Writing—original draft, J.X.; Writing—review & editing, R.S., J.P.K., D.T., R.F. and F.S. All authors have read and agreed to the published version of the manuscript.

Funding: This research received no external funding

Acknowledgments: The present work was performed in partial fulfilment of the requirements for obtaining the degree "Dr. med." (J.X). This research was made possible out of research cooperation between members of the Musculoskeletal Biomechanics Network (www.msb-net.org) of the DGOOC (German Association of Orthopedics and Orthopedic Surgery).

Conflicts of Interest: The authors declare no conflict of interest.

Appendix A

Table A1. Head–taper implant features and case number.

	Group S	Group M	Group L	
Head Material	Ceramic	Ceramic	Ceramic	
Taper Material	CoCrMo	CoCrMo	CoCrMo	
Head Diameter	28 mm	28 mm	28 mm	
Head Neck Length	S	M	L	
Taper	12/14	12/14	12/14	
CCD Angle	126 deg	126 deg	126 deg	
Numbers	26	13	6	Sum: 45

Table A2. Case number in each follow up time.

Follow Up	Group			Sum
	S	M	L	
1.5 months	18	7	5	30
3 months	21	9	5	35
6 months	22	11	6	39
12 months	23	12	6	41
24 months	21	11	6	38
60 months	11	8	4	23
120 months	7	8	3	18

Table A3. Pearson correlation coefficient of intermethod difference and head–taper translation on each follow up time. * Denote a significant correlation with a significance level of 0.05.

		Medial-Lateral	Cranial–Caudal	Anterior–Posterior
Follow up [months]	3	0.42 *	0.66 *	0.58 *
	6	0.49 *	0.74 *	0.52 *
	12	0.39 *	0.68 *	0.60 *
	24	0.27	0.55 *	0.45 *
	60	0.44 *	0.65 *	0.58 *
	120	0.28	0.61 *	0.44 *

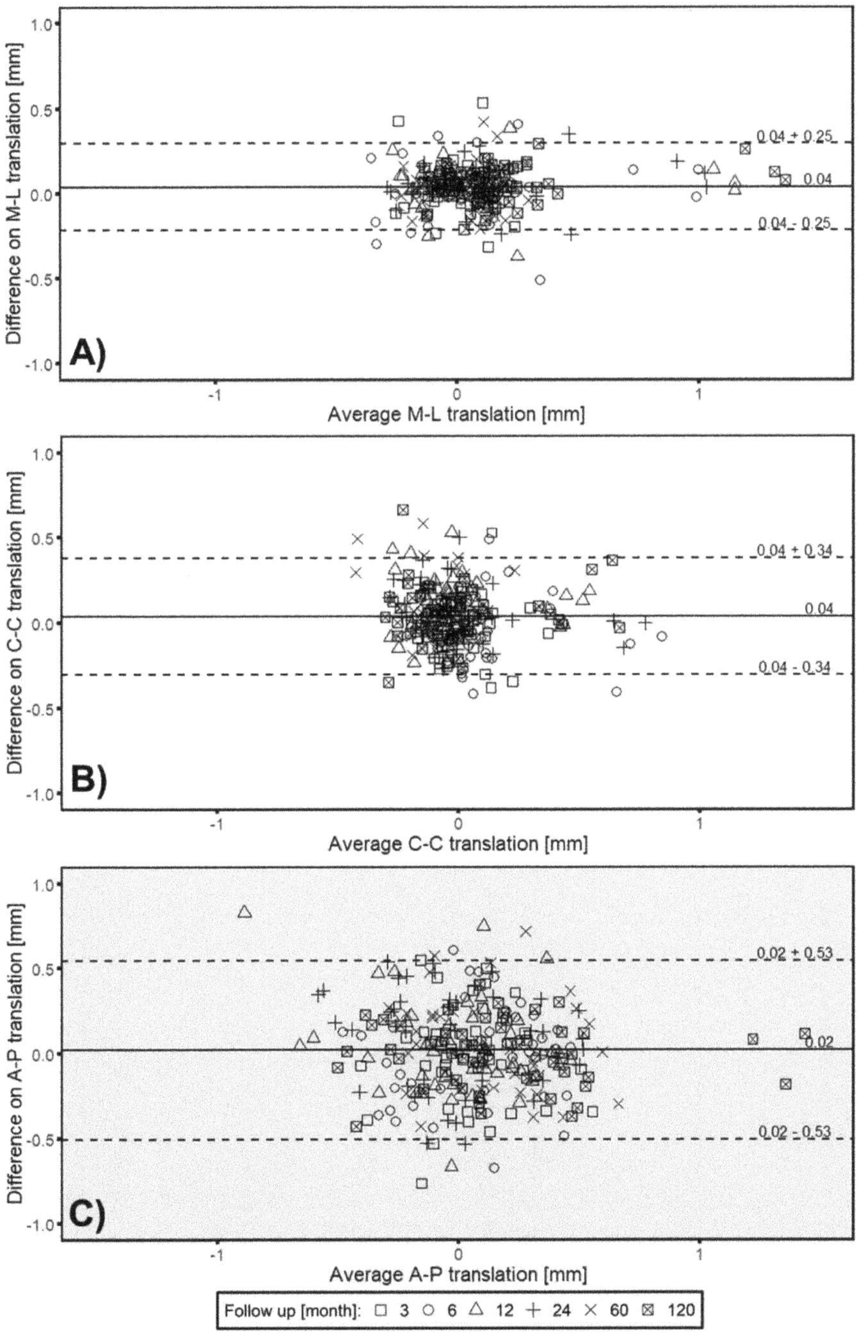

Figure A1. Bland–Altman plots presenting of the calculated average and difference between marker-based RSA and model-based RSA (EGS) for translational measurements along (**A**) Medial–Lateral, (**B**) Cranial–Caudal and (**C**) Anterior–Posterior direction. Solid line indicates the mean difference with LoA (dashed line) for 3 (square), 6 (circle), 12 (triangle), 24 (plus), 60 (cross) and 120 (square with cross) months follow up.

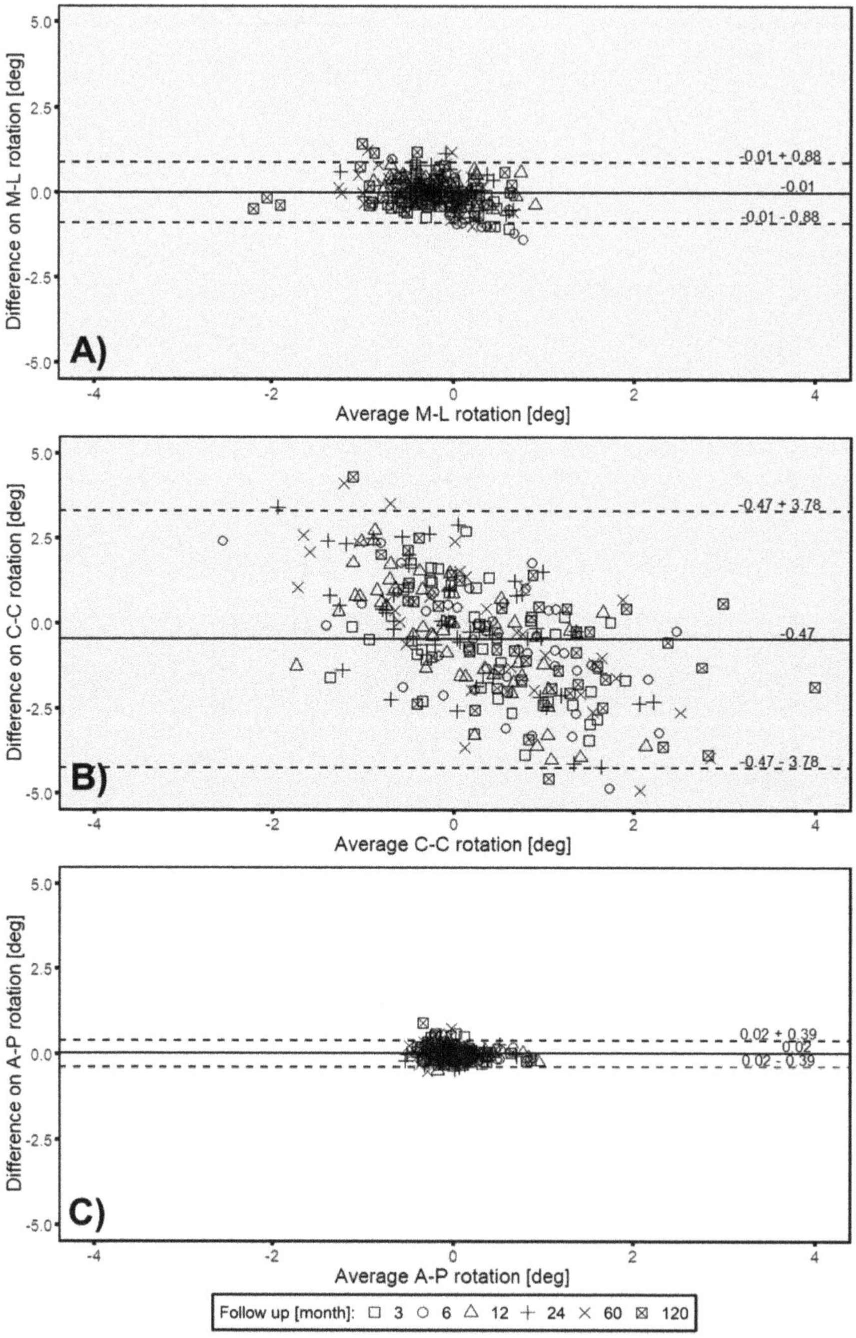

Figure A2. Bland–Altman plots presenting of the calculated average and difference between marker-based RSA and model-based RSA (EGS) for rotational measurements along (**A**) medial–lateral, (**B**) cranial–caudal and (**C**) anterior–posterior direction. Solid line indicates the mean difference with LoA (dashed line) for 3 (square), 6 (circle), 12 (triangle), 24 (plus), 60 (cross) and 120 (square with cross) months follow up.

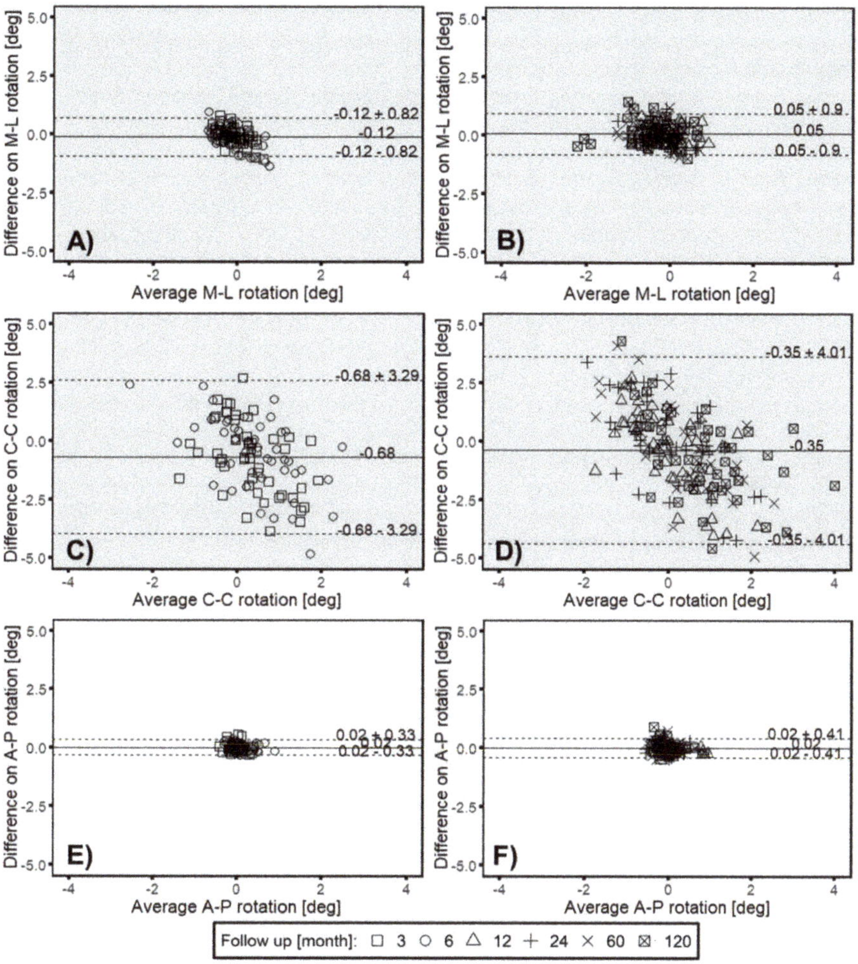

Figure A3. The difference between marker-based RSA and model-based RSA (EGS) showed by Bland & Altman plots on three rotational measurements (M-L: medial–lateral, C-C: cranial–caudal, A-P: anterior–posterior). The results were divided into two groups. (**A**,**C**,**E**): 3 and 6 months; (**B**,**D**,**F**): 12 to 120 months. (Mean difference equal to 0: (**A**): $p < 0.05$; (**B**): $p = 0.12$; (**C**): $p < 0.05$; (**D**): $p < 0.05$; (**E**): $p = 0.17$); (**F**): $p = 0.19$.

References

1. Pivec, R.; Johnson, A.J.; Mears, S.C.; Mont, M.A. Hip arthroplasty. *Lancet* **2012**, *380*, 1768–1777. [CrossRef]
2. Sundfeldt, M.; V Carlsson, L.; B Johansson, C.; Thomsen, P.; Gretzer, C. Aseptic loosening, not only a question of wear: A review of different theories. *Acta Orthop.* **2006**, *77*, 177–197. [CrossRef] [PubMed]
3. MacInnes, S.J.; Gordon, A.; Wilkinson, J.M. Risk factors for aseptic loosening following total hip arthroplasty. In *Recent Advances in Arthroplasty*, 1st ed.; Fokter, S., Ed.; InTech: Rijeka, Croatia, 2012; pp. 275–294. [CrossRef]
4. Valstar, E.R.; Gill, R.H. Radiostereometric analysis in orthopaedic surgery: Editorial comment. *Clin. Orthop. Relat. Res.* **2006**, *448*, 2. [CrossRef] [PubMed]
5. Kaptein, B.; Valstar, E.; Stoel, B.; Rozing, P.; Reiber, J. A new model-based RSA method validated using CAD models and models from reversed engineering. *J. Biomech.* **2003**, *36*, 873–882. [CrossRef]

6. Hurschler, C.; Seehaus, F.; Emmerich, J.; Kaptein, B.L.; Windhagen, H. Comparison of the model-based and marker-based roentgen stereophotogrammetry methods in a typical clinical setting. *J. Arthroplast.* **2009**, *24*, 594–606. [CrossRef] [PubMed]
7. Seehaus, F.; Emmerich, J.; Kaptein, B.L.; Windhagen, H.; Hurschler, C. Experimental analysis of Model-Based Roentgen Stereophotogrammetric Analysis (MBRSA) on four typical prosthesis components. *J. Biomech. Eng.* **2009**, *131*, 041004. [CrossRef]
8. Valstar, E.R.; de Jong, F.W.; Vrooman, H.A.; Rozing, P.M.; Reiber, J.H. Model-based Roentgen stereophotogrammetry of orthopaedic implants. *J. Biomech.* **2001**, *34*, 715–722. [CrossRef]
9. Collier, J.P.; Surprenant, V.A.; Jensen, R.E.; Mayor, M.B.; Surprenant, H.P. Corrosion between the components of modular femoral hip prostheses. *J. Bone Jt. Surg. Br.* **1992**, *74*, 511–517. [CrossRef]
10. Koper, M.; Mathijssen, N.; Witt, F.; Morlock, M.; Vehmeijer, S. Severe Wear and Pseudotumor Formation Due to Taper Mismatch in a Total Hip Arthroplasty: A Case Report. *JBJS Case Connect.* **2015**, *5*, e29. [CrossRef]
11. Morlock, M.; Bünte, D.; Gührs, J.; Bishop, N. Corrosion of the head-stem taper junction—Are we on the verge of an epidemic? *HSS J.* **2017**, *13*, 42–49. [CrossRef]
12. Jennings, J.M.; Dennis, D.A.; Yang, C.C. Corrosion of the head-neck junction after total hip arthroplasty. *J. Am. Acad. Orthop. Surg.* **2016**, *24*, 349–356. [CrossRef]
13. Pastides, P.S.; Dodd, M.; Sarraf, K.M.; Willis-Owen, C.A. Trunnionosis: A pain in the neck. *World J. Orthop.* **2013**, *4*, 161–166. [CrossRef] [PubMed]
14. Sultan, A.A.; Cantrell, W.A.; Khlopas, A.; Berger, R.J.; Sodhi, N.; Molloy, R.M.; Krebs, V.E.; Mont, M.A. Evidence-based management of trunnionosis in metal-on-polyethylene total hip arthroplasty: A systematic review. *J. Arthroplast.* **2018**, *33*, 3343–3353. [CrossRef] [PubMed]
15. Mueller, U.; Braun, S.; Schroeder, S.; Sonntag, R.; Kretzer, J.P. Same Same but Different? 12/14 Stem and Head Tapers in Total Hip Arthroplasty. *J. Arthroplast.* **2017**, *32*, 3191–3199. [CrossRef]
16. Plummer, D.R.; Berger, R.A.; Paprosky, W.G.; Sporer, S.M.; Jacobs, J.J.; Della Valle, C.J. Diagnosis and management of adverse local tissue reactions secondary to corrosion at the head-neck junction in patients with metal on polyethylene bearings. *J. Arthroplast.* **2016**, *31*, 264–268. [CrossRef] [PubMed]
17. Lash, N.; Whitehouse, M.; Greidanus, N.; Garbuz, D.; Masri, B.; Duncan, C. Delayed dislocation following metal-on-polyethylene arthroplasty of the hip due to 'silent'trunnion corrosion. *Bone Jt. J.* **2016**, *98*, 187–193. [CrossRef] [PubMed]
18. McGrory, B.J.; MacKenzie, J.; Babikian, G. A high prevalence of corrosion at the head–neck taper with contemporary Zimmer non-cemented femoral hip components. *J. Arthroplast.* **2015**, *30*, 1265–1268. [CrossRef]
19. Hussey, D.K.; McGrory, B.J. Ten-year cross-sectional study of mechanically assisted crevice corrosion in 1352 consecutive patients with metal-on-polyethylene total hip arthroplasty. *J. Arthroplast.* **2017**, *32*, 2546–2551. [CrossRef]
20. Nawabi, D.H.; Gold, S.; Lyman, S.; Fields, K.; Padgett, D.E.; Potter, H.G. MRI predicts ALVAL and tissue damage in metal-on-metal hip arthroplasty. *Clin. Orthop. Relat. Res.* **2014**, *472*, 471–481. [CrossRef]
21. Nishii, T.; Sakai, T.; Takao, M.; Yoshikawa, H.; Sugano, N. Is ultrasound screening reliable for adverse local tissue reaction after hip arthroplasty? *J. Arthroplast.* **2014**, *29*, 2239–2244. [CrossRef]
22. Sesselmann, S.; Hong, Y.; Schlemmer, F.; Wiendieck, K.; Söder, S.; Hussnaetter, I.; Müller, L.A.; Forst, R.; Wierer, T. Migration measurement of the cemented Lubinus SP II hip stem–a 10-year follow-up using radiostereometric analysis. *Biomed. Tech.* **2017**, *62*, 271–278. [CrossRef] [PubMed]
23. RSAcore. *RSAcore User Manual Model-Based RSA 4.1*; RSAcore: Leiden, The Netherlands, 2015; Volume Version 1.1.
24. Valstar, E.R.; Gill, R.; Ryd, L.; Flivik, G.; Borlin, N.; Karrholm, J. Guidelines for standardization of radiostereometry (RSA) of implants. *Acta Orthop.* **2005**, *76*, 563–572. [CrossRef] [PubMed]
25. Bland, J.M.; Altman, D.G. Statistical methods for assessing agreement between two methods of clinical measurement. *Lancet* **1986**, *1*, 307–310. [CrossRef]
26. Karrholm, J.; Borssen, B.; Lowenhielm, G.; Snorrason, F. Does early micromotion of femoral stem prostheses matter? 4-7-year stereoradiographic follow-up of 84 cemented prostheses. *J. Bone Jt. Surg. Br.* **1994**, *76*, 912–917. [CrossRef]
27. Ryd, L.; Albrektsson, B.; Carlsson, L.; Dansgard, F.; Herberts, P.; Lindstrand, A.; Regner, L.; Toksvig-Larsen, S. Roentgen stereophotogrammetric analysis as a predictor of mechanical loosening of knee prostheses. *J. Bone Jt. Surg. Br.* **1995**, *77*, 377–383. [CrossRef]

28. Pearson, K. Note on regression and inheritance in the case of two parents. *Proc. R. Soc. Lond.* **1895**, *58*, 240–242.
29. Fisher, R.A. Frequency distribution of the values of the correlation coefficient in samples from an indefinitely large population. *Biometrika* **1915**, *10*, 507–521. [CrossRef]
30. Morlock, M.M.; Dickinson, E.C.; Gunther, K.P.; Bunte, D.; Polster, V. Head Taper Corrosion Causing Head Bottoming Out and Consecutive Gross Stem Taper Failure in Total Hip Arthroplasty. *J. Arthroplast.* **2018**, *33*, 3581–3590. [CrossRef]
31. Pennock, A.T.; Schmidt, A.H.; Bourgeault, C.A. Morse-type tapers: Factors that may influence taper strength during total hip arthroplasty. *J. Arthroplast.* **2002**, *17*, 773–778. [CrossRef]
32. Witt, F.; Gührs, J.; Morlock, M.M.; Bishop, N.E. Quantification of the contact area at the head-stem taper interface of modular hip prostheses. *PLoS ONE* **2015**, *10*, e0135517. [CrossRef]
33. Osman, K.; Panagiotidou, A.; Khan, M.; Blunn, G.; Haddad, F. Corrosion at the head-neck interface of current designs of modular femoral components: Essential questions and answers relating to corrosion in modular head–neck junctions. *Bone Jt. J.* **2016**, *98*, 579–584. [CrossRef] [PubMed]
34. McTighe, T.; Brazil, D.; Keppler, L.; Keggi, J.; McPherson, E. Metallic Modular Taper Junctions in Total Hip Arthroplasty. *Reconstr. Rev.* **2015**, *5*, 29–42. [CrossRef]
35. Panagiotidou, A.; Meswania, J.; Osman, K.; Bolland, B.; Latham, J.; Skinner, J.; Haddad, F.; Hart, A.; Blunn, G. The effect of frictional torque and bending moment on corrosion at the taper interface: An in vitro study. *Bone Jt. J.* **2015**, *97*, 463–472. [CrossRef] [PubMed]
36. Natu, S.; Sidaginamale, R.P.; Gandhi, J.; Langton, D.J.; Nargol, A.V. Adverse reactions to metal debris: Histopathological features of periprosthetic soft tissue reactions seen in association with failed metal on metal hip arthroplasties. *J. Clin. Pathol.* **2012**, *65*, 409–418. [CrossRef]
37. Morlock, M.M. The taper disaster-how could it happen? *Hip. Int.* **2015**, *25*, 339–346. [CrossRef] [PubMed]
38. Kurtz, S.M.; Kocagöz, S.B.; Hanzlik, J.A.; Underwood, R.J.; Gilbert, J.L.; MacDonald, D.W.; Lee, G.-C.; Mont, M.A.; Kraay, M.J.; Klein, G.R. Do ceramic femoral heads reduce taper fretting corrosion in hip arthroplasty? A retrieval study. *Clin. Orthop. Relat. Res.* **2013**, *471*, 3270–3282. [CrossRef]
39. Tan, S.C.; Lau, A.C.; Del Balso, C.; Howard, J.L.; Lanting, B.A.; Teeter, M.G. Tribocorrosion: Ceramic and oxidized zirconium vs cobalt-chromium heads in total hip arthroplasty. *J. Arthroplast.* **2016**, *31*, 2064–2071. [CrossRef]
40. Khatod, M.; Cafri, G.; Inacio, M.C.; Schepps, A.L.; Paxton, E.W.; Bini, S.A. Revision total hip arthoplasty: Factors associated with re-revision surgery. *J. Bone Jt. Surg. Am.* **2015**, *97*, 359–366. [CrossRef]
41. Kocagöz, S.B.; Underwood, R.J.; Sivan, S.; Gilbert, J.L.; MacDonald, D.W.; Day, J.S.; Kurtz, S.M. Does taper angle clearance influence fretting and corrosion damage at the head–stem interface? A matched cohort retrieval study. *Semin. Arthroplast.* **2013**, *24*, 246–254. [CrossRef]
42. Del Balso, C.; Teeter, M.; Tan, S.; Lanting, B.; Howard, J. Taperosis: Does head length affect fretting and corrosion in total hip arthroplasty? *Bone Jt. J.* **2015**, *97*, 911–916. [CrossRef]
43. Panagiotidou, A.; Meswania, J.; Hua, J.; Muirhead-Allwood, S.; Hart, A.; Blunn, G. Enhanced Wear and Corrosion in Modular Tapers in Total Hip Replacement Is Associated With the Contact Area and Surface Topography. *J. Orthop. Res.* **2013**, *31*, 2032–2039. [CrossRef] [PubMed]

materials

Article

Retrieval Analysis of Modern Knee Tumor Megaendoprosthesis Shows Considerable Volumetric Metal Wear Generated at the Rotating Hinge

Therese Bormann [1,*], Sebastian Jäger [1], J. Philippe Kretzer [1], Laura Nebel [1], Lucas Clarius [1], Georg Omlor [2], Rudi Bitsch [3] and Burkhard Lehner [2]

1 Laboratory of Biomechanics and Implant Research, Clinic for Orthopedics and Trauma Surgery, Heidelberg University Hospital, 69118 Heidelberg, Germany; Sebastian.Jaeger@med.uni-heidelberg.de (S.J.); philippe.kretzer@med.uni-heidelberg.de (J.P.K.); lauranebel-besigheim@web.de (L.N.); lucas-clarius@t-online.de (L.C.)

2 Clinic for Orthopedics and Trauma Surgery, Heidelberg University Hospital, 69118 Heidelberg, Germany; georg.omlor@med.uni-heidelberg.de (G.O.); burkhard.lehner@med.uni-heidelberg.de (B.L.)

3 ATOS Clinic Heidelberg, 69115 Heidelberg, Germany; rudi.bitsch@atos.de

* Correspondence: therese.bormann@med.uni-heidelberg.de

Received: 28 February 2020; Accepted: 25 March 2020; Published: 26 March 2020

Abstract: Frequently occurring damage, as well as elevated blood metal ion levels, are reported in relation to a tumor and revision system for total knee arthroplasty (TKA), which applies a rotating hinge coupling with a metal-on-metal (MoM) articulation. As the patient collective for this specific system is small, there is no data on wear generated from the couplings. In this study, wear volume and influencing parameters were investigated at 44 retrieved TKAs with MoM couplings. A scoring system rating frequently occurring abrasive wear between 0 (no wear) and 3 (distinct wear) was established. The wear score was correlated to time in vivo, bone resection length, patient weight and polyethylene inlay damage. Volumetric wear was estimated applying coordinate measurements. An elevated wear score of two or higher was found in 43% of cases. The mean wear rate accounted to 7.8 mm^3/year. The main influencing coefficient for the extent of wear is time in vivo. We found a tendency for higher wear scores with higher inlay degradation scores. Patient weight and bone resection length did not impact coupling wear. Assessment of wear damage by a semi-quantitative scoring system has proven to be a reliable option for non-destructive coupling evaluation. The generated wear volume is high.

Keywords: metal wear; retrieval study; metal-on-metal articulation; volumetric wear; megaendoprosthesis; total knee arthroplasty; bone tumor

1. Introduction

Tumors in the distal femur or the proximal tibia often result in the resection of large bone segments. To salvage the limb, the bone and knee joint are usually reconstructed by endoprosthesis systems [1,2]. Modern systems are constructed modularly in order to address different resection lengths and to permit intra-operational flexibility [1]. As together with the bone also soft tissue and ligaments of the leg are removed, a higher degree of constrain in the knee joint is necessary to ensure stability of the joint. For that, tumor and revisions systems usually couple the distal femur and proximal tibia with a hinge or a rotating hinge system. Generally, the outcome of a fully stabilized and hinged prosthesis is worse than that of a less stabilized knee endoprosthesis [3,4]. Revision rates after five years rise from 3.5% for minimally stabilized total knee arthroplasty (TKA) to 8.1% for the coupled systems, with infection being the most common cause [5].

Aside from infections, mechanical complications like aseptic loosening, structural failure and instability are common reasons for a revision [2,6]. The high incidence of mechanical complications is associated with the coupling, as substantial stresses are transmitted via the coupling mechanism [6,7] and—for the hinged systems—these are further transmitted to the implant–bone interface.

As bone tumors often occur between the second and forth decade of life, the patient collective is considerably younger than that of general primary total knee replacement, which has a mean age of 69 years [5]. It has been shown that the probability for a second complication increases after a first complication requiring revision surgery [8]. Bone tumor patients are therefore very likely to have multiple revision surgeries during the course of their life. Not seldom, the affected limb is sacrificed at some point in the patient's life [8,9].

The modular universal tumor and revision system (MUTARS®, Implantcast, Buxtehude, Germany) is a commonly used endoprosthesis system for tumor patients with good functional clinical results [7,9,10]. The femoral and tibial components of this system are joined by a rotating hinge coupling mechanism, sometimes also referred to as bushing. Despite the satisfactory clinical results of this system, structural complications are reported for up to 30% of the patients [7,9–11]. This failure mode includes periprosthetic fracture, breakage of a prosthesis part and wear of the bushing, all requiring a revision operation. The bushing design was changed several times since the introduction of the system in 1995. The latest design uses a metal-on-metal (MoM) coupling mechanism made of cobalt–chromium–molybdenum-alloy. Nevertheless, problems occurring with respect to the coupling mechanism do not seem to be overcome. Bushing wear has still been reported, as well as elevated blood cobalt and chromium levels related to the MoM coupling [9,12]. Even though exchange of the bushing itself is a standard procedure not requiring other parts of the system to be replaced, it affects the patient's life as an additional surgery. Furthermore, the generated metal particles and ions are released into the human body, where they can cause adverse biological reactions that, in turn, can lead to clinical problems such as soft-tissue necrosis, pseudotumors and particle-induced implant loosening [13–15].

In our in-house retrieval registry, we regularly observed (obvious) coupling failures like bushing fracture. Apart from that we noticed marks of abrasive wear on retrieved MUTARS® total knee replacements. As there is no data on wear generated at coupling mechanisms in constrained total knee arthroplasty, we aimed to answer the following research questions in this retrospective study: (1) Is the degree of wear at MoM couplings correlated to implant or patient specific factors? (2) How much wear is actually generated, and can it be quantified?

2. Materials and Methods

2.1. Retrievals

We analyzed 44 retrieved MUTARS® prostheses (Implantcast, Buxtehude, Germany) with a metal-on-metal coupling. Retrievals are from our in-house retrieval registry, which is approved by an ethical committee. Implantation of the 44 prostheses was carried out between 1997 and 2018, with explantation between 2001 and 2018. The mean time in vivo was 31.0 ± 38.7 months (median 17.6 months, range 0.4–172.1 months). The coupling is made of cobalt–chromium–molybdenum-alloy (CoCrMo, as specified in DIN ISO 5832-4 [16]), and consists of a sleeve, inside which a piston with a hemispherical head articulates. Among the retrieved couplings we identified four different designs, as displayed in Figure 1a. In the oldest (design A), the piston is fixed within the tibia plateau by a thread. All following designs connect the piston by a stuck bolt to the tibia plateau. Design B is characterized by a bore in the piston stem. Designs C and D are characterized by a bore on top of the hemispherical piston head instead of the one in the stem. Designs C and D further differ in the dimensions of sleeve and piston. In the latest design from our cohort, design D, the piston stem and head are thicker, and the sleeve is shorter as for the designs A, B and C.

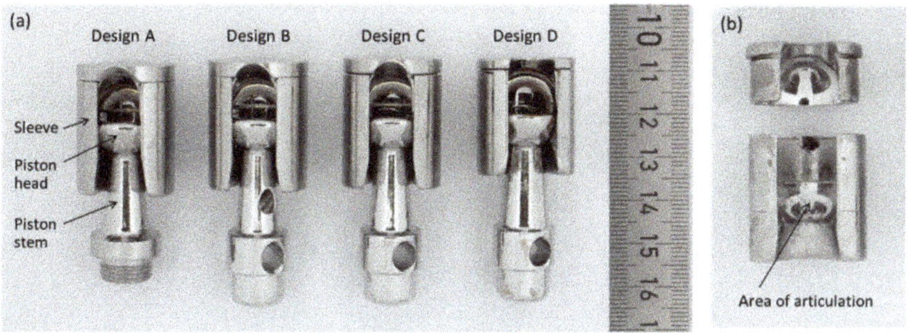

Figure 1. (**a**) Metal-on-metal (MoM) coupling mechanisms consisting of piston and sleeve. The four design variations from our cohort are displayed. Design A refers to the oldest modification, while design D is the most recent. (**b**) The sleeve after separation of the lid for quantitative wear analysis. The arrow indicates the articulating counter face, which was investigated by a coordinate measuring machine (CMM).

2.2. Semi-Quantitative Wear Analysis by Scoring System

Failure modes of these prostheses were classified by the Henderson classification. As wear of the coupling occurs independently from the actual failure mechanism of the device, and appears to the naked eye in almost all retrieved samples, we developed a semi-quantitative scoring system to rate coupling wear. The scoring system is based on the Hood and Goldberg scoring systems, which both are well established visual methods to assess damage of retrieved implant parts [17,18]. Arnholt et al. applied a similar scoring system to assess damage of CoCr femoral condyles [19]. As both parts are prone to wear, we rated them individually. Rating was accomplished on a scale between 0 (no wear) and 3 (distinct wear), and the criteria for the point assignment and examples are displayed in Table 1. For the cylindrical sleeve, the external surface was divided into four quadrants, which were rated individually and subsequently averaged. The piston was divided in two zones, the hemispherical head and the stem. Both zones were rated individually from 0 to 3, and subsequently averaged. The score for the whole coupling was determined by averaging the sleeve and piston scores. Scoring was carried out by two independent investigators in order to verify the assessment of wear by the developed scoring scale. All data presented refer to the evaluation of investigator 1.

Table 1. Criteria for the determination of the coupling wear score.

Score	Sleeve (Each Quadrant)	Piston Head	Piston Stem
0	No damage	No damage	No damage
1	Scratches and/or polishing < 10% of the surface	Scratches visible	Scratches visible
2	Polishing 10–50% of the surface	Planar abrasion noticeable	Planar abrasion < 10% of the surface
3	Polishing > 50% of the surface	Planar abrasion with distinct edge on the surface	Planar abrasion > 10% of the surface

Couplings were ultrasonically cleaned in two steps prior to score determination. In the first step, cleaning was carried out at 60 °C in a detergent-containing solution for one hour. Subsequently, couplings were immersed in Ethanol for 10 min.

Pre-operative radiographs for metering the bone resection lengths of both the tibia and femur, respectively, were available from 35 patients. In case of the resection of the distal femur and proximal tibia, both values were totalized for further analysis.

Damage of polyethylene inlays was assessed on 40 inlays applying a modified Hood score [18]. We rated five damage mechanisms, i.e., delamination, pitting, scratching, burnishing and surface deformation at a scale between 0 (no damage) and 3 (distinct damage) for seven zones of the inlay, as shown in Figure 2. Achieved degradation scores were summed up, which results in a maximal possible score of 105. Determination of the modified Hood score was done by two investigators. The mean difference of the degradation score of the PE-inlays between the two investigators referred to 4.78 ± 3.2. The data presented refers to the evaluation of investigator 1.

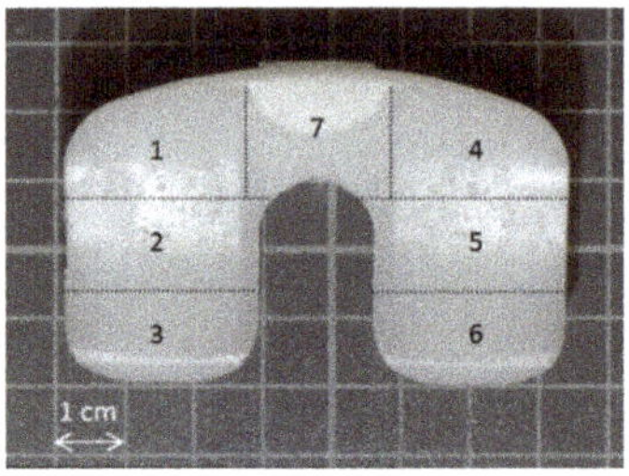

Figure 2. Polyethylene inlay with typical traces of wear. For the determination of the Hood score, the inlay was divided into seven zones.

2.3. *Quantitative Wear Analysis by Coordinate Measurements*

In order to estimate the material loss over the period of implantation, we assessed the geometry of the articulating surfaces of the coupling mechanism by a coordinate measuring machine (CMM; MS222, Mahr, Göttingen, Germany). For that, the sleeve and piston were separated by cutting the welding seam by which the lid was connected to the sleeve (see Figure 1b). Subsequently, the piston head was cut from the piston stem. The articulating faces of the semispherical piston head and the opposing counter face (Figure 1b) were scanned point wise in spherical coordinates with angle steps of 4° and 6° for theta (θ, polar angle) and phi (φ, azimuthal angle), respectively, as demonstrated for a cut off piston head in Figure 3a. Point clouds were further processed by Imageware^{TM} (UGS Corporation, Plano, TX, USA). As the original geometry of the measured parts is unknown, we did the analysis applying a best-case scenario. For this, we considered as much as possible of the spherical cap as un-worn and reasonable, and derived the referencing surface by fitting a sphere to these areas. Normal distances between each point and the fitted surface were derived (illustrated in Figure 3b,c) and further processed to determine the volumetric material loss. The latter was done using Matlab (Version 7.10.0, The MathWorks, Inc., Natick, MA, USA).

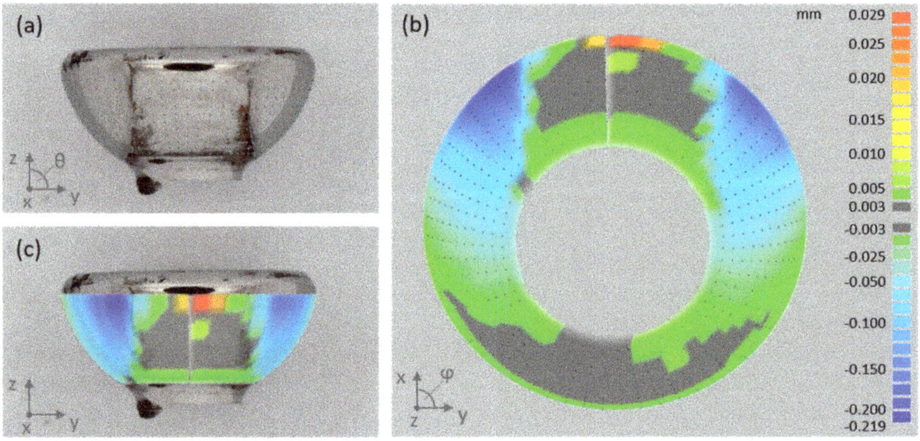

Figure 3. (**a**) Piston head (cut from the piston stem) with point cloud measured by CMM. (**b**) Color plot of linear wear, i.e., normal deviation between measured data and referencing surface. Gray areas refer to linear wear < 3 μm, which can be considered as unworn. (**c**) Overlay of linear wear plot and piston head image illustrates the unworn and most heavily abraded areas, respectively.

2.4. Statistical Analysis

Correlations of wear score and volumetric material loss, respectively, and time in vivo, bone resections length, patient weight and inlay degradation score, were assessed by Spearman's rank order-correlation. The correlation of derived coupling wear scores from two independent investigators was assessed by the Pearson correlation coefficient. Differences in mean values ± standard deviation were tested for statistical significance by student's t-test on unpaired samples with a level of significance of $p \leq 0.05$. Statistical analyses were carried out applying SPSS software (IBM Corp., Version 22.0., Armonk, NY, USA).

3. Results

3.1. Failure Modes of Retrieved Implants

Figure 4 represents the classification of the reasons for revision according to Henderson et al. [6]. 52% of the implants were revised because of infection (Henderson type IV). Aseptic loosening (Henderson type II) caused revision in 27% of the cases. Structural failure (Henderson type III), like breakage or joint instability due to worn couplings, occurred in 18% of the cases, while 2% were revised because of soft tissue complications (Henderson type I). No revisions were necessary because of local recurrences (Henderson type V). We found one case of piston breakage and four cases of sleeve breakage (see Figure 5a) in our collective. Three out of four explants with broken sleeves were revised for aseptic loosening. In one case, the piston was bent (see Figure 5b).

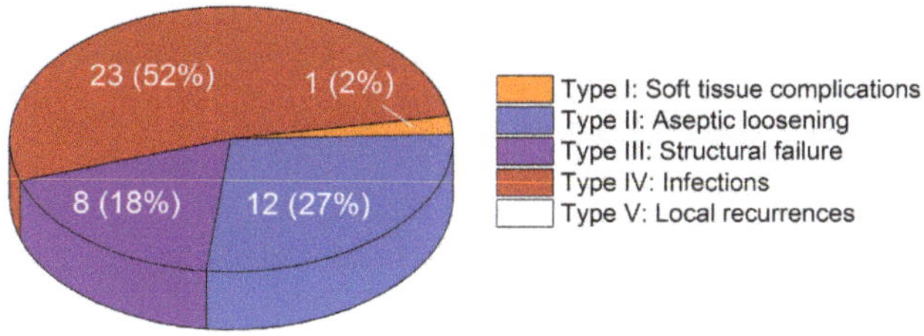

Figure 4. Distribution of failure modes according to the Henderson classification in our collective of 44 retrievals.

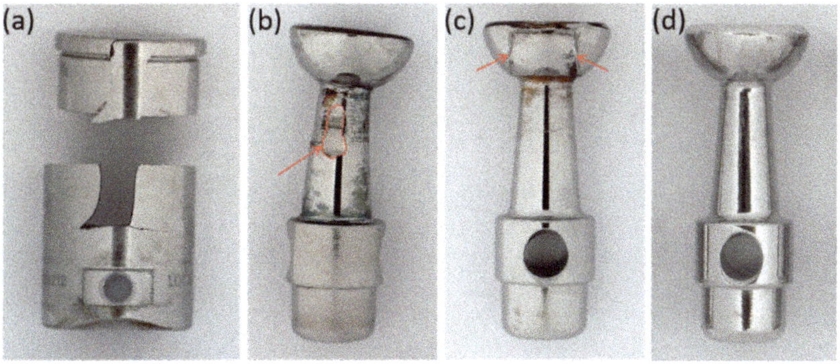

Figure 5. Damage occurring at couplings: (**a**) broken coupling sleeve, (**b**) bent piston with area of abrasive wear at piston stem (encirculated area, arrow), (**c**) piston head with zones of planar abrasion, the arrows mark the created edge between the worn and the unworn zones, (**d**) piston without noticeable planar abrasion.

3.2. Evaluation of Wear by Damage Score

Almost all retrieved MUTARS® coupling mechanisms exhibit abrasive wear to some extent. The outside of the sleeves shows abrasion in the form of highly polished areas. Wear at the piston appears in the form of brush marks at the stem and head, as demonstrated in Figure 5. At the heads, abraded zones can be easily distinguished from unworn areas, as enough material is lost in the articulating zone to create an edge that separates the worn from the unworn zones at the originally semispherical surface (Figure 5c). According to the evaluation by damage score, 43% (n = 19) of the couplings were rated with a score of two or higher; i.e., they showed clear signs of planar abrasion. The mean wear score was 1.79 ± 0.62. The score increased with time in vivo (see Figure 6), with a Spearman's rank-order correlation of $r_s = 0.65$, $p < 0.01$.

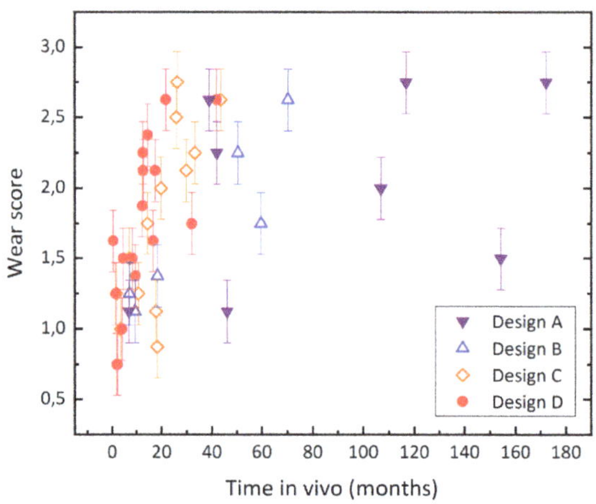

Figure 6. Coupling wear score increases with time in vivo.

We analyzed the correlation between the wear score of the coupling and patient weight, bone resection length and inlay degradation (scatter plots are displayed in Figure 7). Patient weight or bone resection length did not correlate with the damage score of the coupling ($r_s = -0.02$, $p = 0.9$ and $r_s = -0.32$, $p = 0.06$, respectively). The coupling wear showed a trend to increase with the degradation of the polyethylene inlay, as can be seen in Figure 7c; the Spearman's rank-order correlation coefficient r_s amounted to 0.49 ($p < 0.01$).

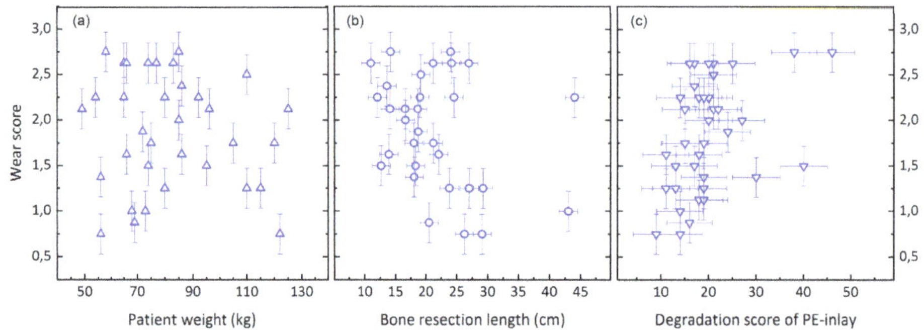

Figure 7. Scatter plots of wear score over (**a**) patient weight, (**b**) bone resection length and (**c**) degradation score of the polyethylene inlays.

Assessment of coupling wear by the two investigators showed a high correlation (Pearson r = 0.89). The mean difference in the coupling wear score of both investigators referred to 0.22 ± 0.19, while the median difference referred to 0.13, the minimum to 0, and the maximum to 0.75.

3.3. *Quantification of Wear by Coordinate Measurements*

Quantitative wear analysis of 20 retrievals estimated the volumetric material loss of the articulating surfaces of the coupling to range from 0.5 mm^3 to 30 mm^3. The total mean value accounted to 7.9 ± 8.1 mm^3. The mean wear rate was 7.8 ± 8.6 mm^3/year. Volumetric wear increased with time in vivo (Figure 8a), with a Spearman's-rank order correlation of $r_s = 0.69$, $p < 0.01$. Wear score and volumetric material loss are correlated, as illustrated in Figure 8b ($r_s = 0.867$, $p < 0.01$).

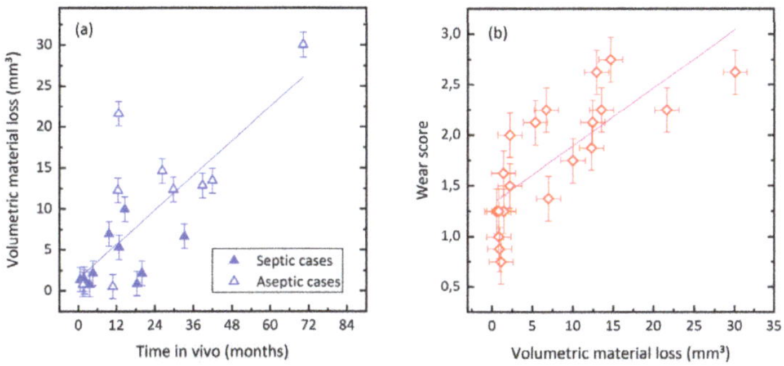

Figure 8. (**a**) Volumetric material loss increases with time in vivo. (**b**) Relationship between wear score and volumetric material loss.

Cases retrieved for joint infection showed 3.5 ± 3.2 mm^3 of cumulated material loss, while the material loss of aseptic cases amounted to 13.2 ± 9.2 mm^3. This can be attributed to the significantly shorter in-vivo periods of septic revisions, as shown in Table 2. Wear rates, i.e., volumetric material loss per time in vivo of both groups, are similar.

Table 2. Total wear volume, time in vivo and wear rates of MoM couplings, subclassified into septic and aseptic revisions.

	Number of Cases	Time in Vivo (months)	Total Wear Volume (mm^3)	Wear Rate (mm^3/year)
	n	mean (SD/range)	mean (SD/range)	mean (SD/range)
Septic revision	11	10.9 (10.1/0.4–33.2)	3.5 (3.2/0.8–10.0)	8.2 (10.7/0.6–39.1)
Aseptic revision	9	34.3 (30.0/1.4–70.2)	13.2 (9.2/0.5–30.0)	6.7 (6.1/0.6–20.6)
p	–	<0.05	<0.05	0.7
All cases	20	19.4 (18/0.4–70.2)	7.9 (8.1/0.5–30.0)	7.8 (8.6/0.6–39.1)

4. Discussion

The study investigated damage at a MoM rotating hinge mechanism due to abrasive wear with two different techniques: A semi-quantitative examination by which a coupling wear score is visually determined, and a CMM-based analysis that estimates actual volumetric wear. Quantitative measures on wear volume showed a relevant mean wear rate of 7.8 mm^3/year. Both applied methods showed a clear relation between the extent of wear at the coupling and the time of in vivo service.

From Figure 6 it appears that the older designs (A and B) exhibited a better in vivo performance, because they show the longest periods of implantation, and the inclination of the wear score over in vivo service time seems to increase in the order of the design A, B, C and D. The shorter in vivo service time is related to the more recent implantations of the newer designs. Design D is implanted since 2015, design C was implanted between 2011 and 2014, design B between 2010 and 2012, and the retrievals with design A were implanted between 1997 and 2011. The available data, however, does not allow us to compare the long-term performance of designs B, C and D with the performance of design A, because there are no retrievals with coupling designs B, C or D that were implanted for more than 80 months in the investigated retrieval collective. Also, we do not know the number of well working implants that are currently implanted. A possible explanation for the difference in inclination could be that the highest wear rate occurs in the beginning of the in-vivo service. However, the quantitative analysis rather showed a linear relation, while the semi-quantitative correlation could also be interpreted as a logarithmic relation. We believe that this is mainly related to the fact that the semi-quantitative evaluation grades the extent of damage in finer steps in the beginning of the abrasive process. Brush marks at the piston head can be seen and were rated with a score of 1, while the actual volumetric material loss determined by the CMM evaluation can be still close to zero. As soon as the abrasive process led to the formation of the distinct edge between worn and unworn areas, the piston heads were rated with a score of 3, independent from the severity of the created edge.

The semi-quantitative analysis also rates those zones of the coupling that cannot be assessed by coordinate measurements, such as the outside of the sleeve. Here, abrasion is noticed visually in the form of areas with a highly polished appearance. Another advantage of the semi-quantitative wear analysis is its non-destructiveness. Coordinate measurements as quantitative alternative require dissociation of lid and bushing of the sleeve and of the piston head from stem.

It has been shown in simulator and retrieval studies, that wear and corrosion of metallic parts does occur, even in conventional, i.e., minimally stabilized, total knee arthroplasty. Considerable levels of metal ions are released during cyclic wear simulation [20]. Clinically, this leads to elevated metal ion levels in the patient's blood [21–23]. In addition, the released metal ions accumulate in tissue surrounding the joint, and can lead to adverse local tissue reactions like metallosis [19,24]. This in turn can lead to aseptic loosening, which is always followed by a revision operation. However, quantitative data on metal wear from total knee arthroplasty is rare. Kretzer et al. report the total release of Co- Cr- and Mo-ions to be about 2.5 mg after 5 million cycles [20]. The wear rate of 7.8 mm^3/year determined for the MoM couplings would correspond to a metal release into the human body of about 65 mg/year, which would be at least about 50 times the metal release from the articulating parts. Generally, clinical problems like adverse reactions to metallic debris (ARMD) or pseudotumor formation related to wear of CoCr-articulating surfaces are well known from MoM total hip arthroplasty (THA) [15,16,25]. Volumetric wear of retrieved MoM THAs has been examined in several studies. Lord et al. analyzed 22 retrieved cup-head MoM pairs, whereof the majority of patients had been revised for ARMD. The median of the determined wear rates was 9.4 mm^3/year [25]. Glyn-Jones et al. found a significantly increased mean volumetric wear rate of 3.3 and 2.5 mm^3/year for femoral heads and acetabular cups, respectively, in retrievals revised for pseudotumors compared to a no pseudotumor control group [26].

Grammatopoulos et al. confirmed these findings, which determined the total mean wear rate in the pseudotumor group to be 5.5 mm^3/year, whereas the control group exhibited 0.4 mm^3/year [16]. Gascoyne et al. reported a total median rate for metal loss from joint articulation and head-stem taper of 1.5 mm^3/year in a retrieval study on 24 MoM hip replacements of a single design, with 16 of the 24 patients revised for MoM-related reasons, such as adverse reaction to metallic debris (ARMD) or high blood metal ion levels. The mean wear rate from articulating parts only referred to 4.6 mm^3/year ranging from 0.01 to 66.4 mm^3/year [27]. The mean amount of metallic wear of about 7.8 mm^3/year created at MoM couplings in the MUTARS prosthesis is thus in the same order of magnitude than wear created from MoM hip TEPs that have been revised for metal-related causes. It is very likely that the debris created in the rotating-hinge system leads to similar metal-related problems, as it has been reported for failing MoM artificial hip joints. This hypothesis is supported by several studies that show considerably elevated metal ion levels related to hinged TKA: Laitinen et al. reported elevated Co- and Cr-ion levels of > 5 ppb in whole-blood samples of 19/22 patients with a MUTARS® prosthesis applying a MoM coupling. In the control group with a MUTARS® prosthesis using a metal-on-polyetheretherketone (PEEK) coupling, only 1 of 12 patients showed a similarly elevated

Co-ion level [12]. Klasan et al. found > 5 ppb Co- and Cr-ions in the blood serum of 16/23 patients with a MoM hinge knee design [28]. Friesenbichler et al. showed elevated Co- and Cr-ion levels in serum blood for two hinged TKA systems (different from the MUTARS®-prosthesis), but not for a standard rotating hinge TKA [29].

A difficulty in the determination of material loss from retrievals is that the original geometry of the worn parts is unknown. Determining volumetric and linear wear therefore always requires estimation of the original geometry by a referencing surface or volume. The MUTARS® rotating hinges are based on a ball joint, so the articulating faces can be represented by a spherical cap. We chose the referencing spherical face in such a way that as much of the measured articulation area as possible could be considered unworn, so we analyzed the best possible case for the coupling in terms of wear. In this way, resulting wear plots usually showed two unworn areas: The unworn, dorsally-oriented face not in contact with the bushing and the opposing ventrally-oriented face. As the latter actually is part of the joint articulation, it is very likely that it experienced abrasive wear to some extent, too. In addition, we determined the quantitative wear volume just from the articulating surfaces. Visually, we detected abrasion also on other parts of the hinge, such as the piston stem, the lid of the sleeve, the outside of the sleeve and on the rims of the openings in the sleeve. The calculated wear volumes therefore represent rather a minimum level of occurring metal abrasion.

In order to estimate how many revisions were caused by worn or broken bushings, we had a closer look on the reasons for revision: Seven prostheses out of 8 revised for 'structural failure' were explicitly revised for mechanical complications related to the coupling, such as instability due to bushing wear and material breakage. From the 12 cases that were revised for aseptic loosening, three cases showed broken coupling sleeves, and in four cases, metallosis was described in the operation report. Two of the metallosis-annotated cases coincided with the broken coupling sleeves. In the case of aseptic loosening with a diagnosed metallosis, we attributed implant loosening to the generated metallic wear. This would sum up to 12 out of 44 revisions (27%), which were probably related to the wear of the rotating hinge mechanism. In addition, five retrievals exhibited bushing or stem breakage, which results in a fracture rate of 11% in the investigated retrieval collective. About half of the investigated retrievals were revised due to an infection, which we did not consider to be related to metal wear. However, there are studies that suggest that metallic debris might be able to promote joint infections, as the tissue damaged by metallic wear products is an optimal environment for bacterial growth [30,31]. It should be noted that after a bone tumor resection at the proximal tibia or distal femur, which requires knee reconstruction by an endoprosthesis, musculature and soft tissue around the joint must be removed, and are therefore missing to stabilize the knee. The tumor endoprostheses systems are therefore subjected to much higher loads than the less stabilized TKA systems. The complications with the rotating hinge mechanism therefore have their reason at least partially in the underlying disease. However, an improved coupling design might be able to diminish occurring coupling wear and damage to some extent.

The study has several limitations. First of all, it is a retrieval study, meaning that we draw our conclusions only based on explants that clinically failed without any control group. Wear volume was estimated on the basis of a fitted referencing spherical surface, as the actual geometry of the measured parts is unknown. In addition, the spherical fit was done on rather small areas of the measured spherical caps. However, the total wear volume we calculated can be considered a minimum value, as we analyzed the measured parts under the estimation of a best-case scenario. Evaluation of coupling wear by damage score instead of quantitative analysis is subjective, and thus to some extent observer-dependent. Also, it does not scale linearly with the amount of abrasive wear.

5. Conclusions

This study investigated damage on the rotating hinge of a tumor and revision system (MUTARS®) for total knee replacements. Pronounced wear on the coupling mechanism occurs frequently, and was assessed by two different methods. Applying a scoring system has proven to be a reliable, non-destructive option to rate the degree of wear. It was shown that the extent of damage increases with time in vivo, but not with patient weight or bone resection length. Volumetric wear was estimated using coordinate measurements, and revealed high wear rates of about 8 mm^3/year.

Author Contributions: Conceptualization, S.J.; methodology, T.B., S.J. and J.P.K.; software, T.B., J.P.K.; validation, T.B.; formal analysis, T.B.; investigation, L.N., L.C. and T.B.; resources, S.J., J.P.K., B.L. and G.O.; data curation, T.B.; writing—original draft preparation, T.B.; writing—review and editing, S.J., J.P.K., G.O., R.B. and B.L.; visualization, T.B.; supervision, B.L., J.P.K. and G.O.; project administration, S.J.; All authors have read and agreed to the published version of the manuscript.

Funding: This research received no external funding.

Acknowledgments: We acknowledge financial support by the Baden-Württemberg Ministry of Science, Research and the Arts and by Ruprecht-Karls-Universität Heidelberg.

Conflicts of Interest: The authors declare no conflict of interest related to this study.

References

1. Gosheger, G.M.D.; Gebert, C.M.D.; Ahrens, H.M.D.; Streitbuerger, A.M.D.; Winkelmann, W.M.D.; Hardes, J.M.D. Endoprosthetic reconstruction in 250 patients with sarcoma. *Clin. Orthop. Relat. Rese.* **2006**, *450*, 164–171. [CrossRef]
2. Myers, G.J.C.; Abudu, A.T.; Carter, S.R.; Tillman, R.M.; Grimer, R.J. Endoprosthetic replacement of the distal femur for bone tumours. *J. Bone Jt. Surg. Br. Vol.* **2007**, *89*, 521–526. [CrossRef] [PubMed]
3. Heisel, C.; Kinkel, S.; Bernd, L.; Ewerbeck, V. Megaprostheses for the treatment of malignantbone tumours of the lower limbs. *Int. Orthop.* **2006**, *30*, 452–457. [CrossRef]
4. Toepfer, A.; Harrasser, N.; Schwarz, P.-R.; Pohlig, F.; Lenze, U.; Mühlhofer, H.M.L.; Gerdesmeyer, L.; von Eisenhart-Rothe, R.; Suren, C. Distal femoral replacement with the MML system: A single center experience with an average follow-up of 86 months. *BMC Musculoskelet. Disord.* **2017**, *18*, 206. [CrossRef] [PubMed]
5. Australian Orthopaedic Association National Joint Replacement Registry (AOANJRR). *Hip, Knee & Shoulder Arthroplasty: 2019 Annual Report*; AOANJRR: Adelaide, Australia, 2019.
6. Henderson, E.R.; Groundland, J.S.; Pala, E.; Dennis, J.A.; Wooten, R.; Cheong, D.; Windhager, R.; Kotz, R.I.; Mercuri, M.; Funovics, P.T.; et al. Failure mode classification for tumor endoprostheses: Retrospective review of five institutions and a literature review. *JBJS* **2011**, *93*, 418–429. [CrossRef] [PubMed]
7. Kinkel, S.; Lehner, B.; Kleinhans, J.A.; Jakubowitz, E.; Ewerbeck, V.; Heisel, C. Medium to long-term results after reconstruction of bone defects at the knee with tumor endoprostheses. *J. Surg. Oncol.* **2010**, *101*, 166–169. [CrossRef] [PubMed]
8. Theil, C.; Röder, J.; Gosheger, G.; Deventer, N.; Dieckmann, R.; Schorn, D.; Hardes, J.; Andreou, D. What is the likelihood that tumor endoprostheses will experience a second complication after first revision in patients with primary malignant bone tumors and what are potential risk factors? *Clin. Orthop. Relat. Res.* **2019**, *477*, 2705–2714. [CrossRef]
9. Hardes, J.; Henrichs, M.-P.; Gosheger, G.; Guder, W.; Nottrott, M.; Andreou, D.; Bormann, E.; Eveslage, M.; Hauschild, G.; Streitbürger, A. Tumour endoprosthesis replacement in the proximal tibia after intra-articular knee resection in patients with sarcoma and recurrent giant cell tumour. *Int. Orthop.* **2018**, *42*, 2475–2481. [CrossRef]
10. Bus, M.P.A.; van de Sande, M.A.J.; Fiocco, M.; Schaap, G.R.; Bramer, J.A.M.; Dijkstra, P.D.S. What are the long-term results of MUTARS® modular endoprostheses for reconstruction of tumor resection of the distal femur and proximal tibia? *Clin. Orthop. Relat. Res.* **2017**, *475*, 708–718. [CrossRef]
11. Hardes, J.; Henrichs, M.P.; Hauschild, G.; Nottrott, M.; Guder, W.; Streitbuerger, A. Silver-coated megaprosthesis of the proximal tibia in patients with sarcoma. *J. Arthroplast.* **2017**, *32*, 2208–2213. [CrossRef]
12. Laitinen, M.; Nieminen, J.; Reito, A.; Pakarinen, T.-K.; Suomalainen, P.; Pamilo, K.; Parkkinen, J.; Lont, T.; Eskelinen, A. High blood metal ion levels in 19 of 22 patients with metal-on-metal hinge knee replacements. *Acta Orthop.* **2017**, *88*, 269–274. [CrossRef] [PubMed]
13. Lützner, J.; Günther, K.-P.; Postler, A.; Morlock, M. Metallionenfreisetzung nach Hüft- und Kniegelenkendoprothetik—Mechanismen, biologische Wirkungen und notwendige Diagnostik. *ZS Orthounfall* **2019**. [CrossRef]

14. Paulus, A.C.; Ebinger, K.; Cheng, X.; Haßelt, S.; Weber, P.; Kretzer, J.P.; Bader, R.; Utzschneider, S. Local biological reactions and pseudotumor-like tissueformation in relation to metal wear in a murine in vivo model. *BioMed Res. Int.* **2019**, *2019*, 10. [CrossRef] [PubMed]
15. Grammatopoulos, G.; Pandit, H.; Kamali, A.; Maggiani, F.; Glyn-Jones, S.; Gill, H.S.; Murray, D.W.; Athanasou, N. The correlation of wear with histological features after failed hip resurfacing arthroplasty. *JBJS* **2013**, *95*, e81. [CrossRef]
16. German Standard DIN ISO 5832-4. *Implants for Surgery—Metallic Materials—Part 4: Cobalt-chromium-molybdenum Casting Alloy (ISO 5332-4:2014)*; Deutsches Institut für Normung e.V., Beuth Verlag GmbH: Berlin, Germany, 2015.
17. Goldberg, J.R.; Gilbert, J.L.; Jacobs, J.J.; Bauer, T.W.; Paprosky, W.; Leurgans, S. A Multicenter retrieval study of the taper interfaces of modular hip prostheses. *Clin. Orthop. Relat. Res.* **2002**, *401*, 149–161. [CrossRef]
18. Hood, R.W.; Wright, T.M.; Burstein, A.H. Retrieval analysis of total knee prostheses: A method and its application to 48 total condylar prostheses. *J. Biomed. Mater. Res.* **1983**, *17*, 829–842. [CrossRef]
19. Arnholt, C.M.; MacDonald, D.W.; Malkani, A.L.; Klein, G.R.; Rimnac, C.M.; Kurtz, S.M.; Kocagoz, S.B.; Gilbert, J.L. Corrosion damage and wear mechanisms in long-term retrieved cocr femoral components for total knee arthroplasty. *J. Arthroplast.* **2016**, *31*, 2900–2906. [CrossRef]
20. Kretzer, J.P.; Reinders, J.; Sonntag, R.; Hagmann, S.; Streit, M.; Jeager, S.; Moradi, B. Wear in total knee arthroplasty—Just a question of polyethylene? *Int. Orthop.* **2014**, *38*, 335–340. [CrossRef]
21. Luetzner, J.; Krummenauer, F.; Lengel, A.M.; Ziegler, J.; Witzleb, W.-C. Serum metal ion exposure after total knee arthroplasty. *Clin. Orthop. Relat. Res.* **2007**, *461*, 136–142. [CrossRef]
22. Lons, A.; Putman, S.; Pasquier, G.; Migaud, H.; Drumez, E.; Girard, J. Metallic ion release after knee prosthesis implantation: A prospective study. *Int. Orthop.* **2017**, *41*, 2503–2508. [CrossRef]
23. Oğurel, T.; Serbest, S.; Oğurel, R.; Tiftikçi, U.; Ölmez, Y. Blood chromium-cobalt levels in patients after total knee arthroplasty and their effect on the retinal nerve fiber layer and macular ganglion cell complex. *Retina* **2019**. [CrossRef]
24. Arnholt, C.M.; White, J.B.; Lowell, J.A.; Perkins, M.R.; Mihalko, W.M.; Kurtz, S.M. Postmortem retrieval analysis of metallosis and periprosthetic tissue metal concentrations in total knee arthroplasty. *J. Arthroplast.* **2020**, *35*, 569–578. [CrossRef] [PubMed]
25. Lord, J.K.; Langton, D.J.; Nargol, A.V.F.; Joyce, T.J. Volumetric wear assessment of failed metal-on-metal hip resurfacing prostheses. *Wear* **2011**, *272*, 79–87. [CrossRef]
26. Glyn-Jones, S.; Roques, A.; Taylor, A.; Kwon, Y.-M.; McLardy-Smith, P.; Gill, H.S.; Walter, W.; Tuke, M.; Murray, D. The in vivo linear and volumetric wear of hip resurfacing implants revised for pseudotumor. *JBJS* **2011**, *93*, 2180–2188. [CrossRef] [PubMed]
27. Gascoyne, T.C.; Turgeon, T.R.; Burnell, C.D. Retrieval analysis of large-head modular metal-on-metal hip replacements of a single design. *J. Arthroplast.* **2018**, *33*, 1945–1952. [CrossRef] [PubMed]
28. Klasan, A.; Meine, E.; Fuchs-Winkelmann, S.; Efe, T.; Boettner, F.; Heyse, T.J. Are serum metal ion levels a concern at mid-term followup of revision knee arthroplasty with a metal-on-metal hinge design? *Clin. Orthop. Relat. Res.* **2019**, *477*, 2007–2014. [CrossRef] [PubMed]
29. Friesenbichler, J.; Sadoghi, P.; Maurer-Ertl, W.; Szkandera, J.; Glehr, M.; Ogris, K.; Wolf, M.; Weger, C.; Leithner, A. Serum metal ion concentrations in paediatric patients following total knee arthroplasty using megaprostheses. *BioMed Res. Int.* **2014**, *2014*, 1–7. [CrossRef]
30. Favetti, F.; Mazzotta, G.; Papalia, M.; Panegrossi, G.; Casella, F.; Falez, F. Contamination of revision procedures in patients with adverse tissues reaction to metal on metal implant. *Eur. Rev. Med. Pharmacol. Sci.* **2019**, *23*, 86–93.
31. Wyles, C.C.; Van Demark, R.E.; Sierra, R.J.; Trousdale, R.T. High rate of infection after aseptic revision of failed metal-on-metal total hip arthroplasty. *Clin. Orthop. Relat. Res.* **2014**, *472*, 509–516. [CrossRef]

materials

MDPI

Article

Antimicrobial Silver Multilayer Coating for Prevention of Bacterial Colonization of Orthopedic Implants

Martin Fabritius [1,*], Amir Andreas Al-Munajjed [2], Christiane Freytag [3], Henriette Jülke [3], Markus Zehe [4], Thomas Lemarchand [5], Jacobus J. Arts [6], Detlef Schumann [1], Volker Alt [7] and Katrin Sternberg [1]

1. Aesculap AG, Research and Development, Am Aesculap-Platz, 78532 Tuttlingen, Germany; detlef.schumann@aesculap.de (D.S.); katrin.sternberg@aesculap.de (K.S.)
2. Bio-Gate AG, Neumeyerstraße 28-34, 90411 Nuremberg, Germany; amir.al-munajjed@bio-gate.de
3. FREY_TOX GmbH, 04916 Herzberg, Germany; Christiane.Freytag@web.de (C.F.); FREY-TOX@t-online.de (H.J.)
4. QualityLabs, Neumeyerstr. 46a, 90411 Nuremberg, Germany; markus.zehe@qualitylabs-bt.de
5. TPL Path Labs, Sasbacher Straße 10, 79111 Freiburg, Germany; lemarchand@tpl-path-labs.com
6. Department of Orthopedic Surgery, Research School CAPHRI, Maastricht University Medical Centre, 6202 Maastricht, The Netherlands; j.arts@mumc.nl
7. Department of Trauma Surgery, University of Regensburg, Franz-Josef-Strauss-Allee 11, 93053 Regensburg, Germany; volker.alt@ukr.de

* Correspondence: martin.fabritius@aesculap.de; Tel.: +49-7461-952100

Received: 20 February 2020; Accepted: 12 March 2020; Published: 20 March 2020

check for updates

Abstract: Due to increasing rates of periprosthetic joint infections (PJI), new approaches are needed to minimize the infection risk. The first goal of this study was to modify a well-established infection model to test surface-active antimicrobial systems. The second goal was to evaluate the antimicrobial activity of a silver multilayer (SML) coating. In vitro tests with SML items showed a >4 Log reduction in a proliferation assay and a 2.2 Log reduction in an agar immersion test (7 d). In the in vivo model blank and SML coated K-wires were seeded with ~2×10^4 CFU of a methicillin-sensitive Staphylococcus epidermidis (MSSE) and inserted into the intramedullary tibial canal of rabbits. After 7 days, the animals were sacrificed and a clinical, microbiological and histological analysis was performed. Microbiology showed a 1.6 Log pathogen reduction on the surface of SML items ($p = 0.022$) and in loosely attached tissue ($p = 0.012$). In the SML group 7 of 12 SML items were completely free of pathogens (cure rate = 58%, $p = 0.002$), while only 1 of 12 blank items were free of pathogens (cure rate = 8%, $p = 0.110$). No silver was detected in the blood or urine of the SML treated animals and only scarcely in the liver or adjacent lymph nodes. In summary, an in vivo infection model to test implants with bacterial pre-incubation was established and the antimicrobial activity of the SML coating was successfully proven.

Keywords: periprosthetic joint infections; infection prophylaxis; *Staphylococcus epidermidis*; in vivo osteomyelitis model

1. Introduction

Periprosthetic joints infection (PJI) is a severe complication for patients undergoing a joint replacement procedure that can lead to a straining month-long treatment process of early implant revision, multiple re-revisions or even an amputation of the infected limb. While infection rates after primary implantation are relatively low (0.5% to 2%) [1–3], they increase dramatically in the case of an implant revision (>10%) [4,5]. If the implant is revised due to a previous infection, the risk to develop a subsequent infection is even higher (~26%) [6]. Besides the high risk for the patient, PJI is also a tremendous economic burden to the healthcare system. Kasch et al. found that in Germany the direct hospital care costs for the management of a septic revision total knee arthroplasty (TKA) are about twice as high as for an aseptic failure [7]. Other papers conclude that the costs for a septic revision are 3 to 4 times that of an primary implantation [8,9].

Early (<3 month after surgery) or delayed infections (3–24 month after surgery) are caused by exogenic pathogens entering the surgical wound during surgery [4]. While there is a controversial discussion on which pathogens are the most relevant to cause PJI, it is generally agreed that Staphylococci species are predominant [4,10–14]. The main focus of scientific as well as public interest is given to *Staphylococcus aureus* (*S. aureus*) due to its high virulence and an increased awareness of antibiotic resistances that renders methicillin resistant *Staphylococcus aureus* (MRSA) the biggest threat to develop PJI. Nevertheless, there are various papers and clinical case studies that highlight the importance of coagulase-negative Staphylococci (Co-NS) e.g., *Staphylococcus epidermidis* (*S. epi*) to be the predominant pathogens causing PJI [4,11–15].

While early infections are mostly caused by very virulent pathogens like *S. aureus*, *Escherichia Coli or Pseudomonas aeruginosa*, delayed infection are caused by less virulent pathogens like Co-NS or Propionibacterium species [16–19]. Early infections are relatively easy to detect by swelling, increased temperature or pain. In contrast, delayed infections are mostly clinically unobtrusive, with delayed and/or nonspecific signs [17,19,20]. If an infection is detected in less than three weeks after implantation, there is a reasonable chance to treat it with the least invasive treatment option of debridement, antibiotic treatment and implant retention [17]. In the case of delayed or chronic infections, the implant most likely has to be revised. This means that delayed infections caused by Co-NS pathogens like a *Staphylococcus epidermidis* are even more dangerous and harder to treat than early infections caused by *S. aureus* [4,12–14,19,21].

Even though operations are carried out under strict hygiene measures, a perioperative contamination from the air or the patients skin can occur [22–24]. The immune system is well capable to address high loads of over 10^6 pathogens. However, in the presence of a foreign body material like an implant, as little as 100 pathogens suffices to cause a severe infection [25]. This is caused by the biofilm formation of pathogens on the artificial surface, which renders the bacteria practically immune to host immune attacks or antibiotic treatment [22]. Therefore, it is of great interest to prevent the biofilm formation on the implants surface. To minimize the infection risk, the thorough implementation of prevention guidelines is essential [26]. In addition to that, technical solutions that protect the implant surface against bacterial colonization should be developed and transferred into clinical practice.

Many techniques and antimicrobial systems have been reported in literature, ranging from active antibiotic release devices to contact killing surfaces [27–31]. Silver is a long known antimicrobial substance, which is successfully applied in various clinically implemented and currently available implant systems on the market [32].

The MUTARS® tumor prosthesis is galvanically silver coated (m(Ag) = 0.33–2.89 g) and is widely used in Europe, Australia and various Asian countries [33,34]. Hardes et al. Donati et al. and Zajonz et al. reported a successful treatment and a reduced infection rate with the silver coated MUTARS® prosthesis compared to a standard implant [35–38]. The Agluna® technology dopes a titanium surfaces with silver ions by an electrochemical process [39]. In a case control study with 170 patients, Wafa et al. reported lower rates of early PJI when Agluna®-treated tumor implants were used [15]. The PorAg® coating is a dual layer system with a silver base layer (1 µm thick) and a rigid top layer of TiAg20N (0.1 µm thick) [40]. Scoccianti et al. reported the successful use of PorAg® coated tumor prosthesis in 33 patients without negative side-effects like argyria [41]. The AgPROTEX® coating is a Hydroxyapatite/Ag_2O system which is applied to metal surfaces by flame spraying (T = 2700 °C) [42]. The coating is approved for the use in primary total hip arthroplasty (THA) in Japan [43]. According to

Eto et al. no adverse reactions were detected in the first clinical application of AgPROTEX® coated primary THA implants (m(Ag) = 1.9 to 2.9 mg) with 20 patients [44]. However, all these silver-based systems are only suitable for metal implants and cannot be applied to polymer surfaces. This leaves the Polyethylene (PE) liner unprotected, even though it is known that the PE components are most often affected by PJI and carry the highest bacterial load [45].

The silver multilayer coating (SML) (HyProtect™, Bio-Gate, Nuernberg, Germany) can be applied to both metal and polymer components. Silver clusters are embedded in a polysiloxane (SiOxCy) matrix and act as a reservoir for the release of silver ions that are anti-microbially active on the coating surface. Therefore, elementary silver itself is not in direct contact with the surrounding bone or tissue. Due to its ultra-thin layer (~90 nm), SML maintains the porosity of nano-structured/porous surfaces intact and does not seal them, which is important for osseointegration. The combination of osteoconductive and osteoinductive biomaterials like calcium phosphate, hydroxyapatite, bisphosphonate and silicates in combination with nanoscale therapeutics like BMP-2 have also shown to support bone regeneration, which proves beneficial for secondary implant stability [46,47].

A paper by Khalilpour et al. reported on various successful tests of the SML coating like the in vitro antimicrobial activity, no cytotoxicity according to ISO 10993-5 and ex vivo antimicrobial activity [48]. In a recent case study, the SML coating was used in a successful knee arthrodesis after recurrent periprosthetic knee infection, and silver levels in the drainage fluid and blood were evaluated. Silver blood concentrations after 48 h remained under the detection limit of 2 ppb, whereas the silver concentrations in the wound drainage fluid reached 170 and 57 ppb 24 and 48 h post-operatively, respectively [49].

In most established in vivo osteomyelitis models, an implant is placed in the tibia medullary canal, and a bacteria suspension is injected afterwards [27,28,42,50–52]. This leads to a localized high concentration of pathogens with no uniform distribution along the tibia canal. In case of drug release systems, this is of minor importance as the released drug is able to target present pathogens in a larger vicinity of the implant. In contrast, surface-active coatings are unable to target these pathogens, which can subsequently grow in the more distant tissues. Once an infection is established, high numbers of pathogens are released into the surrounding tissues, which leads to an overpowering of the surface-active system, and therefore, no antimicrobial activity can be proven with such models.

The objective of this study was (i) to establish a suitable in vivo osteomyelitis model in rabbits and (ii) to evaluate the antimicrobial activity of a silver multilayer coating (SML) under realistic pre-clinical conditions. We hypothesized that the SML coating can significantly reduce the CFU count on the K-wire surface at explantation by a minimum of 1 Log reduction compared to the initial CFU count.

2. Materials and Methods

2.1. Implant Items

Gamma-sterilized pure titanium K-wires with a diameter of 2.0 mm (MEDE Technik GmbH, Emmingen-Liptingen, Germany) and a length of 150 mm were used for the study. To provide an inert surface, all K-wires were coated with the Advanced Surface® ceramic multilayer coating (AS®, Aesculap AG, Tuttlingen, Germany) over a length of ~140 mm. [53,54] The test items were coated with the SML coating by Bio-Gate AG (Nuremberg, Germany) (Figure 1).

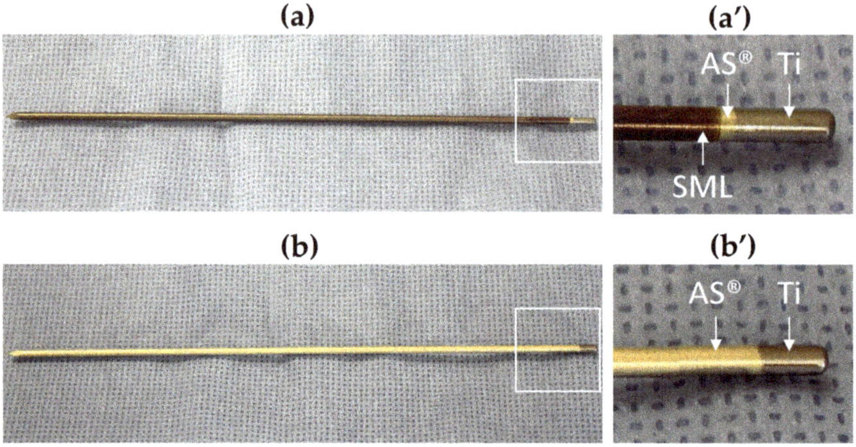

Figure 1. (**a**) Test item: SML-coated AS®/titanium K-wire, (**a′**) shows the two coatings in magnification. SML = bronze, AS® = golden. (**b**) Blank item: AS®-coated titanium K-wire, (**b′**) shows the AS® coating in magnification. The silver color at the blunt end of each K-wire shows the uncoated titanium surface.

SML Coating

Test items were coated with the SML coating in a three-step process. In a first step, a SiO_xC_y base layer was deposited on the respective surface by chemical vapor deposition (CVD). In a second step, silver clusters (~2.7 μg/cm^2) were deposited on the base layer by physical vapor deposition (PVD), and in a third step covered with a SiO_xC_y top layer. This resulted in a coating with a thickness of ~90 nm. The coated items were packed individually, and gamma sterilized (BBF Sterilisationsservice GmbH, Kernen-Rommelshausen, Germany). Previous publications give more detail on the SML coating [48,49].

Every SML coating batch was characterized after production by various test methods on planar surface aluminum coated PET foil (dummy items). The chemical structure of the SML coating was analyzed by FTIR spectrometry (Tensor 27; Bruker Optik GmbH, Ettlingen, Germany). The silver content was determined by inductive coupled plasma-optical emission spectroscopy (ICP-OES) according to EN ISO 11885 (Seibersdorf Labor GmbH, Seibersdorf, Austria). The coating thickness was measured by spectral ellipsometry (IFAM Fraunhofer Institut, Bremen, Germany). Each of these tests were performed in triplicate on non-sterilized dummy items. Additionally, titanium test plates were coated simultaneously and analyzed according to ISO 10993-5 to prove non-cytotoxic behavior.

2.2. *In Vitro Antimicrobial Activity*

Prior to the in vivo tests various in vitro tests were performed to re-verify the antimicrobial activity of the SML coating on surfaces relevant to orthopedic implants.

2.2.1. QualiScreen® Tests

The in vitro antimicrobial activity of blank and test items was evaluated using a proliferation assay described previously [55,56]. In brief, four replicates of SML coated test items and 4 blank items were incubated in 10% human plasma for 1 h and subsequently washed in 1 × phosphate buffered saline (PBS) for 10 min. Afterwards, the items were incubated in a cell suspension of 5×10^6 culture forming units (CFU)/mL MSSE (RKI 10-00621) at 37 °C for 1 h to allow bacterial cells to adhere to the item surface. Loosely attached bacteria were then removed by rinsing. Subsequently, the remaining cells were incubated for 18 h at 37 °C (challenge time). After removal of the test items, 50 μL of tryptic soy broth (TPS) was added to each well. The bacterial growth of the remaining daughter cells was monitored with a microplate reader over a period of 48 h.

2.2.2. Agar Immersion Test

To mimic the in vivo situation, we subsequently tested the antimicrobial activity in an agar immersion test. SML coated items and blank items (AS® coated titanium) were incubated with MSSE (RKI 10-00621) as described in Section 2.3. The seeded K-wires were immersed in a pre-prepared agar slurry (1% Agar and 0.1% TSB) and incubated for 24 and 72 h as well as 7 days. Afterwards, the K-wires were sonicated (3 min) and vortexed (30 s) in PBS to detach adherent bacteria from the item surface. The number of colonies was determined by agar plate count. Each measurement was performed in triplicate.

2.3. *In Vivo Study Design*

To determine the antimicrobial activity of the SML an existing in vivo model published by Alt et al. was adapted and an MSSE was used as contaminant [28]. The modification of this model consisted of the us of pre-incubated implants with bacteria in order to assess the effect of the silver coating on the implant surface compared to the inoculation of bacteria into the intramedullary canal after implantation of the K-wire in the referenced model. SML coated and blank K-wires were used as implants. Both implants were loaded with MSSE before implantation (see below). The studies were approved by the German regional authority of Brandenburg (2347-A-4-10-2014) in compliance with the EU principles for animal care.

2.3.1. Pilot Studies

In order to establish and validate the in vivo model, two pilot studies with 6 SPF New Zealand White Rabbits (Envigo, 5800 Venray, The Netherlands) each with a body weight from 3.6 to 4.0 kg where performed. Three animals were treated with a blank K-wire and three animals with an SML coated K-wire. The operation procedure, microbial contamination and subsequent analysis were identical with the one of the main study described below. The silver analysis in organs, blood and urine was only performed in the main study.

2.3.2. Main Study

The study included 27 SPF New Zealand White Rabbits (Envigo, 5800 Venray, The Netherlands) with a body weight from 3.7 to 4.4 kg. As the SML is only surface-active and has no widespread release of antimicrobial substance, we seeded the implants directly with pathogens, instead of injecting bacteria inoculum into the tibia canal. With this approach, we were able to guarantee a contact of pathogens with SML and simultaneously avoid an uncontrolled distribution of bacteria in the tibia canal, which might lead to false negative results.

The K-wires were contaminated in vitro with ~2×10^4 colony-forming units (CFUs) of MSSE RKI 10-00621 and implanted into the intramedullary canal of the tibia in the rabbit. The method to determine the bacterial load on the K-wire surface is described in Section 2.4.2.

The test group of twelve animals received a SML coated K-wire (test item), while the control group received blank K-wires (blank item). Both were contaminated with identical loads of bacteria. The remaining three animals were implanted with test items without any microbial load. These three animals served as control to differentiate between implantation process-related and bacterial contamination-related lesions in the histological examination.

2.4. Bacteria

2.4.1. Bacterial Strains

MSSE RKI 10-00621 was used as a contaminant in this study. RKI 10-00621 is a clinical isolate from a patient with PJI. It was obtained from the national reference center for Staphylococci and Enterococci of the Robert Koch Institute (RKI) in Wernigerode, Germany.

2.4.2. Bacteria Cultivation and Pre-Incubation of Implants

The bacterium was grown in 20% Tryptic soy broth (TSB) for 4 h at 37 °C. Prior to implantation, two SML K-wires (per test tube) were contaminated in a test tube over a length of 9 cm by incubation in a bacterial solution of ~1×10^6 CFU/mL for 30 to 60 min under dynamic conditions. In pretests, it was proven that the bacterial load level on the item surface can be preserved for this time period. The bacteria were freshly prepared and in logarithmic phase for the controlled contamination of the implant. No biofilm was present at the time of surgery. Subsequently to incubation, non-adherent bacteria were removed by rinsing the test item in PBS for 10 min. One K-wire was implanted into the tibia canal, and the other K-wire was used to determine the pathogen load on the surface at the time of implantation. The pathogens were removed by sonication and vortexing, and the bacterial count was determined by agar plate count.

2.5. Surgery

Intramuscular anesthesia was performed using ketamine (40 mg kg^{-1} body weight) and xylazine (6 mg kg^{-1} body weight). Perioperative analgesia was applied subcutaneously (Butorphanol, 0.5 mg kg^{-1} body weight).

Surgery was carried out under aseptic conditions according to the model published by Alt et al. and no systemic antibiotics were given [28]. The left hind leg was shaven, fixated, disinfected and draped in sterile covering. The tibia was approached via an infrapatellar skin incision and subsequent preparation of Ligamentum patellae and Tuberositas tibiae. The Tuberositas tibiae was sharply penetrated and the medullary channel opened with a 2.0 mm K-wire. A template K-wire was inserted along the medial cortical bone until the proximal part of Malleolus medialis for channel preparation. After removal of the template, one implant per animal was applied. The implant was inserted into the prefabricated channel at a length of approx. 10 cm and fixated in the proximal part of the Malleolus medialis. The K-wires were contaminated over a length of 9 cm to make sure that no pathogens enter the knee joint. Table 1 lists the group arrangement. Group 1 (test item) and Group 2 (blank item) were seeded with 2×10^4 CFU MSSE. Group 3 (test item) served as a negative control for histology to determine the lesions induced by the implantation process. The protruding part of the implant was removed, and the implant site was rinsed with Octenisept® and NaCl 0.9%. Afterwards the wound was closed, and post-operative X-ray control was performed (Figure 2).

Table 1. Main study-group setup.

Group	Implant	Contamination	Animals	Time of Implantation Post Operation
1	Test item (SML item) K-wire (Titanium/AS®) + SML	2×10^4 CFU MSSE RKI 10-00621	12	7 days
2	Blank item K-wire (Titanium/AS®)	2×10^4 CFU MSSE RKI 10-00621	12	7 days
3	Negative control: Test item (SML item) K-wire (Titanium/AS®) + SML	Without contamination	3	7 days

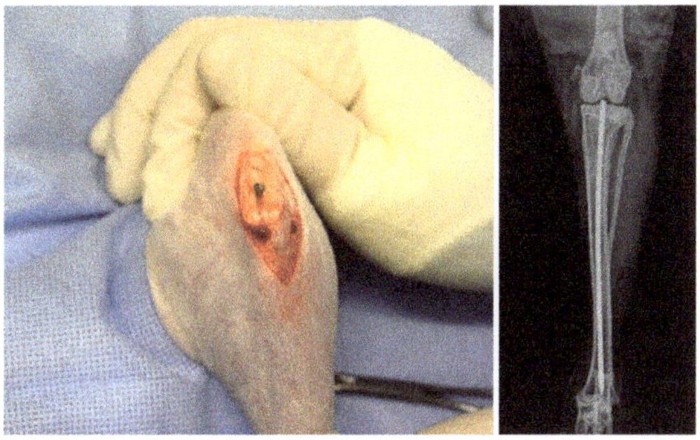

Figure 2. (**a**) Image of SML coated K-wire implanted in intramedullary canal, (**b**) Post-operative X-ray image of the rabbit tibia with the K-wire placement.

After 7 days post operation, the animals were put under general anesthesia (ketamine 40 mg kg^{-1} body weight, xylazine 6 mg kg^{-1} body weight) prior to euthanasia (T61 intravenously). The implantation site and the surrounding tissue were examined macroscopically, and samples were harvested. For microbiological examination, the implant and bone marrow were collected. The proximal tibia was transversally cut open for implant removal. Bone marrow was extracted from the central tibial bone using sterile instruments. It was manually mixed and halved for microbiological analysis and silver level measurements. For histological evaluation, the tibial bone was segmented into three parts by two transversal cuts. The proximal cut was set distally to the Tuberositas tibiae, and the distal cut was placed proximal to the Malleoli (Figure S1). Both proximal and distal segments were immersed in neutral buffered formalin for subsequent histopathological examination.

2.6. Sample Collection for Silver Measurement

For pharmacokinetic examinations, samples of whole blood and urine were collected from groups 1 and 2 at defined time points. Samples for determination of zero levels were gathered 7 days prior to surgery at maximum. Subsequently, samples were taken on days 1 and 7 post operation. Approximately 1.5 mL whole blood was taken using a lithium-heparin-tube. Urine samples were harvested as 24-h-samples prior to surgery and on day 1 post operation, whereas on day 7 post operation, urine was favorable collected as punctate. For extended determination of silver levels, situs associated lymph nodes (Lnn. poplitei and Lnn. inguinales), and liver were collected.

2.7. Evaluation Methods

2.7.1. Clinical Assessment for Infection

Before harvesting of the bone, the knee joint and the surrounding of the implantation site were evaluated for clinical signs of inflammation or swelling. Therefore, areas surrounding the insertion site being directly net to the knee joint, adjacent soft tissue as well as the external structure of the tibia were examined regarding swelling, edema, excessive fluids and pus.

After implant removal and transversal opening of the tibia, the bone marrow was assessed for signs of infection. The quality of the bone marrow was defined as follows. Physiological quality: grey-white–pinkish–light red coloration, not washed-out, no watery phase, formed structure. Inflamed quality: reddish–red–bright red–dark red coloration, washed-out, with aqueous phase and unformed structure. In case of multiple findings, the most severe grade was taken into account for evaluation. In order to standardize qualitative description, macroscopic evaluation was performed in a blinded manner by the same investigator in all cases.

2.7.2. Microbiological Assessment for Infection

At explantation, it was noted that varying amounts of bone marrow or tissue adhered to the K-wires. To avoid a distortion of the CFU count, the implants were gently rinsed in phosphate buffered saline (PBS) to remove the attached tissue (rinsing solution = "rinsing sol.") Subsequently the K-wires were sonicated (3 min) and vortexed (30 s) in PBS to detach adherent bacteria from the item surface ("K-wire"). The bone marrow was collected as described in Section 2.5 and was sonicated in PBS to harvest containing pathogens. Bacterial contamination in the bone marrow was standardized to 1 g ("BMstd"). The three bacterial suspensions (K-wire, rinsing sol., BMstd) were diluted (1:1, 1:10), plated out on agar plates (1000 μL) and incubated over night at 37 °C. The number of colonies was determined by visual agar plate count.

2.7.3. Histological Evaluation

Tibia parts (proximal + distal) for histological examination were fixed in 4% phosphate-buffered formaldehyde solution for 72 h. The formalin-fixed bones were cut with a band saw sagittally, para-sagittally and longitudinally into a total of five ~3 mm slabs, according to Figure S1, 1 transverse slab and 2 longitudinal slabs for the proximal segment and 2 longitudinal slabs for the distal segment. The slabs were then demineralized for 1 to max 2 weeks in a solution containing formic acid (5% formic acid in distilled water) and dehydrated and embedded in paraffin. Formalin-fixed paraffin-embedded tissues were cut into 3 μm thick tissue sections and stained with haematoxylin and eosin (HE) or Gram stain according to a method derived from Brown and Brenn [57]. In order to maximize the chances to detect bacterial colonies and inflammatory lesions, step sections at least 50 μm apart were prepared from each block. The presence or absence and if present, an evaluation of the extent of osteomyelitis and bacterial colonies at the site of the contaminated K-wire insertion was carried out at the proximal and distal ends of the tibia. A qualitative ordinal scoring approach (ordinal non-continuous categorical response variables) (usually improperly referred to as "semi-quantitative scoring") was used for the histological evaluation following current animal disease model literature and pharmaceutical development pathological safety/efficacy investigations, as well as Annex E of DIN EN ISO 10993-6:2014-12 [58,59]. It was based on the presence or absence of a microscopic change, finding or lesion and a scoring of extent and magnitude using a relative or absolute scale in a 6-category system: none, minimal, slight (mild), moderate, marked and very marked (severe). No morphological change quantification with digital section and image analysis was carried out, as most often the change was no longer present in a majority of sections.

2.7.4. Silver Levels

For silver level analysis, samples collected according to Section 2.6 were used. The urine samples were centrifuged, and an aliquot of each sample was diluted and measured with inductively coupled plasma mass spectrometry (ICP-MS) according to ÖNORM EN ISO 17294-2 [60]. Blood, liver and lymph nodes were digested in an UV-digestion apparatus by using nitric acid and hydrogen peroxide and analyzed as described above. The detection limit of the ICP-MS depends on the sample quantity, which is analyzed. The following detection limits apply: blood (d0, d1, d7) < 6 μg/kg; urine (d0, d1, d7) < 0.6 μg/kg; liver < 3–6 μg/kg. Due to the varying sample quantity, the detection limit for the lymph nodes fluctuated considerably.

2.7.5. Statistical Analysis

Statistical analysis was performed with Minitab R 17 (Minitab LLC, PA, USA). Due to a significant deviation from a normal distribution of the SML group, a Mann–Whitney test was used to evaluate whether there is a difference between the contamination at the beginning (implantation) and the end (explantation) of the experiment. This was done for the three respective bacterial suspensions (K-wire, rinsing sol., BMstd). We also compared the CFU count of SML coated and blank items after explantation. A Mann–Whitney test was used to evaluate whether there is a difference between the bacterial counts in the bacterial suspensions K-wire and rinsing sol. Differences were considered as significant for $p < 0.05$.

3. Results

3.1. *In Vitro Antimicrobial Activity*

Proliferation assay: The SML coated items and blank items were tested against MSSE (RKI 10-00621). The blank items showed a brutto Onset-OD of 10.3 ± 0.8 h, while the SML items showed a brutto Onset-OD of 21.5 ± 11.0 h. This results in an average netto Onset-OD of 11.2 ± 7.3 h, which relates to a >4 Log reduction [55,56] (Figure 3).

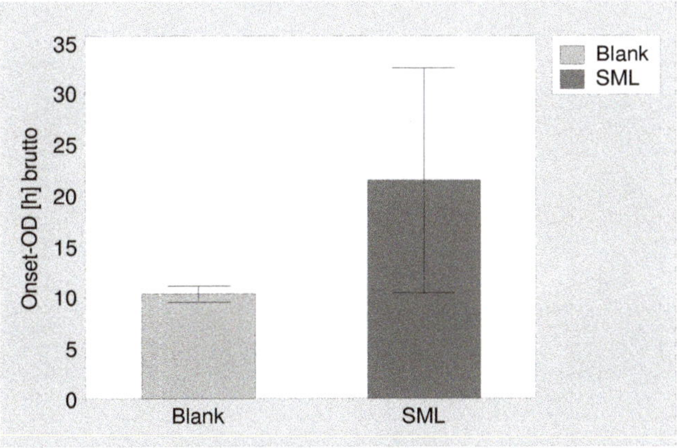

Figure 3. Results of the proliferation assay. Brutto Onset-OD time and 95% confidence interval for blank items and SML coated items. The netto Onset-OD of 11.2 h relates to a >4 Log reduction.

Agar immersion test: SML and blank items were challenged with MSSE (RKI 10-00621) and incubated for 24, 72 and 168 h in agar slurry. At the respective times the SML coated items showed a CFU reduction of 1.4 ± 0.2, 1.3 ± 0.3 and 2.2 ± 0.2 Log compared to uncoated blank items (Figure 4).

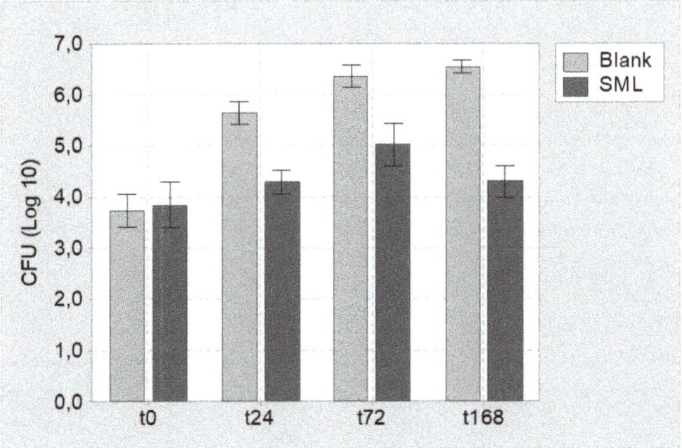

Figure 4. Results of the agar immersion test. Mean CFU count and 95% confidence interval for blank and SML coated K-wires at time point t = 0 h (t0), t = 24 h (t24), t = 72 h (t72) and t = 168 h (t168).

3.2. In Vivo Experiments

3.2.1. Clinical Assessment

In general, the established infection model was considered to be mild, as septic arthritis of the knee joint was not detected by clinical observation in any case (Figure 5a,b). In 4 of 12 animals of the SML group signs of osteomyelitis were found in the bone marrow. This was marked by an increased red and/or washed-out coloration, as well as an unformed structure of the bone marrow and partial presence of an aqueous phase. In the other 8 cases of the SML group, the bone marrow was evaluated as physiological or "cured" (SML–cured 8/12 = 67%). On the contrary, bone marrow of animals from the blank group showed signs of osteomyelitis in 11 of 12 cases and only one animal was documented as physiologic or cured (Blank–cured 1/12 = 8%) (Figure 5).

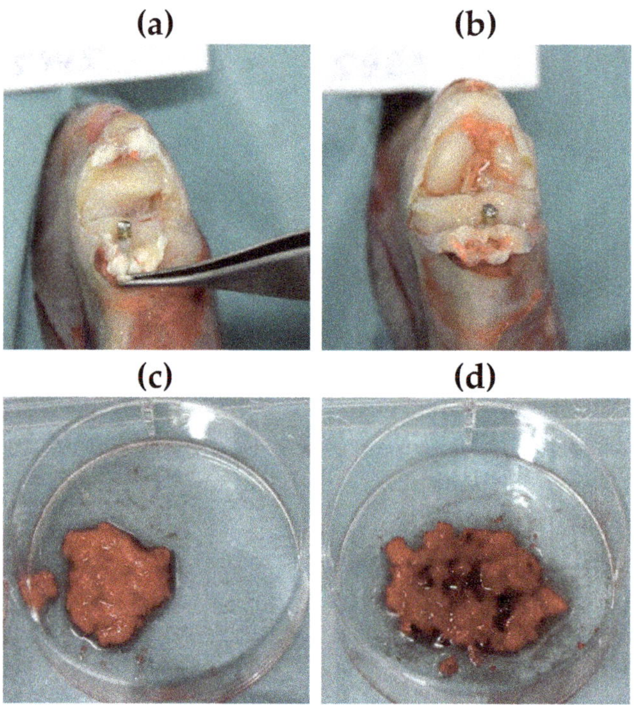

Figure 5. (**a**,**b**) Exemplary images of the knee joint postmortem of animals treated with (**a**) SML item and (**b**) blank item. (**c**,**d**) Exemplary images of bone marrow after explantation of (**c**) SML item and (**d**) blank item. (**c**) Physiological bone marrow was found in 11 of 12 animals. This equals a cure rate = 8%. (**d**) Fragmented and hemorrhagic aspects indicating osteomyelitis were found in 4 of 12 animals. This equals to a cure rate of 67%.

3.2.2. Microbiological Assessment for Infection

In the microbiological evaluation, 7 of 12 SML coated K-wires were free of pathogens and the remaining 5 K-wires showed a distinct CFU reduction compared to implantation. This equals a cure rate of 58% ($p = 0.002$). The mean pathogen count for the whole 12 SML K-wires was 353 ± 529 CFU. In the control group, only 1 of 12 K-wires was free of pathogens and the CFU count on the whole twelve blank K-wires was 9.282 ± 10.585 CFU (Figure 6). This equals a cure rate of 8% ($p = 0.110$).

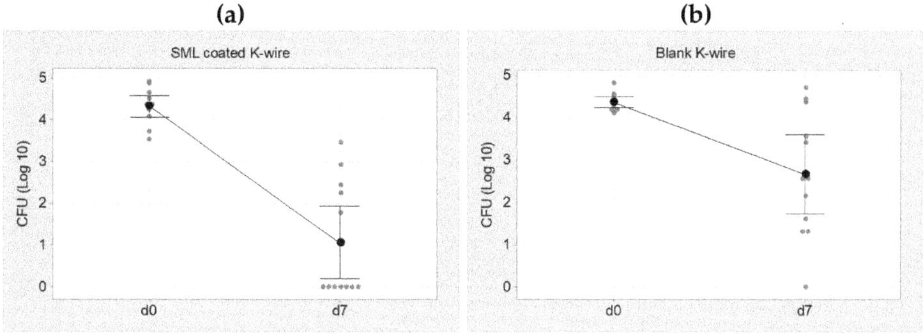

Figure 6. Individual value plot of CFU at d0 (implantation, ~2×10^4 CFU) and d7 (explantation). (**a**) SML coated K-wires, mean CFU = 353 ± 529 CFU. This equals a cure rate of 58% (**b**) Blank K-wires, mean CFU = 9.282 ± 10.585 CFU. This equals a cure rate of 8%. The black dots represent the mean value of each group with a 95% confidence interval.

The results of the Mann–Whitney test for the three bacterial suspensions (K-wire, rinsing sol., BMstd) are listed in Table 2. A comparison of the CFU count at implantation and explantation showed a decrease in all bacterial suspensions and the reduction is significant for both blank and SML coated items ($p < 0.05$, Table 2). The CFU reduction was always higher for SML coated K-wires than for blank K-wires.

Table 2. Results from the Mann–Whitney test of Log (Implantation) compared to Log (Explantation).

Type	Response	Mean	Difference from Log (Implantation)	*p*-Value Mann–Whitney Test
blank	Log(Implantation)	4.376		
	Log(K-wire)	2.670	−1.706	0.007
	Log(rinsing sol.)	3.393	−0.983	0.069
	Log(BMstd)	2.649	−1.727	0.112
SML	Log(Implantation)	4.336		
	Log(K-wire)	1.073	−3.263	0.000
	Log(rinsing sol.)	1.797	−2.539	0.000
	Log(BMstd)	2.148	−2.188	0.001

Sol. is solution and BMstd is bone marrow standardized.

Comparing the CFU count at explantation for SML and blank group showed a 1.6 Log reduction for the SML coated items on the K-wire surface ($p = 0.022$) and in the rinsing solution ($p = 0.012$, Table 3). The bone marrow of the SML group also exhibited less pathogens than the blank group (0.5 Log), yet the effect was less pronounced (Figure 7).

Table 3. Results from the Mann–Whitney test of Blank and SML items after explantation.

Type	Difference	*p*-Value Mann–Whitney Test
Log(K-Wire) Blank vs. SML	1.597	0.022
Log(rinsing sol.) Blank vs. SML	1.596	0.012
Log(BMstd) Blank vs. SML	0.501	0.362

Sol. is solution and BMstd is bone marrow standardized.

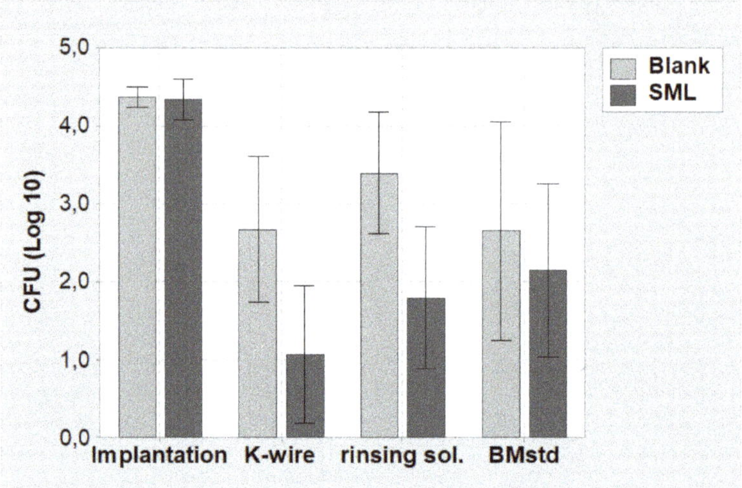

Figure 7. Mean values and 95% confidence interval of the CFU count of the in vivo study. (Implantation) = CFU count at implantation on the K-wire surface. (K-wire) = Bacterial suspension derived from the K-wire surface at explantation. (rinsing sol.) = Bacterial suspension derived from rising the K-wire surface to remove attached tissue. (BMstd) = Bacterial suspension derived from the bone marrow (normalized to 1 g). Light grey = Blank items, dark grey = SML items.

3.2.3. Histological Evaluation

Histology showed less heterophilic infiltration/pus and fibroplasia in the distal tibia of animals with SML items, compared to the distal tibia of animals with blank items (Figures 8 and 9). Similar degrees of inflammation and associated repair were noted in the proximal tibia of SML group and blank group animals. In general, the induced inflammation was very mild and barely above the one induced by the surgery alone. Very importantly, the new bone formation around the implant was very active in this disease model and comparable for test item and blank item. The negative control group (SML item with no bacterial contamination) showed no suppuration and excellent implant stabilization by means of fibrous connective tissue and recent new bone formation (Figure 10).

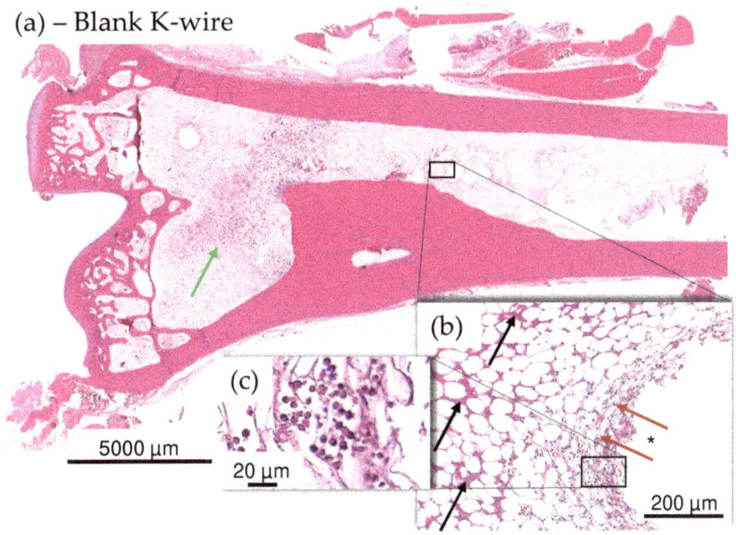

Figure 8. (**a**–**c**) Histology images of distal tibia of blank item animal. (**a**) There was a higher incidence of mild focal ongoing osteomyelitis in the distal tibia of blank K-wire implanted animals (**b**). Green arrows indicate new fibrous tissue (fibroplasia). Osteomyelitis was focal and showed evidence of several days old pus (black arrows) at periphery of implant, in the bone marrow and more recent exudate of intact and degenerated heterophils (red arrows and (**c**)) in the bone marrow immediately adjacent to the implant imprint (*), indicating active suppurative inflammation.

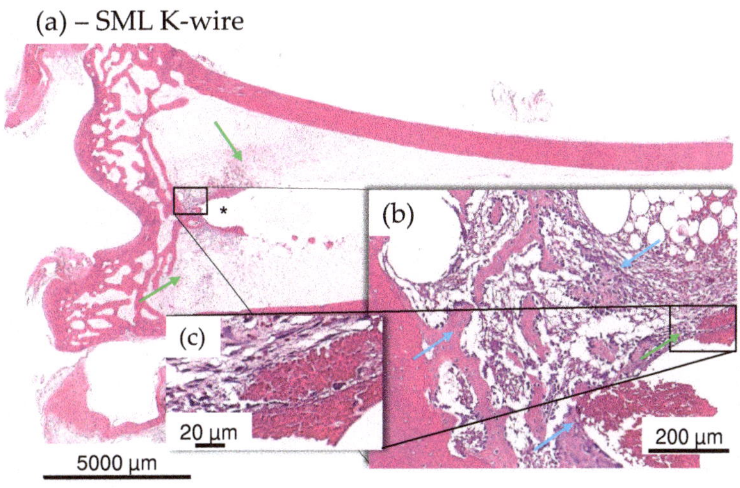

Figure 9. (**a**–**c**) Histology images of distal tibia of test item animal. No evidence of mild ongoing osteomyelitis along the K-wire in most SML K-wire implanted animal and stabilization was often seen to be more significant in the test item at tip of the K-wire imprint (*). Integration was by means of fibroplasia (green arrows) and new fibrous bone formation at host-implant interface ((**b**), blue arrows), devoid of any heterophilic infiltration (**c**).

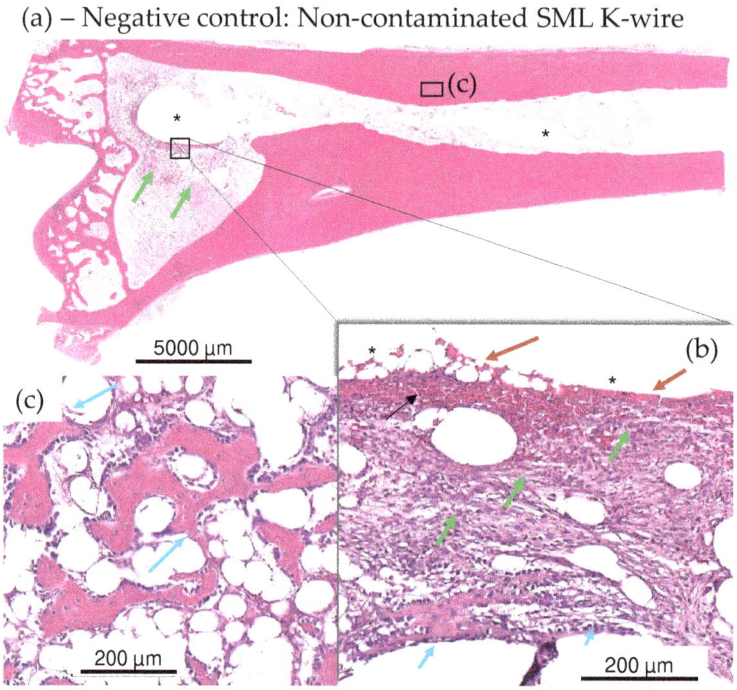

Figure 10. (**a–c**): Test item SML K-wire-implanted rabbits with no bacterial contamination, showing lack of suppuration and excellent implant stabilization by means of fibrous connective tissue (green arrows inset (**b**)) and recent new bone formation (see magenta arrows in insets (**b**,**c**)). Very recent new fibrous bone formation at host-implant interface (see short magenta arrows in inset (**b**)), devoid of any heterophilic infiltration, and more mature and older new fibrous bone (see longer magenta arrows in panel (**c**) from adjacent step section in metaphysis region adjacent the epiphysis EP). At the interface of the removed K-wire and the new connective tissue, there is limited red blood cell extravasation or hemorrhage (red arrows in inset (**b**)).

3.2.4. Blood and Urine Analysis

No silver was detected in the blood and urine of all SML group animals at days 0, 1 and 7. In 1 of 12 animals, silver was detected in the liver (5 μg/kg), while in 11 of 12 animals, silver levels were below the detection limit. In Lnn poplitei, silver was detectable in 3 of 12 animals and in Lnn inguinales in 6 of 12 animals while in all other cases silver levels were below the detection limit (Figure 11).

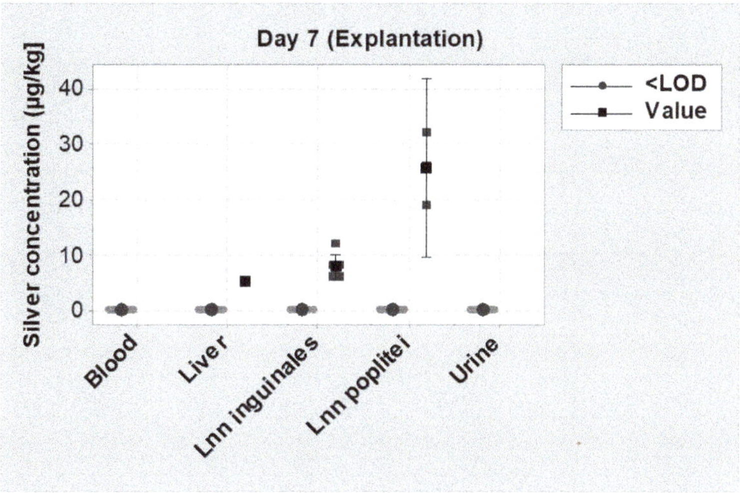

Figure 11. Individual value plot of silver levels measured by inductively coupled plasma mass spectrometry (ICP-MS) on day 7 (explantation) in blood, urine, liver and two lymph nodes. <LOD = Silver concentration was below the detection limit; Value = Silver concentration could be measured. The big symbols represent the mean value of each group with a 95% confidence interval.

4. Discussion

In the development of PJI, the initial step of bacteria adhering to the implant surface is of utmost importance. It initiates the cascade of bacterial proliferation and subsequent biofilm formation, which protects the bacteria from the host defense system. This contributes to inferior results in treatment options, as the susceptibility to antibiotics is dramatically decreased at this location. Therefore, the protection of the implant surface from bacterial colonization is potentially the most important step to prevent PJI.

The objective of this study was (i) to establish a suitable in vivo osteomyelitis model with pre-incubated implants with *Staphyloccocus epidermidis* in rabbits and (ii) to evaluate the antimicrobial activity of a silver multilayer coating (SML) under realistic pre-clinical conditions. Our study has some limitations that need to be taken into account when interpreting the results. Hischebeth et al. showed that MRSE infections are more difficult to cure than MRSA infections [21]. However, in our study, we used MSSE instead of MRSE due to safety aspects and to avoid unnecessary use of MRSE. Aspects such as antibiotic resistance, no differences in several bacterial properties such as proliferation, biofilm development and adherence are to be expected between MSSE and MRSE. The limitation of a relatively short test period of 7 days when compared to the timeframe of delayed infections (3–10 weeks) was accepted mainly for animal-welfare reasons. The test period was sufficient to identify the microbiological significant difference in the infection course. To test for complete clearance of the infection, further research with a longer test interval could be performed.

In literature, various models have been established to determine the antimicrobial activity of antimicrobial coatings most of which use MRSA as contaminant [52]. Recent findings emphasize that in clinical reality its "little brother" methicillin-resistant *Staphyloccocus epidermidis* (MRSE) is even harder to cure and spreads around the globe undetected [21,61]. In our study design, we therefore tested the antimicrobial activity of the SML coating against MSSE, which is associated with a strong biofilm forming capacity [62].

Most of the mentioned animal models are designed to test active release systems that address pathogens in the larger tissue region around the coated implant and therefore could actively treat an infection [27,28,50,51]. For example, Suhardi et al. reported the successful treatment of a PJI with an antibiotic releasing PE [27]. However, this approach has a major drawback for cementless endoprosthesis as they have the necessity of bone ingrowth which is essential to guaranty secondary implant stability. Antimicrobial substances like antibiotics or silver can have a cytotoxic effect on osteoblasts and thus may impair bone ingrowth around the implant [63,64]. Therefore, it is important to minimize the exposure to these substances while still guaranteeing an antimicrobial activity. Our measurements showed no silver in blood and urine and very low silver concentrations in the liver and adjacent lymph nodes, which proves a very limited systemic silver exposure. These findings are in line with the histopathological evaluation, which found very active new bone formation for both test item and blank item.

Another relevant aspect to take into account is the clinical approach currently used to treat PJI. The surgical techniques involve the thorough debridement of infected tissue, systemic antibiotic therapy and wound irrigation to reduce the pathogen load in the wound as much as possible. In clinical practice, an implant is never placed in an infected wound but in an environment with as little pathogens as possible. However, a contamination of the implants' surface with a few pathogens from whichever source can always occur, and as little as 100 bacteria are enough to induce a periprosthetic infection [25]. An antimicrobial coating, which is surface active and capable to kill the bacteria trying to adhere to the implant surface adds an additional safety characteristic to the implant while avoiding the potential negative effects of high dose drug release. The remaining planktonic pathogens in the soft tissue can be attacked by the host immune system.

To address the aspects mentioned above, we deviated from established animal models and adapted ours to enable the testing of surface-active coatings without typical release properties and drug efficacy of antibiotics.

The SML coating tested in this study showed a distinct activity against MSSE with a statistically significant reduction of pathogens on the implants surface. Due to the missing or mild infection signs in the clinical observation and histology as well as no systemic effects, the model was considered to be mild, with only a minimal to mild impact on the animal health. The decrease in CFU count of both blank and SML item shows that the immune system of the animals was able to fight the pathogen concentration of ~2×10^4 CFU of MSSE. When compared with the initial pathogen dose, a statistically significant CFU reduction of over 3 Log after explantation on the surface of SML coated items and an over 2 Log reduction in the corresponding rinsing solution and bone marrow was detected. This also significant CFU reduction on the AS® coated blank items was unexpected. Recent papers report similar findings, yet to date, there is no explanation for this effect. This will need to be evaluated further [65].

Comparing the CFU count after explantation, the SML coated items showed a 1.6 Log reduction on both the K-wire surface and the rinsing solution compared to the blank items. In the SML group, 7 of 12 test items were completely free of pathogens, which equals to a cure rate of 58% (p = 0.002). On the contrary, only 1 of 12 blank items were free of pathogens, which equals a cure rate of 8% ((p = 0.110). These results clearly prove the antimicrobial activity of the SML coating. In addition, the results from the main study are backed up by the results from the two pilot studies, which also showed a clear reduction of pathogens. As the difference in CFU count of blank and SML items was far less pronounced in the bone marrow, this indicates the localized antimicrobial effect of the SML coating.

As no reference values exists that indicate how high the in vivo CFU reduction of an antimicrobial system has to be to prove a certain and clinically relevant antimicrobial activity, it might be hard for the reader to interpret the measured 1.6 Log CFU reduction. To put this into perspective, the clinical reality is to be considered in this context. Each K-wire was implanted with a load of ~20,000 CFU on the K-wire surface. This was done to induce an infection in the animals and subsequently detect a difference in CFU count between SML and blank items. Lower contamination doses were quickly eradicated by the animal's immune system, and therefore, no effect of the antimicrobial coating could be detected. A contamination of over 20,000 CFU per K-wire equals over 3500 CFUs per cm^2, which is several orders of magnitud higher than the contamination of a hospital toilet door handle (7.97 ± 0.68 CFU/cm^2) or a hospital washroom floor (20 CFU/cm^2) [66,67]. In an orthopedic setting with adequate hygiene management, such high pathogen loads will enter the wound neither through the orthopedic implant, which is delivered sterile, nor through the patient's skin, which is disinfected and draped before the operation.

At explantation, the mean pathogen count on the surface of a blank K-wire was 9.282 ± 10.585 CFU, and only one of twelve K-wires was pathogen free. On the contrary, on SML coated items, the mean pathogen count was 353 ± 529 CFU, and seven of twelve K-wires were pathogen free, which shows a decrease in pathogen count by the SML coating of roughly 9000 CFU. This is over 3 times the pathogen count that is present at the surface of an PE liner, which is explanted due to PJI (2768 CFUs) [45]. Therefore, the 1.6 Log reduction of pathogens when SML is used is considered to be a substantial safety benefit.

To our knowledge, this is the first established mild osteomyelitis model that works with pre-incubated implants. In this model, we successfully proved an antimicrobial activity of the SML coating.

To date, various orthopedic implants with silver-based surface coatings are approved and used successfully in clinical practice. Within the European Union, to date, these systems are limited to large revision and tumor prosthesis and not used on primary or standard revision implants for total joint arthroplasty. However, as they can only be applied to metal, the PE liner, which is most prone to infection, remains unprotected. A technology that can be applied to both metal and polymer surfaces and that has been proven to reduce adherent bacterial load broadens the possibilities to fight PJI. The SML coating is not designed to be wear resistant, nor has it been tested for wear under continuous loads in this study. In areas with high friction like the joint articulation, the SML coating could be sheared off. It is therefore recommended that for orthopedic implants, the majority of the surfaces be coated to protect against bacterial colonization, but that areas exposed to high wear remain uncoated. Further studies should be conducted to address the osseointegration and wear behavior of the SML coating.

5. Conclusions

There is a great need for new infection prophylaxis systems that can improve the safety of patients undergoing joint replacement surgery. We have successfully (i) established a mild osteomyelitis model in rabbits with pre-incubated implants and (ii) demonstrated excellent antimicrobial activity of the presented SML coating. The performed in vitro and in vivo experiments both showed a statistically significant CFU reduction in a clinically relevant scale. The local and systemic silver release remained close to detection limits. Its broad applicability renders the SML coating a promising candidate as an infection prophylaxis system for orthopedic applications.

Supplementary Materials: The following are available online at http://www.mdpi.com/1996-1944/13/6/1415/s1, Figure S1: Schematics of band saw cutting of un-demineralized specimens for block preparation. From each of the 5 blocks depicted, 2 section levels spaced at least 50 μm were cut.

Author Contributions: Conceptualization, M.F., A.A.A.-M., C.F., H.J., M.Z. and K.S.; investigation, M.F., A.A.A.-M., C.F., H.J., M.Z. and T.L.; methodology, M.F., A.A.A.-M., C.F., H.J. and M.Z.; visualization M.F., C.F., H.J. and T.L.; supervision D.S., K.S; writing—original draft preparation, M.F. and A.A.A.-M.; writing—review and editing, J.J.A., V.A. and K.S. All authors have read and agreed to the published version of the manuscript.

Funding: The study was funded by Aesculap AG (Tuttlingen, Germany) and supported by Bio-Gate AG (Nuremberg, Germany).

Conflicts of Interest: Three of the authors (M.F., D.S., K.S.) are employees of Aesculap AG Tuttlingen, a manufacturer of orthopedic implants. One author (A.A.A.-M.) is employee of Bio-Gate AG Nuremberg, a manufacturer of antimicrobial materials. Four authors (C.F., H.J., M.Z., T.L.) are employees of the respective labs (C.F., H.J., FreyTox Gmbh Herzberg; M.Z., QualityLabs Nuremberg; T.L. TPL Path Labs Freiburg). Two of the authors (J.J.A., V.A.) are scientific advisors in Aesculap R&D projects. V.A. is member of the Supervisory Board of Bio-Gate AG (Nuremberg, Germany).

References

1. National Joint Registry for England, Wales, Northern Ireland and the Isle of Man. *15th Annual Report 2018*; National Joint Registry for England: London, UK, 2018.
2. Australian Orthopaedic Association. *Australian Orthopaedic Association National Joint Replacement Registry: Annual Report 2019*; Australian Orthopaedic Association: Adelaide, Australia, 2019.
3. Trauma Register, D.G.U. *Jahresbericht 2019*; Endoprothesenregister Deutschland (EPRD): Berlin, Germany, 2019.
4. Zimmerli, W. Clinical presentation and treatment of orthopaedic implant-associated infection. *J. Intern. Med.* **2014**, *276*, 111–119. [CrossRef]
5. Kärrholm, J.; Garellick, G.; Rogmark, C.; Herberts, P. *Swedish Hip Arthroplasty Register: Annual Report 2017*; Sahlgrenska University Hospital: Gothenburg, Sweden, 2018.
6. Cochran, A.R.; Ong, K.L.; Lau, E.; Mont, M.A.; Malkani, A.L. Risk of Reinfection After Treatment of Infected Total Knee Arthroplasty. *J. Arthroplast.* **2016**, *31*, 156–161. [CrossRef] [PubMed]
7. Kasch, R.; Merk, S.; Assmann, G.; Lahm, A.; Napp, M.; Merk, H.; Flessa, S. Comparative Analysis of Direct Hospital Care Costs between Aseptic and Two-Stage Septic Knee Revision. *PLoS ONE* **2017**, *12*, e0169558. [CrossRef] [PubMed]
8. Klouche, S.; Sariali, E.; Mamoudy, P. Total hip arthroplasty revision due to infection: A cost analysis approach. *Orthop. Traumatol. Surg. Res.* **2010**, *96*, 124–132. [CrossRef] [PubMed]
9. Hernandez-Vaquero, D.; Fernandez-Fairen, M.; Torres, A.; Menzie, A.M.; Fernandez-Carreira, J.M.; Murcia-Mazon, A.; Guerado, E.; Merzthal, L. Treatment of periprosthetic infections: An economic analysis. *Sci. World J.* **2013**, *2013*, 821650. [CrossRef]
10. Arciola, C.R.; Campoccia, D.; Ehrlich, G.D.; Montanaro, L. Biofilm-based implant infections in orthopaedics. *Adv. Exp. Med. Biol.* **2015**, *830*, 29–46. [CrossRef]
11. Montanaro, L.; Speziale, P.; Campoccia, D.; Ravaioli, S.; Cangini, I.; Pietrocola, G.; Giannini, S.; Arciola, C.R. Scenery of Staphylococcus implant infections in orthopedics. *Future Microbiol.* **2011**, *6*, 1329–1349. [CrossRef]
12. Dapunt, U.; Radzuweit-Mihaljevic, S.; Lehner, B.; Haensch, G.M.; Ewerbeck, V. Bacterial Infection and Implant Loosening in Hip and Knee Arthroplasty: Evaluation of 209 Cases. *Materials* **2016**, *9*, 871. [CrossRef]
13. Zimmerli, W.; Trampuz, A.; Ochsner, P.E. Prosthetic-Joint Infections. *N. Engl. J. Med.* **2004**, 1645–1654. [CrossRef]
14. Stefánsdóttir, A.; Johansson, D.; Knutson, K.; Lidgren, L.; Robertsson, O. Microbiology of the infected knee arthroplasty: Report from the Swedish Knee Arthroplasty Register on 426 surgically revised cases. *Scand. J. Infect. Dis.* **2009**, *41*, 831–840. [CrossRef]
15. Wafa, H.; Grimer, R.J.; Reddy, K.; Jeys, L.; Abudu, A.; Carter, S.R.; Tillman, R.M. Retrospective evaluation of the incidence of early periprosthetic infection with silver-treated endoprostheses in high-risk patients: Case-control study. *Bone Jt. J.* **2015**, *97*, 252–257. [CrossRef] [PubMed]
16. Zimmerli, W. *Bone and Joint Infections: From Microbiology to Diagnostics and Treatment*; John Wiley & Sons: Hoboken, NJ, USA, 2015.
17. Sendi, P.; Zumstein, M.A.; Zimmerli, W. Protheseninfektionen—Eine Übersichtsarbeit für die Praxis. *Praxis* **2011**, *100*, 787–792. [CrossRef]
18. Li, C.; Renz, N.; Trampuz, A. Management of Periprosthetic Joint Infection. *Hip Pelvis* **2018**, *30*, 138–146. [CrossRef] [PubMed]
19. Boyle, K.K.; Wood, S.; Tarity, T.D. Low-Virulence Organisms and Periprosthetic Joint Infection-Biofilm Considerations of These Organisms. *Curr. Rev. Musculoskelet. Med.* **2018**, *11*, 409–419. [CrossRef] [PubMed]
20. Vasso, M.; Schiavone Panni, A. Low-grade periprosthetic knee infection: Diagnosis and management. *J. Orthop. Traumatol.* **2015**, *16*, 1–7. [CrossRef]
21. Hischebeth, G.T.; Randau, T.M.; Ploeger, M.M.; Friedrich, M.J.; Kaup, E.; Jacobs, C.; Molitor, E.; Hoerauf, A.; Gravius, S.; Wimmer, M.D. Staphylococcus aureus versus Staphylococcus epidermidis in periprosthetic joint infection-Outcome analysis of methicillin-resistant versus methicillin-susceptible strains. *Diagn. Microbiol. Infect. Dis.* **2019**, *93*, 125–130. [CrossRef] [PubMed]
22. Busscher, H.J.; van der Mei, H.C.; Subbiahdoss, G.; Jutte, P.C.; van den Dungen, J.J.A.M.; Zaat, S.A.J.; Schultz, M.J.; Grainger, D.W. Biomaterial-associated infection: Locating the finish line in the race for the surface. *Sci. Transl. Med.* **2012**, *4*, 1–11. [CrossRef]

23. Cristina, M.L.; Sartini, M.; Schinca, E.; Ottria, G.; Spagnolo, A.M. Operating room environment and surgical site infections in arthroplasty procedures. *J. Prev. Med. Hyg.* **2016**, *57*, E142.
24. Suzuki, A.; Namba, Y.; Matsuura, M.; Horisawa, A. Airborne contamination in an operating suite: Report of a five-year survey. *J. Hyg.* **1984**, *93*, 567–573. [CrossRef]
25. Elek, S.D.; Conen, P.E. The Virulence of Staphylococcus pyogenes for Man.: A Study of the Problems of Wound Infection. *Br. J. Exp. Pathol.* **1957**, *38*, 573–586.
26. Berríos-Torres, S.I.; Umscheid, C.A.; Bratzler, D.W.; Leas, B.; Stone, E.C.; Kelz, R.R.; Reinke, C.E.; Morgan, S.; Solomkin, J.S.; Mazuski, J.E.; et al. Centers for Disease Control and Prevention Guideline for the Prevention of Surgical Site Infection, 2017. *JAMA Surg.* **2017**, *152*, 784–791. [CrossRef] [PubMed]
27. Suhardi, V.J.; Bichara, D.A.; Kwok, S.J.J.; Freiberg, A.A.; Rubash, H.; Malchau, H.; Yun, S.H.; Muratoglu, O.K.; Oral, E. A fully functional drug-eluting joint implant. *Nat. Biomed. Eng.* **2017**, *1*, 80. [CrossRef] [PubMed]
28. Alt, V.; Kirchhof, K.; Seim, F.; Hrubesch, I.; Lips, K.S.; Mannel, H.; Domann, E.; Schnettler, R. Rifampicin-fosfomycin coating for cementless endoprostheses: Antimicrobial effects against methicillin-sensitive Staphylococcus aureus (MSSA) and methicillin-resistant Staphylococcus aureus (MRSA). *Acta Biomater.* **2014**, *10*, 4518–4524. [CrossRef] [PubMed]
29. Coraça-Huber, D.C.; Putzer, D.; Fille, M.; Hausdorfer, J.; Nogler, M.; Kühn, K.-D. Gentamicin palmitate as a new antibiotic formulation for mixing with bone tissue and local release. *Cell Tissue Bank.* **2014**, *15*, 139–144. [CrossRef]
30. Cao, Y.; Su, B.; Chinnaraj, S.; Jana, S.; Bowen, L.; Charlton, S.; Duan, P.; Jakubovics, N.S.; Chen, J. Nanostructured titanium surfaces exhibit recalcitrance towards Staphylococcus epidermidis biofilm formation. *Sci. Rep.* **2018**, *8*, 1071. [CrossRef]
31. Campoccia, D.; Montanaro, L.; Arciola, C.R. A review of the biomaterials technologies for infection-resistant surfaces. *Biomaterials* **2013**, *34*, 8533–8554. [CrossRef]
32. Alt, V. Antimicrobial coated implants in trauma and orthopaedics-A clinical review and risk-benefit analysis. *Injury* **2017**, *48*, 599–607. [CrossRef]
33. Hardes, J.; Ahrens, H.; Gebert, C.; Streitbuerger, A.; Buerger, H.; Erren, M.; Gunsel, A.; Wedemeyer, C.; Saxler, G.; Winkelmann, W.; et al. Lack of toxicological side-effects in silver-coated megaprostheses in humans. *Biomaterials* **2007**, *28*, 2869–2875. [CrossRef]
34. Bus, M.P.A.; van de Sande, M.A.J.; Fiocco, M.; Schaap, G.R.; Bramer, J.A.M.; Dijkstra, P.D.S. What Are the Long-term Results of MUTARS(R) Modular Endoprostheses for Reconstruction of Tumor Resection of the Distal Femur and Proximal Tibia? *Clin. Orthop. Relat. Res.* **2017**, *475*, 708–718. [CrossRef]
35. Hardes, J.; Henrichs, M.P.; Hauschild, G.; Nottrott, M.; Guder, W.; Streitbuerger, A. Silver-Coated Megaprosthesis of the Proximal Tibia in Patients with Sarcoma. *J. Arthroplast.* **2017**, *32*, 2208–2213. [CrossRef]
36. Hardes, J.; von Eiff, C.; Streitbuerger, A.; Balke, M.; Budny, T.; Henrichs, M.P.; Hauschild, G.; Ahrens, H. Reduction of periprosthetic infection with silver-coated megaprostheses in patients with bone sarcoma. *J. Surg. Oncol.* **2010**, *101*, 389–395. [CrossRef] [PubMed]
37. Donati, F.; Di Giacomo, G.; D'Adamio, S.; Ziranu, A.; Careri, S.; Rosa, M.; Maccauro, G. Silver-Coated Hip Megaprosthesis in Oncological Limb Savage Surgery. *Biomed. Res. Int.* **2016**, *2016*, 9079041. [CrossRef] [PubMed]
38. Zajonz, D.; Birke, U.; Ghanem, M.; Prietzel, T.; Josten, C.; Roth, A.; Fakler, J.K.M. Silver-coated modular Megaendoprostheses in salvage revision arthroplasty after periimplant infection with extensive bone loss–A pilot study of 34 patients. *BMC Musculoskelet. Disord.* **2017**, *18*, 383. [CrossRef] [PubMed]
39. Accentus Medical. How Agluna Works. Available online: http://www.accentus-medical.com/products-agluna.asp (accessed on 18 March 2020).
40. Schmidt-Braekling, T.; Streitbuerger, A.; Gosheger, G.; Boettner, F.; Nottrott, M.; Ahrens, H.; Dieckmann, R.; Guder, W.; Andreou, D.; Hauschild, G.; et al. Silver-coated megaprostheses: Review of the literature. *Eur. J. Orthop. Surg. Traumatol.* **2017**, *27*, 483–489. [CrossRef] [PubMed]
41. Scoccianti, G.; Frenos, F.; Beltrami, G.; Campanacci, D.A.; Capanna, R. Levels of silver ions in body fluids and clinical results in silver-coated megaprostheses after tumour, trauma or failed arthroplasty. *Injury* **2016**, *47* (Suppl. 4), S11–S16. [CrossRef] [PubMed]

42. Akiyama, T.; Miyamoto, H.; Yonekura, Y.; Tsukamoto, M.; Ando, Y.; Noda, I.; Sonohata, M.; Mawatari, M. Silver oxide-containing hydroxyapatite coating has in vivo antibacterial activity in the rat tibia. *J. Orthop. Res.* **2013**, *31*, 1195–1200. [CrossRef]
43. PMDA. PMDA_Products Approved in FY 2015 New Medical Devices. Available online: https://www.pmda.go.jp/english/review-services/reviews/approved-information/devices/0001.html (accessed on 18 March 2020).
44. Eto, S.; Kawano, S.; Someya, S.; Miyamoto, H.; Sonohata, M.; Mawatari, M. First Clinical Experience with Thermal-Sprayed Silver Oxide-Containing Hydroxyapatite Coating Implant. *J. Arthroplast.* **2016**, *31*, 1498–1503. [CrossRef]
45. Holinka, J.; Pilz, M.; Hirschl, A.M.; Graninger, W.; Windhager, R.; Presterl, E. Differential bacterial load on components of total knee prosthesis in patients with prosthetic joint infection. *Int. J. Artif. Organs* **2012**, *35*, 735–741. [CrossRef]
46. Kerativitayanan, P.; Tatullo, M.; Khariton, M.; Joshi, P.; Perniconi, B.; Gaharwar, A.K. Nanoengineered Osteoinductive and Elastomeric Scaffolds for Bone Tissue Engineering. *ACS Biomater. Sci. Eng.* **2017**, *3*, 590–600. [CrossRef]
47. Barry, M.; Pearce, H.; Cross, L.; Tatullo, M.; Gaharwar, A.K. Advances in Nanotechnology for the Treatment of Osteoporosis. *Curr. Osteoporos. Rep.* **2016**, *14*, 87–94. [CrossRef]
48. Khalilpour, P.; Lampe, K.; Wagener, M.; Stigler, B.; Heiss, C.; Ullrich, M.S.; Domann, E.; Schnettler, R.; Alt, V. Ag/SiO(x)C(y) plasma polymer coating for antimicrobial protection of fracture fixation devices. *J. Biomed. Mater. Res. B Appl. Biomater.* **2010**, *94*, 196–202. [CrossRef] [PubMed]
49. Alt, V.; Heiss, C.; Rupp, M. Treatment of a Recurrent Periprosthetic Joint Infection with an Intramedullary Knee Arthrodesis System with Low-Amount Metallic Silver Coating. *J. Bone Jt. Infect.* **2019**, 111–114. [CrossRef] [PubMed]
50. Alt, V.; Bitschnau, A.; Osterling, J.; Sewing, A.; Meyer, C.; Kraus, R.; Meissner, S.A.; Wenisch, S.; Domann, E.; Schnettler, R. The effects of combined gentamicin-hydroxyapatite coating for cementless joint prostheses on the reduction of infection rates in a rabbit infection prophylaxis model. *Biomaterials* **2006**, *27*, 4627–4634. [CrossRef] [PubMed]
51. Lucke, M.; Schmidmaier, G.; Sadoni, S.; Wildemann, B.; Schiller, R.; Haas, N.P.; Raschke, M. Gentamicin coating of metallic implants reduces implant-related osteomyelitis in rats. *Bone* **2003**, *32*, 521–531. [CrossRef]
52. Reizner, W.; Hunter, J.G.; O'Malley, N.T.; Southgate, R.D.; Schwarz, E.M.; Kates, S.L. A systematic review of animal models for Staphylococcus aureus osteomyelitis. *Eur. Cells Mater.* **2015**, 196–212. [CrossRef]
53. Reich, J.; Hovy, L.; Lindenmaier, H.-L.; Zeller, R.; Schwiesau, J.; Thomas, P.; Grupp, T.M. Präklinische Ergebnisse beschichteter Knieimplantate für Allergiker. *Orthopade* **2010**, *39*, 495–502. [CrossRef]
54. Grupp, T.M.; Giurea, A.; Miehlke, R.K.; Hintner, M.; Gaisser, M.; Schilling, C.; Schwiesau, J.; Kaddick, C. Biotribology of a new bearing material combination in a rotating hinge knee articulation. *Acta Biomater.* **2013**, *9*, 7054–7063. [CrossRef]
55. Bechert, T.; Steinrücke, P.; Guggenbichler, J.P. A new method for screening anti-infective biomaterials. *Nat. Med.* **2000**, *6*, 1053–1056. [CrossRef]
56. Bruenke, J.; Roschke, I.; Agarwal, S.; Riemann, T.; Greiner, A. Quantitative Comparison of the Antimicrobial Efficiency of Leaching versus Nonleaching Polymer Materials. *Macromol. Biosci.* **2016**, *16*, 647–654. [CrossRef]
57. Taylor, R.D. Modification of the Brown and Brenn gram stain for the differential staining of gram-positive and gram-negative bacteria in tissue sections. *Am. J. Clin. Pathol.* **1966**, *46*, 472–474. [CrossRef]
58. Schafer, K.A.; Eighmy, J.; Fikes, J.D.; Halpern, W.G.; Hukkanen, R.R.; Long, G.G.; Meseck, E.K.; Patrick, D.J.; Thibodeau, M.S.; Wood, C.E.; et al. Use of Severity Grades to Characterize Histopathologic Changes. *Toxicol. Pathol.* **2018**, *46*, 256–265. [CrossRef] [PubMed]
59. Klopfleisch, R. Multiparametric and semiquantitative scoring systems for the evaluation of mouse model histopathology—A systematic review. *BMC Vet. Res.* **2013**, *9*, 123. [CrossRef] [PubMed]
60. Austrian Standards. *Water Quality—Application of Inductively Coupled Plasma Mass Spectrometry (ICP-MS). Part. 2: Determination of Selected Elements Including Uranium Isotopes*; Austrian Standards: Wien, Austria, 2016; EN ISO 17294-2:2016.
61. Lee, J.Y.H.; Monk, I.R.; da Silva, A.G.; Seemann, T.; Chua, K.Y.L.; Kearns, A.; Hill, R.; Woodford, N.; Bartels, M.D.; Strommenger, B.; et al. Global spread of three multidrug-resistant lineages of Staphylococcus epidermidis. *Nat. Microbiol.* **2018**, *3*, 1175–1185. [CrossRef] [PubMed]

62. Heilmann, C.; Ziebuhr, W.; Becker, K. Are coagulase-negative staphylococci virulent? *Clin. Microbiol. Infect.* **2019**, *25*, 1071–1080. [CrossRef] [PubMed]
63. Rathbone, C.R.; Cross, J.D.; Brown, K.V.; Murray, C.K.; Wenke, J.C. Effect of various concentrations of antibiotics on osteogenic cell viability and activity. *J. Orthop. Res.* **2011**, *29*, 1070–1074. [CrossRef]
64. Pauksch, L.; Hartmann, S.; Rohnke, M.; Szalay, G.; Alt, V.; Schnettler, R.; Lips, K.S. Biocompatibility of silver nanoparticles and silver ions in primary human mesenchymal stem cells and osteoblasts. *Acta Biomater.* **2014**, *10*, 439–449. [CrossRef]
65. Pilz, M.; Staats, K.; Tobudic, S.; Assadian, O.; Presterl, E.; Windhager, R.; Holinka, J. Zirconium Nitride Coating Reduced Staphylococcus epidermidis Biofilm Formation on Orthopaedic Implant Surfaces: An In Vitro Study. *Clin. Orthop. Relat. Res.* **2019**, *477*, 461–466. [CrossRef]
66. Claro, T.; O'Reilly, M.; Daniels, S.; Humphreys, H. Surface microbial contamination in hospitals: A pilot study on methods of sampling and the use of proposed microbiologic standards. *Am. J. Infect. Control.* **2015**, *43*, 1000–1002. [CrossRef]
67. Best, E.; Parnell, P.; Couturier, J.; Barbut, F.; Le Bozec, A.; Arnoldo, L.; Madia, A.; Brusaferro, S.; Wilcox, M.H. Environmental contamination by bacteria in hospital washrooms according to hand-drying method: A multi-centre study. *J. Hosp. Infect.* **2018**, *100*, 469–475. [CrossRef]

materials

MDPI

Article

Comparison of Different Locking Mechanisms in Total Hip Arthroplasty: Relative Motion between Cup and Inlay

Sebastian Jaeger [1,*,†], Maximilian Uhler [1,*,†], Stefan Schroeder [1], Nicholas A. Beckmann [2,3] and Steffen Braun [1]

1 Laboratory of Biomechanics and Implant Research, Clinic for Orthopedics and Trauma Surgery, Heidelberg University Hospital, Heidelberg University, 69118 Heidelberg, Germany; stefan.schroeder@med.uni-heidelberg.de (S.S.); Steffen.Braun@med.uni-heidelberg.de (S.B.)

2 Clinic for Orthopedics and Trauma Surgery, Heidelberg University Hospital, Heidelberg University, 69118 Heidelberg, Germany; nick.beckmann@gmail.com

3 Department of Orthopaedic Surgery and Traumatology, Inselspital, Bern University Hospital, 3010 Bern, Switzerland

* Correspondence: sebastian.jaeger@med.uni-heidelberg.de (S.J.); Maximilian.Uhler@med.uni-heidelberg.de (M.U.)

† Both authors contributed equally to the work and share first authorship.

Received: 21 February 2020; Accepted: 16 March 2020; Published: 19 March 2020

check for updates

Abstract: The resulting inflammatory reaction to polyethylene (PE) wear debris, which may result in osteolysis, is still considered to be a main reason for aseptic loosening. In addition to the primary wear in hip joint replacements caused by head-insert articulation, relative motions between the PE liner and the metal cup may cause additional wear. In order to limit this motion, various locking mechanisms were used. We investigated three different locking mechanisms (Aesculap, DePuy, and Zimmer Biomet) to address the resulting relative motion between the acetabular cup and PE liner and the maximum disassembly force. A standardized setting with increasing load levels was used in combination with optically based three-dimensional measurements. In addition the maximum disassembly forces were evaluated according to the ASTM F1820-13 standard. Our data showed significant differences between the groups, with a maximum relative motion at the maximum load level (3.5 kN) of 86.5 ± 32.7 μm. The maximum axial disassembly force was 473.8 ± 94.6 N. The in vitro study showed that various locking mechanisms may influence cup-inlay stability.

Keywords: cup-inlay stability; total hip arthroplasty; disassembly forces; relative motion

1. Introduction

Total hip arthroplasty (THA) is one of the most successful orthopedic procedures in joint replacement [1,2]. The THA shows survival rates between 82% and 96% after 10 years, depending on the age of the patient and their physical functions [3]. However, the resulting inflammatory reaction to polyethylene (PE) wear debris, which may result in osteolysis [4], is still considered to be a main reason for aseptic loosening [5–7]. Therefore, wear debris production is thought to be the main factor limiting long-term survival of THA. It is known that not only can primary wear processes at the articulation sliding surfaces of hip cup and femoral head play an essential role, but in addition, secondary PE wear also takes also place at the backside of the PE liners (backside wear) due to relative motion between cup and insert [8–11]. A non-conforming fit between cup and PE liner, in combination with increased relative motion, could lead to burnishing, gouging, scratching or third-body wear on the PE backside surface [12]. In order to limit the motion between the metal shell and the liner, various locking mechanisms are used. It is assumed that a firm connection of the components will reduce the relative motion and the particle migration inside the cup system [13,14]. Another rather rare reason for revision is the dislocation between cup and PE liner, which presumably can be attributed to the acetabular locking mechanism.

The aim of this study was to evaluate specific locking mechanisms regarding the resulting relative motion between the acetabular cup and the PE liner. In addition, the maximum dissembling forces were investigated. For this purpose, three implant designs were tested, to determine if different types of locking mechanisms affect the magnitude of relative motion and dissembling force.

2. Materials and Methods

2.1. Acetabular Hip Cup Systems

Three acetabular titanium hip cup designs, with different inner designs to fixate the PE liner, were investigated (Figure 1): (A) Allofit®-S Alloclassic® with a cross-linked Durasul®-PE liner (Zimmer Biomet, Warsaw, IN, USA), (B) Pinnacle®-Multihole combinates with a cross-linked Marathon®-PE liner (DePuy Synthes, Warsaw, IN, USA), (C) Plasmafit® Plus7 with a conventional ultra-high-molecular-weight polyethylene (UHMWPE) liner (Aesculap, Tuttlingen, BW, Germany). All hip cup systems were non-cemented with an outer cup diameter of 52 mm and a corresponding 32 mm PE liner. In each group, six acetabular hip cups with corresponding PE liners were used.

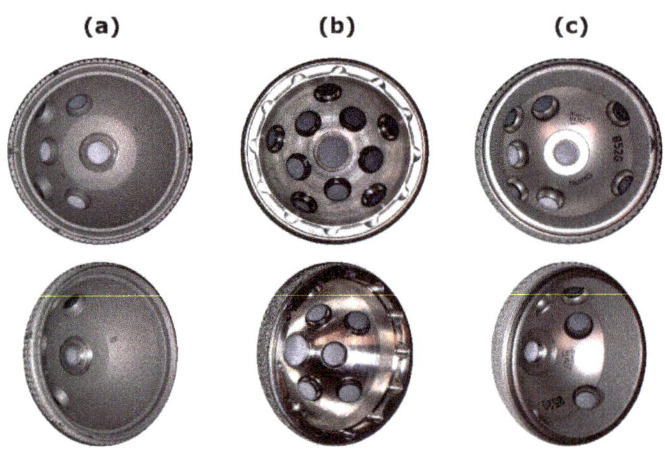

Figure 1. Acetabular titanium hip cup components. (**a**) Zimmer Biomet, (**b**) DePuy Synthes, (**c**) Aesculap.

The PE liner of the hip cup system Allofit®-S Alloclassic® is fixed by a circular press fit mechanism. In addition, two spikes in the inner polar region of the cup on which the PE liner is inserted improve rotational stability. The inner anchoring surface of the cup is spherical, and has a smooth surface (Ra = 1 μm; Rz = 6–8 μm) [15].

The anchoring of the PE liner of the hip cup system Pinnacle®-Multihole is realised by a central dome region and a taper lock mechanism. Additionally, 12 grooves are placed at the rim of the shell. Corresponding to the grooves, six tabs on the liner were found to improve rotational stability. The inner anchoring surface of the cup showed values of Ra = 0.2 μm and Rz = 1 μm.

The PE liner of the hip cup system Plasmafit® Plus7 is fixed by a conical locking mechanism. The cup shows a rough inner surface (Ra = 4 μm, Rz = 25 μm) [9,15].

An overview of the locking mechanisms and the surface specifications of the different cup systems are shown in Figure 2.

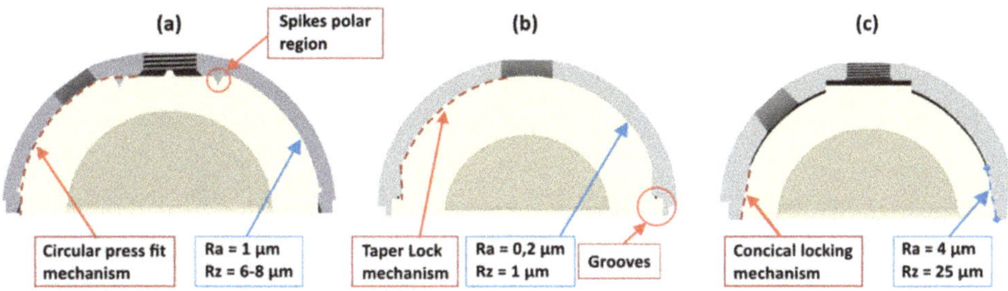

Figure 2. Locking mechanisms and surface specifications of the different cup systems. (**a**) Zimmer Biomet, (**b**) DePuy Synthes, (**c**) Aesculap.

2.2. Test Setup

In order to carry out a standardized PE liner stability test, the cups were fixed according to Braun et al. in polyurethane (RenCast FC 53 A/B, Gößl & Pfaff GmbH, Karlskron, Germany) at an angle of 30 degrees to the vertical load axis [9]. A loading scenario was implemented using a uniaxial servo hydraulic testing machine (Bosch Rexroth, Lohr am Main, Germany). The components were loaded with an increasing dynamic load at a frequency of 1 Hz and a sinus shaped wave form. The initial load level started at 0.5 kN and increased incrementally with each load level by 1 kN, up to the maximum load of 3.5 kN. The cyclic load at each load level was maintained over 1000 cycles. The fixed acetabular cups were integrated into the testing machine and mounted on a rocker (Figure 3b). The rocker was placed on a free linear bearing to prevent transverse forces [9,15].

A rigid connection was generated between femoral head and PE liner. This, in combination with a 2 mm offset (Figure 3b L) between the centre of the implant and the rotational axis (Figure 3b green) of the rocker, initiated torque in the cup-system (Figure 3b) [9,15]. In combination with the axial compression a theoretically torque of 1, 3, 5 and 7 Nm were applied.

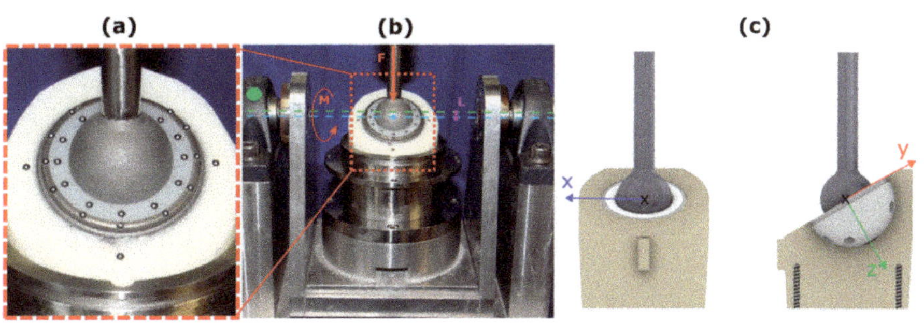

Figure 3. (**a**) Arrangement of the optical marker; (**b**) Relative motion measurement setup; (**c**) Camera coordinate system.

2.3. Determination of Relative Motion

The determination of relative motion was performed by placing optical markers (uncoded passive white markers, diameter 0.8 mm, GOM Item Number: 21874; GOM Gmbh, Braunschweig, Germany) along the rim of the hip cup and PE liner (Figure 3a). These markers were detected using a stereo camera system and a triangulation algorithm to calculate the 3D marker position (x-, y-, and z-axes). Based on the measured marker motion in the respective spatial direction (x, y and z direction), the resulting vector was calculated using the equation $XYZ = \sqrt{(X^2 + Y^2 + Z^2)}$. In a next step, the cup and PE liner markers were separated and the resultant relative motion XYZ between both components was calculated. This calculation was carried out for every measurement time point. In a further step, the maximum of the resulting relative motion (XYZ) was determined for each individual load cycle in order to calculate the mean value of the maximum resulting relative motion for each load level (0.5, 1.5, 2.5 and 3.5 kN) [16]. The coordinate system was defined as shown in Figure 3c. The relative motion between the components was measured at cycles 10, 100, 300, 500, 800 and 995 for each load level.

2.4. Disassembly Test

In addition, the maximum disassembly force of the PE Liner was determined according to the ASTM F1820-13 standard. The PE liners of the investigated cup designs were assembled in the shell with a mean peak load of 2002.2 N ± 1.1 N using a material testing machine (Zwick Roell, Z005, Ulm, Germany). Therefore, a displacement control rate of 0.04 mm/s was used. The force was applied by a femoral head coincident with the polar axis of the liner. The assembled liner shell constructs were placed in a fixture frame, with continuous support of the cup as shown in Figure 4. The axial force was applied through the centre hole of the cup with a 6 mm rod and a rate of 51 mm/min. The maximum disassembly force was recorded [17].

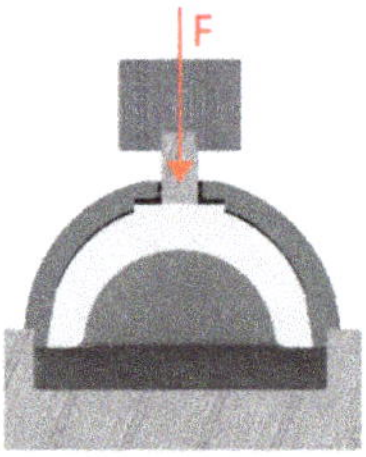

Figure 4. Test Setup Disassembly Test.

2.5. Statistical Analysis

A repeated measures analysis of variance (ANOVA) was conducted to test for significant differences in relative motion depending on the load level for all investigated cup designs. Additionally, we conducted a one-way ANOVA to assess the effect of cup design on the maximum pull out force and relative motion. To identify which particular differences between pairs of means were significant, a post hoc analysis was performed. We used a Bonferroni Test as post hoc analysis to explore differences between the three group means while controlling the experiment wise error rate. Pre-analysis, the normal distribution of the data was evaluated using a Shapiro–Wilk-test and the homogeneity of variance was verified using the Levene-test. The results allowed for the use of the ANOVA test. For the repeated measures ANOVA, a Greenhouse–Geisser adjustment was used to correct for violations of sphericity.

Additionally, the data were evaluated descriptively using the arithmetic mean, standard deviation, minimum and maximum. The data were analysed using SPSS 25 (IBM, Armonk, NY, USA) with a significance level of $p < 0.05$.

3. Results

3.1. Relative Motion

The mean resulting maximum relative motion in XYZ-direction for each load level and all investigated cup designs are shown in Table 1 and Figure 5.

Table 1. Mean Relative Motion XYZ ± (μm).

Load Level (kN)	Zimmer Biomet	DePuy	Aesculap
0.5	58.8 ± 32.6	15.2 ± 8.3	6.2 ± 1.1
1.5	74.3 ± 33.7	43.9 ± 14.5	12.5 ± 6.4
2.5	82.8 ± 33.5	66.3 ± 16.1	25.1 ± 8.8
3.5	86.5 ± 32.7	81.1 ± 18.5	38.2 ± 9.9

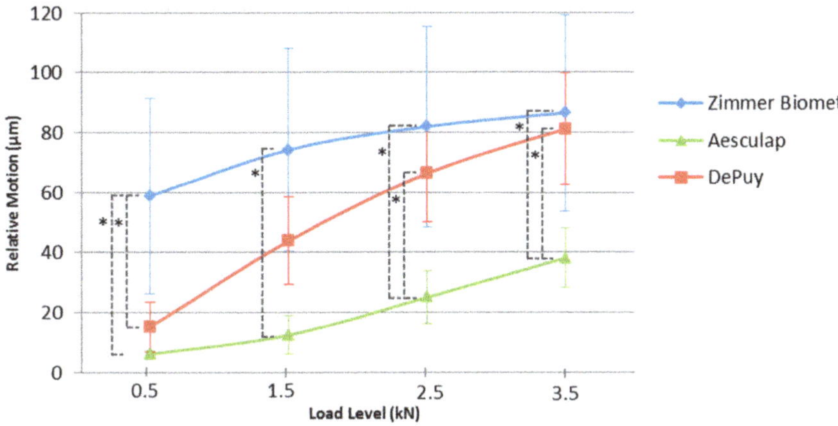

Figure 5. Resulting maximum relative motion, * showed significant difference.

For the load level of 0.5 kN the Bonferroni-adjusted post-hoc analysis revealed a significant difference ($p = 0.001$) for mean resulting maximum relative motion of the Zimmer Biomet and the Aesculap group (52.67, 95%-CI [82.85, 22.49]) and also between the Zimmer Biomet and DePuy group ($p = 0.004$) (43.61, 95%-CI [13.43, 73.79]). There was no significant difference ($p = 1.0$) for the relative motion between the Aesculap and DePuy groups. The load level of 1.5 kN showed a significant difference ($p < 0.001$) for the relative motion between the Zimmer Biomet and the Aesculap group (61.75, 95%-CI [28.31, 95.18]). There was no significant difference between the Zimmer Biomet and DePuy groups ($p = 0.082$) or between the Aesculap and DePuy groups ($p = 0.070$). The load level of 2.5 kN showed a significant difference ($p = 0.001$) between the Zimmer Biomet and Aesculap groups (56.85, 95%-CI [22.52, 91.18]) and between the Aesculap and DePuy groups ($p = 0.017$) (41.23, 95%-CI [6.90, 75.56]). There was no statistically significant difference for the relative motion between the Zimmer Biomet and DePuy groups ($p = 0.718$). For the load level of 3.5 kN, the Bonferroni-adjusted post-hoc analysis revealed a significant difference ($p = 0.006$) for the mean resulting maximum relative motion between the Zimmer Biomet and Aesculap groups (48.28, 95%-CI [13.41, 83.14]) and between the Aesculap and DePuy groups ($p = 0.014$) (42.92, 95%-CI [8.06, 77.79]). There was no statistically significant difference for the relative motion between the Zimmer Biomet and DePuy groups ($p = 1.0$). The repeated measures ANOVA with a Greenhouse–Geisser correction determined that the load levels showed a significant influence on the relative motion, $F\ (1.264, 18.964) = 309.886$, $p < 0.001$.

3.2. Disassembly Test

The mean maximum axial disassembling forces showed the highest values for the Aesculap group at 473.7 ± 94.6 N, followed by the Zimmer Biomet group at 294.8 ± 48.2 N and the DePuy group at 146.8 ± 49.8 N (Figure 6). The mean maximum force differed statistically significantly for the investigated cup designs, $F\ (2, 15) = 29.25$, $p < 0.001$.

The Bonferroni-adjusted post-hoc analysis revealed a significant difference for the mean maximum disassembling force between the Aesculap and Zimmer Biomet groups ($p = 0.002$), (178.90, 95%-CI

[63.60, 294.20]) and between the Aesculap and DePuy groups ($p < 0.001$), (326.90, 95%-CI [211.60, 442.20]). There was also a significant difference between the Zimmer Biomet and DePuy groups ($p = 0.011$), (148.00, 95%-CI [32.70, 263.30]).

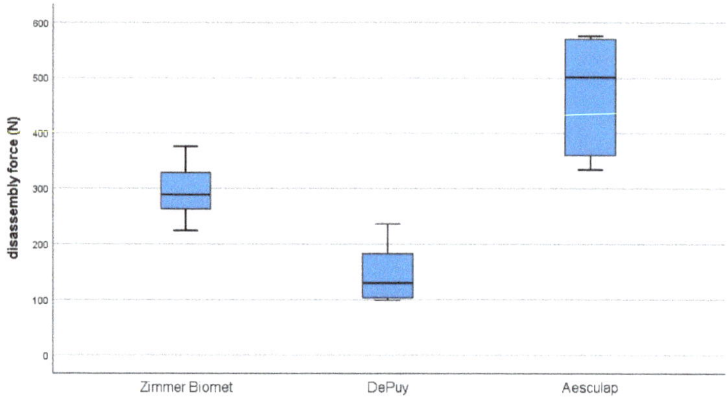

Figure 6. Resulting maximum axial disassembling forces.

4. Discussion

Osteolysis-induced bone destruction and the following aseptic loosening is one of the major complications of prosthetic hip replacement [5–7]. PE wear particles can cause osteolysis due to an inflammatory reaction. Existing publications proved that, additionally to the PE wear due to the main articulation, wear could be a consequence of the backside wear of the convex side of the PE liner [18]. The design and material could have an impact on the damage of the backside of polyethylene in modern modular acetabular cups [19]. The motion between the PE liner and the cup of a modular acetabular component exacerbated the polyethylene wear as well. This kind of micromotion, respectively, relative motion between cup and liner, is influenced by several factors. Factors could include a reduced conformity, screw holes, and different locking mechanisms and geometries [12,14,20,21].

Kurtz et al. determined the relative motion between cup and liner using a three-dimensional finite element model [12]. The focus was on considering the influence of nonconforming between liner and metal cup. Furthermore, the effects of rim and equatorial restraints were considered. The result of the study stated that backside nonconformity and locking restraints have a relevant influence on backside relative motion. Systems with a conforming cup exhibited an axial motion between 8.5 and 12.8 μm [12]. Nonconforming models offered up to 63% higher motion [12]. An in vitro study was performed by Williams et al. to determine the fixation of PE liners in metal cups [14]. Therefore, the rotational and axial motion was measured using linearly variable differential transducers. Axial loads from 0.272 kN to 2.720 kN and torsional loads from ± 7.5 Nm over 10 million cycles were applied. In the evaluation of the results, it could be derived that the rim, rotational and dome micromotion decreased as the cycles increased. Additionally an influence of the different locking mechanisms was identified. Williams et al. achieved micromotions between 3.35 ± 2.50 μm and 164.7 ± 112.5 μm after 1 million cycles [14].

Our data also showed considerable differences between cup and PE liner. Regardless of the implant design and loading situation, the range was between 6.2 ± 1.1 μm and 86.5 ± 32.7 μm. The Zimmer Biomet hip cup system with a circular snap-fit mechanism, in combination with two spikes in the inner polar region of the cup, showed relative motion between 58.8 ± 32.6 μm and 86.5 ± 32.7 μm (0.5 and 3.5 kN). The DePuy system with a central dome region and a taper lock mechanism plus grooves at the rim of the shell showed relative motion between 15.2 ± 8.3 μm and 81.1 ± 18.5 μm (0.5 and 3.5 kN). For the Aesculap system with a conical locking mechanism and a locking free contact with the base of the cup we found relative motion between 6.2 ± 1.1 μm and 38.2 ± 9.9 μm (0.5 and 3.5 kN).

For the initial load level of 0.5 kN, the DePuy and the Aesculap cups showed no significant differences ($p = 1.0$). The Zimmer Biomet cup showed significantly higher values compared to the Aesculap and DePuy cups ($p = 0.001$, $p = 0.004$). At the highest load level of 3.5 kN, the Zimmer Biomet group and DePuy system no longer showed any significant difference ($p = 1.0$). However, the Aesculap system still showed significantly lower values compared to the Zimmer Biomet and DePuy cups ($p = 0.006$, $p = 0.014$). The locking mechanism of the Aesculap system showed the lowest relative motion between cup and PE liner for the investigated load levels compared to the Zimmer Biomet and DePuy cup. A retrieval study conducted by Wasielewski et al. evaluated the backside of 55 polyethylene liners [22]. At this juncture, a distinction was made between micromotion caused by rotational instability of the insert in the shell and micromotion induced through elastic or plastic deformation. A total of 52 of the 55 inserts showed polyethylene wear on the backside in varying degrees. The most severe wear occurred in the posterior-superior and anterior-superior areas. In addition, five of six patients with acetabular osteolysis showed a high grade of backside wear, which was mainly attributed to elastic or plastic deformation related micromotion [22].

Another aspect besides the relative motion between cup and PE liner was the maximum disassembly force with regard to the risk of dislocation. Therefore, the maximum disassembly force was determined according to ASTM F1820-13 standard [17]. The maximum disassembling force showed the highest values for the Aesculap group with 473.7 ± 94.6 N, followed by the Zimmer Biomet with 294.8 ± 48.2 N and DePuy with 146.8 ± 49.8 N. In comparison, the Aesculap group showed significantly higher forces compared to the Zimmer Biomet and DePuy groups ($p = 0.002$, $p = 0.001$). There was also a significant difference between the Zimmer Biomet and DePuy groups ($p = 0.011$). In a comparative study of eight different cup systems, Tradonsky et al. determined a disassembly force between 129 N and 2950 N [23]. Our data are in the lower third compared to Tradosky et al. However, acetabular liner dissociation is a rare complication following [24].

The results of our investigation showed that a tapered locking mechanism and non-locking contact with the bottom of the cup (Aesculap) showed a significant improvement in relative motion and disassembly force compared to a system with a central dome region and a taper lock mechanism (smooth surface) plus grooves at the rim (DePuy) and a hip cup system with a circular snap-fit mechanism in combination with two spikes in the inner polar region of the cup (Zimmer Biomet).

This relative motion between cup and PE liner could be a predictor of backside wear. However, features such as the surface structure, specifically roughness of the inner geometry of the acetabular cup, could also have a major impact on backside wear behaviour. Braun et al. showed 45% higher backside wear particles for the Aesculap system (Plasmafit® Plus7) compared to the Zimmer Biomet system (Allofit®-S Alloclassic®) using an identical experimental setup [9].

In addition, Braun et al. verified that backside wear particles can occur in clinically established cup systems [15]. Furthermore, Reyna et al. was able to detect backside wear caused by micromotion and poor conformity between cup and liner in an in vitro study [25]. As already mentioned, these particles could increase the potential of a resulting inflammatory reaction, which may result in osteolysis. Compared to the in vitro data from Braun et al. and Reyna et al., similar effects could be demonstrated in an in vivo study from Long et al. [26]. In this study, a series of early aseptic loosenings of acetabular cups were identified, an analysis of x-ray images was realized and an optical analysis of the inlays was carried out. The osteolysis that occurred behind the cup could be attributed to backside wear particles in conjunction with optical wear marks on the backside of the PE inlay [26]. Due to the fact that backside wear particles are also significantly smaller than articulating wear particles, they may additionally cause stronger biological responses [9].

However, experimental and clinical studies can show that backside wear in combination with relative motion between the cup and PE liner may have an impact on the osseointegration of the acetabular cup. Furthermore, subsequent studies have to show the characteristics (particle size, roundness and aspect ratio) of the PE particles that may arise and prove their influence on the human body.

From these findings, and the measured results in our study, it could be concluded that there is not a one dimensional relation between a tight fit of the inlay inside the cup, which is reflected in a low relative motion and increased disassembly forces, and the amount of backside wear. In addition, a closer look will be needed at the inner structure of the cup. It is necessary to distinguish between a smooth and a rough surface. A smooth surface with a high rate of relative motion could generate fewer backside wear particles than a rough surface with less motion.

Limitations

Initially, it should be noted that the fixation of the cups by fixating them in polyurethane does not correspond to the physiological fixation in vivo. However, the advantage of this connection was standardization of the fixation and the focus on the locking mechanisms of the cup systems. Other influencing factors could be excluded. There was a defined impaction force applied on the inlay, but no peak force as it occurs at the impaction by the surgeon.

In the analysis of motion, this work deals with relative motion. The relative motion between cup and inlay, however, could be a combination of plastic and elastic deformation, migration, rotation, creeping and many other factors.

5. Conclusions

This experimental study showed significant differences in relative motion at the cup–liner interface and in the disassembly forces for the three investigated locking mechanisms. How relative motion and assembly force affect backside wear in combination with different roughnesses should be analyzed in more detail in a further investigation.

Author Contributions: Conceptualization, S.J., M.U. and S.B.; Data curation, S.J. and M.U.; Formal analysis, S.J., M.U., S.S. and N.A.B.; Methodology, S.B.; Project administration, S.J., M.U. and S.B.; Writing—original draft, S.J. and M.U.; Writing—review & editing, S.J., M.U., S.S., N.A.B. and S.B. All authors have read and agreed to the published version of the manuscript.

Funding: This research received no external funding.

Acknowledgments: We acknowledge financial support for publication by the Baden-Wuerttemberg Ministry of Science, Research and the Arts and by Ruprecht-Karls-Universitaet Heidelberg.

Conflicts of Interest: The authors declare no pertinent conflict of interest. S.J. report grants from B Braun Aesculap, Johnson & Johnson Depuy Synthes, Heraeus Medical, Waldemar Link, Peter Brehm and Zimmer Biomet that are not related to the current study. M.U., S.S. and S.B. report no conflict of interest. N.A.B. reports research grants from Johnson & Johnson DePuy Synthes that are not related to the current study.

References

1. Crawford, R.W.; Murray, D.W. Total hip replacement: Indications for surgery and risk factors for failure. *Ann. Rheum. Dis.* **1997**, *56*, 455–457. [CrossRef] [PubMed]
2. Pivec, R.; Johnson, A.J.; Mears, S.C.; Mont, M.A. Hip arthroplasty. *Lancet* **2012**, *380*, 1768–1777. [CrossRef]
3. Makarewich, C.A.; Anderson, M.B.; Gililland, J.M.; Pelt, C.E.; Peters, C.L. Ten-year survivorship of primary total hip arthroplasty in patients 30 years of age or younger. *Bone Jt. J.* **2018**, *100-B*, 867–874. [CrossRef] [PubMed]
4. Hallab, N.J.; Jacobs, J.J. Biologic effects of implant debris. *Bull. NYU Hosp. Jt. Dis.* **2009**, *67*, 182–188.
5. Dumbleton, J.H.; Manley, M.T.; Edidin, A.A. A literature review of the association between wear rate and osteolysis in total hip arthroplasty. *J. Arthroplast.* **2002**, *17*, 649–661. [CrossRef]
6. Harris, W.H. Wear and periprosthetic osteolysis: The problem. *Clin. Orthop. Relat. Res.* **2001**, *393*, 66–70. [CrossRef]
7. Prokopovich, P. Interactions between mammalian cells and nano- or micro-sized wear particles: Physico-chemical views against biological approaches. *Adv. Colloid Interface Sci.* **2014**, *213*, 36–47. [CrossRef]
8. Akbari, A.; Roy, M.E.; Whiteside, L.A.; Katerberg, B.J.; Schnettgoecke, D.J. Minimal backside surface changes observed in retrieved acetabular liners. *J. Arthroplast.* **2011**, *26*, 686–692. [CrossRef]

9. Braun, S.; Sonntag, R.; Schroeder, S.; Mueller, U.; Jaeger, S.; Gotterbarm, T.; Kretzer, J.P. Backside wear in acetabular hip joint replacement. *Acta Biomater.* **2019**, *83*, 467–476. [CrossRef]
10. Kretzer, J.P.; Zietz, C.; Schröder, C.; Reinders, J.; Middelborg, L.; Paulus, A.; Sonntag, R.; Bader, R.; Utzschneider, S. Grundlagen zur tribologischen analyse von endoprothesen. *Der Orthopäde* **2012**, *41*, 844–852. [CrossRef]
11. McKellop, H.A. The lexicon of polyethylene wear in artificial joints. *Biomaterials* **2007**, *28*, 5049–5057. [CrossRef] [PubMed]
12. Kurtz, S.M.; Ochoa, J.A.; White, C.V.; Srivastav, S.; Cournoyer, J. Backside nonconformity and locking restraints affect liner/shell load transfer mechanisms and relative motion in modular acetabular components for total hip replacement. *J. Biomech.* **1998**, *31*, 431–437. [CrossRef]
13. Kyle, R.F.; Riddle, C.D.; Kyle, M.; Rockswold, S.; Bourgeault, C. Factors influencing the initial micromotion between polyethylene acetabular cups and titanium alloy shells. *J. Arthroplast.* **2006**, *21*, 443–448. [CrossRef]
14. Williams, V.G., 2nd; Whiteside, L.A.; White, S.E.; McCarthy, D.S. Fixation of ultrahigh-molecular-weight polyethylene liners to metal-backed acetabular cups. *J. Arthroplast.* **1997**, *12*, 25–31. [CrossRef]
15. Braun, S.; Vardag, S.; Mueller, U.; Schroeder, S.; Sonntag, R.; Bormann, T.; Gotterbarm, T.; Kretzer, J.P. Backside wear, particle migration and effectiveness of screw hole plugs in acetabular hip joint replacement with cross-linked polyethylene. *Acta Biomater.* **2019**, *97*, 239–246. [CrossRef]
16. Beckmann, N.A.; Bitsch, R.G.; Gondan, M.; Schonhoff, M.; Jaeger, S. Comparison of the stability of three fixation techniques between porous metal acetabular components and augments. *Bone Jt. Res.* **2018**, *7*, 282–288. [CrossRef]
17. ASTM F1820-13. *Standard Test Method for Determining the Forces for Disassembly of Modular Acetabular Devices;* ASTM International: West Conshohocken, PA, USA, 2013.
18. Barrack, R.L.; Folgueras, A.; Munn, B.; Tvetden, D.; Sharkey, P. Pelvic lysis and polyethylene wear at 5-8 years in an uncemented total hip. *Clin. Orthop. Relat. Res.* **1997**, *335*, 211–217. [CrossRef]
19. Bali, K.; McCalden, R.W.; Naudie, D.D.; MacDonald, S.J.; Teeter, M.G. Backside wear is not dependent on the acetabular socket design in crosslinked polyethylene liners. *Clin. Orthop. Relat. Res.* **2016**, *474*, 374–382. [CrossRef] [PubMed]
20. Kligman, M.; Furman, B.D.; Padgett, D.E.; Wright, T.M. Impingement contributes to backside wear and screw-metallic shell fretting in modular acetabular cups. *J. Arthroplast.* **2007**, *22*, 258–264. [CrossRef] [PubMed]
21. Lieberman, J.R.; Kay, R.M.; Hamlet, W.P.; Park, S.H.; Kabo, J.M. Wear of the polyethylene liner-metallic shell interface in modular acetabular components. An in vitro analysis. *J. Arthroplast.* **1996**, *11*, 602–608. [CrossRef]
22. Wasielewski, R.C.; Jacobs, J.J.; Arthurs, B.; Rubash, H.E. The acetabular insert-metal backing interface: An additional source of polyethylene wear debris. *J. Arthroplast.* **2005**, *20*, 914–922. [CrossRef] [PubMed]
23. Tradonsky, S.; Postak, P.D.; Froimson, A.I.; Greenwald, A.S. A comparison of the disassociation strength of modular acetabular components. *Clin. Orthop. Relat. Res.* **1993**, *296*, 154–160. [CrossRef]
24. O'Neill, C.K.; Napier, R.J.; Diamond, O.J.; O'Brien, S.; Beverland, D.E. Acetabular liner dissociation following total hip arthroplasty: A rare but serious complication that may be easily misinterpreted in the emergency department. *Case Rep. Emerg. Med.* **2015**, *2015*, 802753. [CrossRef] [PubMed]
25. Puente Reyna, A.L.; Jager, M.; Floerkemeier, T.; Frecher, S.; Delank, K.S.; Schilling, C.; Grupp, T.M. Backside wear analysis of retrieved acetabular liners with a press-fit locking mechanism in comparison to wear simulation in vitro. *BioMed Res. Int.* **2016**, *2016*, 8687131. [CrossRef] [PubMed]
26. Long, W.J.; Nayyar, S.; Chen, K.K.; Novikov, D.; Davidovitch, R.I.; Vigdorchik, J.M. Early aseptic loosening of the tritanium primary acetabular component with screw fixation. *Arthroplast. Today* **2018**, *4*, 169–174. [CrossRef]

materials

MDPI

Article

Influence of the Acetabular Cup Material on the Shell Deformation and Strain Distribution in the Adjacent Bone—A Finite Element Analysis

Danny Vogel *, Matthias Klimek, Michael Saemann and Rainer Bader

Biomechanics and Implant Technology Research Laboratory, Department of Orthopaedics, Rostock University Medical Center, Doberaner Straße 142, 18057 Rostock, Germany; matthias.klimek@gmx.de (M.K.); michael.saemann@med.uni-rostock.de (M.S.); rainer.bader@med.uni-rostock.de (R.B.)

* Correspondence: danny.vogel@med.uni-rostock.de; Tel.: +49-381-494-9375

Received: 16 January 2020; Accepted: 13 March 2020; Published: 18 March 2020

Abstract: In total hip arthroplasty, excessive acetabular cup deformations and altered strain distribution in the adjacent bone are potential risk factors for implant loosening. Materials with reduced stiffness might alter the strain distribution less, whereas shell and liner deformations might increase. The purpose of our current computational study was to evaluate whether carbon fiber-reinforced poly-ether-ether-ketones with a Young´s modulus of 15 GPa (CFR-PEEK-15) and 23 GPa (CFR-PEEK-23) might be an alternative shell material compared to titanium in terms of shell and liner deformation, as well as strain distribution in the adjacent bone. Using a finite element analysis, the press-fit implantation of modular acetabular cups with shells made of titanium, CFR-PEEK-15 and CFR-PEEK-23 in a human hemi-pelvis model was simulated. Liners made of ceramic and polyethylene were simulated. Radial shell and liner deformations as well as strain distributions were analyzed. The shells made of CFR-PEEK-15 were deformed most (266.7 μm), followed by CFR-PEEK-23 (136.5 μm) and titanium (54.0 μm). Subsequently, the ceramic liners were radially deformed by up to 4.4 μm and the polyethylene liners up to 184.7 μm. The shell materials slightly influenced the strain distribution in the adjacent bone with CFR-PEEK, resulting in less strain in critical regions (<400 μm/m or >3000 μm/m) and more strain in bone building or sustaining regions (400 to 3000 μm/m), while the liner material only had a minor impact. The superior biomechanical properties of the acetabular shells made of CFR-PEEK could not be determined in our present study.

Keywords: modular acetabular cup; poly-ether-ether-ketone (PEEK); titanium; ceramics; ultra-high-molecular-weight polyethylene (UHMW-PE); implant deformation; strain distribution; bone stock

1. Introduction

The aseptic loosening of the acetabular cup due to stress shielding and altered strain distributions within the adjacent bone stock is a common cause for the failure of a total hip replacement [1]. If the modular cups are stiffer than the adjacent bone cavity, the strain distribution is altered, leading to high strains in some regions and low strains in others. It results in a potential risk of bone fracture on the one hand and bone atrophy on the other [2–4]. To avoid these effects, the stiffness of the acetabular cup should be similar to the adjacent bone [5]. Dickinson et al. investigated the influence of the material of monolithic acetabular cups on the pelvis cortex surface strains using a composite bone model. It was shown that monolithic cups made of ultra-high molecular weight polyethylene (UHMWPE) have less influence on the load transfer into the bone stock, when compared to metallic cups, as the reduced elastic modulus of UHMWPE bears more similarity to the bone stock [2]. In terms of the liner material, Kim et al. observed no significant differences in bone remodeling after five years postoperatively when ceramic and UHMWPE liners combined with structurally identical metallic shells were compared [6]. Therefore, the shell material of modular acetabular cups might have a stronger influence on the strain distribution than the liner material and shells made of polymers might have a lower influence on strain distribution.

Besides UHMWPE, poly-ether-ether-ketone (PEEK), often applied with carbon fibers for reinforcement (CFR-PEEK), is a biocompatible polymer that has been introduced as a material for monolithic cups and liners, but not as a material for acetabular shells [7–11].

However, PEEK and CFR-PEEK might be suitable alternatives to metal shells in modular acetabular cups as well, in order to avoid adverse effects on stress and strain distribution within the bone stock. However, it was shown that the reduced stiffness of shells made of PEEK and CFR-PEEK leads to increased shell deformations in the case of press-fit fixation in a finite element analysis (FEA) [12]. In particular, shells made of pure PEEK without reinforcement were deformed excessively, resulting in strong deformations of the liners, which might cause further problems in vivo. Excessive liner deformations can lead to a reduced clearance between the liner and femoral ball head and subsequently to increased frictional torques and increased wear rates [13–16]. Moreover, the seating of the liner might be jeopardized by strong shell deformations [17,18], which might lead to increased peak stresses within ceramic liners and, therefore, an increased risk of fracture [19–21]. Thus, shells made of PEEK without carbon fiber reinforcement do not seem to be a suitable alternative, whereas the suitability of shells made of CFR-PEEK needs to be further evaluated.

Hence, the aim of this computational study was to evaluate whether acetabular shells made of CFR-PEEK can serve as suitable alternatives to shells made of titanium in terms of strain distribution in the adjacent bone and whether the stiffness of such shells is sufficient to withstand adverse shell and liner deformations when they are pushed into an under-reamed bone cavity.

2. Materials and Methods

In the present numerical study, the insertion of a modular acetabular cup into a hemi-pelvis was simulated using ABAQUS/CAE (v 6.12-2, Dassault Systèmes Simulia Corp., Providence, RI, USA). The modular cup was designed based on a commercially available cup consisting of a titanium shell for press-fit fixation and a ceramic liner. The shell was flattened and spherical, with an outer diameter of 54 mm and a wall thickness of approximately 5.8 mm in the region of press-fit contact. The liner had an inner diameter of 36 mm and a maximum wall thickness of approximately 3.9 mm. In terms of simplification and comparability, the liner geometry was identical for both liner materials (ceramic and polyethylene), even so the real parts would have geometric differences in the dependency of the material. To apply realistic joint loads, a ceramic ball head was simulated in the form of an ideal sphere, with a diameter of 36 mm.

The hemi-pelvis included an acetabulum, which was equipped with a bone cavity, which was adapted to the shells outer diameter to simulate a diametric press-fit of 1 mm. The modular acetabular cup was oriented at 45° of inclination and 15° of anteversion, relative to the pelvic plane.

Due to the complex geometry, the hemi-pelvis was meshed using tetrahedral elements (141,877 elements), while the shell, liner, and ball head were meshed with 5987, 6162, and 14,400 hexahedral elements, respectively. To avoid excessive computational time, the components had to be meshed using linear elements. The element length at the contact area of the hemi-pelvis was set to 1.0 mm and the global element length of the acetabular shell, liner and head were set to 1.5 mm, 1.5 mm and 1.25 mm, respectively. A convergence analysis was carried out previously in order to determine the sensitivity of the results in relation to the mesh density, the element length of each component was varied between 1 mm and 4 mm. To exclude the influence of contact definitions on the results of the convergence analysis, each component was evaluated separately by means of deformation analyses. Thereby, the individual components were deformed by a defined amount and the deforming force was evaluated. The deforming forces determined with the chosen mesh densities changed less than 6% compared to the finest mesh.

The shell material was varied between a titanium alloy and two different kinds of CFR-PEEK (CFR-PEEK-15 and CFR-PEEK-23), while the liner material was varied between alumina toughened zirconia (ATZ) ceramic and UHMWPE. All the chosen materials were defined to be homogenous and isotropic (Table 1). As the yield strengths of the polymers were not exceeded, all the materials were simulated to be linear elastic.

Table 1. Overview of the Young's modulus and Poisson's ratio defined for the various simulated materials.

Part	Material	Young's Modulus (MPa)	Poisson's Ratio
Acetabular shell	Titanium	110,000	0.40
	Carbon fiber-reinforced poly-ether-ether-ketone (CFR-PEEK-15)	15,000	0.40
	Carbon fiber-reinforced poly-ether-ether-ketone (CFR-PEEK-23)	23,000	0.40
Liner	Alumina toughened zirconia (ATZ) ceramic	261,000	0.27
	Ultra-high-molecular-weight polyethylene (UHMW-PE) [22]	945	0.45
Ball head	Alumina toughened zirconia (ATZ) ceramic	261,000	0.27

The numerically simulated hemi-pelvis was reconstructed from computed tomography (CT) data of a human cadaveric hemi-pelvis (Ethics Committee of the University of Rostock; Reg. No. A 2009 38), using an algorithm introduced by Kluess et al. [23]. The local X-ray attenuation from the CT resulted in a distribution of Hounsfield units (HU) in the CT slices, which directly correlate with bone density and were therefore mapped onto the FE mesh using a previously described approach [24]. For this, the HU of the CT dataset were treated as temperatures of a temperature-dependent material model assigned to the FE nodes. This results in a realistic, heterogeneous distribution of material characteristics throughout the bone geometry. To assign a corresponding Young's modulus to the HU, the HU from the CT were correlated with the local apparent densities, and the apparent densities were correlated with the Young's modulus. Due to the fact that the existing CT data provided no scanned bone mineral density phantom, an apparent density of 1.8 g/cm^3 was assigned to the maximal HU value, as per Taddei et al. [25]. The HU values of the cancellous bone were averaged, and the averaged value was assigned with an apparent density of 0.425 g/cm^3, which represents a reasonable value for cancellous pelvic bone [26]. A linear correlation between the HU and apparent density was assumed using the following equation:

$$\rho_{app}\ (\text{g/cm}^3) = 0.0007918{*}\text{HU} + 0.4718988 \quad (1)$$

To calculate the Young's modulus from the apparent density, the following equation based on the equation of Carter and Hayes was used [27]:

$$\text{E (GPa)} = 3.79\text{e}^{0.06}{\rho_{app}}^3 \quad (2)$$

Forty equally spaced reference points were selected between the maximum and minimum HU, at which the Young's modulus was calculated and assigned. Between the reference points, a linear interpolation was used. Moreover, a lower limit of 500 MPa and an upper limit of 20,000 MPa were defined as thresholds to adequately represent the stiffness limits of the cancellous and cortical bones [28]. The mapping of the HU values and correlation to local stiffness was achieved with the self-developed software script AbaCTMat, based on Python 2.6.2 and implemented as plug-in in ABAQUS/CAE.

The model included three sliding contact formulations, which were described by normal and tangential contact. A penalty friction model with Coulomb friction formulation was chosen. The first contact was defined between the outer surface of the shell and the hemi-pelvis. The friction coefficient was set to 0.6, which represents a reasonable value for contacts between bones and porous coatings [29]. The second contact was defined between the shell and the liner, where a friction coefficient of 0.16

was applied [30]. The third contact was defined between the liner and the ball head, with a friction coefficient of 0.05 [13].

The simulation was executed in subsequent steps. Initially, the constraints for the hemi-pelvis were applied at the sacroiliac joint and the pubic symphysis, whereby the translational degrees of freedom at the sacroiliac joint were completely fixed (BC-1), while the motion in the sagittal plane was enabled at the pubic symphysis (BC-2) [13] (Figure 1). In the first step, the acetabular shell was pushed under displacement-control into the cavity, until a predefined shell overhang was reached using a kinematic coupling definition (Figure 1b). In the following step no forces or displacements were applied, to enable an elastic relaxation of the shell and cavity. Thereafter, the liner was also moved in a displacement-controlled manner, until a first contact between the shell rim and the liner was achieved (Figure 1c), followed by a force-controlled insertion with 500 N using a kinematic coupling definition again (Figure 1d). The coupling was applied to the distal inner surface of the liner, to not bias the radial deformation. Subsequently, the femoral ball head was inserted in a displacement-controlled manner into the liner to initiate contact (Figure 1e), followed by the application of a realistic hip joint load (Figure 1f). A load during normal walking (resulting load: 784.8 N) was chosen in accordance with Bergmann et al. [31] and applied via a reference point to the ball head using a kinematic coupling, again. The load was divided into the different load compounds in the x-, y-, and z-directions (F_x = 502.3 N, F_y = −78.5 N and F_z = 2048.3 N).

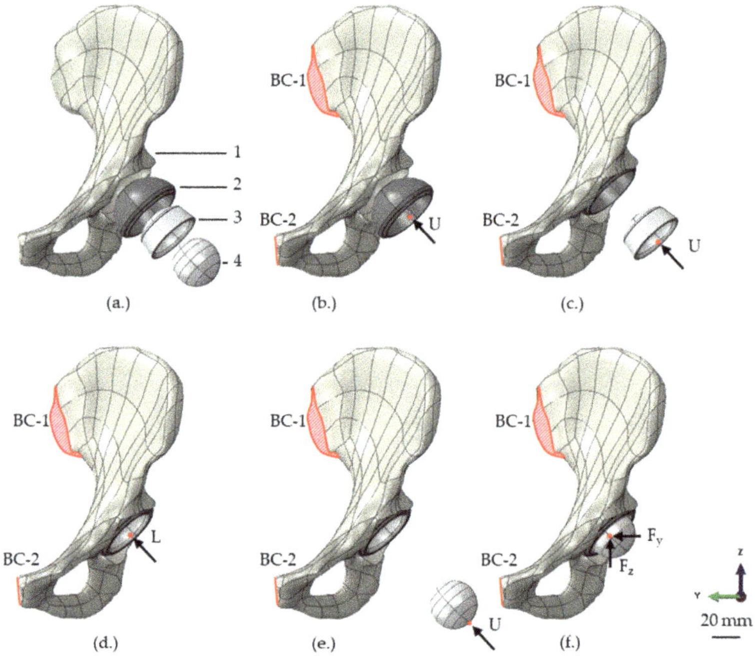

Figure 1. Depiction of the assembly (**a**) of the 3D finite element model consisting of a hemi-pelvis with a prepared bone cavity (1), acetabular shell (2), liner (3) and femoral ball head (4). The simulation was executed in several steps, consisting of the displacement-controlled insertion of the shell (**b**), displacement-controlled motion of the liner until the first contact to the shell (**c**), force-controlled insertion of the liner (**d**), displacement-controlled insertion of the femoral ball head (**e**) and the final loading, by applying a realistic hip joint load in x-, y- and z-direction (F_x, F_y and F_z) via the femoral ball head (**f**). During all simulation steps, the translational degrees of freedom at the sacroiliac joint were completely fixed (BC-1) and the pubic symphysis was fixed in the y-direction (BC-2).

The numerical simulations were performed as non-linear static calculations with incremental loading conditions. The FEA was evaluated in terms of the maximum radial shell and liner deformation after insertion and subsequent relaxation. Therefore, polar coordinate systems were created along the rotational axis of the shell and liner, respectively, and circumferential node paths along the inner rim of the acetabular shell (62 nodes) and liner (47 nodes) were defined and evaluated.

Moreover, the strain distribution over the full surface of the hemi-pelvis was investigated at a total of 29,923 node points (Figure 2c). It should be noted that only the magnitudes of the strains were considered and therefore no distinction was made between the compression and tensile strains.

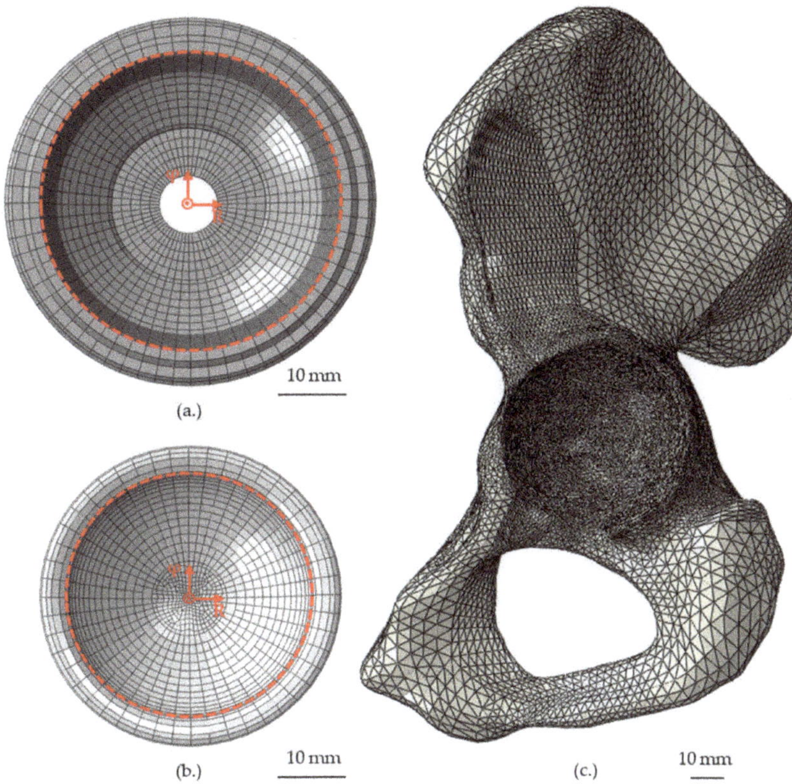

Figure 2. Depiction of the node paths at the inner rim of the shell (**a**) and liner (**b**), consisting of 62 and 47 nodes, to analyze the radial deformation in polar coordinate systems. Moreover, the analyzed node region in terms of strain distribution at the surface of the hemi-pelvis is depicted (**c**).

The determined logarithmic strains were divided into different strain regions, dependent on the bone response, based on studies by Biewener [32] and Frost [33], as summarized in Table 2.

Table 2. Values set for bone thresholds and response to particular strain rates [32,33].

Strain (μm/m)	Bone Response
<400	Atrophy
400–3000	Bone preserving and building
3000–20,000	Yielding
>20,000	Fracture

3. Results

3.1. Shell and Liner Deformation

After the insertion into the hemi-pelvis and subsequent relaxation of the shells, the shell made of CFR-PEEK-15 showed the highest deformation (266.7 µm), followed by the shell made of CFR-PEEK-23 (136.5 µm) and the shell made of titanium, which exhibited the smallest maximum deformation (54.0 µm). Therefore, the deformation of shells made of CFR-PEEK-15 was approx. two and five times the deformation of shells made of CFR-PEEK-23 and titanium, respectively, while the deformation of a shell made of CFR-PEEK-23 was approx. 2.5 times higher than the deformation of the titanium shell. The shells made of CFR-PEEK were compressed in the region of the Os ilium and Os ischium and expanded in the direction of the acetabular notch (Figure 3a).

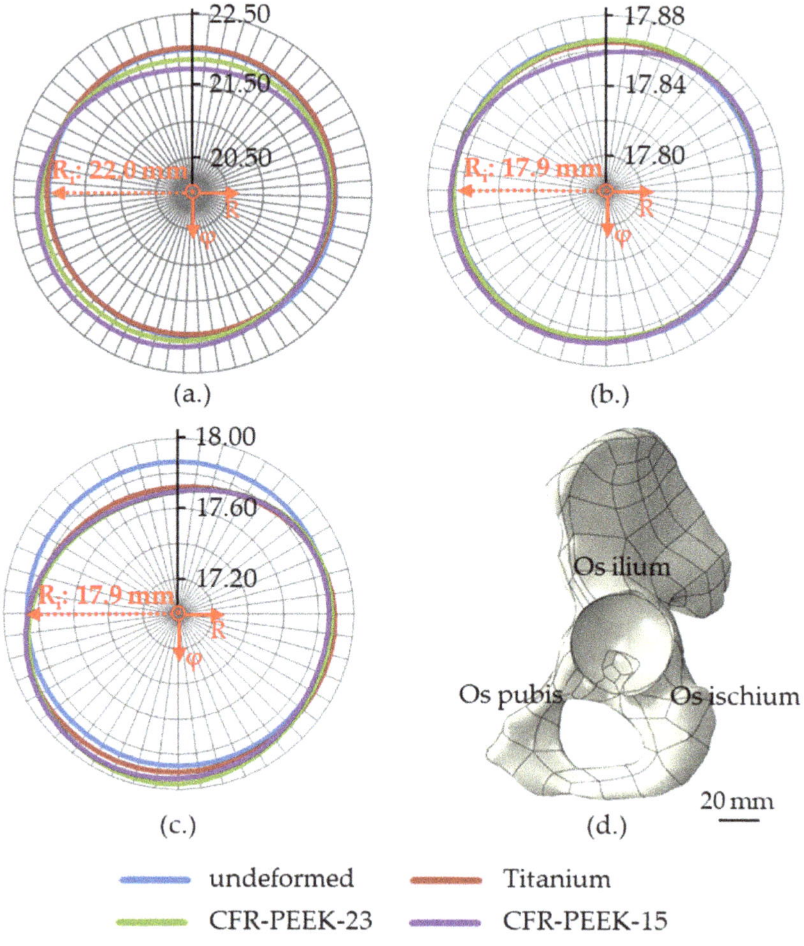

Figure 3. Exaggerated radial deformation (mm) at the inner rim of the shell (**a**) and liners made of ATZ ceramic (**b**) and UHMWPE (**c**) dependent on the chosen shell material and the orientation of the hemi-pelvis (**d**). The deformations are depicted in a polar coordinate system and in relation to the initial radii (R_i).

The deformation of the shells influenced the radial deformation of the liners, with the largest liner deformation resulting in combination with the shell made of CFR-PEEK-15, followed by CFR-PEEK-23. The UHMWPE liners displayed greater deformation compared to ceramic liners. The maximum radial liner deformations have been summarized in Table 3.

Table 3. Maximum radial liner deformations dependent on the acetabular shell and liner material.

Shell Material	Titanium		CFR-PEEK-23		CFR-PEEK-15	
Liner material	Ceramic	UHMWPE	Ceramic	UHMWPE	Ceramic	UHMWPE
Maximum deformation (μm)	2.0	149.0	1.8	184.7	4.4	180.6

Like the shells, the liners were compressed in the region of the Os ilium and Os ischium and expanded in the direction of the acetabular notch (Figure 3b,c).

3.2. Strain Distribution

The resulting strains in the hemi-pelvis were influenced by the shell material (Table 4). The strongest influence could be seen for strains beneath 3000 μm/m. The percentage of nodes in the region of expected bone loss (<400 μm/m) was higher in shells made of CFR-PEEK, while the percentage of nodes in a region of bone preservation and building (400–3000 μm/m) was lower. On the other hand, the percentage of nodes with strains above 20,000 μm/m was higher in shells made of titanium.

Table 4. Comparison of the percentage of analyzed nodes, categorized into the different strain regions after shell relaxation and dependent on the shell material (Titanium, CFR-PEEK-23 or CFR-PEEK-15).

Strain Region (μm/m)	Titanium	CFR-PEEK-23	CFR-PEEK-15
<400 (Atrophy)	17.2%	18.2%	18.3%
400 < x < 3000 (Bone preserving and building)	40.6%	39.7%	39.5%
3000 < x < 20,000 (Yielding)	39.8%	40.2%	40.5%
>20,000 (Fracture)	2.4%	1.8%	1.7%

On the macroscopic scale, there were almost no differences of the visible strain distribution (Figure 4). The strongest alteration of the strain was seen at the acetabular fossa, where the strain increased when CFR-PEEK was used instead of titanium. Minor deviations were seen at the Os ischium and Os pubis.

The liner material had a small influence on the strains in the hemi-pelvis after insertion and subsequent relaxation, with more nodes with a strain beneath 400 μm/m and less nodes between 400 μm/m and 20,000 μm/m, when UHMWPE was used (Table 5). Due to the joint load, the overall strains increased and the influence of the liner material was compensated (Table 5 and Figure 5). The percentage of nodes with strains beneath 400 μm/m was reduced up to 73.0%, due to the hip load, while the percentage of nodes with strains in the range from 400 μm/m to 3000 μm/m and from 3000 μm to 20,000 μm/m increased up to 21.0% and 19.8%, respectively.

After the insertion of the liner, no relevant differences in strain distribution in the dependency of the shell or liner material were visible on the macroscopic scale. Only small areas near the acetabular rim showed altered strain distribution due to the shell material (Figure 5a,b). After loading with a hip joint load, the strains in the acetabulum and the wing of ilium increased strongly (Figure 5c,d). A small effect of the shell material was seen in the same regions as in the unloaded condition. In Figure 5, only the shells made of titanium and CFR-PEEK-15 with a liner made of UHMWPE were compared as an example. The results of the other simulations showed similar strain patterns.

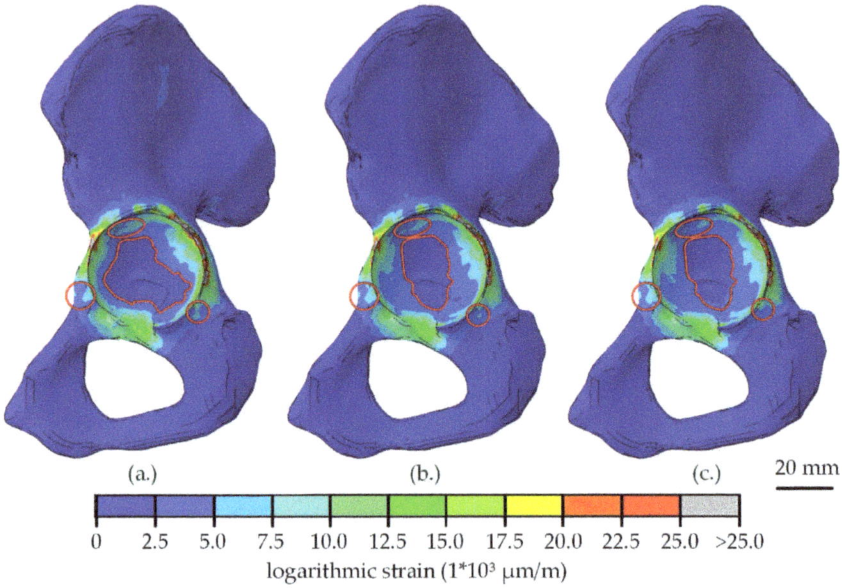

Figure 4. Strain distribution in the adjacent bone of the hemi-pelvis after relaxation of the acetabular shell, dependent on the chosen shell material (**a**) titanium, (**b**) CFR-PEEK-23 and (**c**) CFR-PEEK-15. Differences visible on the macroscopic scale are highlighted in red.

Table 5. Comparison of the percentage of analyzed nodes, categorized into the different strain regions after liner insertion and after loading in dependency of the acetabular shell and liner material.

Strain Regions (μm/m)	Load Scenario	Titanium		CFR-PEEK-23		CFR-PEEK-15	
		Ceramic	UHMWPE	Ceramic	UHMWPE	Ceramic	UHMWPE
<400 (Atrophy)	Liner Insertion	17.3%	19.3%	18.9%	20.2%	19.3%	20.6%
	Loading	5.3%	5.3%	5.6%	5.5%	5.4%	5.6%
400 < x < 3000 (Bone preserving and building)	Liner Insertion	41.1%	39.1%	41.2%	39.9%	41.3%	40.8%
	Loading	47.2%	47.3%	48.5%	48.1%	48.0%	48.5%
3000 < x < 20,000 (Yielding)	Liner Insertion	39.3%	39.3%	38.1%	38.0%	37.7%	36.9%
	Loading	45.2%	45.1%	44.1%	44.5%	44.7%	44.2%
>20,000 (Fractures)	Liner Insertion	2.3%	2.3%	1.8%	1.9%	1.7%	1.7%
	Loading	2.3%	2.3%	1.8%	1.9%	1.9%	1.8%

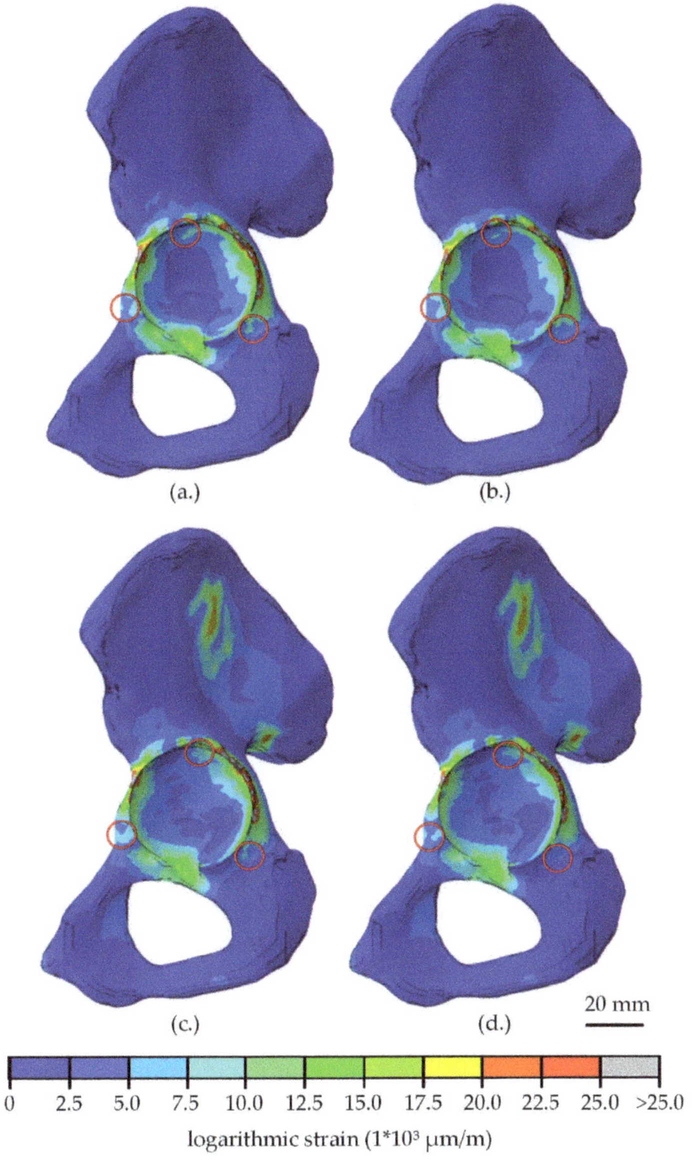

Figure 5. Comparison of strain distribution in the adjacent bone of the hemi-pelvis after the UHMWPE liner insertion into the titanium (**a**) and CFR-PEEK-15 (**b**) shell, followed by a relaxation. Additional the alteration of strain distribution after the application of a hip joint load in dependency of the shell material (**c**) titanium, (**d**) CFR-PEEK-15.

4. Discussion

In total hip arthroplasty, the stiffness of modular acetabular cups should be similiar to the adjacent bone, to avoid stress and strain shielding, that could otherwise lead to subsequent atrophy in the adjacent bone [5,34]. Compared to titanium, polymer materials are characterized by Young's modulus closer to bone and, therefore, might alter the stress and strain distribution to a lesser degree. Particularly, reinforced polymers, such as CFR-PEEK, might be a suitable alternative for acetabular cups, as their material properties can be altered to fit the bones' properties by varying the proportion of carbon fibers. However, the reduced stiffness compared to titanium shells might lead to increased acetabular shell and liner deformation, that can subsequently lead to a reduced clearance, possibly resulting in higher frictional torques and wear rates [13–16,19,20].

CFR-PEEK has previously been tested in numerical, experimental and in vivo studies as a material for monolithic cups and liners, but barely as a shell material [7–11,35,36]. Therefore, the purpose of our current computational study was to evaluate whether the stiffness of two different CFR-PEEK materials is sufficient to avoid excessive shell and liner deformation when they are used as shell materials for modular acetabular cups and whether these shells may reduce the strains in the adjacent bone.

Our study is restricted by certain limitations. First of all, only one hemi-pelvis was used in the current study and, thus, anatomical differences and differences in terms of age-related bone properties could not be taken into account. In particular, the bone density might be crucial in terms of shell deformation, even though previous studies concluded that altered bone properties do not influence the stress distribution greatly [36]. Moreover, the hemi-pelvis was CT-scanned without an additional bone mineral density phantom. Therefore, the equation to correlate the HU and apparent density had to be set manually, using additional data from the literature [25,26]. To calculate the Young's modulus from the apparent density, an equation based on an equation by Carter and Hayes [27], which was originally set up to calculate the compressive modulus depending on the apparent density and strain rate, was used as this equation can be used for cancellous as well as cortical bones. The numerically simulated hemi-pelvis showed unreasonably small HU in a region in the Os ilium; therefore, a minimum Young's modulus of 500 MPa was established, to avoid areas with excessively low modulus. The maximum Young's modulus of the cortical bone was set to 20,000 MPa, which is high compared to other studies, in which the maximal Young's modulus for cortical bones was set to between 8500 MPa and 18,000 MPa [20,23,36,37]. This is applicable, because the highest HU was defined as the point of maximum stiffness, but this point still includes naturally occurring deviations in x-ray attenuation which, however, are compensated over several HU. This still may lead to a slight overestimation of the bone stiffness, but is favored in order to consider a disadvantageous scenario in terms of acetabular shell and liner deformations. Moreover, these deformations were not experimentally validated in the current study, as the simulated hemi-pelvis was not accessible for experimental analysis. It is assumed that a qualitative comparison between the shells is legitimate, as the evaluated feature for shell and liner deformation, as well as strain distribution, was the material definition of the shell. Additionally, the material behavior of the CFR-PEEK was simplified. The CFR-PEEK was simulated as elastic with isotropic behavior, despite CFR-PEEK being an anisotropic material with its mechanical properties depending on the carbon fiber orientation. By adjusting the fiber orientation in the direction of strongly loaded areas, the shell could be strengthened, possibly leading to reduced shell deformations in these areas. Hence, an advanced material definition should be considered for future investigations. Moreover, it is known that implants made of CFR-PEEK have to be coated to enable osseointegration in vivo [38–40]. Even so, it was previously shown that the coating procedure can influence the mechanical properties (e.g., yield strength) of PEEK materials [41], the material properties of uncoated CFR-PEEK were used in the current study in terms of simplification.

Despite these limitations, the current study was suitable for providing an insight into whether CFR-PEEK can be used as an alternative shell material in the context of acetabular shell and liner deformation, as well as strain distribution in the adjacent bone. In terms of shell deformation, a correlation between the shell material stiffness and deformation value was determined, as shown previously [12]. The determined maximum radial deformation of the titanium shell after insertion into the bone cavity was 54.0 µm. When the shells made of CFR-PEEK-23 or CFR-PEEK-15 with a reduced stiffness were simulated, the radial shell deformation increased by about 2.5 times and 5 times, respectively. In a previous study, the same shell was numerically inserted into bone substitute cavities, with a diametric press-fit of 2 mm. The deformation of the titanium shell in the present study is about three times, and the deformation of the shell made of CFR-PEEK-23 is still about 1.5 times higher compared to the data of the previous study, even so the diametric press-fit was only 1 mm, indicating that the simulated hemi-pelvis is stiffer compared to the bone substitute material used before [12]. In another study, Everitt et al. inserted monolithic cups made of CFR-PEEK with a Young's modulus of 15 GPa into cavities made of polyurethane foams (20 pcf or 30 pcf), with a diametric press-fit of 1 mm or 2 mm [42]. The monolithic cup was deformed by about 250 μm when inserted into a 1 mm under-reamed cavity made of 20 pcf polyurethane foams, which is in good agreement with the present deformation in the case of the shell made of CFR-PEEK-15. When the monolithic cups were inserted in 2 mm under-reamed cavities or cavities made of 30 pcf polyurethane foams, the deformation doubled [42]. In case of modular cups, an increased shell deformation results in greater liner deformations and stresses, harboring the risk of reduced clearance, increased frictional torques, incomplete seating and misalignment, as well as higher wear rates at the articulating surfaces [13–16]. Therefore, a diametric press-fit above 1 mm or the insertion of acetabular shells made of CFR-PEEK in young patients with high bone mineral density could be critical in terms of deformation, and the applicability of CFR-PEEK shells has to be cautiously questioned in these cases. In further investigations, the effect of an increased press-fit and increased bone mineral density on the implant deformation should be considered. On the other hand, one advantage of CFR-PEEK is the ability to adjust the material properties for a specific application. For example, the orientation of the carbon fibers could be altered to create an anisotropic material that can be used for acetabular shells. It would be conceivable to radially reinforce the shell in order to avoid strong deformations, while remaining flexible in the direction of the load application. Another opportunity to alter the mechanical properties of PEEK is by adding various additives (like TiO2) to the PEEK powder before molding, which can increase the bioactivity and therefore osseointegration at the same time [43].

The shells were mainly compressed in the contact area at the Os ilium and Os ischium, and expanded in the direction of the Os pubis and acetabular notch. Therefore, the shell deformation was consistent with the distribution of the bone density, as the deformation was initiated from the stiffest region. This is also in line with previous studies, in which similar deformation patterns were found [13,20,44]. In the present study, only the initial deformation of the shells was determined, even though creep deformation over time is a well-known problem of polymers. The creep deformation of shells made of polymers might lead to liner migration and, subsequently, to altered liner deformations and stresses. In worst-case scenarios, this could lead to increased rates of ceramic liner fractures, as reported for sandwich liners [45–51]. Therefore, the creep deformation of acetabular shells made of CFR-PEEK should be considered in future long-term experimental investigations.

The radial liner deformation in the present study was in a range between 149.0 μm (titanium) and 184.7 μm (CFR-PEEK-23) in case of the UHMWPE liner, and 1.8 μm (CFR-PEEK-23) and 4.4 mm (CFR-PEEK-15) in case of the ceramic liner after the load-controlled insertion and subsequent relaxation. Therefore, a linear correlation between shell and liner deformation was not shown, although previous studies have shown that increased shell deformations lead to the increased deformation of the inserted liners [12,13,46]. The non-uniform liner deformations might occur due to other factors, like the liner's initial seating depth after load-controlled insertion, and a varying bounce-back after the liner relief.

Compared to the above mentioned study [12] in which the same acetabular cups were numerically inserted into bone substitute cavities, the deformation of the UHMWPE liner was deformed six times more in case of the titanium shell and 1.5 times in case of the CFR-PEEK-23 shell during the load-controlled liner insertion. After subsequent relaxation, the liner deformation decreased, but was still three times and 1.5 times higher, respectively. However, when the determined liner deformations are compared to typical radial clearance of 300 μm for hard-on-soft bearings [52], the articulation between a femoral ball head and an UHMWPE liner would not be hampered in combination with the simulated shell materials. The deformations of the ceramic liner were also influenced by the chosen shell material, but were less critical regarding the clearance. The maximum radial deformation of the ceramic liner ranged from 1.8–4.4 μm and, thus, was approximately 20% of a typical radial clearance for ceramic-on-ceramic bearings of 20 μm [53]. The maximum determined peak tensile stress of the ceramic liner (129.9 MPa), arising during load-controlled liner insertion, was nowhere near a critical value in terms of ceramic fracture as well [54]. However, under unfavorable conditions, such as an increased press-fit, a decreased wall thickness of the shell, a high bone mineral density, or increased loads, the radial deformation of the shell would increase, leading to increased liner deformation, increased stresses and a further decrease of the clearance, especially in the case of the UHMWPE liners [12–14,17,18,42,44,55–57]. Decreased clearances are known to increase the wear rates [15] and frictional torques at the articulating surfaces and therefore, might hamper the initial fixation of acetabular cups [16].

The magnitudes of the strains in the hemi-pelvis were slightly influenced by the stiffness of the shell material. With the decreasing stiffness of the shells, the proportion of nodes with a strain of less than 400 μm/m increased, indicating that more nodes are located in a bone resorbing range in the case of shells made of CFR-PEEK as compared to titanium, with the most affected region detected at the acetabular fossa. At the regions where the shell is in direct contact with the bone, the highest strains occurred (overloading), indicating a possible risk of bone fracture during shell insertion. In these regions, the strains were the lowest when the acetabular shell made of CFR-PEEK-15 was simulated. However, it is assumed that the determined strains would be reduced in vivo over time, due to the viscoelastic response of the adjacent bone stock, allowing the re-expansion of the shells [14,56].

Overall, only small effects of the shell material on the strains were determined, supporting the conclusions of other studies that a simple change in material stiffness is not sufficient to avoid the loss of acetabular bone [36,58]. The chosen liner material had nearly no impact on the strain distribution under loading, which is also in line with the findings of other studies [6,59,60].

The strain distribution pattern at the surface of the bone cavity was similar to previous studies [36,59]. The peak strains occurred at the rim of the reamed bone cavity, and there was nearly no strain at the acetabular fossa, as shown previously [36,59,61]. In our present FE study, only the magnitudes were evaluated. We found that the strains in the acetabulum were mainly tensile strains, except for the surfaces in direct contact to the shell, where compressive strains arose.

In the present study, only three loading states were analyzed: either directly after the shell and liner insertion with subsequent relaxation (no external loading), or at the peak load of a normal gait cycle [31,62]. We showed that nodes in the bone sustaining or building region of 400–3000 μm/m increased up to 21.0% during the loading, compared to the unloaded cup. Therefore, the loading condition was crucial in terms of strain distribution, as expected, and more loading conditions, such as stair-climbing, sitting down and standing up, or stumbling, with higher loads and altered loading directions should be considered in further investigations [63].

5. Conclusions

In conclusion, the deformation of acetabular shells made of CFR-PEEK is more pronounced as compared to titanium shells during intraoperative insertion, resulting in the increased radial deformation of liners. However, liner deformations were not critical in terms of the radial clearance in our present computational study. Particularly, the combination of shells made of CFR-PEEK and liners made of ATZ ceramic showed only minor differences in implant deformations, as compared to titanium shells. In terms of strain distribution, only small effects were observed to be dependent on the shell and liner material, with a greater area of bone being in a bone-sustaining or building region when CFR-PEEK was used. Therefore, clear advantages in terms of strain distribution were not seen in our computational study. Acetabular shells made of CFR-PEEK are not likely to replace titanium shells in case of standard THA. However, as the strain distribution was not impaired and the shell and liner deformations were not critical, shells made of CFR-PEEK might be a possible alternative in special cases. However, in our present study only shell and liner deformations as well as strain distribution in the adjacent bone were considered, but even so, more parameters have to be taken into account before a final statement can be made about the suitability of acetabular shells made of CFR-PEEK. Therefore, further experimental investigations are needed, e.g., to determine the creep and wear properties of acetabular shells made of CFR-PEEK.

Author Contributions: Conceptualization: D.V. and R.B.; methodology: D.V., M.K., M.S. and R.B.; software: D.V., M.K. and M.S.; data curation: D.V. and M.K.; formal analysis: D.V. and M.K.; investigation: D.V. and M.K.; resources: R.B.; writing—original draft preparation: D.V.; writing—review and editing: D.V., M.K., M.S. and R.B.; visualization: D.V. and M.K.; supervision: D.V. and R.B. All authors have read and agreed to the published version of the manuscript.

Funding: This research received no external funding.

Conflicts of Interest: The authors declare no conflict of interest.

References

1. Herberts, P.; Malchau, H. Long-term registration has improved the quality of hip replacement: A review of the Swedish THR Register comparing 160,000 cases. *Acta Orthop. Scand.* **2000**, *71*, 111–121. [CrossRef]
2. Dickinson, A.S.; Taylor, A.C.; Browne, M. The influence of acetabular cup material on pelvis cortex surface strains, measured using digital image correlation. *J. Biomech.* **2012**, *45*, 719–723. [CrossRef]
3. Kim, Y.S.; Callaghan, J.J.; Ahn, P.B.; Brown, T.D. Fracture of the acetabulum during insertion of an oversized hemispherical component. *J. Bone Jt. Surg.* **1995**, *77*, 111–117. [CrossRef]
4. Wright, J.M.; Pellicci, P.M.; Salvati, E.A.; Ghelman, B.; Roberts, M.M.; Koh, J.L. Bone density adjacent to press-fit acetabular components. A prospective analysis with quantitative computed tomography. *J. Bone Jt. Surg.* **2001**, *83*, 529–536. [CrossRef]
5. Meneghini, R.M.; Ford, K.S.; McCollough, C.H.; Hanssen, A.D.; Lewallen, D.G. Bone remodeling around porous metal cementless acetabular components. *J. Arthroplast.* **2010**, *25*, 741–747. [CrossRef]
6. Kim, Y.-H.; Yoon, S.-H.; Kim, J.-S. Changes in the bone mineral density in the acetabulum and proximal femur after cementless total hip replacement: Alumina-on-alumina versus alumina-on-polyethylene articulation. *J. Bone Jt. Surg.* **2007**, *89*, 174–179. [CrossRef]
7. Nakahara, I.; Takao, M.; Bandoh, S.; Bertollo, N.; Walsh, W.R.; Sugano, N. In vivo implant fixation of carbon fiber-reinforced PEEK hip prostheses in an ovine model. *J. Orthop. Res. Off. Publ. Orthop. Res. Soc.* **2013**, *31*, 485–492. [CrossRef]
8. Pace, N.; Marinelli, M.; Spurio, S. Technical and histologic analysis of a retrieved carbon fiber-reinforced poly-ether-ether-ketone composite alumina-bearing liner 28 months after implantation. *J. Arthroplast.* **2008**, *23*, 151–155. [CrossRef] [PubMed]
9. Scholes, S.C.; Inman, I.A.; Unsworth, A.; Jones, E. Tribological assessment of a flexible carbon-fibre-reinforced poly(ether-ether-ketone) acetabular cup articulating against an alumina femoral head. *Proc. Inst. Mech. Eng. Part H J. Eng. Med.* **2008**, *222*, 273–283. [CrossRef]
10. Wang, A.; Lin, R.; Polineni, V.K.; Essner, A.; Stark, C.; Dumbleton, J.H. Carbon fiber reinforced polyether ether ketone composite as a bearing surface for total hip replacement. *Tribol. Int.* **1998**, *31*, 661–667. [CrossRef]
11. Wang, Q.Q.; Wu, J.J.; Unsworth, A.; Briscoe, A.; Jarman-Smith, M.; Lowry, C.; Simpson, D.; Collins, S. Biotribological study of large diameter ceramic-on-CFR-PEEK hip joint including fluid uptake, wear and frictional heating. *J. Mater. Sci. Mater. Med.* **2012**, *23*, 1533–1542. [CrossRef]
12. Vogel, D.; Schulze, C.; Dempwolf, H.; Kluess, D.; Bader, R. Biomechanical behavior of modular acetabular cups made of poly-ether-ether-ketone: A finite element study. *Proc. Inst. Mech. Eng. Part H J. Eng. Med.* **2018**, *232*, 1030–1038. [CrossRef]
13. Goebel, P.; Kluess, D.; Wieding, J.; Souffrant, R.; Heyer, H.; Sander, M.; Bader, R. The influence of head diameter and wall thickness on deformations of metallic acetabular press-fit cups and UHMWPE liners: A finite element analysis. *J. Orthop. Sci. Off. J. Jpn. Orthop. Assoc.* **2013**, *18*, 264–270. [CrossRef]
14. Meding, J.B.; Small, S.R.; Jones, M.E.; Berend, M.E.; Ritter, M.A. Acetabular cup design influences deformational response in total hip arthroplasty. *Clin. Orthop. Relat. Res.* **2013**, *471*, 403–409. [CrossRef]
15. Ong, K.L.; Rundell, S.; Liepins, I.; Laurent, R.; Markel, D.; Kurtz, S.M. Biomechanical modeling of acetabular component polyethylene stresses, fracture risk, and wear rate following press-fit implantation. *J. Orthop. Res. Off. Publ. Orthop. Res. Soc.* **2009**, *27*, 1467–1472. [CrossRef]

16. Schmidig, G.; Patel, A.; Liepins, I.; Thakore, M.; Markel, D.C. The effects of acetabular shell deformation and liner thickness on frictional torque in ultrahigh-molecular-weight polyethylene acetabular bearings. *J. Arthroplast.* **2010**, *25*, 644–653. [CrossRef]
17. Bone, M.C.; Dold, P.; Flohr, M.; Preuss, R.; Joyce, T.J.; Aspden, R.M.; Holland, J.; Deehan, D. The influence of the strength of bone on the deformation of acetabular shells: A laboratory experiment in cadavers. *Bone Jt. J.* **2015**, *97*, 473–477. [CrossRef]
18. Howcroft, D.W.J.; Qureshi, A.; Graham, N.M. Seating of ceramic liners in the uncemented trident acetabular shell: Is there really a problem? *Clin. Orthop. Relat. Res.* **2009**, *467*, 2651–2655. [CrossRef]
19. McAuley, J.P.; Dennis, D.A.; Grostefon, J.; Hamilton, W.G. Factors affecting modular acetabular ceramic liner insertion: A biomechanical analysis. *Clin. Orthop. Relat. Res.* **2012**, *470*, 402–409. [CrossRef]
20. Qiu, C.; Wang, L.; Li, D.; Jin, Z. The influence of metallic shell deformation on the contact mechanics of a ceramic-on-ceramic total hip arthroplasty. *Proc. Inst. Mech. Eng. Part H J. Eng. Med.* **2016**, *230*, 4–12. [CrossRef]
21. Sharkey, P.F.; Hozack, W.J.; Callaghan, J.J.; Kim, Y.S.; Berry, D.J.; Hanssen, A.D.; LeWallen, D.G. Acetabular fracture associated with cementless acetabular component insertion: A report of 13 cases. *J. Arthroplast.* **1999**, *14*, 426–431. [CrossRef]
22. Kluess, D.; Martin, H.; Mittelmeier, W.; Schmitz, K.-P.; Bader, R. Influence of femoral head size on impingement, dislocation and stress distribution in total hip replacement. *Med. Eng. Phys.* **2007**, *29*, 465–471. [CrossRef]
23. Kluess, D.; Souffrant, R.; Mittelmeier, W.; Wree, A.; Schmitz, K.-P.; Bader, R. A convenient approach for finite-element-analyses of orthopaedic implants in bone contact: Modeling and experimental validation. *Comput. Methods Programs Biomed.* **2009**, *95*, 23–30. [CrossRef]
24. Schultze, C.; Klüss, D.; Martin, H.; Hingst, V.; Mittelmeier, W.; Schmitz, K.-P.; Bader, R. Finite-Elemente-Analyse einer zementierten, keramischen Femurkomponente unter Berücksichtigung der Einbausituation bei künstlichem Kniegelenkersatz. Biomedizinische Technik. *Biomed. Eng.* **2007**, *52*, 301–307. [CrossRef] [PubMed]
25. Taddei, F.; Pancanti, A.; Viceconti, M. An improved method for the automatic mapping of computed tomography numbers onto finite element models. *Med. Eng. Phys.* **2004**, *26*, 61–69. [CrossRef]
26. Helgason, B.; Perilli, E.; Schileo, E.; Taddei, F.; Brynjólfsson, S.; Viceconti, M. Mathematical relationships between bone density and mechanical properties: A literature review. *Clin. Biomech.* **2008**, *23*, 135–146. [CrossRef] [PubMed]
27. Carter, D.R.; Hayes, W.C. The compressive behavior of bone as a two-phase porous structure. *J. Bone Jt. Surg.* **1977**, *59*, 954–962. [CrossRef]
28. Grifka, J.; Kuster, M. *Orthopädie und Unfallchirurgie*; Springer: Berlin/Heidelberg, Germany, 2011.
29. Shirazi-Adl, A.; Dammak, M.; Paiement, G. Experimental determination of friction characteristics at the trabecular bone/porous-coated metal interface in cementless implants. *J. Biomed. Mater. Res.* **1993**, *27*, 167–175. [CrossRef]
30. Kurtz, S.M.; Edidin, A.A.; Bartel, D.L. The role of backside polishing, cup angle, and polyethylene thickness on the contact stresses in metal-backed acetabular components. *J. Biomech.* **1997**, *30*, 639–642. [CrossRef]
31. Bergmann, G.; Bender, A.; Dymke, J.; Duda, G.; Damm, P. Standardized Loads Acting in Hip Implants. *PLoS ONE* **2016**, *11*, e0155612. [CrossRef]
32. Biewener, A.A. Safety factors in bone strength. *Calcif. Tissue Int.* **1993**, *53*, S68–S74. [CrossRef]
33. Frost, H.M. Bone's mechanostat: A 2003 update. *Anat. Rec. Part A Discov. Mol. Cell. Evol. Biol.* **2003**, *275*, 1081–1101. [CrossRef]
34. Cho, Y.J.; Bae, C.I.; Yoon, W.K.; Chun, Y.S.; Rhyu, K.H. High incidence of early subtrochanteric lateral cortical atrophy after hip arthroplasty using bone-conserving short stem. *Int. Orthop.* **2018**, *42*, 303–309. [CrossRef]
35. Field, R.E.; Rajakulendran, K.; Eswaramoorthy, V.K.; Rushton, N. Three-year prospective clinical and radiological results of a new flexible horseshoe acetabular cup. *Hip Int. J. Clin. Exp. Res. Hip Pathol. Ther.* **2012**, *22*, 598–606. [CrossRef]
36. Manley, M.T.; Ong, K.L.; Kurtz, S.M. The potential for bone loss in acetabular structures following THA. *Clin. Orthop. Relat. Res.* **2006**, *453*, 246–253. [CrossRef]
37. Anderson, A.E.; Peters, C.L.; Tuttle, B.D.; Weiss, J.A. Subject-specific finite element model of the pelvis: Development, validation and sensitivity studies. *J. Biomech. Eng.* **2005**, *127*, 364–373. [CrossRef]

38. Devine, D.M.; Hahn, J.; Richards, R.G.; Gruner, H.; Wieling, R.; Pearce, S.G. Coating of carbon fiber-reinforced polyetheretherketone implants with titanium to improve bone apposition. *J. Biomed. Mater. Res. Part B Appl. Biomater.* **2013**, *101*, 591–598. [CrossRef]
39. Han, C.-M.; Lee, E.-J.; Kim, H.-E.; Koh, Y.-H.; Kim, K.N.; Ha, Y.; Kuh, S.-U. The electron beam deposition of titanium on polyetheretherketone (PEEK) and the resulting enhanced biological properties. *Biomaterials* **2010**, *31*, 3465–3470. [CrossRef]
40. Walsh, W.R.; Bertollo, N.; Christou, C.; Schaffner, D.; Mobbs, R.J. Plasma-sprayed titanium coating to polyetheretherketone improves the bone-implant interface. *Spine J. Off. J. N. Am. Spine Soc.* **2015**, *15*, 1041–1049. [CrossRef]
41. Vogel, D.; Dempwolf, H.; Baumann, A.; Bader, R. Characterization of thick titanium plasma spray coatings on PEEK materials used for medical implants and the influence on the mechanical properties. *J. Mech. Behav. Biomed. Mater.* **2018**, *77*, 600–608. [CrossRef]
42. Everitt, H. Deformation on Impaction of a Large Diameter Carbon Fibre Reinforced Poly-Ether-Ether-Ketone Acetabular Cup and Its Effects on the Tribology of the Bearing. Ph.D. Thesis, University Cardiff, Cardiff, UK, August 2011.
43. Wu, X.; Liu, X.; Wei, J.; Ma, J.; Deng, F.; Wei, S. Nano-TiO_2/PEEK bioactive composite as a bone substitute material: In Vitro and in vivo studies. *Int. J. Nanomed.* **2012**, *7*, 1215–1225.
44. Lin, Z.M.; Meakins, S.; Morlock, M.M.; Parsons, P.; Hardaker, C.; Flett, M.; Isaac, G. Deformation of press-fitted metallic resurfacing cups. Part 1: Experimental simulation. *Proc. Inst. Mech. Eng. Part H J. Eng. Med.* **2006**, *220*, 299–309.
45. Park, Y.-S.; Hwang, S.-K.; Choy, W.-S.; Kim, Y.-S.; Moon, Y.-W.; Lim, S.-J. Ceramic failure after total hip arthroplasty with an alumina-on-alumina bearing. *J. Bone Jt. Surg.* **2006**, *88*, 780–787.
46. Ha, Y.-C.; Kim, S.-Y.; Kim, H.J.; Yoo, J.J.; Koo, K.-H. Ceramic liner fracture after cementless alumina-on-alumina total hip arthroplasty. *Clin. Orthop. Relat. Res.* **2007**, *458*, 106–110. [CrossRef]
47. Hasegawa, M.; Sudo, A.; Hirata, H.; Uchida, A. Ceramic acetabular liner fracture in total hip arthroplasty with a ceramic sandwich cup. *J. Arthroplast.* **2003**, *18*, 658–661. [CrossRef]
48. Lopes, R.; Philippeau, J.M.; Passuti, N.; Gouin, F. High rate of ceramic sandwich liner fracture. *Clin. Orthop. Relat. Res.* **2012**, *470*, 1705–1710. [CrossRef]
49. Viste, A.; Chouteau, J.; Desmarchelier, R.; Fessy, M.-H. Fractures of a sandwich ceramic liner at ten year follow-up. *Int. Orthop.* **2012**, *36*, 955–960. [CrossRef]
50. Okita, S.; Hasegawa, M.; Takahashi, Y.; Puppulin, L.; Sudo, A.; Pezzotti, G. Failure analysis of sandwich-type ceramic-on-ceramic hip joints: A spectroscopic investigation into the role of the polyethylene shell component. *J. Mech. Behav. Biomed. Mater.* **2014**, *31*, 55–67. [CrossRef]
51. Wang, T.; Sun, J.-Y.; Zha, G.-C.; Dong, S.-J.; Zhao, X.-J. Mid term results of total hip arthroplasty using polyethylene-ceramic composite (Sandwich) liner. *Indian J. Orthop.* **2016**, *50*, 10–15.
52. Hua, X.; Li, J.; Wang, L.; Jin, Z.; Wilcox, R.; Fisher, J. Contact mechanics of modular metal-on-polyethylene total hip replacement under adverse edge loading conditions. *J. Biomech.* **2014**, *47*, 3303–3309. [CrossRef]
53. Mabuchi, K.; Sakai, R.; Ota, M.; Ujihira, M. Appropriate radial clearance of ceramic-on-ceramic total hip prostheses to realize squeeze-film lubrication. *Clin. Biomech.* **2004**, *19*, 362–369. [CrossRef]
54. Begand, S.; Glien, W.; Oberbach, T. ATZ—A New Material with a High Potential in Joint Replacement. *KEM* **2005**, *284*, 983–986. [CrossRef]
55. Dold, P.; Pandorf, T.; Flohr, M.; Preuss, R.; Bone, M.C.; Joyce, T.J.; Holland, J.; Deehan, D. Acetabular shell deformation as a function of shell stiffness and bone strength. *Proc. Inst. Mech. Eng. Part H J. Eng. Med.* **2016**, *230*, 259–264. [CrossRef]
56. Markel, D.; Day, J.; Siskey, R.; Liepins, I.; Kurtz, S.; Ong, K. Deformation of metal-backed acetabular components and the impact of liner thickness in a cadaveric model. *Int. Orthop.* **2011**, *35*, 1131–1137. [CrossRef]
57. Yew, A.; Jin, Z.M.; Donn, A.; Morlock, M.M.; Isaac, G. Deformation of press-fitted metallic resurfacing cups. Part 2: Finite element simulation. *Proc. Inst. Mech. Eng. Part H J. Eng. Med.* **2006**, *220*, 311–319. [CrossRef]
58. Thompson, M.S.; Northmore-Ball, M.D.; Tanner, K.E. Effects of acetabular resurfacing component material and fixation on the strain distribution in the pelvis. *Proc. Inst. Mech. Eng. Part H J. Eng. Med.* **2002**, *216*, 237–245. [CrossRef]

59. Chatterjee, S.; Kobylinski, S.; Basu, B. Finite Element Analysis to Probe the Influence of Acetabular Shell Design, Liner Material, and Subject Parameters on Biomechanical Response in Periprosthetic Bone. *J. Biomech. Eng.* **2018**, *140*, 101014. [CrossRef]
60. Pitto, R.P.; Bhargava, A.; Pandit, S.; Munro, J.T. Retroacetabular stress-shielding in THA. *Clin. Orthop. Relat. Res.* **2008**, *466*, 353–358. [CrossRef]
61. Widmer, K.-H.; Zurfluh, B.; Morscher, E.W. Load transfer and fixation mode of press-fit acetabular sockets. *J. Arthroplast.* **2002**, *17*, 926–935. [CrossRef]
62. Bergmann, G.; Graichen, F.; Rohlmann, A.; Bender, A.; Heinlein, B.; Duda, G.N.; Heller, M.O.; Morlock, M.M. Realistic loads for testing hip implants. *Bio-Med. Mater. Eng.* **2010**, *20*, 65–75. [CrossRef]
63. Ghosh, R.; Gupta, S. Bone remodelling around cementless composite acetabular components: The effects of implant geometry and implant-bone interfacial conditions. *J. Mech. Behav. Biomed. Mater.* **2014**, *32*, 257–269. [CrossRef] [PubMed]

materials

MDPI

Article

Third Body Wear of UHMWPE-on-PEEK-OPTIMA™

Raelene M. Cowie [1], Naveen Manikya Pallem [1], Adam Briscoe [2] and Louise M. Jennings [1,*]

[1] Institute of Medical and Biological Engineering, School of Mechanical Engineering, University of Leeds, Leeds LS2 9JT, UK; r.cowie@leeds.ac.uk (R.M.C.); speak2manikya@gmail.com (N.M.P.)

[2] Invibio Ltd., Invibio Global Technology Centre, Hillhouse International, Thornton-Cleveleys, Lancashire FY5 4QD, UK; abriscoe@invibio.com

* Correspondence: l.m.jennings@leeds.ac.uk

Received: 18 February 2020; Accepted: 9 March 2020; Published: 11 March 2020

Abstract: PEEK-OPTIMA™ is being considered as an alternative to cobalt chrome (CoCr) in the femoral component of total knee replacements. To date, investigations of ultra-high molecular weight polyethylene (UHMWPE)-on-PEEK have shown an equivalent wear rate to conventional implant materials under standard conditions. In this study, the third body wear performance of UHMWPE-on-PEEK was directly compared to UHMWPE-on-CoCr in a series of pin-on-plate studies using two approaches for third body damage. Damage simulation with particles of bone cement showed a significant ($p < 0.001$), four-fold increase in the mean surface roughness of PEEK plates compared to CoCr. However, wear simulation against the damaged plates showed no significant difference in the wear of UHMWPE pins against the different materials ($p = 0.59$), and a polishing effect by the pin against the PEEK plates was observed. Scratching PEEK and CoCr counterfaces with a diamond stylus to create scratches representative of severe third body damage (4 μm lip height) resulted in a significantly higher ($p = 0.01$) wear of UHMWPE against CoCr compared to PEEK and again, against PEEK plates, polishing by the UHMWPE pin led to a reduction in scratch lip height. This study shows that in terms of its wear performance under third body wear/damage conditions, UHMWPE-on-PEEK differs from conventional knee replacement materials.

Keywords: biomaterials; arthroplasty; orthopaedic tribology; experimental simulation; total knee replacement; PEEK-OPTIMA™; UHMWPE; third body wear

1. Introduction

Total knee replacement (TKR) is a highly successful procedure with >90% survival at 10 years [1]; however, up to 20% of patients may be dissatisfied with their procedure [2]. When considering how to improve patient satisfaction, the materials from which the implant is manufactured is one potential variable to be considered. PEEK-OPTIMA™ (poly-ether-ether-ketone) has recently been investigated as an alternative to cobalt chrome (CoCr) for use in the femoral component of TKR [3–5]. There are a number of potential advantages of using PEEK in this application. For example, the Young's modulus of the PEEK femoral component (~3.7 GPa) is more similar to that of bone (0.001–20 GPa) than a CoCr femoral component (~210 GPa), giving the potential for a reduction in stress shielding, which can lead to bone resorption particularly behind the anterior flange of the implant and, hence, increase the risk of implant loosening [6]. In addition, an all-polymer implant would be lighter weight than conventional materials and more similar to the weight of the natural joint.

Several in vitro pin-on-plate and whole joint simulation studies have been carried out investigating the wear performance of the PEEK-on-ultra-high molecular weight polyethylene (UHMWPE) bearing couple and have demonstrated an equivalent rate of wear for this bearing couple compared to CoCr-on-UHMWPE for a well-positioned implant tested in a clean environment under standard conditions [3,7,8]. In addition, an in vivo large animal study has also investigated the potential to use PEEK in an all-polymer TKR [9,10]. Despite this animal study being relatively short-term, no problems were encountered with fracture or fixation of the device; there was, however, an inflammatory response but, due to the lack of a control group (metal-on-polyethylene implant), it is not known whether a similar response would occur if conventional TKR materials were implanted in this animal model. Having obtained promising initial results, prior to the clinical adoption of the device, it is important to consider how the implant will respond under a wider envelope of clinically relevant conditions [11]. In this study, the influence of third body wear on UHMWPE-on-PEEK-OPTIMA™ in a simple geometry pin-on-plate configuration has been considered.

Third body wear can occur when hard particles such as bone cement particles, bone fragments or other particulate materials become trapped between the articulating surfaces of an implant causing damage to the joint surfaces and accelerating implant wear. Two approaches have been taken for the simulation of third body wear of arthroplasty bearing materials. One approach has been to introduce clinically relevant particles such as polymethyl methacrylate (PMMA) cement into the articulating interface [12]; an alternative approach has been to inflict third body damage directly to the surface(s) either by scratching the surface of the implant directly using a diamond stylus, which gives close control of the position and geometry of the scratches [13], or by abrading the component bearing surface producing a random orientation of scratches [14]. There are advantages and limitations to these approaches. Protocols using particles to replicate third body damage can more closely represent the clinical situation; however, controlling the particles trapped between the articulating surfaces can be difficult especially in the knee where the low conforming nature of the articulating surfaces may lead to particles being ejected from the joint. In bearing couples containing polymeric implant materials, if particles become embedded in the polymer, assessing wear of the polymer gravimetrically can be unreliable. Scratching the surface of the implant directly can give more consistent and reproducible damage. Controlling the lip height of the scratches, which is the variable that most influences polyethylene wear in metal-on-UHMWPE configurations, using a method such as a diamond stylus allows the scratch geometry to be optimised to more closely replicate observations of retrieved implants [15]. However, this approach could be considered a less clinically relevant method for simulating third body damage than using particles.

Whilst carrying out whole joint wear simulation studies is optimal, valuable information can be gained from pin-on-plate studies. By simplifying component geometries and applied loads and motions, the influence of individual variables can be investigated [15]. The aim of this study was to investigate the influence of third body damage on the wear of UHMWPE-on-PEEK-OPTIMA™ in simple geometry pin-on-plate wear simulation. Third body damage was simulated using two approaches: (1) using particles of PMMA cement and (2) scratching the implants directly using a diamond stylus. When considered together, the outcomes of these two approaches can give a better understanding both of how third body particles can damage the articulating surfaces, and how different magnitudes of damage can influence wear. For all the studies, the wear of UHMWPE-on-PEEK was directly compared to conventional knee implant materials, UHMWPE-on-CoCr, which were tested in parallel. It was hypothesised that, because of the different material properties of PEEK compared to CoCr, the third body wear behaviour of the two material combinations would differ.

2. Materials and Methods

The studies were split into two phases, first carrying out damage simulation (using each method) before determining the wear factors of UHMWPE against the damaged surfaces. This two-phase approach has been adopted in previous third body wear simulation studies [15,16].

2.1. Materials

The pins used were GUR 1020 UHMWPE (conventional, non-sterile) machined into a truncated cone with either 3 mm or 8 mm flat contact face for damage or wear simulation, respectively. The plates were either injection moulded, implant grade, unfilled (natural) PEEK-OPTIMA™ (Invibio Ltd, UK), initial surface roughness (Ra) ~0.02 μm or CoCr, polished to an initial Ra < 0.01 μm. To create damage with particles, Palacos R + G PMMA cement (Heraeus, Germany) was mixed and cured in a block as per the manufacturers' instructions before turning and crushing with a mortar and pestle to create particles, which were sieved within a size range of 500–1000 μm diameter. The particle size range used was within a clinically relevant range [17].

2.2. Methods

2.2.1. Damage Simulation: Third Body Damage with PMMA Cement Particles

The protocol used for damage simulation with particles was adapted from previously described studies [15,16]. In brief, the PEEK or CoCr plate was clamped onto a sliding table mounted on the platen of an Instron materials testing machine (Instron, MA, USA). The PMMA cement particles described above were trapped (in excess) between a UHMWPE pin (3 mm diameter contact face) and the plate; before a load of 120 N was applied axially through the pin. Then, using the Instron materials testing machine, the plate was pulled beneath the pin for 15 mm at a speed of 8 mm/s to create third body damage. Five regions of damage were created on each plate with a spacing of 3 mm, the particles were passed over the plate 5 times in each region of damage. Third body damage was created perpendicular to the direction of the subsequent wear test (Figure 1a).

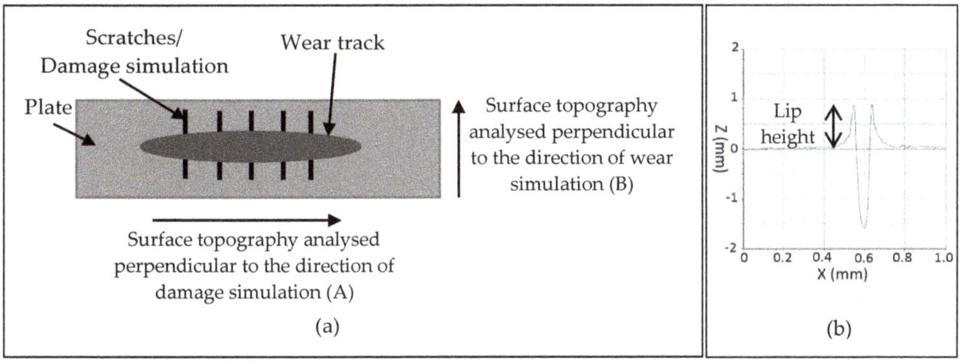

Figure 1. (**a**) Schematic showing the respective directions of the damage simulation and wear simulation; (**b**) schematic showing the profile of a 1 μm scratch created with a diamond stylus and the lip height measurement taken.

2.2.2. Damage Simulation: Third Body Damage Created Using a Diamond Stylus

To create a scratch in the plates with a reproducible geometry, the plate was set up on a sliding table as described for the particle method, a stylus with a 200 μm radius diamond tip was axially loaded and the plate pulled beneath the stylus to create scratches. On each plate, 5 scratches were created running perpendicular to the direction of the wear study (Figure 1a). The lip height of the scratches was adjusted to either 1, 2 or 4 μm by changing the load applied to the stylus (Table A1 in Appendix A) [18].

2.2.3. Pin-on-Plate Wear Simulation

Wear simulation was carried out using a six-station multi-axial pin-on-plate reciprocating rig and aimed to replicate the average contact pressure and cross-shear ratio in a total knee replacement [19]. The plate was mounted in a lubricant-containing bath, which reciprocated over a length of 20 mm at 1Hz, and as the bath reciprocated, the UHMWPE pin rotated (±20°) via a rack and pinion mechanism. a constant axial load of 160 N was applied through the pin (8 mm flat contact face) to give an average contact pressure of 3.18 MPa consistent with previous wear simulation of the UHMWPE-on-PEEK bearing couple [8]. All wear simulation studies were carried out using 25% bovine serum supplemented with 0.03% sodium azide solution as a lubricant and were carried out under rig running (room) temperature conditions as previously described and optimised for the UHMWPE-on-PEEK bearing couple [8]. Prior to the start of the study, the UHMWPE pins were soaked for a minimum of 2 weeks to maximise water uptake prior to cleaning and weighing using an XP26 digital microbalance (Mettler Toledo Inc., OH, USA) with a resolution of 1 μg. Measurements of each pin were taken until 5 consecutive measurements were within ± 5 μg and an average of these 5 measurements taken. Studies were carried out for 1 million cycles (MC) with gravimetric analysis of the UHMWPE pins every 0.3 MC. The weight loss of the pins was converted to a volume loss (V) using two unloaded soak controls to compensate for the uptake of moisture by the polyethylene and a density of 0.934 mg/mm^3 for GUR 1020 UHMWPE. The wear factor (k) was calculated using the sliding distance for the test (X) and the applied load (P) as shown in Equation (1).

$$k = V/PX, \tag{1}$$

For both methods of damage simulation, the wear of the UHMWPE pins against the plates which had undergone damage simulation was compared to that against undamaged, polished plates which served as a control. In the second approach (scratching using a diamond stylus), these plates are referred to as 0 μm lip height.

The surface topography of the plates was assessed pre- and post-test using a contacting Form Talysurf (Taylor Hobson, Leicester, UK) with a 2 μm conical tipped stylus. To analyse the surface roughness, a Gaussian filter was applied to the measurements and a 0.25 mm upper cut-off used as described in the ISO standard [20]. The surface roughness parameter of interest was the mean surface roughness (Ra). Measurements were taken both perpendicular to the direction of damage simulation (A) and perpendicular to the direction of the wear test (B) as shown in Figure 1a. The lip height (Figure 1b) of the scratches was also assessed by carrying out LS line form removal and primary analysis. For third body damaged plates, the density of the scratches within a given lip height range was calculated and expressed as the number of scratches per mm both after damage simulation and after the subsequent wear test. For this analysis, a threshold was applied whereby scratches with a lip height of <0.1 μm were not measured as they were considered too indistinct from the background topography of the plate to be reliably measured. For the plates scratched with a diamond stylus, the lip height of each scratch was measured at 5 points, scratches with a lip height less than 0.2 μm were not measured.

At the conclusion of the studies, images of the wear scars on the PEEK and CoCr plates were taken using an optical profiler, Alicona G5 IF (Graz, Austria) with 5× magnification.

A minimum of 3 sets of samples were used for each material configuration for each of the two approaches. The mean wear factor of the UHMWPE pins, the mean surface roughness (Ra) and the mean lip height of the scratches on the plates was determined with 95% confidence limits. Statistical analysis was carried out using a one-way ANOVA to compare the two configurations of UHMWPE-on-PEEK to UHMWPE-on-CoCr with significance taken at $p < 0.05$.

The data associated with this article is openly available from the University of Leeds Data Repository [18].

3. Results

3.1. Third Body Damage with PMMA Cement Particles

Having carried out third body damage simulation with PMMA cement particles, on the surface of all 3 PEEK plates, there was a high density of linear scratching visible in the region of damage; for CoCr, on one plate, scratching was visible, and on the other plates, there was little evidence of damage (Figure 2). Surface roughness measurements taken perpendicular to the direction of damage simulation

(A) showed a significant increase of more than four-fold ($p < 0.001$) in mean surface roughness of the PEEK plates compared to CoCr (Table 1). Inspection of the plates showed no evidence of PMMA cement particles becoming embedded in the surface of the PEEK.

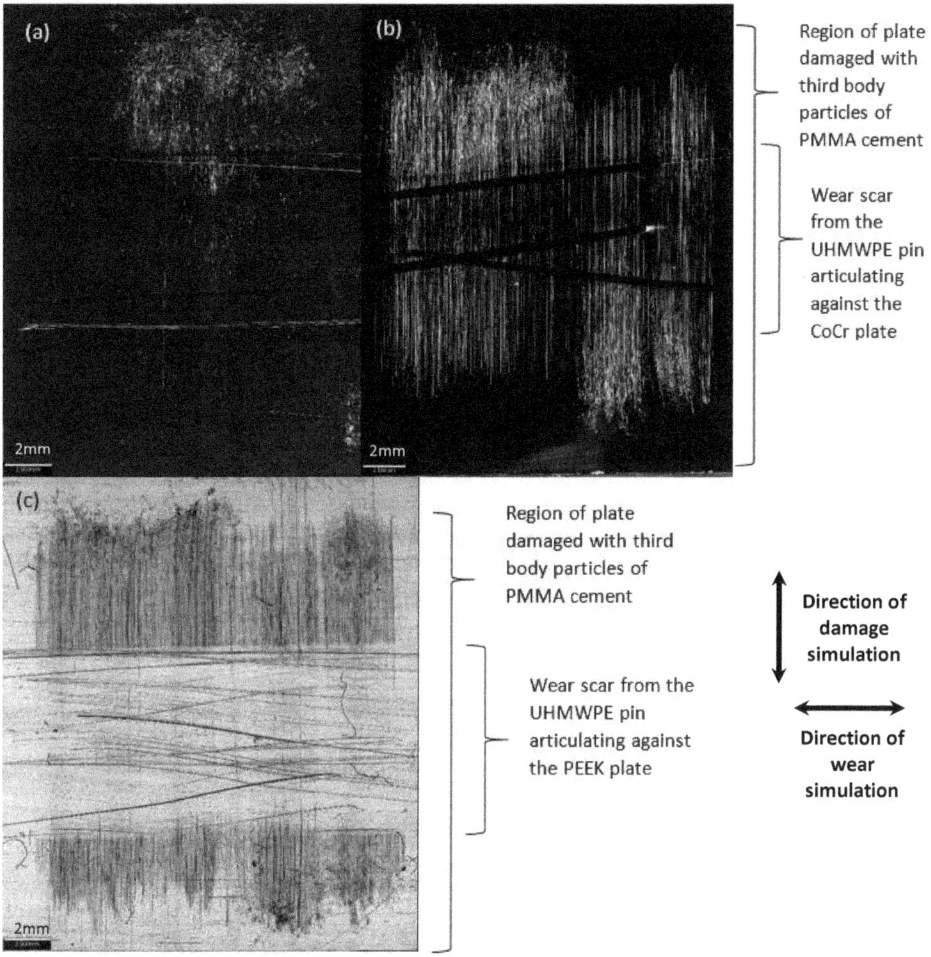

Figure 2. Representative images of the CoCr and PEEK plates following damage simulation with PMMA cement particles and wear simulation taken using an Alicona G5 IF with 5× magnification. (**a**,**b**) CoCr plates. In (**b**), scratches caused by the PMMA cement particles are more clearly visible; in both images, isolated scratches from the wear simulation can be seen. (**c**) a PEEK plate. The polishing and scratching from the wear simulation is visible running perpendicular to the direction of damage simulation. Scale bars represent 2 mm, arrows denote directions of damage simulation and wear simulation.

Table 1. Mean surface roughness (Ra), µm (± 95% confidence limits) of cobalt chrome and PEEK plates, pre-test, after damage simulation with polymethyl methacrylate (PMMA) cement particles and after wear simulation. Measurements were taken either perpendicular to or parallel to the direction of damage simulation (N = 3). Statistical analysis to compare ultra-high molecular weight polyethylene (UHMWPE)-on-PEEK to UHMWPE-on-CoCr, * denotes $p < 0.05$.

Direction		Cobalt Chrome			PEEK-OPTIMA™		
		Pre-Test	After Damage Simulation with Particles	After Wear Simulation	Pre-Test	After Damage Simulation with Particles	After Wear Simulation
Control test	Perpendicular to damage simulation (A)	0.013 ± 0.001		0.014 ± 0.004	0.014 ± 0.001 *		0.019 ± 0.006
	Perpendicular to wear test (B)	0.013 ± 0.000		0.028 ± 0.030	0.019 ± 0.001 *		0.092 ± 0.159 *
Damage simulation	Perpendicular to damage simulation (A)	0.011 ± 0.014	0.012 ± 0.015	0.011 ± 0.009	0.006 ± 0.002	0.054 ± 0.006 *	0.018 ± 0.008
	Perpendicular to wear test (B)	0.011 ± 0.015	0.012 ± 0.016	0.014 ± 0.004	0.018 ± 0.001	0.023 ± 0.002 *	0.098 ± 0.164

In the control tests, after one MC wear simulation, the articulation of the pin against the polished plates caused discrete linear scratches on the CoCr plates and a higher density of scratches on the PEEK plates, resulting in a significantly higher post-test mean surface roughness of PEEK compared to CoCr ($p = 0.03$) (Table 1). Following wear simulation against the damaged plates, burnishing could be seen on the PEEK plates, where the scratches in the PEEK as a result of the damage simulation had been polished out leaving a defined wear scar (Figure 2c). However, within the wear scar, there was evidence of scratching from the articulation of the UHMWPE pin against the PEEK plate. On the CoCr plates, some discrete scratching was evident as a result of the wear simulation. At the conclusion of the study, there was no significant difference ($p = 0.07$) in the mean surface roughness of the PEEK and CoCr plates when measured either perpendicular to the damage simulation (A) or perpendicular to the direction of the wear test (B) ($p = 0.09$). However, despite there being no significant difference in the measurements, when measured perpendicular to the wear simulation, the mean surface roughness of the PEEK was much higher than that of CoCr (more than six-fold) with greater variability in the measurements.

In the control test, the mean wear factor with 95% confidence limits of the UHMWPE pins against smooth CoCr and PEEK plates was similar ($p = 0.84$) at $1.88 \times 10^{-7} \pm 0.92 \times 10^{-7}$ mm^3/Nm and $1.97 \times 10^{-7} \pm 1.52 \times 10^{-7}$ mm^3/Nm, respectively (Figure 3). Against third body damaged CoCr plates, the wear of the UHMWPE pins was $2.24 \times 10^{-7} \pm 1.41 \times 10^{-7}$ mm^3/Nm; against third body damaged PEEK plates, wear was $1.93 \times 10^{-7} \pm 1.82 \times 10^{-7}$ mm^3/Nm. There was no significant difference ($p = 0.59$) in the wear factors of the UHMWPE against the different material types and for all tests, the wear of the UHMWPE pins was linear.

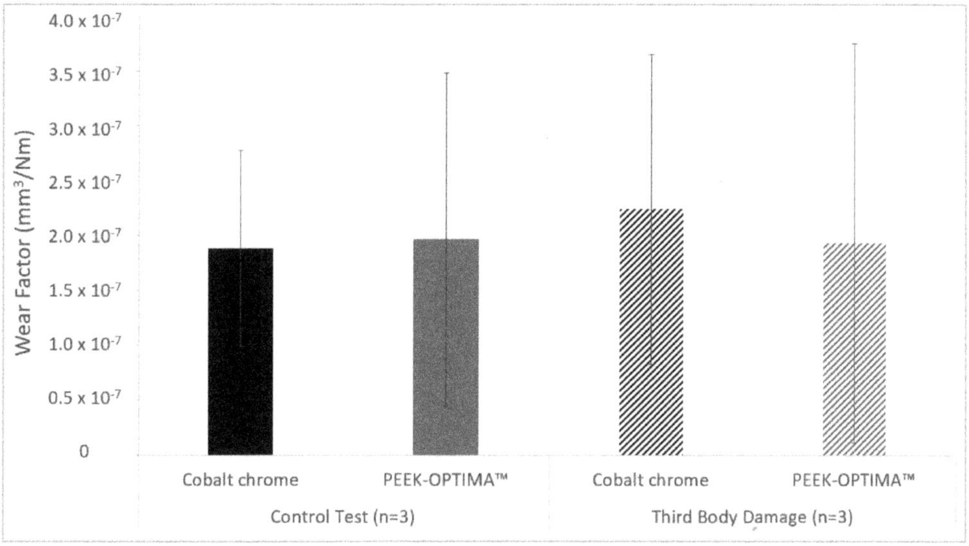

Figure 3. Mean wear factor (mm^3/Nm) of UHMWPE pins (± 95% confidence limits) against smooth (control) CoCr and PEEK-OPTIMA™ plates and plates which had undergone damage simulation with particles of PMMA cement (n = 3).

Analysis of the lip height of the scratches (Figure 4) showed that on the CoCr plates, lip heights were primarily in the 0.1–0.2 µm range both before and after damage simulation. For the PEEK plates, after damage simulation, most scratch lip heights were between 0.2 and 0.4 µm but some were as large as 0.8 µm, then after wear simulation, scratch lip heights were smaller and predominantly in the range of 0.1–0.2 µm.

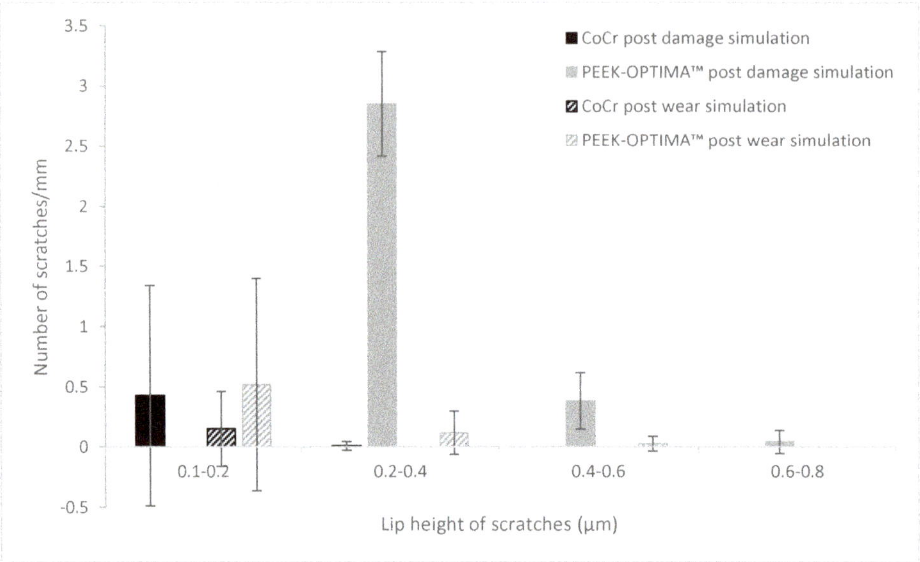

Figure 4. Mean number of scratches per mm (± 95 % confidence limits) within given lip height ranges on cobalt chrome and PEEK plates. Measurements taken post damage simulation and post wear test, (n = 3).

3.2. Third Body Damage Created Using a Diamond Stylus

To adjust the lip height of the scratches created, the load applied to the diamond stylus was varied. To create scratches in the PEEK plates of the same magnitude as the CoCr, a lower load, approximately one-quarter of that required to create scratches in the CoCr was applied to the stylus (Table A1 in Appendix A) [18]. The magnitude of the lip heights created in the PEEK plates was more difficult to control than in the CoCr so variability in the lip heights was greater (Figure 5 and Table 2).

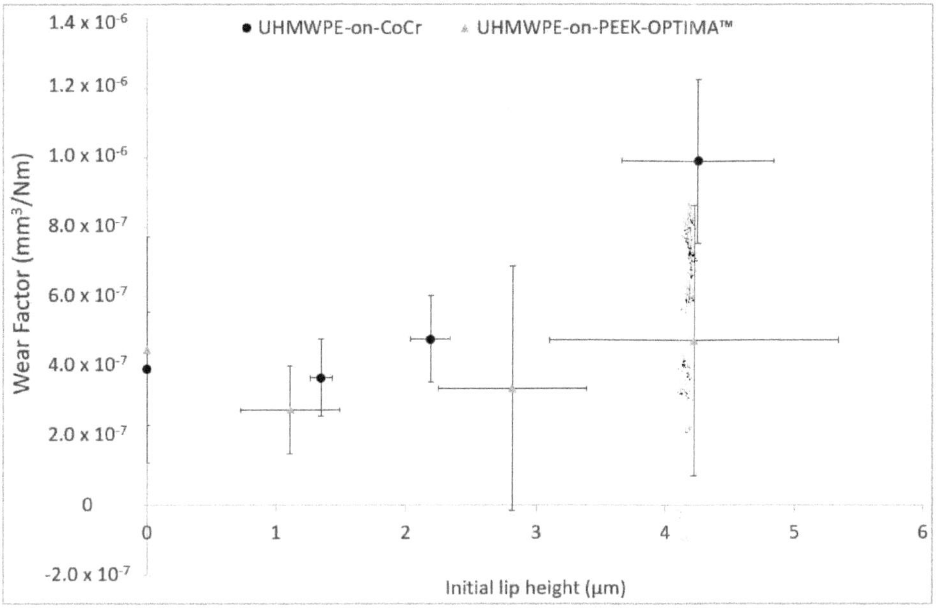

Figure 5. Mean wear factor (mm^3/Nm) ± 95% confidence limits of UHMWPE pins tested against PEEK-OPTIMA™ and cobalt chrome plates scratched with a diamond stylus as a function of pre-test (initial) lip height, μm (± 95% confidence limits).

Table 2. Mean lip height, μm (± 95% confidence limits) of scratches on PEEK-OPTIMA™ and cobalt chrome plates taken pre- and post- wear simulation (minimum N = 3). Statistical analysis to compare PEEK to CoCr, * denotes $p < 0.05$.

Lip Height (μm)	Cobalt Chrome		PEEK-OPTIMA™	
	Pre-Test	**Post-Test**	**Pre-Test**	**Post-Test**
1	1.35 ± 0.09	1.23 ± 0.16	1.11 ± 0.38	0.66 ± 0.66 *
2	2.19 ± 0.15	1.99 ± 0.16	2.82 ± 0.57 *	1.31 ± 0.70 *
4	4.25 ± 0.59	4.13 ± 0.44	4.22 ± 1.12	1.79 ± 1.02 *

After 1 MC wear simulation, discrete scratches were visible in the wear scars on the CoCr plates; whereas a polished region was evident and there was a visible reduction in the lip height of the scratches in the PEEK plates (Figure 6).

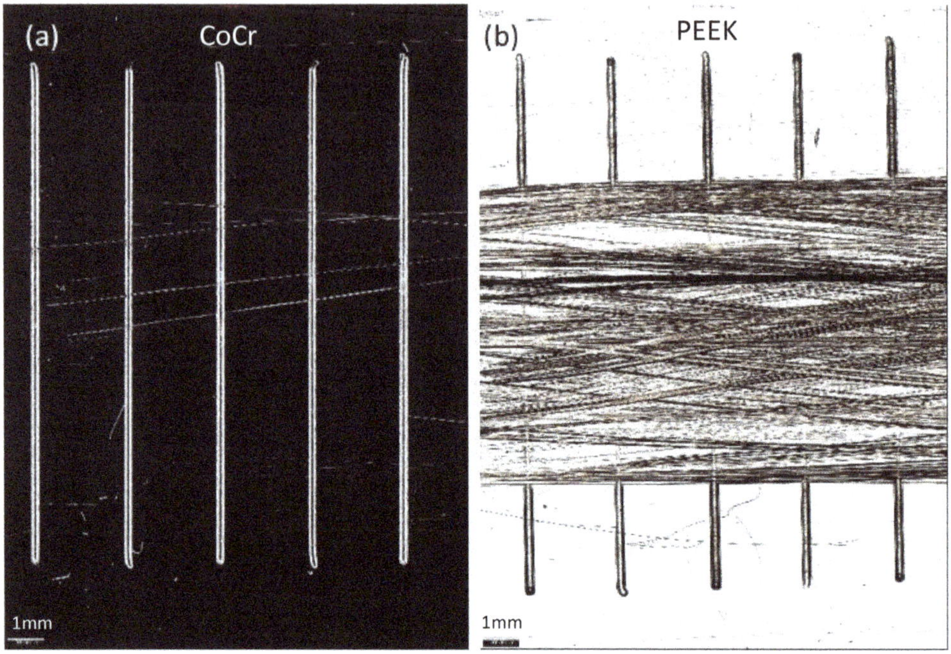

Figure 6. Representative images of CoCr (**a**) and PEEK (**b**) plates taken using an Alicona G5 IF with 5× magnification. The plates were scratched with a diamond stylus to create a scratch lip height of 4 μm, then 1 million cycles (MC) wear simulation was carried out perpendicular to the direction of the scratches. Scale bar represents 1 mm.

The mean wear factors of the UHMWPE pins against the scratched plates are shown in Figure 5. With no scratches on the plates (0 μm lip height), the wear of the UHMWPE against CoCr ($3.4 \times 10^{-7} \pm 8.2 \times 10^{-7}$ mm^3/Nm) and PEEK ($3.9 \times 10^{-7} \pm 5.3 \times 10^{-7}$ mm^3/Nm) was similar ($p = 0.64$); when scratches were created in the plates with a lip height of 4 μm, the wear of UHMWPE was significantly higher ($p = 0.01$) against CoCr than against PEEK at $9.7 \times 10^{-7} \pm 4.3 \times 10^{-7}$ mm^3/Nm and $3.8 \times 10^{-7} \pm 4.4 \times 10^{-7}$ mm^3/Nm, respectively, there were no significant differences ($p > 0.05$) in wear factor for the 1 and 2 μm lip height conditions.

The lip heights of the scratches were assessed pre- and post-test. For the pre-test measurements, there was no significant difference ($p < 0.05$) between the lip height of the scratches in PEEK and CoCr for 1 or 4 μm. Post-test, the lip height of the scratches in the PEEK was significantly lower ($p < 0.05$) than in the CoCr for all scratch lip heights investigated due to the polishing effect of the pin against the plate (Table 2).

Surface roughness measurements taken perpendicular to the direction of wear simulation (B) between the scratches created with a diamond stylus are shown in Table 3. Pre-test, the mean surface roughness of the PEEK plates was significantly higher ($p > 0.05$) than the CoCr plates. After one MC wear simulation, there was a high density of linear scratching on the PEEK plates, and discrete scratches visible on the CoCr plates, resulting in a significantly higher ($p < 0.05$) mean surface roughness of the PEEK plates compared to CoCr.

Table 3. Mean surface roughness (Ra), µm (± 95% confidence limits) of PEEK-OPTIMA™ and cobalt chrome plates taken perpendicular to the direction of wear simulation (B) pre- and post- wear testing. Statistical analysis to compare PEEK to CoCr, * denotes $p < 0.05$.

Lip Height (µm)	Cobalt Chrome		PEEK-OPTIMA™	
	Pre-Test	Post-Test	Pre-Test	Post-Test
0	0.004 ± 0.001	0.027 ± 0.031	0.019 ± 0.004 *	0.540 ± 0.236 *
1	0.004 ± 0.001	0.008 ± 0.004	0.030 ± 0.012 *	0.729 ± 0.327 *
2	0.004 ± 0.001	0.019 ± 0.016	0.025 ± 0.007 *	0.616 ± 0.506 *
4	0.004 ± 0.001	0.019 ± 0.026	0.025 ± 0.008 *	0.839 ± 0.519 *

In both studies, the PEEK plates were weighed at each measurement point; however, despite extensive pre-test soaking (>90 days in sterile water [21]), due to inconsistencies in moisture uptake by the PEEK [22], it was not possible to assess wear and the data was considered unreliable.

4. Discussion

The aim of this study was to investigate the influence of third body damage on the wear of UHMWPE-on-PEEK in a simple geometrical configuration. Two approaches were used to simulate third body damage: using particles of PMMA cement to create damage and scratching the implant surfaces directly with a diamond stylus prior to carrying out multi-directional pin-on-plate wear simulation against the damaged counterfaces. The wear of UHMWPE-on-PEEK was compared to UHMWPE-on-CoCr in studies carried out in parallel. It was hypothesised that the third body wear performance of UHMWPE-on-PEEK would differ to that of UHMWPE-on-CoCr.

Simulating third body damage with particles of PMMA cement caused light scratching on one of the three CoCr plates, which highlights the difficulty in controlling the particles trapped between the pin and plate to create reproducible third body damage using a particle method. Following damage simulation, the mean surface roughness of the plates was twice that reported by Cowie et al. [16] where third body damage simulation was carried out using a similar method. This higher surface roughness may be attributed to differences in the composition of the cement used. In the previous study, the cement used contained 10% barium sulphate as a radiopacifier, whilst in this study, the cement contained 14.75% zirconium dioxide. Whilst both barium sulphate and zirconium dioxide are commonly used additives to improve the visibility of bone cement under x-ray, previous third body wear simulation studies have shown the harder zirconium dioxide agglomerates to cause more damage to metallic counterfaces than barium sulphate [23]. This may account for the higher surface roughness of the plates following damage simulation in this study compared to previous work. Simulating third body wear with cement particles against PEEK plates gave a significantly higher mean surface roughness of the PEEK plates, which, therefore, indicates a reduced scratch resistance of PEEK compared to CoCr. Previous wear simulation studies of scratched metallic counterfaces have shown that it is the lip height of scratches in the metal that influences the wear of UHMWPE and that, to increase polyethylene wear, the scratch lip height must exceed a critical value [15,24]. When compared to control tests carried out against smooth plates, the damage simulation protocol used in this study did not generate scratches with a lip height of sufficient magnitude to influence the wear of UHMWPE against either damaged PEEK or CoCr plates. The wear factor of the UHMWPE pins in the control study was similar to that previously reported under comparable conditions [8]. It is interesting to note that the variability in the data was consistently larger in the UHMWPE-on-PEEK bearing couple compared to UHMWPE-on-CoCr and that following third body damage simulation, the variability in the data was further increased for both material combinations. This higher variability suggests that larger sample sizes and further testing under a wider range of conditions may need to be considered to fully understand the wear performance of the bearing couple.

Scratching the counterfaces with a diamond stylus allowed the geometry of the scratches to be more closely controlled. Previous studies have shown the lip height of scratches on retrieved hip and knee implants to be in the range of 0.01–4.1 μm [25,26] and experimental studies have shown larger scratch lip heights, up to 4 μm, to increase the wear of both hip [27] and knee implants [28]. To create a scratch of the same lip height in PEEK, a lower axial load was applied to the diamond stylus against the PEEK plates compared to CoCr due to the lower hardness of the PEEK polymer (Table A1) [18]. For both materials, there was a linear relationship between the load applied to the stylus and the lip height of the scratch created. Greater control of the lip height could be achieved in CoCr plates, where the scratch lip heights also had a more consistent geometry.

Against polished plates, there was no significant difference in the wear of UHMWPE against PEEK or CoCr. Scratches of 1 or 2 μm lip height in CoCr did not increase the wear of UHMWPE compared to polished control plates; however, with lip heights of 4 μm, the wear of UHMWPE was higher. This exponential relationship between scratch lip height and wear has previously been reported. Minakawa et al. showed scratches with lip heights below 0.5 μm to have a wear factor similar to that of polished plates; with larger lip heights (1 μm), a dramatic increase in wear factor was seen [15], and Lancaster et al. showed a similar relationship between Ra and wear factor [24]. In this study, a larger lip height (4 μm) was necessary to significantly increase the wear rate of the polyethylene, perhaps due to the differences in the grade of polyethylene used. When investigating the influence of scratch lip height in PEEK on the wear of UHMWPE, a different trend was seen. In the PEEK plates, a polishing effect of the pin against the plate was seen and the wear factor of the UHMWPE did not increase with lip height. When a 4 μm lip height was created in PEEK plates, the wear factor of the UHMWPE was significantly lower ($p < 0.05$) than when a scratch of similar magnitude was created in CoCr. This difference in wear performance can be attributed to the different hardness of the PEEK and CoCr materials.

Independent control tests were carried out for each of the damage simulation methodologies used and the different wear factors obtained may be attributable to the studies being carried out using different experimental rigs, slight differences in the composition of the cobalt chrome, the polymers used being from different batches and/or the studies being carried out by different researchers. It is, however, a strength of the study that independent control tests were carried out for each method.

There were a number of limitations that should be considered in the interpretation of this study. Firstly, this was a fundamental study carried out in a pin-on-plate configuration using simplified geometries of components and whilst the input kinematics replicated the average cross-shear and contact pressure in a total knee replacement [29]. The loading and motion profiles were simplified and consistent with previous studies [8], and it is not known whether the results would be replicated in a clinical setting. The components used were not sterilised or cross-linked; however, sterilisation by ethylene oxide, which is commonly used for TKRs, has been shown not to influence mechanical properties or induce crosslinking in polyethylene and, therefore, does not influence wear [30]. The sample size was small. In some cases, three repeats were carried out; however, it was deemed that a sufficiently large number of samples were tested to show the polishing effect of the UHMWPE pin against the PEEK plates scratched either with particles or a diamond stylus. This polishing effect did not occur in the UHMWPE-on-CoCr studies. It is also unclear whether the scratching on the PEEK plates after damage simulation or the polishing of the scratched regions of the plate by UHMWPE pins would contribute to an increase in PEEK wear or wear debris and whether in vivo this would accelerate osteolysis, as the inconsistent moisture uptake of PEEK made gravimetric analysis unreliable [8,22]. It is also likely that the relationship between scratch lip height and wear would be different for other grades or compositions of UHMWPE, particularly when articulating against CoCr or with the introduction of different kinematic conditions or geometries of components. Finally, this was a short-term study. When investigating damage simulation with particles, the particles were passed over the plate five times in each region of damage; and in the wear simulation, studies were limited to 1 MC.

5. Conclusions

Simulating third body damage with PMMA particles did not influence mean surface roughness enough to increase the wear of UHMWPE against either CoCr or PEEK. On the PEEK plates, a polishing effect of the UHMWPE pin against the scratched plate was observed.

When scratches with a lip height of greater magnitude were created directly on PEEK and CoCr plates, an exponential increase in UHMWPE wear factor was observed with increasing lip height against CoCr plates; whereas against PEEK plates, increasing lip height did not have a similar influence on the wear factor of UHMWPE.

Author Contributions: Conceptualisation, R.M.C., A.B. and L.M.J.; data curation, R.M.C. and N.M.P.; formal analysis, R.M.C., N.M.P. and L.M.J.; funding acquisition, L.M.J.; investigation, R.M.C. and N.M.P.; methodology, R.M.C. and L.M.J.; project administration, L.M.J.; resources, A.B. and L.M.J.; supervision, L.M.J.; writing—original draft, R.M.C.; writing—review and editing, N.M.P., A.B. and L.M.J. All authors have read and agreed to the published version of the manuscript.

Funding: This research was supported by Invibio Knees Ltd. and the Innovation and Knowledge Centre in Regenerative Therapies and Devices funded by the EPSRC, TSB and BBSRC under grant number EP/J017620/1. It was partially funded through WELMEC, a centre of Excellence in Medical Engineering funded by the Wellcome Trust and EPSRC under grant number WT 088908/Z/09/Z and supported by the EPSRC Centre for Innovative Manufacturing in Medical Devices under grant number EP/K029592/1.

Acknowledgments: The PEEK-OPTIMA™ plates were provided by Invibio Knees Ltd. Thanks to Phil Wood and his team for technical assistance.

Conflicts of Interest: A.B. is a paid employee of Invibio Ltd. and had a role in the study design and writing the manuscript; L.M.J. has received research funding support from Invibio Ltd.

Appendix A

Table A1. Load applied to a 200 μm radius diamond stylus to create scratches of a known geometry in PEEK and cobalt chrome plates.

Intended Lip Height of Scratches	PEEK-OPTIMA™		Cobalt Chrome	
	Weight Applied to Diamond Stylus (Kg)	Actual Lip Height of Scratches (μm)	Weight Applied to Diamond Stylus (Kg)	Actual Lip Height of Scratches (μm)
1 μm	0.51	1.11 ± 0.38	2.08	1.35 ± 0.09
2 μm	0.71	2.82 ± 0.57	2.99	2.19 ± 0.15
4 μm	1.02	4.22 ± 1.12	4.90	4.25 ± 0.59

References

1. National Joint Registry 16th Annual Report. 2018. Available online: https://reports.njrcentre.org.uk/Portals/0/PDFdownloads/NJR%2016th%20Annual%20Report%202019.pdf (accessed on 10 March 2020).
2. Bourne, R.B.; Chesworth, B.M.; Davis, A.M.; Mahomed, N.N.; Charron, K.D.J. Patient satisfaction after total knee arthroplasty: Who is satisfied and who is not? *Clin. Orthop. Relat. Res.* **2010**, *468*, 57–63. [CrossRef] [PubMed]
3. Cowie, R.M.; Briscoe, A.; Fisher, J.; Jennings, L.M. PEEK-OPTIMA™ as an alternative to cobalt chrome in the femoral component of total knee replacement: a preliminary study. *Proc. Inst. Mech. Eng. Part H J. Eng. Med.* **2016**, *230*, 1008–1015. [CrossRef] [PubMed]
4. Rankin, K.E.; Dickinson, A.S.; Briscoe, A.; Browne, M. Does a PEEK Femoral TKA Implant Preserve Intact Femoral Surface Strains Compared with CoCr? a Preliminary Laboratory Study. *Clin. Orthop. Relat. Res.* **2016**, *474*, 2405–2413. [CrossRef] [PubMed]
5. de Ruiter, L.; Janssen, D.; Briscoe, A.; Verdonschot, N. The mechanical response of a polyetheretherketone femoral knee implant under a deep squatting loading condition. *Proc. Inst. Mech. Eng. Part H J. Eng. Med.* **2017**, *231*, 1204–1212. [CrossRef] [PubMed]
6. De Ruiter, L.; Janssen, D.; Briscoe, A.; Verdonschot, N. a preclinical numerical assessment of a polyetheretherketone femoral component in total knee arthroplasty during gait. *J. Exp. Orthop.* **2017**, *4*, 3. [CrossRef]

7. East, R.H.; Briscoe, A.; Unsworth, A. Wear of PEEK-OPTIMA® and PEEK-OPTIMA®-Wear Performance articulating against highly cross-linked polyethylene. *Proc. Inst. Mech. Eng. Part H J. Eng. Med.* **2015**, *229*, 187–193. [CrossRef]
8. Cowie, R.M.; Briscoe, A.; Fisher, J.; Jennings, L.M. Wear and Friction of UHMWPE-on-PEEK OPTIMA™. *J. Mech. Behav. Biomed. Mater.* **2019**, *89*, 65–71. [CrossRef]
9. Du, Z.; Zhu, Z.; Yue, B.; Li, Z.; Wang, Y. Feasibility and Safety of a Cemented PEEK-on-PE Knee Replacement in a Goat Model: a Preliminary Study. *Artif. Organs* **2018**, *42*, E204–E214. [CrossRef]
10. Meng, X.; Du, Z.; Wang, Y. Characteristics of wear particles and wear behavior of retrieved PEEK-on-HXLPE total knee implants: a preliminary study. *RSC Adv.* **2018**, *8*, 30330–30339. [CrossRef]
11. Jennings, L.M.; Al-Hajjar, M.; Brockett, C.L.; Williams, S.; Tipper, J.L.; Ingham, E.; Fisher, J. (iv) Enhancing the safety and reliability of joint replacement implants. *Orthop. Trauma* **2012**, *26*, 246–252. [CrossRef]
12. Wang, A.; Essner, A. Three-body wear of UHMWPE acetabular cups by PMMA particles against CoCr, alumina and zirconia heads in a hip joint simulator. *Wear* **2001**, *250*, 212–216. [CrossRef]
13. Barbour, P.; Stone, M.; Fisher, J. a hip joint simulator study using new and physiologically scratched femoral heads with ultra-high molecular weight polyethylene acetabular cups. *Proc. Inst. Mech. Eng. Part H J. Eng. Med.* **2000**, *214*, 569–576. [CrossRef] [PubMed]
14. Ries, M.D.; Salehi, A.; Widding, K.; Hunter, G. Polyethylene wear performance of oxidized zirconium and cobalt-chromium knee components under abrasive conditions. *JBJS* **2002**, *84*, 129–135. [CrossRef]
15. Minakawa, H.; Stone, M.; Wroblewski, B.; Lancaster, J.; Ingham, E.; Fisher, J. Quantification of third-body damage and its effect on UHMWPE wear with different types of femoral head. *J. Bone Jt. Surg. Br. Vol.* **1998**, *80*, 894–899. [CrossRef]
16. Cowie, R.M.; Carbone, S.; Aiken, S.; Cooper, J.J.; Jennings, L.M. Influence of third-body particles originating from bone void fillers on the wear of ultra-high-molecular-weight polyethylene. *Proc. Inst. Mech. Eng. Part H J. Eng. Med.* **2016**, *230*, 775–783. [CrossRef]
17. Niki, Y.; Matsumoto, H.; Otani, T.; Tomatsu, T.; Toyama, Y. How much sterile saline should be used for efficient lavage during total knee arthroplasty? Effects of pulse lavage irrigation on removal of bone and cement debris. *J. Arthroplast.* **2007**, *22*, 95–99. [CrossRef]
18. Cowie, R.M.; Jennings, L.M. *Data Associated with 'Third Body Wear of UHMWPE-on-PEEK-OPTIMA™'*; University of Leeds: Leeds, UK, 2020. [CrossRef]
19. Galvin, A.; Kang, L.; Tipper, J.; Stone, M.; Ingham, E.; Jin, Z.; Fisher, J. Wear of crosslinked polyethylene under different tribological conditions. *J. Mater. Sci. Mater. Med.* **2006**, *17*, 235–243. [CrossRef]
20. ISO 4288:1996. *Geometrical Product Specifications (GPS)—Surface Texture: Profile Method—Rules and Procedures for the Assessment of Surface Texture*; ISO: Geneva, Switzerland, 1996.
21. Grupp, T.M.; Meisel, H.-J.; Cotton, J.A.; Schwiesau, J.; Fritz, B.; Blömer, W.; Jansson, V.J.B. Alternative bearing materials for intervertebral disc arthroplasty. *Biomaterials* **2010**, *31*, 523–531. [CrossRef]
22. Brockett, C.L.; John, G.; Williams, S.; Jin, Z.; Isaac, G.H.; Fisher, J. Wear of ceramic-on-carbon fiber-reinforced poly-ether ether ketone hip replacements. *J. Biomed. Mater. Res. Part B Appl. Biomater.* **2012**, *100*, 1459–1465. [CrossRef]
23. Isaac, G.; Atkinson, J.; Dowson, D.; Kennedy, P.; Smith, M. The causes of femoral head roughening in explanted Charnley hip prostheses. *Proc. Inst. Mech. Eng. Part H J. Eng. Med.* **1987**, *16*, 167–173. [CrossRef]
24. Lancaster, J.; Dowson, D.; Isaac, G.; Fisher, J. The wear of ultra-high molecular weight polyethylene sliding on metallic and ceramic counterfaces representative of current femoral surfaces in joint replacement. *Proc. Inst. Mech. Eng. Part H J. Eng. Med.* **1997**, *211*, 17–24. [CrossRef] [PubMed]
25. Tipper, J.; Ingham, E.; Hailey, J.; Besong, A.; Fisher, J.; Wroblewski, B.; Stone, M. Quantitative analysis of polyethylene wear debris, wear rate and head damage in retrieved Charnley hip prostheses. *J. Mater. Sci. Mater. Med.* **2000**, *11*, 117–124. [CrossRef] [PubMed]
26. Jones, V.; Williams, I.; Auger, D.; Walsh, W.; Barton, D.; Stone, M.; Fisher, J. Quantification of third body damage to the tibial counterface in mobile bearing knees. *Proc. Inst. Mech. Eng. Part H J. Eng. Med.* **2001**, *215*, 171–179. [CrossRef] [PubMed]
27. Bowsher, J.; Shelton, J. a hip simulator study of the influence of patient activity level on the wear of crosslinked polyethylene under smooth and roughened femoral conditions. *Wear* **2001**, *250*, 167–179. [CrossRef]
28. Cowie, R.M.; Aiken, S.S.; Cooper, J.J.; Jennings, L.M. The Influence of a Calcium Sulfate Bone Void Filler on the Third Body Damage and Polyethylene Wear of Total Knee Replacements. *Bone Jt. Res.* **2019**, *8*, 65–72. [CrossRef]

29. Fisher, J.; McEwen, H.M.; Tipper, J.L.; Galvin, A.L.; Ingram, J.; Kamali, A.; Stone, M.H.; Ingham, E. Wear, debris, and biologic activity of cross-linked polyethylene in the knee: Benefits and potential concerns. *Clin. Orthop. Relat. Res.* **2004**, *428*, 114–119. [CrossRef]
30. Kurtz, S.M. Chapter 3—Packaging and Sterilization of UHMWPE. In *UHMWPE Biomaterials Handbook*, 2nd ed.; Kurtz, S.M., Ed.; Academic Press: Boston, MA, USA, 2009; pp. 21–30.

materials

MDPI

Article

How to Proceed with Asymptomatic Modular Dual Taper Hip Stems in the Case of Acetabular Revision

Thomas M. Grupp [1,2,*,†], Marc Baxmann [1,†], Volkmar Jansson [2], Henning Windhagen [3], Karl-Dieter Heller [4], Michael M. Morlock [5] and Hanns-Peter Knaebel [6]

1 Aesculap AG Research & Development, 78532 Tuttlingen, Germany; marc.baxmann@gmx.de
2 Department of Orthopaedic Surgery, Physical Medicine & Rehabilitation, Campus Grosshadern Ludwig Maximilians University Munich, 81377 Munich, Germany; volkmar.jansson@med.uni-muenchen.de
3 Department of Orthopaedic Surgery, Hannover Medical School, 30625 Hannover, Germany; henning.windhagen@diakovere.de
4 Department of Orthopaedic Surgery, Herzogin-Elisabeth-Hospital, 38124 Braunschweig, Germany; kd.heller@heh-bs.de
5 Institute of Biomechanics, TUHH Hamburg University of Technology, 21073 Hamburg, Germany; morlock@tuhh.de
6 Roechling SE & Co. KG, 68165 Mannheim, Germany; hp.knaebel@roechling.com
* Correspondence: thomas.grupp@aesculap.de; Tel.: +49-7461-95-2667; Fax: +49-7461-95-382667
† Both primary authors contributed equally to the study.

Received: 28 November 2019; Accepted: 18 February 2020; Published: 2 March 2020

Abstract: How to proceed with a clinically asymptomatic modular Metha®Ti alloy stem with dual taper CoCr neck adapter in case of acetabular revision? To systematically answer this question the status of research and appropriate diagnostic methods in context to clinically symptomatic and asymptomatic dual taper stem-neck couplings has been evaluated based on a systematic literature review. A retrieval analysis of thirteen Metha®modular dual taper CoCr/Ti alloy hip stems has been performed and a rational decision making model as basis for a clinical recommendation was developed. From our observations we propose that in cases of acetabular revision, that for patients with a serum cobalt level of > 4 μg/L and a Co/Cr ratio > 3.6, the revision of the modular dual taper stem may be considered. Prior to acetabular revision surgery a systematic diagnostic evaluation should be executed, using specific tests such as serum metal (Co, Cr) ion analysis, plain antero-posterior and lateral radiographs and cross-sectional imaging modalities (Metal Artefact Reduction Sequence Magnetic Resonance Imaging). For an asymptomatic Metha®dual taper Ti alloy/CoCr stem-neck coupling at the stage of acetabular revision careful clinical decision making according to the proposed model should be followed and overreliance on any single examination should be avoided, considering the complete individual differential diagnosis and patient situation.

Keywords: total hip arthroplasty; dual taper modular hip stem; acetabular revision; asymptomatic stem modularity; decision making model; threshold

1. Introduction

Failure of total hip arthroplasty (THA) is a relatively rare condition and the most common causes for revision are aseptic loosening, dislocation, septic loosening and peri-prosthetic fractures [1,2]. For a better adaptation to different diaphyseal and extra-diaphyseal anatomical conditions a modular dual taper neck design was clinically introduced for primary THA by Toni et al. [3] in 1995. Using an anatomic cementless stem design in combination with a modular rectangular tapered neck adapter in a consecutive series of 216 hip arthroplasties they reported a survival rate of 98.6% at 5 years including all implant related complications [3]. Clinical benefits of modular dual taper neck adapter hip stems are the restoration of the hip centre of rotation in combination with an adequate soft tissue balancing in complex anatomical, muscular and ligamenteous patient situations [4–6]. For anatomic differences between specific patients (e.g., dysplasia), as well as different hip morphologies in males and females in regard to femur size, Caput-Collum-Diaphysis (CCD) angle, femoral offset and anteversion [7], these designs allow to adjust the native femoral anteversion and offset to restore an adequate abductor muscles lever arm [5], limb length [7] and patients' original biomechanics [8], independently of the stem size and femoral fixation [6,9]. In an exploratory study on 684 mono-bloc and 125 THAs with a modular neck in patients dedicated to a rapid recovery programme based on selected cases with primary arthritis in an otherwise anatomically normal hip joint, Gerhardt et al. [10] did not find a clear benefit in restoring hip geometry and dislocation rate—only the abductor moment arm which is closely correlated to the femoral offset was better reconstructed in the modular neck adapter cohort. For this "straightforward" THA the exclusion criteria were profound acetabular dysplasia, discrepancy in leg length or anatomical deformities of the proximal femur [10].

The mid- to long-term clinical results for some modular dual taper stem designs are comparable to the implant survivorship of mono-bloc cementless stem designs [7,9,11–15]. Examining the effectiveness of neck modularity in THA considering sex differences in hip morphology, Traina et al. [7] retrospectively reviewed the clinical results of 2131 modular stems implanted in 1051 men and 1080 women. They reported an estimated Kaplan-Meier survival rate at 11 years of 97.6% for men and 96.0% for women without any modular Ti neck failures [7]. In a series of demanding THA procedures with developmental dysplasia of the hip in 47 patients with 61 modular neck stems with an average follow-up of 117.2 months (range 57–167) Traina et al. [9] reported a cumulative survival of 97.5% at 11 years (one ceramic liner fracture). Analysing the long-term survivorship and complication rate of a cementless modular neck primary stem design in the data base of the Emilia-Romagna registry of orthopaedic prosthetic implants, Fitch et al. [12] found for 692 modular THAs implanted in 26 orthopaedic units an overall Kaplan-Meier survivorship of 95.8% at 12 years follow-up, similar to the 96.1% for all mono-bloc cementless stem designs implanted during the same period.

A major disadvantage of dual taper modular hip stems is the additional interface at the neck-stem junction, increasing the risk for fatigue fracture, fretting, crevice corrosion and metal particle and ion release [6,16–22]. For Ti alloy hip stem designs with Ti alloy neck adapters, the main implant-related clinical failure mode is fatigue fracture of the neck taper in the highly stressed neck-stem junction due to fretting corrosion, local stress concentration and crack initiation by multi-directional bending and torsion [16,23–27]. For CoCr neck adapters only few fatigue failures have been reported clinically for a Profemur Z titanium alloy stem (Profemur Hip System, Wright, Arlington, TN, USA) with long 8° varus type CoCr alloy modular necks in overweight or obese patients with considerable physical activity [28–30]. For CoCr neck adapters, the main clinical failure mechanism is the generation of particulate Co and Cr wear debris and metal ion release from the neck adapter due to tribo-corrosion, that may cause adverse local tissue reactions (ALTR) [17,31–34] like osteolysis, metallosis and pseudo-tumor formation in the surrounding tissue [19,35,36] as well as elevated serum ion levels [17,33,37–39] and systemic toxicity [40,41]. This failure mode is predominantly occurs in mixed CoCr neck/Ti alloy stem junctions [31,33,39,40,42–44], but has also been described for Ti alloy/Ti alloy dual taper stems [37,38] and single alloy CoCr couplings [45,46].

In a previous study [16] clinical fatigue failures for a short modular Ti alloy stem in combination with a Ti alloy neck adapter and the influence of the implant material on the endurance behaviour have been described. The current study is focused on the evaluation of possible clinical failure modes related to the effects of particulate debris and ion release due to mechanically assisted fretting and crevice corrosion from the same stem in combination with a CoCr neck adapter in order to derive a clinical decision making model.

2. Objectives

The objectives of our study therefore to attempt to answer the following questions:

1) Which suggestions are given in literature how to proceed with clinically symptomatic and asymptomatic dual taper stem-neck couplings?—Status of research and appropriate diagnostic methods.
2) What are the relevant findings based on neck adapter retrievals of the Metha®dual taper CoCr/Ti alloy hip stem design?
3) How to proceed with a clinically asymptomatic dual taper modular hip stem in case of acetabular revision?—Definition of a rational decision making model as basis for a clinical recommendation.

3. Literature on Symptomatic and Asymptomatic Dual Taper Hip Stems

Searching for total hip arthroplasty with dual taper or bi-modular neck adapters in combination with crevice or tribo-corrosion, particle release, metallosis or adverse local tissue reactions we performed a systematic review in PubMed and EMBASE to present an actual literature overview (time frame 1 January 2006 to 31 January 2020). We found after removal of duplicates n = 281 publications, within n = 166 are not related to the topic of dual taper hip stems. Undergoing a systematic full text review based on n = 115 publications we found n = 75 suitable publications and identified additional n = 20 publications (registry reports, conference proceedings, et al.) from our internal database including in total n = 95 studies into the meta-analysis (Scheme 1).

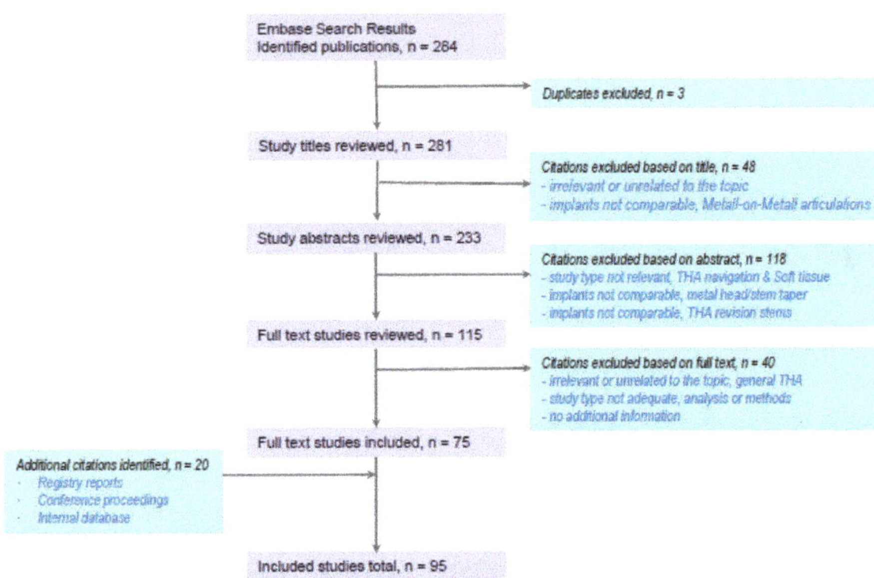

Scheme 1. Systematic literature review (Embase n = 284 publications) and studies included in meta-analysis (n = 95).

In a review, Mistry et al. [19] postulated etiologies like modular head-neck or neck-stem junctions wear, corrosion damage and metal ion release as "trunnionosis", a cause of failed total hip arthroplasties [47]. They described the effects of the femoral head size as well as the trunnion design and localized biological reactions associated with trunnionosis [19]. Jacobs et al. [17] described a variety of factors, like head size, taper geometry, material composition, metallurgical processing, surface finish, neck offset and length, design-related factors and a contamination of the taper interface during assembly which may contribute to mechanically assisted crevice corrosion (MACC) in modular junctions involving at least one CoCr component. They found adverse local tissue reactions (ALTR) with clinically symptoms in eleven patients due to MACC with 19-fold elevated mean serum cobalt levels for a beta-titanium alloy ($TiMo_{12}Zr_6Fe_2$) modular stem with CoCr neck compared to well-functioning primary THA patients with metal-on-polyethylene articulations at a relatively early post-operative period (mean 7.9 months) [17]. Dimitriou et al. [33] reported about the early outcomes of revision surgery for taper corrosion of two dual taper THA designs (Rejuvenate & ABG II) with CoCr neck and beta-titanium alloy stem based on 198 revision surgeries in 187 patients. They described a significant decline of patients serum ion levels for cobalt from 5.3 μg/L (range 2.3 to 48.5 μg/L) to 1.4 μg/L (range 0.2 to 8.8 μg/L) and for chromium from 2.6 μg/L (range 0.2 to 64 μg/L) to 0.7 μg/L (range 0.1 to 3.9 μg/L) after revision surgery, with a half-life of 3.2 months for cobalt and of 5 months for chromium. The cobalt/chromium ratio also significantly decreased from 4.7 (range 2.1 to 35) to 2.2 (range 0.4 to 8.8) [33].

For the Rejuvenate dual taper modular neck stem consisting of a CoCr neck and a beta-titanium alloy stem, Bernstein et al. [44] described a revision rate of 86% (63 of 73 hip stems) at a mean follow-up of 4.2 years (range 3 to 5.5) with indications for revision surgery being a serum cobalt ion level > 4 μg/L, persistent pain or abnormal MRI (Magnetic Resonance Imaging) findings. They reported mean serum cobalt and chromium ion levels prior to revision of 10.0 μg/L (range 0.3 to 40.0 μg/L) and of 2.3 μg/L (range 1.0 to 7.4 μg/L), respectively and found a substantial decrease of cobalt levels post-operatively. The unrevised group had a serum cobalt level of 2.1 ± 2.0 μg/L and chromium of 1.2 ± 0.4 μg/L, whereas the patients with abnormal MRI findings had 8.1 μg/L (0.3 to 28.9) and 2.0 μg/L (1.0 to 7.0), respectively [44]. Meftah et al. [31] examined the rate of corrosion-related failure and survivorship of the dual taper Rejuvenate stem (n = 97) and correlated implant and patient factors with serum cobalt ion levels and revisions. The Kaplan-Meier survivorship was 40% at four years and the mean cobalt and chromium levels related to metal corrosion in symptomatic patients were 8.1 μg/L (range 0.4 to 31 μg/L) and 2.5 μg/L (range 0.2 to 4.3 μg/L), respectively.

In contrast to that Vundelinckx et al. [39] described for the ABG II dual taper stem design, which was also made of beta-titanium alloy with CoCr neck, only one revision out of a cohort of 306 THAs consisting of a ceramic-on-polyethylene or a ceramic-on-ceramic bearing interface implanted between 2007 and 2011. The patient undergoing a revision developed intermittent pain in the trochanteric area 2 years post-operatively and showed after 4 years peri-prosthetic fluid accumulation and a soft tissue mass around the proximal stem and neck region and an increased serum level of 7.4 μg/L for cobalt. Taking a randomized sample of 19 asymptomatic patients, 9 patients presented a cobalt level > 4 μg/L with a maximum of 7.5 μg/L.

Walsh et al. [48] studied 103 THA revision cases (78 Rejuvenate; 25 ABG II) with dual taper modular neck at a mean time of 2.4 years (range 0.66 to 5) from index surgery to revision and found a mean serum cobalt level of 7.6 μg/L (range 1.1 to 23 μg/L) and a mean serum chromium level of 1.8 μg/L (range 0.1 to 6.8 μg/L). They performed an aspiration of the hip for synovial fluid metal content prior revision THA in 40 patients and found a mean cobalt level of 916 μg/L (range 12 to 3900 μg/L) and a mean chromium level of 599 μg/L (range 3.4 to 3300 μg/L) in the synovial fluid [48].

Barlow et al. [49] reported about 54 patients undergoing revision surgery with 59 revised Rejuvenate dual taper stems based on a cohort of 199 THAs implanted between 2010 and 2012 by a senior surgeon. They analysed the serum ion levels prior to revision and the decline of serum levels at 6 weeks and 12 weeks after revision surgery. For 49 patients with unilateral THA they found a mean serum cobalt ion level of 8.19 ± 5.54 μg/L, which significantly decreased in all patients to 2.68 ± 2.67 μg/L at 6 weeks and to 1.58 ± 1.57 μg/L at 3 months. In five patients with bilateral modular Rejuvenate hip arthroplasty they measured a pre-revision serum cobalt level of 13.33 ± 6.45 μg/L and also a significant drop to 3.73 ± 2.19 μg/L at six weeks post-revision. An overview of additional studies analyzing serum levels of cobalt, chromium and Ti alloy is given for modular dual taper hip designs with $TiAl_6V_4/TiAl_6V_4$ and $CoCr_{29}Mo_6/TiMo_{12}Zr_6Fe_2$ neck stem couplings [50–54] (Table 1).

Table 1. Literature overview about clinical cohorts with dual taper neck adapter.

Study	Prosthesis	Neck/ Stem Material	Mean Age (Years)	Follow-up (Years)	Number of Patients	Study Group	Serum Level (µg/L) Co	Cr	Ti	Co/Cr Ratio	MRI Findings
Omlor et al. 2013 [37]	Profemur E	Ti6Al4V/Ti6Al4V	66.0	9.0	67	Modular (Uni-/bilateral)	-	-	3.0/6.0	-	-
	CLS	Ti6Al4V	71.0		11	Non-modular (Uni-/bilateral)	-	-	2.7/6.2	-	-
Molloy et al. 2014 [42]	ABGII	Vitallium/TMZF	64.3	3.4	7	Revised	6.3	1.3	-	6.4	86%
				3.6	8	Non-revised	3.9	1.1	-	3.7	13%
Silverton et al. 2014 [50]	Profemur Z	Ti6Al4V/Ti6Al4V	59.5	4.5	137	Follow-up	2.4	1.7	4.2	-	-
Gofton et al. 2015 [38]	Profemur TL	Ti6Al4V/Ti6Al4V	60.2	2.0	25	MoM articulation	2.5	2.1	2.8	-	-
					25	MoP articulation	0.3	0.3	2.9	-	-
Lanting et al. 2015 [51]	Rejuvenate	Vitallium/TMZF	65.0	1.7	19	Revised	5.5	0.8	3.0	8.2	83%
Restrepo et al. 2014 [52]	ABGII	Vitallium/TMZF	61.0	2.0	85	Symptomatic	4.0	1.2	-		46%
					110	Asymptomatic	3.4	1.2	-		11%
Barlow et al. 2016 [55]	Rejuvenate	Vitallium/TMZF	65.8	2.4	90	Revised	6.6	1.3	-	-	81%
Chillemi et al. 2017 [53]	ABGII	Vitallium/TMZF	69.4	>1.0	5	Symptomatic	3.0	0.4	-	7.4	-
			65.1		17	Asymptomatic	3.6	0.7	-	9.6	-
Kwon et al. 2017 [36]	Rejuvenate/ABGII	Vitallium/TMZF	59.0	1.4	90	Pseudotumor	5.0	0.8	-	6.0	100%
			58.0	1.3	58	No pseudotumor	3.7	0.8	-	3.7	0 %
Liow et al. 2016 [54]	Rejuvenate/ABGII	Vitallium/TMZF	57.4	2.3	31	Revised	3.8	1.0	-	3.8	100 %

In addition, Hussey and McGrory [56] performed a ten years cross-sectional study including 1352 consecutive patients with metal-on-polyethylene THA combined with 12/14 trunnion Ti alloy stem types and found symptomatic MACC present in 43 cases (3.2%). A dual taper Ti alloy neck and Ti alloy stem design (M/L Taper) showed a higher prevalence (4.9%) of MACC than all other Ti alloy stems combined (1.2%) of the same manufacturer. For these stem design (M/L Taper Kinectiv) with a bi-modular Ti alloy neck on a Ti alloy stem and a CoCr 40 mm head, Canham et al. [57] described in a case study a characteristic peri-prosthetic pseudo-tumor formation as an ALTR and an elevated serum cobalt level of 12.3 μg/L and chromium of 1.8 μg/L whereby the only potential source was the head-neck junction.

For symptomatic patients with dual taper hip arthroplasty Kwon et al. [58] propose a differential diagnosis of the temporal onset, duration, severity and location of pain and recommend that also additional symptoms as a swelling or feeling of fullness around the hip should be elicited. To evaluate dual-taper modular implants with corrosion related metal debris contamination, the analysis of serum inflammatory markers as erythrocyte sedimentation rate and C-reactive protein as well as hip aspiration for synovial fluid counts are described to differentiate and exclude periprosthetic joint infection [59]. The elevation of metal ion serum levels and an increase in the cobalt/chromium ratio is also observed [18,40,59].

Adverse local tissue reactions associated with metal ion and debris released by a dual taper Rejuvenate stem were identified in 36 revised stems out of a cohort of 118 THAs by Ghanem et al. [43]. The symptoms of the 36 THAs which were considered as failed began at a mean post-operative time of 14.8 months (range 2.8–34.8) and the average time to revision was 24.1 months (range 8.8–50.2). The authors described higher cobalt serum levels in the failed group of 9.5 ± 6.8 μg/L (range 1.9–24.7) compared to the asymptomatic group of 4.9 ± 3.6 μg/L (range 0.1–15.7) and higher cobalt/chromium ratios of 5.2 ± 3.2 for the failed compared to 3.6 ± 2.3 for the asymptomatic group. However, they reported no correlation in diagnostic accuracy for ALTR, while MRI scan considering pseudo-tumor size was more sensitive [43].

Kwon et al. [35] evaluated 97 consecutive dual taper modular stem THAs in a retrospective study by MARS-MRI and stratified 83 of these patients in pseudo-tumor absent (n = 53) and pseudo-tumor present (n = 30). In the pseudo-tumor present group they found no substantial difference between symptomatic (n = 21) and asymptomatic patients (n = 9) in the serum cobalt level (sympt.: 7.6 μg/L (range 3.3–14.4) versus asympt. 6.2 μg/L (range 3.4–11.7)) or cobalt/chromium ratio (sympt.: 8.25 (range 4.5–68) versus asympt. 10.6 (range 4.8–29.5)). For the THA patient group with a pseudo-tumor the cobalt serum level and cobalt/chromium ratio (8.0 μg/L; 10.3) were significantly higher than for those without a pseudo-tumor (2.0 μg/L; 2.4) and based on this the authors suggest cross-sectional imaging (MARS-MRI) for THA patients with elevated metal ion serum levels [35]. In a nano-analysis of wear particles from retrieved peri-prosthetic tissue, Xia et al. [60] could clearly distinguish between metal-on-metal (MoM) surface replacement (n = 12; implantation time 51.6 months), MoM large head THA (n = 18; implantation time 59.9 months) and non-MoM dual modular neck hip arthroplasty (Rejuvenate) (n = 23; implantation time 30.9 months). They found that the particle physical characteristics and metal composition are consistent in each implant category and concluded that substantial differences in size, shape and element composition of the metallic particles correlate with the histological features of severity of ALTR and variability in implant performance, indicating that the immunogenicity and toxicity of the released particles is a leading factor in the specific onset and severity of the reaction [60,61].

4. Clinical Case Presentations and Retrieval Analysis

4.1. Metha®Dual Taper CoCr/Ti Alloy Couplings with Adverse Local Tissue Reactions

Data on all known cases (n = 5) with soft tissue reactions due to debris and metal ion release following THA with the dual modular short stem hip prosthesis Metha®(Aesculap AG, Tuttlingen,

Germany; Figure 1) with CoCr neck adapter in the period 2007 to 2019 (Table 2). In single cases (#3 & #5) microbiology and histological analysis was performed and showed a wear particle induced peri-prosthetic interface membrane type I according to the classification of Krenn and Morawietz [62,63]. Average duration between index procedure and revision was 92.6 months (61–128 months).

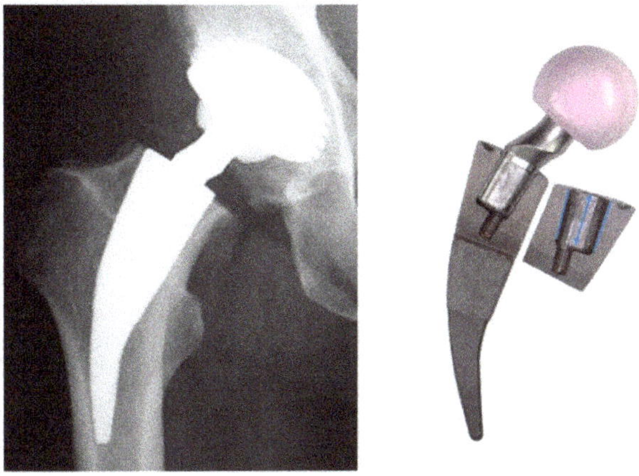

Figure 1. Modular Ti alloy short hip stem with a dual taper CoCr neck—Metha® modular.

All retrieved CoCr neck components were characterized by means of light microscopy (M165c, Leica Microsystems GmbH, Wetzlar, Germany), scanning electron microscopy (SEM, EVO 50, Carl Zeiss Microscopy GmbH, Jena, Germany) and energy dispersive X-ray analysis (EDS, X-Max 50, Oxford Instruments plc, Abingdon, UK). Volumetric material loss of the neck adapters was quantified according to a previously published algorithm by Buente et al. [64] using a tactile co-ordinate measurement machine (Figures 2–4).

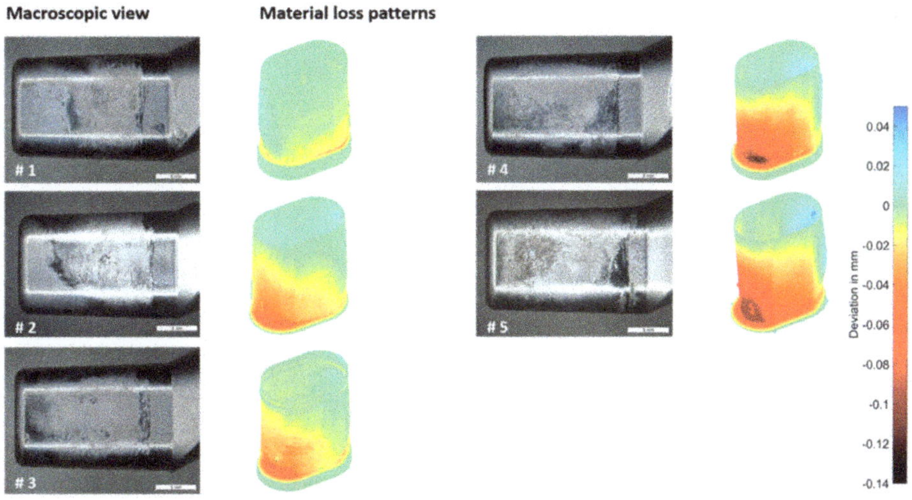

Figure 2. Macroscopic view (left) and volumetric material loss (right) of 5 symptomatic CoCr dual taper neck adapter retrieved after 61–128 months in vivo. The absence of oxygen in the energy dispersive X-ray measurements on characteristic taper surface points indicated that electrochemical processes due to contact, crevice and fretting corrosion affected the ability of the CoCr alloy to form a passive oxide layer in the physiological environment, resulting in increased material loss and metal ion release.

Table 2. Overview of five symptomatic clinical cases with retrieved Metha® CoCr dual taper neck adapter.

Patient (n)	Indication & Revision Procedure	Time In Situ (mths)	Sex	Age (years)	BMI (kg/m^2)	Stem Size & Side	CCD-Angle Version	Head Size & Offset	Head Material	Serum Level (μg/L)	Co/Cr Ratio	Max. Wear Depth (μm)	Material Loss (mm^3)
1	Soft tissue reactions progressive pain & swelling, ALTR & pseudo-tumor formation	61	Female	60	-	2 right	135° 0°	40 +0	Ceramic/Ti-Sleeve	Co=31.3 Cr=0.3	R=104	37	2.6
2	Groin pain, hip aspiration—cloudy-yellow synovial fluid, local tissue infiltrations, thickened capsular tissue	84	Female	67	-	3 left	135° 0°	36 −4.0	Ceramic	-	-	72	5.0
3	Soft tissue reactions, persistent hip pain ALTR, positive MRI findings, peri-prosthetic joint infection, extensive debridement of capsular tissue & metal debris	94	Female	76	32.0	2 left	130° 7.5° AV	32 +4.0	Ceramic	Co=8.3 Cr=0.8	R=10.37	82	6.5
4	Soft tissue reactions, persistent hip pain & swelling, elevated Co level, pseudo-tumor resection & debridement of metal debris, severe loosening of ScrewCup® SC	96	Female	58	24.6	2 -	135° 0°	32 −4.0	Ceramic	-	-	128	9.7
5	CT hip scan, migrated Hofer cup, suspicion cup loosening, exploration oft he hip, accumulation of metal debris, muddy-yellow fluid accumulation within joint capsula	128	Female	79	-	2 left	135° 7.5° AV	36 +0	Ceramic	-	-	100	12.0

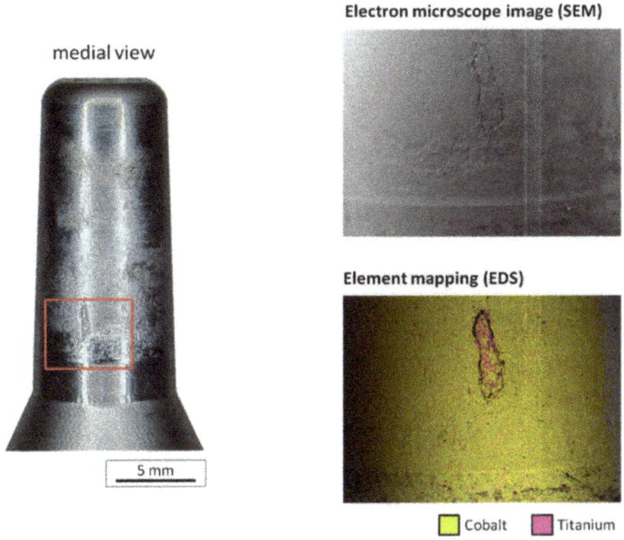

Figure 3. SEM and EDS analysis of the medial contact area of one symptomatic retrieved CoCr neck adapter (Case #3). The SEM and EDS-analysis indicated that the local damage of the protective oxide layer and the suppressed repassivation process led to fretting wear and contact corrosion with metal dissolution in the medio-proximally and latero-proximally contact area of the retrieved neck adapter. Furthermore, local material deposition on the neck adapter with a high element concentration of titanium was observed at the circular medial taper interface caused by local adhesion and cold welding in the mixed CoCr/Ti neck-stem junction.

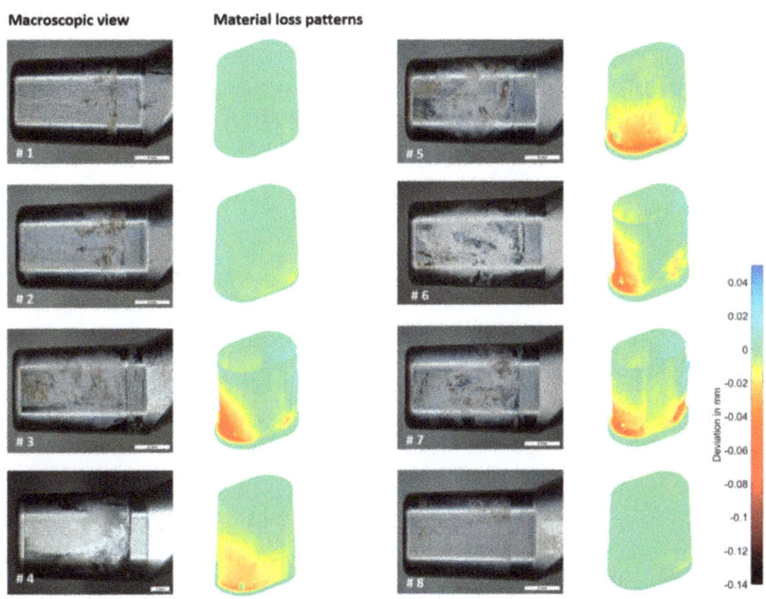

Figure 4. Macroscopic view (left) and volumetric material loss (right) of eight asymptomatic CoCr dual taper neck adapter retrieved after 13–111 months in vivo..

Statistics were performed with SPSS 24 (IBM Inc., Armonk, NY USA). Normality was checked using Shapiro-Wilk test. The homogeneity of variances between the groups was checked using Levene's test. Univariate analysis of variance (ANOVA) was performed to compare the two cohorts linear regression analysis to investigate the dependency of material loss with time in situ. The type 1 error level was set to 5%.

Case 1 (exemplified):

In March 2011, a fifty-one years old woman underwent an uncomplicated total hip arthroplasty of the right hip and received a Metha®(Ti_6Al_4V, stem size 2) with a CoCr neck adapter (CCD-angle of 135°, 0° neutral version) and a 40 mm ceramic-on-ceramic articulation with a titanium sleeve (DeltaMotion®, Finsbury Orthopaedics, Leatherhead, UK). At 61 months after this index procedure, she presented with progressive pain and swelling localized to the right proximal thigh. The clinical and radiographical evaluation showed no pathologic findings with regard to the range of motion, component position and osseointegration. Magnetic resonance imaging (MRI) demonstrated a large heterogeneous fluid collection (7 cm × 3 cm × 3.4 cm) in the ventral aspect of the iliopsoas muscle. Serum cobalt ion level was elevated at 31.3 μg/L and chromium level was normal at 0.3 μg/L. The positive MRI findings and the high Co/Cr-ratio (ratio = 104) were in keeping with adverse local tissue reactions and pseudo-tumor formation and revision total hip arthroplasty was performed in April 2016. The retrieved modular neck exhibited local surface changes and black deposits on the taper interface. Material analysis showed characteristic fretting and corrosion wear patterns concentrated on the medial and lateral contact region of the neck adapter (Figure 2, neck adapter #1). The largest amount of material loss was observed proximal at the medial taper interface with a maximum wear depth of 37 μm. The total volumetric material loss of the neck adapter was 2.6 mm^3 (see also Table 2 cases #2–5 and Figure 2, symptomatic neck adapter #2–5).

4.2. Retrieval Analysis of CoCr Neck Adapters Revised for Other Reasons than Adverse Local Tissue Reactions

A total of eight asymptomatic Metha®modular CoCr neck adapters revised in the period of 2007 to 2019 for other reasons than adverse local tissue reactions, listed in our database (Table 3).

The reasons for revision were insufficient osseointegration with migration, cup malpositioning, patient discomfort, cup loosening, luxation, acetabular fracture and periprosthetic fracture. The mean age at time of revision was 68.7 years (63 to 77 years) and the mean period of implantation was 62 months (13 to 111 months). The time to revision for the asymptomatic group is comparably shorter than for the symptomatic group (92.6 months).

Table 3. Overview of eight asymptomatic clinical cases with retrieved CoCr dual taper neck adapter.

Patient (n)	Revision	Time In Situ (mths)	Sex	Age (Years)	BMI (kg/m²)	Stem Size & Side	CCD-Angle Version	Head Size (mm)	Head Offset (mm)	Head Material	Max. Wear Depth (µm)	Material Loss (mm³)
1	Osseointegration insufficient	13	Male	73	29.7	5 right	130° 0°	36	+4.0	Ceramic	7	0.1
2	Malposition cup	26	Female	77	31.6	1 right	130° 0°	32	−4.0	Ceramic	16	0.5
3	Patient discomfort	36	Female	63	20.2	2 right	135° 0°	32	−4.0	Ceramic	78	5.1
4	Cup loosening	54	Female	69	-	2 left	135° 0°	36	−4.0	Ceramic	39	3.0
5	Patient discomfort	72	Female	67	-	3 right	130° 0°	32		Ceramic/Ti-Sleeve	86	6.3
6	Luxation	80	Female	63	29.1	3 left	130° 0°	32	−4.0	Ceramic	61	5.3
7	Acetabulum fracture	104	Female	63	25.2	2 right	135° 0°	28	+5.0	Ceramic	90	5.5
8	Periprosthetic fracture	111	Female	74	-	3 right	135° 7.5° RV	28	+3.5	Ceramic	14	0.3

Visual inspection of the modular neck adapters of the asymptomatic group (Figure 4) by light microscopy and SEM/ EDS-analysis, demonstrated the same characteristic signs of corrosion and debris concentrated on the medial and lateral surface of the neck/stem interface, but qualitatively less pronounced than for the symptomatic group (Figure 2). One patient (Table 3, #4) with signs of a cup loosening revised after 54 months in situ showed some local tissue reactions, which, however, were classified as asymptomatic due to the low neck adapter wear and material loss (3.0 mm^3) (Figure 4, neck adapter #4).

Only minor traces of wear and corrosion were seen at the neck adapter of patient #8 (Table 3), based on 111 months of implant in service. The maximum wear depth was 14 μm observed at the proximal lateral taper interface, resulting in a very low total volumetric material loss of 0.3 mm^3 (Figure 4, neck adapter #8). This may be related to parameters like a low demanding biomechanical loading of the hip, low tribo-corrosion due to a less aggressive joint fluid composition or unknown co-diseases or morbidity of the patient. The adapter was classified as an outlier and not included into the analysis.

The volumetric material loss of the CoCr neck adapters increased for both cohorts with time in situ (Figure 5; adjusted $r^2 = 0.73$, $p < 0.001$). A strong trend for a lower wear rate for the asymptomatic cohort compared to the symptomatic cohort was observed ($p = 0.059$). The total volume loss for the symptomatic group was 7.16 ± 3.73 mm^3 and for the asymptomatic 3.69 ± 2.52 mm^3 ($p = 0.082$). Time in situ tended to be longer for the symptomatic group (92.6 ± 24.2 months) compared to the asymptomatic group (55.0 ± 32.4 months) ($p = 0.054$). The two cohorts did not differ with respect to gender distribution or any other patient or implant specific variable shown in Tables 2 and 3 ($p > 0.1$ for all analysis).

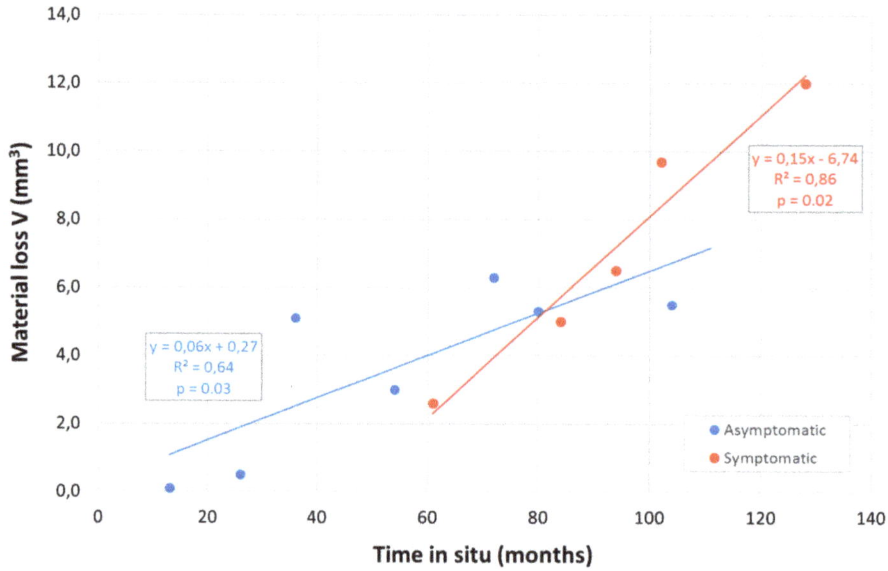

Figure 5. The volumetric material loss versus time in situ of seven asymptomatic and five symptomatic retrieved modular CoCr neck components. Note: asymptomatic neck adapter #8 was excluded from the regression analysis.

Due to multiple factors of influence like the patient specific loading situation, weight, activity level, implant orientation and muscular situation, as well as varying physiological surrounding lubricant conditions the comparably low number of only thirteen neck adapters does not allow to estimate a specific patient profile or phenotype. In a descriptive manner of the symptomatic and asymptomatic cases the patient age at revision was 57 to 79 years, their were 12 females and one male, BMI was between 20.2 and 32 (unknown in 6 cases), neck adapter CCD angles were 130° and 135°, and the head material was ceramic in all cases (2/13 ceramic heads with Ti-Sleeve).

5. Decision Making Model for Asymptomatic Dual Taper Stems in Case of Acetabular Revision

In the present study, metal ion concentrations were not determined. Therefore the available data from the literature were utilized to develop a decision making model for the case of acetabular revision for pre- and intra-operative decision making, as an orientation how to decide whether to maintain or revise an asymptomatic dual taper stem (Scheme 2).

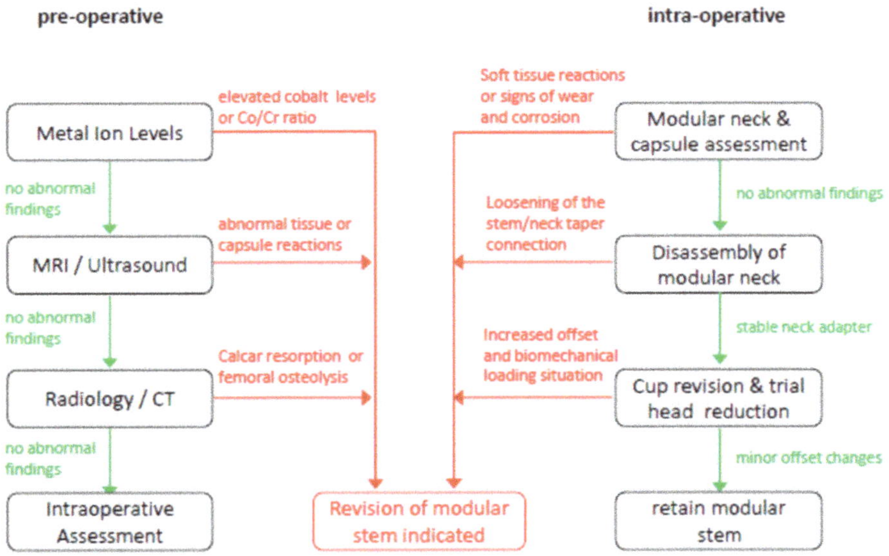

Scheme 2. Practical guide how to proceed with an asymptomatic dual taper modular hip stem in case of acetabular revision.

Cobalt and chromium ions are released from modular dual taper stem connections as consequence of mechanically assisted crevice corrosion [17]. In several retrieval studies elevated cobalt ion levels were associated with adverse local tissue reactions in THA patients with dual taper stems [18,39,45] and a cobalt value of ≥ 8 µg/L serum concentration has been documented for patients having a symptomatic dual taper stem and or a pseudo-tumor present [31,35,39,43,44,49,65,66]. An additional important diagnostic parameter is differential elevation of cobalt relative to chromium [58,65], originated by a predominant cobalt ion release at modular taper connections related to a chemical corrosion process that involves more soluble cobalt dissipating as free ions [40,45,67].

In a series of 447 consecutive patients tested for serum levels Fillingham et al. [65] identified 64 patients with a metal-on-polyethylene THA bearing (12 with a dual taper modular neck), whereas 44 were positive for an adverse local tissue reaction. The diagnosic measures showed a mean serum cobalt level of 8.58 µg/L and a Co/Cr ratio of 11.56. Kwon et al. [35] performed a retrospective study of 97 consecutive patients with a dual taper femoral stem and found substantially elevated cobalt levels of 8.0 µg/L (3.3–14.4) and an elevated Co/Cr ratio of 10.3 (4.5–68.0) in their pseudo-tumor present group. Ghanem et al. [43] identified 107 patients who underwent 118 THAs (11 bilateral cases) with a Rejuvenate dual taper femoral stem and proposed a decision tree to detect whether or not symptoms were present. They found that patients with a serum cobalt level < 6.25 µg/L had a chance of 82% to stay without symptoms, while those with ≥ 18.5 µg/L had a very high risk of failure. Patients in the failure group had a mean cobalt concentration of 9.5 μg/L and a mean Co/Cr ratio of 5.2, whereas the asymptomatic group had a concentration of 4.9 μg/L and a ratio of 3.6.

In a recent study 148 patients with dual taper modular THA (110 Rejuvenate, 38 ABG II) were examined for pseudo-tumors (n = 90) on MARS-MRI and the severity of intra-operative tissue reactions was correlated with pre-operative cobalt ion levels [36]. The occurrence of pseudo-tumors was associated with significantly elevated cobalt levels (5.0 mg/L vs 3.7 mg/L), a higher Co/Cr ratio (6.0 vs 3.7) and also higher intra-operative tissue damage grades demonstrated substantially elevated Co/Cr ratios (8.6 vs 3.4). These findings on cobalt values and Co/Cr ratios were also underligned by a comparison of cobalt and chromium level diagnostic measurements between positive and negative ALTR groups for a specific metal-on-polyethylene head-neck modularity THA design without dual-taper neck (cohort n = 62) [68]. In 43 THA patients with ALTR a mean cobalt level of 8.92 μg/L and a Co/Cr ratio of 5.91 were found.

In a consensus statement of the American Association of Hip and Knee Surgeons, the American Academy of Orthopaedic Surgeons and the Hip Society, Kwon et al. [58] defined a "high risk" group stratification combining a serum cobalt level > 5 μg/L and a Co/Cr ratio > 5 as factors for diverse modular taper junctions associated with adverse local tissue reactions.

Investigating a cohort of 123 Rejuvenate dual taper THAs, Meftah et al. [31] described a comparably high revision free probability for patients with cobalt serum levels < 4 μg/L. In addition their patients with a Co/Cr ratio of < 3.6 had a likelihood to stay within the asymptomatic group as it has been similarly analysed by Ghanem et al. [43] and Kwon et al. [36].

Due to the fact that serum cobalt ion levels and Co/Cr ratio are confounded in patients having a contra-lateral head-neck-trunnionosis, a bilateral dual taper stem, a metal-on-metal bearing or another joint replacement (knee, shoulder, ankle), surgeons should not solely rely on ion serum concentration factors to determine a clinical recommendation for stem revision [58,65].

An aspiration of the hip to rule out peri-prosthetic infection [69] and to perform intra-articular synovial fluid collection for cobalt and chromium ion content analyses may be a considerable diagnostic option to detect symptomatic MACC prior acetabular revision THA [48,70]. McGrory et al. [70] examined the relationship between serum and intra-articular (IA) cobalt and chromium levels in a cohort of 16 patients with symptomatic MACC undergoing hip revision and they concluded that intra-articular joint fluid levels (IA cobalt 940 μg/L; IA chromium 491 μg/L) positively correlated with serum levels (cobalt 5.1 μg/L; chromium 1.3 μg/L), but intra-articular levels were on average 100-fold higher.

As an important diagnostic tool in detection of adverse tissue reactions due to dual-taper fretting wear and corrosion, cross sectional imaging modality by metal artifact reduction sequence magnetic resonance imaging (MARS-MRI) [35,48,58,59,71–75] and musculoskeletal ultrasound (US) [76] have been qualified.

Walsh et al. [75] described the incidence of different pathologies based on MARS-MRI images in a retrospective cohort of 312 THAs in 272 patients with a dual taper CoCr neck and a beta-titanium alloy stem implanted between 2007 and 2012. They noted synovitis in 167 hips (53.5%), osteolysis in 18 hips (5.8%) and an effusion or fluid collection in 194 hips (62.3%), whereas 52 (29.1%) of these fluid collections were noted to contain debris. Solely intra-capsular effusion and fluid collection was found in 127 hips (40.7%) and combined intra- and extra-capsular in 52 hips (16.7%) [48]. Tendinopathy of one of the related muscle groups (glutaeus minimus, glutaeus medius, iliopsoas or hamstrings) was seen in 250 (80.1%) of the hips and in 87 (27.9%) some tendon disruption occurred.

The presence of a thickened capsule in association with an effusion is a common MARS-MRI abnormality often accompanied by findings like iliopsoas and abductor tendinopathy, peri-tendinous collections and also the presence of metallic debris [59]. For the detection of adverse local tissue reactions like solid or cystic pseudo-tumors MARS-MRI is a highly sensitive modality [59].

Barlow et al. [71] performed in a revised cohort of 90 THA patients with 98 Rejuvenate modular femoral neck stems MRI and serum cobalt and chromium ion level analysis before revision and used histologic samples from revision surgery to score for synovial lining, inflammatory infiltrate and tissue organization according to Campbell et al. [77]. They found that MRI enables to accurately describe ALTR in dual taper modular neck THA patients and they predicted histologic severity particularly based on maximal synovial thickness and synovitis volume.

To identify femoral osteolysis, loosening and erosions in trochanteric or calcar regions possibly associated with taper corrosion, a focused review of a series of plain antero-posterior and lateral radiographs is proposed [59,69].

Werner et al. [78] reported about adverse inflammatory soft tissue reactions as a consequence of enhanced wear and corrosion of a dual taper neck-stem interface. At time of revision they describe extensive debridement of pseudo-tumor and necrotic bursal and capsular tissue encapsulating the hip joint, as well as corrosion at the neck-stem interface with significant black corrosive debris throughout the modular neck and the soft tissues [78]. Walsh et al. [48] published a study on a cohort of 99 patients including 103 revisions of a dual taper modular neck stem (78 Rejuvenate, 25 ABG II) at a mean time of 2.4 years from primary surgery. They reported intra-operative findings of the 103 revised hips, observing a black metallic sludge associated from corrosion and wear debris in all hips (100%), bony calcar erosion in 88/103 (85.4%), pseudo-tumor formation in 26/103 (25.2%), peri-capsular necrosis in 84/103 (81.6%), tissue necrosis in 80/103 (77.7%) and synovitis in 101/103 (98.1%) of the cases.

Dimitriou et al. [33] evaluated 198 revision surgeries of a dual taper modular femoral stem in 187 THA patients, by an intra-operative tissue damage grading system and observed adverse tissue reactions in 178 (89%), a large amount of fluid entering the capsule in 160 (81%) and particulate wear debris in 103 (52%) of the hips, whereby in all cases a black metallic corrosion debris was found.

If a disassembling of the modular neck takes place intra-operatively during removal of the femoral head within an acetabular revision procedure, the modular hip stem shall be considered for revision. The reason for this is, that a not firmly fixed modular neck/stem taper connection is of high risk for mechanically assisted crevice corrosion. A loosened neck/stem taper connection can be originated during index surgery by an insufficient assembling force or can be caused by macroscopic visible material loss at the medial neck taper interface due to corrosion resulting in a toggling of the modular CoCr neck relatively to the Ti alloy stem [79].

During acetabular cup revision, it may be necessary to place the new cup in an anatomically different position compared to index surgery and this may impact the necessary offset and neck length of the femoral head. In addition, the restoration of the centre of rotation and related ligament balancing during trial head reduction possibly requires a higher offset or longer neck length [4,7,80,81]. Increasing the offset or neck length to adapt for a different cup orientation or sufficient soft tissue balancing may create a more demanding biomechanical loading situation [7,16,32,82–84] for the so far asymptomatic dual taper neck adapter, possibly triggering MACC [17,32,79]. Therefore an indication may be given for revision of the modular stem.

6. Discussion

In an attempt to find an answer to the question—How to proceed with a clinically asymptomatic modular Metha®stem with dual taper CoCr neck adapter in case of acetabular revision?—following systematic methods have been applied:

1) The status of research and appropriate diagnostic methods in context to clinically symptomatic and asymptomatic dual taper stem-neck couplings was evaluated based on a systematic literature review.
2) A retrieval analysis of thirteen Metha®dual taper CoCr/ Ti alloy hip stems was performed.
3) A rational decision making model as basis for a clinical recommendation was developed.

A limitation may arise by the fact that the literature review about serum ion levels, radiographic and clinical findings and the retrieval analysis was based in the vast majority on two recalled dual taper stems in the material combination $CoCr_{29}Mo_6$ (neck) and $TiMo_{12}Zr_6Fe_2$ (stem), and hence a generalization of the findings and transfer to other implant designs and materials of dual taper stems may be limited [64,85,86]. Meftah et al. [31] used persistent pain and high cobalt levels as predictors for revision surgery of the dual taper Rejuvenate stem and reported a Kaplan-Meier survivorship of 40% at four years with revision related to neck taper corrosion as the end point. They calculated a revision-free probability of 93% for patients with cobalt levels of less than 4.0 μg/L compared with 45% for those with cobalt levels of 4.0 μg/L and found that significantly higher metal ion levels correlated with younger age and a higher femoral offset [31]. Bernstein et al. [44] described a corrosion-related revision rate of 28% at a mean follow-up of 2.7 years in a cohort of 81 Rejuvenate modular hip stems. They prospectively followed this cohort of patients with elevated serum cobalt ion levels (> 4 μg/L), persistent pain, or abnormal MRI findings as indications for revision and observed a clinical failure rate of 86% at a mean of 4.2 years [44]. Koziara et al. [87] reported a study group of 66 out of a cohort of 156 patients who underwent modular Rejuvenate THA with an average follow-up of 55 months (range 22–89) with a revision rate of 31.8% (21 of 66 THAs). They found in the non-revision group a mean serum cobalt ion level of 3.48 μg/L and a Visual Analog Scale (VAS) pain score of 2.4, whereas for the revision group a level of Co = 5.05 μg/L and a VAS pain score 5.1 was present. From the revised group, 18 patients were undergoing MARS-MRI. The THA patients who did not have reactive tissue showed a mean serum cobalt ion level of 4.15 μg/L and a VAS pain score 3.8 and in the group with reactive tissue formation cobalt ion level was Co = 5.01 μg/L and a VAS pain score of 5.63 [87].

For a cohort of 36 patients who underwent uni-lateral primary THA with Profemur®Preserve Ti_6Al_4V alloy femoral stems and ceramic-on-ceramic bearings, Barry et al. [88] determined the impact of the modular neck material Ti alloy (n = 22) or CoCr (n = 14) with no patient being revised. With a comparably short-term follow-up of 20 months (range 9–44), they observed significantly higher cobalt ion serum concentrations in the CoCr neck group (0.46 vs 0.26 μg/L) and higher titanium ion serum concentrations in the Ti alloy neck group (1.98 vs 1.59 μg/L), but on a comparably low level. Laurencon et al. [89] reported serum and whole blood metal ion levels of a prospective cohort study on 40 THA patients with a cementless anatomic SPS stem made of Ti_6Al_4V alloy with modular CoCr necks with a mean follow-up of 23 months (range 12–28) and found in 6 of 40 (15%) serum cobalt ion levels > 2 μg/L and in 3 of 40 (7.5%) values > 4 μg/L. Applying MARS-MRI in all THA patients with a serum ion level > 2 μg/L, they detected a pseudo-tumor in one patient having a serum level of 5.21 μg/L for cobalt, 3.51 for chrome and 42 μg/L for titanium [89]. Using a nationwide retrospective cohort of 324,108 THA patients from the French health insurance system, Colas et al. [90] described a cumulative revision incidence of 6.5% for exchangeable neck THAs (n = 8,931) versus 4.7% for fixed neck THAs (n = 315,177) and a significantly increased adjusted hazard ratio of revision of 1.26.

In the National Joint Replacement Registry Report 2018 of the Australian Orthopaedic Association for exchangeable femoral neck adapters the cumulative percent revision for primary THA was reported to be 4.9% at 5 years and 6.8% at 10 years for Ti alloy-Ti alloy stem-neck couplings [91]. For Ti alloy-CoCr modular neck couplings they reported a cumulative percent revision of 9.6% at 5 years and of 16.6% at 10 years [91].

In THA femoral stems with modular exchangeable neck components had significantly lower 10-year survival rates in literature reviews and in registry data compared to primary THA implant survivorship for femoral mono-bloc stems [92,93].

Su et al. [94] performed a retrieval analysis for neck fretting and corrosion on 60 Rejuvenate modular stem designs and compared those to 26 retrieved implants from seven other modular CoCr and Ti alloy neck designs. For the Rejuvenate design they found significantly higher damage and corrosion scores and a 20-fold increased likelihood to show ALVAL based on histologic samples, than for the other designs. As a relevant parameter they stated—beyond design aspects—the lower Youngs modulus of 80 GPa for the $TiMo_{12}Zr_6Fe_2$ stem material (Ti_6Al_4V; 110 GPa), being responsible for increased metal transfer and surface damage in coupling with a CoCr neck, which could account for the higher ALVAL and corrosion scores [94].

Somers et al. [95] stated for the market withdrawn Rejuvenate and ABG II hip systems that the different design features and the stem material $TiMo_{12}Zr_6Fe_2$ show more fretting corrosion and it is possible that the different metal trace elements (molybdenum, zirconium and iron) might lead to a more pronounced toxic local tissue reaction.

Lewinski and Floerkemeier [13] described their 10-year experience with short stem TH based on 1953 Metha®short stem procedures with an overall aseptic stem revision rate of 1.3% and 1.9% including 12 modular Ti alloy neck adapter failures based on 190 modular stems with Ti alloy necks implanted before the product recall in November 2006 [16]. Schnurr et al. [14] recorded data for 1888 Metha®short stem implantations from 2004 to 2014 with three implanted versions: Modular Ti_6Al_4V alloy stems with Ti alloy (n = 314) or CoCr (n = 230) neck adapters and mono-bloc Ti alloy stems (n = 1090) with a mean follow-up of 6 years (range 1–11). They found a 7-year revision rate for mono-bloc of 1.5%, for modular cobalt-chrome of 1.8% and for modular Ti alloy adapter stems of 5.3%, including 15 modular Ti alloy neck fractures.

Apart of the promising 7 to 10 years clinical experiences with the modular Metha®short stem with CoCr necks [13,14], a limitation of our retrieval analysis study is the small number of symptomatic (n = 5) and asymptomatic (n = 8) retrieved CoCr neck adapters out of a cohort of 25,177 dual taper modular stems with CoCr neck, implanted from January 2007 until January 2017. On the other hand side compared to the mostly short-term follow-up in the literature [31,43,44,70,87–89] the retrievals in our study have an average follow-up of 73.8 months and 6 of them have been in patients service for more than 7 years (range 84–128 months).

7. Conclusions

Based on the analysis of the literature it is suggested that in case of required acetabular revision in patients with a serum cobalt level of > 4 µg/L [31,33,35,36,43,44,58,87] and a Co/Cr ratio > 3.6 [33,35,36,42,43,58] revision of the modular dual taper stem may be considered.

Prior acetabular revision surgery in patients with dual taper modular neck stem THA [59], a systematic diagnostic evaluation has to be executed, using specific tests such as serum metal (Co, Cr) ion analysis, plain antero-posterior and lateral radiographs [59,69] and cross-sectional imaging modalities (MARS-MRI, US) [48,55,59,77]. The patient's stated pain level (e.g., VAS pain score) should also be included as an important factor and measurements of IA cobalt and chromium levels may be meaningful [70].

For an asymptomatic Metha®dual taper Ti alloy/CoCr stem-neck coupling at stage of acetabular revision, careful clinical decision making according to the proposed model should be followed and overreliance on any single examination should be avoided, considering the complete individual differential diagnosis and patient situation.

Author Contributions: Conceptualization, T.M.G., M.B., V.J., H.W., K.-D.H., and H.-P.K.; methodology, T.M.G., M.B. and H.-P.K.; validation, T.M.G., M.B., V.J., H.W., K.-D.H., M.M.M. and H.-P.K.; formal analysis, T.M.G., M.B., V.J., H.W., K.-D.H., M.M.M. and H.-P.K.; investigation, T.M.G., M.B. and M.M.M.; data curation, T.M.G., M.B. and M.M.M. All authors have read and agreed to the published version of the manuscript.

Funding: This research received no external funding.

Acknowledgments: The authors would like to thank Sabine Rusch-Zieris for her performance of the systematic literature review in PubMed and EMBASE and Moritz Wente and Thomas Guettler for their fruitful input and critical review of the decision making model.

Conflicts of Interest: Two of the authors (T.M.G, M.B.) are employees of Aesculap AG Tuttlingen a manufacturer of orthopaedic implants. Three of the authors (V.J., H.W. and K.-D.H.) are advising surgeons in Aesculap research projects and three of the authors (V.J., H.W. and M.M.M.) have got institutional support by Aesculap AG Tuttlingen.

Ethical Approval: Not required.

References

1. Evans, J.T.; Evans, J.P.; Walker, R.W.; Blom, A.W.; Whitehouse, M.R. Sayers A: How long does a hip replacement last? A systematic review and meta-analysis of case series and national registry reports with more than 15 years of follow-Up. *Lancet* **2019**, *393*, 647–654. [CrossRef]
2. Sadoghi, P.; Liebensteiner, M.; Agreiter, M.; Leithner, A.; Böhler, N.; Labek, G. Revision surgery after total joint arthroplasty: A complication-Based analysis using worldwide arthroplasty registers. *J. Arthroplast.* **2013**, *28*, 1329–1332. [CrossRef] [PubMed]
3. Toni, A.; Sudanese, A.; Paderni, S.; Guerra, E.; Bianchi, G.; Antonietti, B.; Giunti, A. Cementless hip arthroplasty with a modular neck. *Chir. Organi. Mov.* **2001**, *86*, 73–85.
4. Traina, F.; De Fine, M.; Biondi, F.; Tassinari, E.; Galvani, A.; Toni, A. The influence of the centre of rotation on implant survival using a modular stem hip prosthesis. *Int. Orthop.* **2009**, *33*, 1513–1518. [CrossRef] [PubMed]
5. Archibeck, M.J.; Cummins, T.; Carothers, J.; Junick, D.W.; White, R.E. A comparison of two implant systems in restoration of hip geometry in arthroplasty. *Clin. Orthop. Relat. Res.* **2011**, *469*, 443–446. [CrossRef] [PubMed]
6. Srinivasan, A.; Jung, E.; Levine, B.R. Modularity of the femoral component in total hip arthroplasty. *J. Am. Acad. Orthop. Surg.* **2012**, *20*, 214–222. [CrossRef]
7. Traina, F.; De Clerico, M.; Biondi, F.; Pilla, F.; Tassinari, E.; Toni, A. Sex differences in hip morphology: Is stem modularity effective for total hip replacement? *J. Bone Jt. Surg. Am.* **2009**, *91*, 121–128. [CrossRef]
8. Ollivier, M.; Parratte, S.; Galland, A.; Lunebourg, A.; Flecher, X.; Argenson, J.N. Titanium-Titanium modular neck for primary THA. Result of a prospective series of 170 cemented THA with a minimum follow-Up of 5 years. *Orthop. Traumatol. Surg. Res.* **2015**, *101*, 137–142. [CrossRef]
9. Traina, F.; De Fine, M.; Tassinari, E.; Sudanese, A.; Calderoni, P.P.; Toni, A. Modular neck prostheses in DDH patients: 11-Year results. *J. Orthop. Sci.* **2011**, *16*, 14–20. [CrossRef]
10. Gerhardt, D.M.; Bisseling, P.; de Visser, E.; van Susante, J.L. Modular necks in primary hip arthroplasty without anatomical deformity: No clear benefit on restoration of hip geometry and dislocation rate. An exploratory study. *J. Arthroplast.* **2014**, *29*, 1553–1558. [CrossRef]
11. Wittenberg, R.H.; Steffen, R. Comparative 5-Year results of short hip total hip arthroplasty with Ti- or CoCr-Neck adapters. *Orthopedics* **2015**, *38*, S33–S39. [CrossRef] [PubMed]
12. Fitch, D.A.; Ancarani, C.; Bordini, B. Long-Term survivorship and complication rate comparison of a cementless modular stem and cementless fixed neck stems for primary total hip replacement. *Int. Orthop.* **2015**, *39*, 1827–1837. [CrossRef]
13. Von Lewinski, G.; Floerkemeier, T. 10-Year experience with short stem total hip arthroplasty. *Orthopedics* **2015**, *38*, S51–S56. [CrossRef] [PubMed]
14. Schnurr, C.; Schellen, B.; Dargel, J.; Beckmann, J.; Eysel, P.; Steffen, R. Low short stem revision rates: 1-11 years results from 1888 total hip arthroplasties. *J. Arthroplast.* **2017**, *32*, 487–493. [CrossRef] [PubMed]
15. Toni, A.; Giardina, F.; Guerra, G.; Sudanese, A.; Montalti, M.; Stea, S.; Bordini, B. 3rd generation alumina-On-Alumina in modular hip prosthesis: 13 to 18 years follow-Up results. *HIP Int.* **2017**, *27*, 8–13. [CrossRef] [PubMed]
16. Grupp, T.M.; Weik, T.; Bloemer, W.; Knaebel, H.-P. Modular titanium alloy neck adapter failures in hip replacement–Failure mode analysis and influence of implant material. *BMC Musculoskelet. Disord.* **2010**, *11*, 1–12. [CrossRef] [PubMed]
17. Jacobs, J.J.; Cooper, H.J.; Urban, R.M.; Wixson, R.L.; Della Valle, C.J. What do we know about taper corrosion in total hip arthroplasty? *J. Arthroplast.* **2014**, *29*, 668–669. [CrossRef]
18. Krishnan, H.; Krishnan, S.P.; Blunn, G.; Skinner, J.A.; Hart, A.J. Modular neck femoral stems. *Bone Jt. J.* **2013**, *95*, 1011–1021. [CrossRef]
19. Mistry, J.B.; Chughtai, M.; Elmallah, R.K.; Diedrich, A.; Le, S.; Thomas, M.; Mont, M.A. Trunnionosis in total hip arthroplasty: A review. *J. Orthopaed. Traumatol.* **2016**, *17*, 1–6. [CrossRef]
20. Wodecki, P.; Sabbah, D.; Kermarrec, G.; Semaan, I. New type of hip arthroplasty failure releated to modular femoral components: Breakage at the neck-stem junction. *Orthop. Traumatol. Surg. Res.* **2013**, *99*, 741–744. [CrossRef]
21. Mihalko, W.M.; Assaf, D.; Sungu, M.T. Reproducing the hip center with a femoral neck-Retaining implant. *Orthopedics* **2015**, *38*, S21–S26. [CrossRef] [PubMed]

22. Pour, A.E.; Borden, R.; Murayama, T.; Groll-Brown, M.; Blaha, J.D. High Risk of Failure With Bimodular Femoral Components in THA. *Clin. Orthop. Relat. Res.* **2016**, *474*, 146–153. [CrossRef] [PubMed]
23. Atwood, S.A.; Patten, E.W.; Bozic, K.J.; Pruitt, L.A.; Ries, M.D. Corrosion-Induced fracture of a double-Modular hip prosthesis: A case report. *J. Bone Jt. Surg. Am.* **2010**, *92*, 1522–1525. [CrossRef] [PubMed]
24. Sotereanos, N.G.; Sauber, T.J.; Tupis, T.T. Modular femoral neck fracture after primary total hip arthroplasty. *J. Arthroplast.* **2013**, *28*, 196.e7–196.e9. [CrossRef] [PubMed]
25. Fokter, S.K.; Rudolf, R.; Molicnik, A. Titanium alloy femoral neck fracture–Clinical and metallurgical analysis in 6 cases. *Acta Orthop.* **2016**, *87*, 198–202. [CrossRef] [PubMed]
26. Fokter, S.K.; Molicnik, A.; Kavalar, R.; Pelicon, P.; Rudolf, R.; Gubeljak, N. Why do some titanium-Alloy total hip arthroplasty modular necks fail. *J. Mech. Behav. Biomed. Mater.* **2016**, *69*, 107–114. [CrossRef]
27. Ceretti, M.; Falez, F. Modular titanium alloy neck failure in total hip replacement: Analysis of a relapse case. *SICOT J.* **2016**, *20*, 1–4. [CrossRef]
28. Menciere, M.L.; Amouyel, T.; Taviaux, J.; Bayle, M.; Laterza, C.; Mertl, P. Fracture of the cobalt-Chromium modular femoral neck component in total hip arthroplasty. *Orthop. Traumatol. Surg. Res.* **2014**, *100*, 565–568. [CrossRef]
29. Murena, L.; Maritan, G.; Concina, C.; Scamacca, V.; Ratti, C.; Canton, G. Fracture of cobalt-Crome modular neck in total hip arthroplasty. *Acta Biomed.* **2019**, *90* (Suppl. 12), 187–191.
30. Kovac, S.; Mavcic, B.; Kotnik, M.; Levasic, V.; Sirse, M.; Fokter, S. What Factors Are Associated with Neck Fracture in One Commonly Used Bimodular THA Design? A Multicenter, Nationwide Study in Slovenia. *Clin. Orthop. Relat. Res.* **2019**, *466*, 1324–1332. [CrossRef]
31. Meftah, M.; Halleem, A.M.; Burn, M.B.; Smith, K.M.; Incavo, S.J. Early Corrosion-Relared Failure of the Rejuvenate Modular Total Hip Replacement. *J. Bone Jt. Surg. Am.* **2014**, *96*, 481–487. [CrossRef] [PubMed]
32. De Martino, I.; Assini, J.B.; Elpers, M.E.; Wright, T.M.; Westrich, G.H. Corrosion and Fretting of a Modular Hip System: A Retrieval Analysis of 60 Rejuvenate Stems. *J. Arthroplast.* **2015**, *30*, 1470–1475. [CrossRef] [PubMed]
33. Dimitriou, D.; Liow, M.H.; Tsai, T.Y.; Leone, W.A.; Li, G.; Kwon, Y.M. Early Outcome of Revision Surgery for Taper Corrosion of Dual Taper Total Hip Arthroplasty in 187 Patients. *J. Arthroplast.* **2016**, *31*, 1549–1554. [CrossRef] [PubMed]
34. Kolatat, K.; Perino, G.; Wilner, G.; Kaplowitz, E.; Ricciardi, B.F.; Boettner, F.; Westrich, G.H.; Jerabek, S.A.; Goldring, S.R.; Purdue, P.E. Adverse local tissue reaction (ALTR) associated with corrosion products in metal-On-Metal and dual modular neck total hip replacements is associated with upregulation of interferon gamma-Mediated chemokine signaling. *J. Orthop. Res.* **2015**, *33*, 1487–1497. [CrossRef] [PubMed]
35. Kwon, Y.M.; Khormaee, S.; Liow, M.H.; Tsai, T.Y.; Freiberg, A.A.; Rubash, H.E. Asymptomatic Pseudo-Tumors in Patients with Taper Corrosion of a Dual-Taper Modular Femoral Stem: MARS-MRI and Metal Ion Study. *J. Bone Jt. Surg.* **2016**, *98*, 1735–1740. [CrossRef] [PubMed]
36. Kwon, Y.M.; Tsai, T.Y.; Leone, W.A.; Liow, M.H. Sensitivity and Specificity of Metal Ion Levels in Predicting "Pseudo-tumors" due to Taper Corrosion in Patients with Dual Taper Modular Total Hip Arthroplasty. *J. Arthroplast.* **2017**, *32*, 996–1000. [CrossRef]
37. Omlor, G.W.; Kretzer, J.P.; Reinders, J.; Streit, M.R.; Bruckner, T.; Gotterbarm, T.; Aldinger, P.R.; Merle, C. In vivo serum titanium ion levels following modular neck total hip arthroplasty–10 years results in 67 patients. *Acta Biomater.* **2013**, *9*, 6278–6782. [CrossRef]
38. Gofton, W.; Beaule, P.E. Serum Metal Ions with a Titanium Modular Neck Total Hip Replacement System. *J. Arthroplast.* **2015**, *30*, 1781–1786. [CrossRef]
39. Vundelinckx, B.J.; Verhelst, L.A.; De Schepper, J. Taper Corrosion in Modular Hip Prostheses Analysis of Serum Metal Ions in 19 Patients. *J. Arthroplast.* **2013**, *28*, 1218–1223. [CrossRef]
40. Cooper, H.J.; Urban, R.M.; Wixson, R.L.; Meneghini, R.M.; Jacobs, J.J. Adverse Local Tissue Reaction arising from Corrosion at the Femoral Neck-Body Junction in a Dual-Taper Stem with a Cobalt-Chromium Modular Neck. *J. Bone Jt. Surg.* **2013**, *95*, 865–872. [CrossRef]
41. Cheung, A.C.; Banerjee, S.; Cherian, J.J.; Wong, F.; Butany, J.; Gilbert, C.; Overgaard, C.; Syed, K.; Zywiel, M.G.; Jacobs, J.J.; et al. Systemic cobalt toxicity from total hip arthroplasties. *Bone Jt. J.* **2016**, *98*, 6–13. [CrossRef] [PubMed]

42. Molloy, D.O.; Munir, S.; Jack, C.M.; Cross, M.B.; Walter Wl Walter, W.K. Fretting and Corrosion in Modular-Neck total Hip Arthroplasty Femoral Stems. *J. Bone Jt. Surg. Am.* **2014**, *96*, 488–493. [CrossRef] [PubMed]
43. Ghanem, E.; Ward, D.M.; Robbins, C.E.; Nandi, S.; Bono, J.V.; Talmo, C.T. Corrosion and Adverse Local Tissue Reaction in One Type of Modular Neck Stem. *J. Arthroplast.* **2015**, *30*, 1787–1793. [CrossRef] [PubMed]
44. Bernstein, D.T.; Mefath, M.; Paranilam, J.; Incavo, S.J. Eighty-Six Percent Failure Rate of a Modular-Neck Femoral Stem Design at 3 to 5 Years: Lessons Learned. *J. Bone Jt. Surg.* **2016**, *98*, e491–e497. [CrossRef] [PubMed]
45. Gill, I.P.S.; Webb, J.; Sloan, K.; Beaver, R.J. Corrosion at the neck-stem junction as a cause of metal ion release and pseudotumour formation. *J. Bone Jt. Surg.* **2012**, *94*, 895–900. [CrossRef]
46. Kop, A.M.; Keogh, C.; Swarts, E. Proximal component modularity in THA–At what cost? An implant retrieval study. *Clin. Orthop. Relat. Res.* **2012**, *470*, 1885–1894. [CrossRef]
47. Wooster, B.M.; Lachiewicz, P.F. Trunnionosis: Is it really a problem? *Semin. Arthroplast.* **2017**, *28*, 206–210. [CrossRef]
48. Walsh, C.P.; Hubbard, J.C.; Nessler, J.P.; Markel, D.C. Revision of recalled modular neck Rejuvenate and ABG femoral implants. *J. Arthroplast.* **2015**, *30*, 822–826. [CrossRef]
49. Barlow, B.T.; Assini, J.; Boles, J.; Lee, Y.Y.; Westrich, G.H. Short-Term metal ion trends following removal of recalled modular neck femoral stems. *J. Arthroplast.* **2015**, *30*, 1191–1196. [CrossRef]
50. Silverton, C.D.; Jacobs, J.J.; Devitt, J.W.; Cooper, H.J. Midterm results of a femoral stem with a modular neck design: Clinical outcomes and metal ion analysis. *J. Arthroplast.* **2014**, *29*, 1768–1773. [CrossRef]
51. Lanting, B.A.; Teeter, M.G.; Vasarhelyi, E.M.; Ivanov, T.G.; Howard, J.L.; Naudie, D.D.R. Correlation of corrosion and biomechanics in the retrieval of a single modular neck total hip arthroplasty design: Modular neck total hip arthroplasty system. *J. Arthroplast.* **2015**, *30*, 135–140. [CrossRef] [PubMed]
52. Restrepo, C.; Ross, D.; Restrepo, S.; Heller, S.; Goyal, N.; Moore, R.; Hozack, W.J. Adverse clinical outcomes in a primary modular neck/stem system. *J. Arthroplast.* **2014**, *29*, S173–S178. [CrossRef] [PubMed]
53. Chillemi, M.; Placella, G.; Caraffa, A.; Cerulli, G.; Antinolfi, P. Serologic and radiographic outcome of total hip arthroplasty with CoCr modular neck at mid-term follow-Up. *Musculoskelet. Surg.* **2017**, *101*, 51–58. [CrossRef] [PubMed]
54. Liow, M.H.L.; Urish, K.L.; Preffer, F.I.; Nielson, G.P.; Kwon, Y.M. Metal Ion Levels Are Not Correlated with Histopathology of Adverse Local Tissue Reactions in Taper Corrosion of Total Hip Arthroplasty. *J. Arthroplast.* **2016**, *31*, 1797–1802. [CrossRef]
55. Barlow, B.T.; Ortiz, P.A.; Fiels, K.G.; Burge, A.J.; Potter, H.G.; Westrich, G.H. Magnetic resonance imaging predicts adverse local tissue reaction histologic severity in modular neck total hip arthroplasty. *J. Arthroplast.* **2016**, *31*, 2325–2331. [CrossRef]
56. Hussey, D.K.; McGrory, B.J. Ten-Year cross-Sectional study of mechanically assisted crevice corrosion in 1352 consecutive patients with metal-On-Polyethylene total hip arthroplasty. *J. Arthroplast.* **2017**, *32*, 2546–2551. [CrossRef]
57. Canham, C.D.; Muradov, P.I.; Simpson, J.B.; Incavo, S.J. Corrosion and adverse local tissue reaction after total hip arthroplasty with a modular titanium alloy femoral neck. *Arthoplasty Today* **2017**, *3*, 211–214. [CrossRef]
58. Kwon, Y.M.; Fehring, T.K.; Lombardi, A.V.; Barnes, C.L.; Cabanela, M.E.; Jacobs, J.J. Risk stratification algorithm for management of patients with dual modular taper total hip arthroplasty: Consensus statement of the American Association of Hip and Knee Surgeons, the American Academy of Orthopaedic Surgeons and the Hip Society. *J. Arthroplast.* **2014**, *29*, 2060–2064. [CrossRef]
59. Kwon, Y.M. Evaluation of the Painful Dual Taper Modular Neck Stem Total Hip Arthroplasty: Do They All Require Revision? *J. Arthroplast.* **2016**, *31*, 1385–1389. [CrossRef]
60. Xia, Z.; Ricciardi, B.F.; Liu, Z.; von Ruhland, C.; Ward, M.; Lord, A.; Hughes, L.; Goldring, S.R.; Purdue, E.; Murray, D.; et al. Nano-Analysis of wear particles from metal-On-Metal and non metal-On-Metal dual modular neck hip arthroplasty. *Nanomed. Nanotechnol. Biol. Med.* **2017**, *13*, 1205–1217. [CrossRef]
61. Perino, G.; Ricciardi, B.F.; Jerabek, S.A.; Martignoni, G.; Wilner, G.; Maass, D.; Goldring, S.R.; Purdue, P.E. Implant based differences in adverse local tissue reaction in failed total hip arthroplasties: A morphological and immunohistochemical study. *BMC Clin. Pathol.* **2014**, *14*. [CrossRef] [PubMed]

62. Morawietz, L.; Classen, R.; Schröder, J.H.; Dynybil, C.; Perka, C.; Skwara, A.; Neidel, A.; Gehrke, T.; Frommelt, L.; Hansen, T.; et al. Proposal for a histopathological consensus classification of the periprosthetic interface membrane. *J. Clin. Pathol.* **2006**, *59*, 591–597. [CrossRef] [PubMed]
63. Krenn, V.; Morawietz, L.; Burmester, G.R.; Kinne, R.W.; Mueller-Ladner, U.; Mueller, B.; Haupl, T. Synovitis score: Discrimination between chronic low-grade and high-Grade synovitis. *Histopathology* **2006**, *49*, 358–364. [CrossRef] [PubMed]
64. Buente, D.; Huber, G.; Bishop, N.; Morlock, M. Quantification of material loss from neck piece taper junctions of a bimodular primary hip prosthesis. A retrieval study from 27 failed Rejuvenate bimodular hip arthroplasties. *Bone Jt. J.* **2015**, *97*, 1350–1357. [CrossRef] [PubMed]
65. Fillingham, Y.A.; Della Valle, C.J.; Bohl, D.D.; Kelly, M.P.; Hall, D.J.; Pourzal, R.; Jacobs, J.J. Serum metal levels for diagnosis of adverse local tissue reactions secondary to corrosion in metal-On-Polyethylene total hip arthroplasty. *J. Arthroplast.* **2017**, *32*, S272–S277. [CrossRef] [PubMed]
66. Nawabi, D.H.; Do, H.T.; Ruel, A.; Lurie, B.; Elpers, M.E.; Wright, T.; Potter, H.G.; Westrich, G.H. Comprehensive analysis of a recalled modular total hip system and recommendations for management. *J. Bone Jt. Surg.* **2016**, *98*, 40–47. [CrossRef] [PubMed]
67. Kretzer, J.P.; Jakubowitz, E.; Krachler, M.; Thomsen, M.; Heisel, C. Metal release and corrosion effects of modular neck total hip artrhroplasty. *Int. Orthop.* **2009**, *33*, 1531–1536. [CrossRef]
68. Kwon, Y.M.; MacAuliffe, J.; Arauz, P.G.; Peng, Y. Sensitivity and specificity of metal ion level in predicting adverse local tissue reactions due to head-Neck taper corrosion in primary metal-On-Polyethylene total hip arthroplasty. *J. Arthroplast.* **2018**. [CrossRef]
69. Della Valle, C.J.; Calkins, T.E.; Jacobs, J.J. Diagnosing Taper Corrosion: When is it the taper and when is it something else? *J. Arthroplast.* **2018**, *33*. [CrossRef]
70. McGrory, B.J.; Payson, A.M.; MacKenzie, J.A. Elevated intra-Articular cobalt and chromium levels in mechanically assisted crevice corrosion in metal-on-polyethylene total hip arthroplasty. *J. Arthroplast.* **2017**, *32*, 1654–1658. [CrossRef]
71. Elmallah, R.K.; Cherian, J.J.; Meneghini, R.M.; Hozack, W.J.; Westrich, G.H.; Mont, M.A. How to approach a recalled dual modular hip implant: An update. *J. Arthroplast.* **2016**, *31*, 2646–2652. [CrossRef] [PubMed]
72. Morozov, P.P.; Sana, M.; McGrory, B.J.; Farraher, S.W.; Abrahams, T.G. Comparison of Pre-Revision Magnetic Resonance Imaging and Operative Findings in Mechanically Assisted Crevice Corrosion in Symptomatic Metal-On-Polyethylene Total Hip Arthroplasties. *J. Arthroplast.* **2017**, *32*, 2535–2545. [CrossRef] [PubMed]
73. Burge, A.J.; Gold, S.L.; Lurie, B.; Nawabi, D.H.; Fields, K.G.; Koff, M.F.; Westrich, G.; Potter, H.G. MR imaging of adverse local tissue reactions around rejuvenate modular dual-Taper stems. *Radiology* **2015**, *277*, 142–150. [CrossRef] [PubMed]
74. Koff, M.F.; Esposito, C.; Shah, P.; Miranda, M.; Baral, E.; Fields, K.; Bauer, T.; Padgett, D.E.; Wright, T.; Potter, H.G. MRI of THA Correlates with Implant Wear and Tissue Reactions: A Cross-Sectional Study. *Clin. Orthop. Relat. Res.* **2019**, *477*, 159–174. [CrossRef] [PubMed]
75. Walsh, C.P.; Hubbard, J.C.; Nessler, J.P.; Markel, D.C. MRI Findings Associated with Recalled Modular Femoral Neck Rejuvenate and ABG Implants. *J. Arthroplast.* **2015**, *30*, 2021–2026. [CrossRef] [PubMed]
76. Frisch, N.B.; Wessell, N.M.; Taliaferro, K.; van Holsbeeck, M.; Silverton, C.D. Ultrasound findings in asymptomatic patients with modular metal on metal total hip arthroplasty. *Skelet. Radiol.* **2017**, *46*, 641–649. [CrossRef] [PubMed]
77. Campbell, P.; Ebramzadeh, E.; Nelson, S.; Takamura, K.; DeSmet, K.; Amstutz, H.C. Histological features of pseudo-Tumor-Like tissues from metal-On-Metal hips. *Clin. Orthop. Relat. Res.* **2010**, *468*, 2321–2327. [CrossRef]
78. Werner, S.D.; Bono, J.V.; Nandi, S.; Ward, D.M.; Talmo, C.T. Adverse tissue reactions in modular exchangeable neck implants: A report of two cases. *J. Arthroplast.* **2013**, *28*, 543–545. [CrossRef]
79. Baxmann, M.; Pfaff, A.; Schilling, C.; Grupp, T.M.; Morlock, M.M. Biomechanical Evaluation of the Fatigue Performance, the Taper Corrosion and the Metal Ion Release of a Dual Taper Hip Prosthesis under Physiological Environmental Conditions. *Biotribology* **2017**, *12*, 1–7. [CrossRef]
80. Sporer, S.M. How to do a revision total hip arthroplasty. Revision of the acetabulum. *Instr. Course Lect.* **2012**, *61*, 303–311. [CrossRef]

81. Tohtz, S.W.; Sassy, D.; Matziolis, G.; Preininger, B.; Perka, C.; Hasart, O. CT evaluation of native acetabular orientation and localization: Sex-Specific data comparison on 336 hip joints. *Technol. Health Care Off. J. Eur. Soc. Eng. Med.* **2010**, *18*, 129–136. [CrossRef] [PubMed]
82. Skendzel, J.G.; Blaha, J.D.; Urquhart, A.G. Total hip arthroplasty modular neck failure. *J. Arthroplast.* **2011**, *26*, 338.e1–338.e4. [CrossRef] [PubMed]
83. Traina, F.; Baleani, M.; Erani, P.; Bordini, B.; de Fine, M.; Stea, S.; Toni, A. Failure of Modular Necks in Primary Total Hip Replacement: Can We Prevent It? In Proceedings of the Scientific Exhibit 79th American Academy of Orthopaedic Surgeons, San Francisco, CA, USA, 14–16 April 2012.
84. Benazzo, F.; Perticarini, L. Comment on: Modular titanium alloy neck failure in total hip replacement. *SICOT J.* **2017**, *3*, 23. [CrossRef] [PubMed]
85. Buente, D.; Bryant, M.; Ward, M.; Neville, A.; Morlock, M.; Huber, G. The taper corrosion pattern observed for one bi-modular stem design is related to geometry-Determined taper mechanics. *Med Eng. Phys.* **2017**, *46*, 79–88. [CrossRef]
86. Cafri, G.; Graves, S.E.; Sedrakyan, A.; Fan, J.; Calhoun, P.; de Steiger, R.N.; Cuthbert, A.; Lorimer, M.; Paxton, E.W. Postmarket surveillance of arthroplasty device components using machine learning methods. *Pharmacoepidemiol. Drug Saf.* **2019**, *28*, 1440–1447. [CrossRef]
87. Koziara, C.R.; Lombardo, D.J.; Petersen-Fitts, G.R.; Jildeh, T.R.; Morawa, L. Effects of cobalt and chromium levels following modular hip stem total hip arthroplasty. *Orthopedics* **2016**, *39*, 288–292. [CrossRef]
88. Barry, J.; Kiss, M.O.; Massé, V.; Lavigne, M.; Matta, J.; Venditolli, P.A. Effect of femoral stem modular neck's material on metal ion release. *Open Orthop. J.* **2017**, *11*, 1337–1344. [CrossRef]
89. Laurencon, J.; Augsburger, M.; Faouzi, M.; Becce, F.; Hassani, H.; Rüdiger, H.A. Systemic metal ion levels in patients with modular neck stems: A Prospective cohort study. *J. Arthroplast.* **2016**, *31*, 1750–1755. [CrossRef]
90. Colas, S.; Allalou, A.; Poichotte a Piriou, P.; Dray-Spira, R.; Zureik, M. Exchangeable femoral neck (dual modular) THA prostheses have poorer survivorship than other designs: A nationwide cohort of 324,108 patients. *Clin. Orthop. Relat. Res.* **2017**, *475*, 2046–2059. [CrossRef]
91. Graves, S.; de Steiger, R.; Lewis, P.; Harris, I. Australian Orthopaedic Association National Joint Replacement Registry for Hip, Knee & Shoulder Arthroplasty. *Annu. Rep.* **2018**, 155–156.
92. Mihalko, W.M.; Wimmer, M.A.; Pacione, C.A.; Laurent, M.P.; Murphy, R.F.; Rider, C. How Have Alternative Bearings and Modularity Affected Revision Rates in Total Hip Arthroplasty? *Clin. Orthop. Relat. Res.* **2014**, *472*, 3747–3758. [CrossRef] [PubMed]
93. Graves, S.E.; De Steiger, R.; Davidson, D.; Donnelly, W.; Rainbird, S.; Lorimer, M.F.; Cashman, K.S.; Vial, R. The use of femoral stems with exchangeable necks in primary total hip arthroplasty increases the rate of revision. *Bone Jt. J.* **2017**, *99*, 766–773. [CrossRef] [PubMed]
94. Su, S.L.; Koch, C.N.; Nguyen, T.M.; Burket, J.C.; Wright, T.M.; Westrich, G.H. Retrieval analysis of neck-stem coupling in modular hip prostheses. *J. Arthroplast.* **2017**, *32*, 2301–2306. [CrossRef] [PubMed]
95. Somers, J.F.A.; Dedrye, L.; Goeminne, S. Metal ion levels in ceramic-On-Ceramic THR with cobalt-Chrome modular necks: Analysis of cobalt and chromium serum levels in 23 healthy hip patients. *Hip Int.* **2017**, *27*, 21–25. [CrossRef] [PubMed]

materials

MDPI

Article

Biological Reactions to Metal Particles and Ions in the Synovial Layer of Mice

Xiangyun Cheng [1], Sabine C. Dirmeier [1], Sandra Haßelt [1], Andrea Baur-Melnyk [2], Jan Philippe Kretzer [3], Rainer Bader [4], Sandra Utzschneider [1,*] and Alexander C. Paulus [1,*,†]

1 Department of Orthopedic Surgery, Physical Medicine and Rehabilitation, University Hospital of Munich, Ludwig-Maximilians-University, Campus Großhadern, Marchioninistraße 15, 81377 Munich, Germany; xiangyun-cheng@outlook.com (X.C.); Sabine.Dirmeier@gmx.de (S.C.D.); sandra.hasselt@med.uni-muenchen.de (S.H.)

2 Department of Radiology, University Hospital of Munich, Ludwig-Maximilians-University, Campus Großhadern, Marchioninistraße 15, 81377 Munich, Germany; andrea.baur@med.uni-muenchen.de

3 Laboratory of Biomechanics and Implant Research, Clinic for Orthopedics and Trauma Surgery, Heidelberg University Hospital, Schlierbacher Landstrasse 200a, 69118 Heidelberg, Germany; philippe.kretzer@med.uni-heidelberg.de

4 Biomechanics and Implant Technology Research Laboratory (FORBIOMIT), Department of Orthopaedics, Rostock University Medical Center, Doberaner Straße 142, 18057 Rostock, Germany; rainer.bader@med.uni-rostock.de

* Correspondence: sandra.utzschneider@med.uni-muenchen.de (S.U.); alexander.paulus@med.uni-muenchen.de (A.C.P.)

† Alexander C. Paulus and Sandra Utzschneider contributed equally.

Received: 29 October 2019; Accepted: 24 February 2020; Published: 26 February 2020

check for updates

Abstract: Metal particles and ions released from implants not only have a fundamental effect on the longevity of total joint replacements, but can also be disseminated to remote organs. Periprosthetic tissues harvested during revision surgeries mainly reflect end-stage failure but may not adequately reveal initial biological reactions and systemic side effects. Therefore, primary reactions caused by metal particles and ions were investigated in an established murine model. Left knee joints in three groups, each consisting of ten female BALB/c mice, received injections of metal ions (MI), metal particles (MP) and phosphate-buffered saline (PBS) (control). Seven days after the injection, immunohistochemical analyses of the synovial layer were performed with respect to some biological markers including Tumor necrosis factor -α (TNF-α), Interleukin-6 (IL-6), Interleukin-1β (IL-1β), Cluster of Differentiation 45 (CD45), Cluster of Differentiation 68 (CD68) and Cluster of Differentiation 3(CD3). The MP group showed significantly enhanced proinflammatory cytokine expression (TNF-α, IL-6 and IL-1β) compared with the other groups ($p < 0.05$). Interestingly, CD3, as a marker for T lymphocytes, did not increase in any of the groups. The MI group showed a significantly increased expression of CD45 compared with the control group ($p < 0.05$). Therefore, during the primary process, metal particles have stronger pro-inflammatory potential than metal ions, and T lymphocytes did not seem to be activated in our murine model. Systemic reactions caused by metal particles and ions were found by observing the untreated right knees.

Keywords: inflammation; cytokines; metal particles; metal ions; synovium

1. Introduction

Medical cobalt-chromium-molybdenum (CoCrMo) alloys are mainly classified by the International Organization for Standardization (ISO) or American Society for Testing and Materials (ASTM) [1] and can be of a cast (ASTM F75, ISO 5832-4) or wrought content (ASTM F1537, ISO 5832-12) [2,3].

Due to their excellent biocompatibility and mechanical properties, these alloys have been widely employed in implant devices that replace hard tissue in the human body [4]. Since the mid-1980s, over one million metal-on-metal (MoM) hip endoprostheses made from a CoCrMo alloy have been implanted worldwide [5]. However, during the 2000s, issues of aseptic loosening due to the release of metal particles and ions were found, and the use of MoM replacements was almost completely stopped [6]. Subsequently, in joint arthroplasties, hybrid combinations were mainly recommended, such as a polyethylene inserts and CoCrMo heads (MoP). Unfortunately, metal wear particles and released metal ions have also been found in patients with MoP prostheses due to the mechanically assisted crevice corrosion of modular taper junctions, including the head–neck and neck–stem taper interfaces [7,8]. Some related studies have been conducted, and researchers have found that CoCrMo particles in revision patients are quite small (< 60 nm) and numerous. In general, high levels of metal ions are also generated from MoM implants [9]. Undoubtedly, the generation of degradation products and the subsequent biological reactions to the metal particles and ions have a fundamental effect on the longevity of total joint replacement [10]. However, to date, the effects of small wear particles and metal ions on local or systemic biological reactions remain complex and are still not fully understood in detail.

Commonly, metal debris, numerous inflammatory cells and increased proinflammatory mediators, such as IL-1β, TNF-α and IL-6, are found in periprosthetic tissues from patients with aseptic loosening [11]. Additionally, a synovial-like membrane with aggressive granulomatous lesions, in which many macrophages are infiltrated, is usually observed around aseptically loosened MoM implants [12]. Therefore, many researchers believe that small metal particles and high levels of metal ions cause an aseptic inflammatory response, which probably leads to periprosthetic osteolysis and, eventually, results in aseptic implant loosening [13]. However, most of the histological evidence for aseptic inflammatory reactions in periprosthetic tissues is from tissues of revision surgeries [14,15], which might not reflect the important initial reactions caused by metal particles and ions. Meanwhile, in terms of metal particles and metal ions, it is still unclear which factor plays a more critical role in aseptic inflammatory reactions.

In addition to aseptic inflammatory reactions, it has been speculated for a long time that type IV hypersensitivity reactions via T lymphocyte activation could play a role in aseptic loosening [13,16]. The assumption is mainly based on the presence of T lymphocytes in periprosthetic tissues from some studies and on the ability of metal ions to activate type-IV hypersensitivity by acting as haptens [17]. However, reactions against metal haptens are mainly determined by individual hereditary factors, with only some patients being genetically susceptible to developing type IV hypersensitivity against metals. Additionally, in some studies, patients with metal-on-metal implants displayed a significant decrease in the number of T lymphocytes and a significant increase in the level of metal ions after the hip replacement [18]. Therefore, it is still somewhat controversial whether type IV hypersensitivity reactions contribute to the aseptic loosening induced by metal particles and metal ions, especially in the short term after joint arthroplasties.

Primary biological reactions around periprosthetic tissues need to be precisely elucidated, not only to improve diagnoses but also to provide valid evidence for subsequent treatment [19–21]. Therefore, the objective of the present study was to evaluate and compare the primary aseptic inflammatory reactions to metal ions and wear particles in an established murine model. In addition, to better understand the role of type IV hypersensitivity reactions caused by degeneration products, T lymphocytes were also evaluated in the established model. The first hypothesis is that the metal particle (MP) group and metal ion (MI) group can both induce enhanced proinflammatory cytokine expression (TNF-α, IL-6 and IL-1β), and that T lymphocytes would be recruited when particles and ions were released in vivo. The second hypothesis was that the MI group have a stronger pro-inflammatory potential and recruit more T lymphocytes than the MP group.

2. Materials and Methods

2.1. Metal Particle/Ion Generation

Standard wrought CoCrMo alloys (ISO 5832 -12, ASTM F1537) [20] for hip implants were used to generate metal ions and wear particles in this study.

To obtain metal particles that were similar to clinically produced particles, a newly developed pin-on-disc simulator was used to perform wear tests at a frequency of 1 Hz. The wear volume was determined in the test medium by a high resolution inductively coupled plasma mass spectrometry (HR-ICP-MS) instrument (Thermo Scientific, Bremen, Germany). After performing an analysis via scanning electron microscopy (SEM; Zeiss EVO 50, Carl Zeiss NTS GmbH, Oberkochen, Germany), the following parameters were recorded: the equivalent circular diameter (ECD), the aspect ratio (AR) and the roundness (R) of wear particles.

To induce the release of metal ions, the CoCrMo alloys were anodized in a corrosion chamber and phosphate-buffered saline (PBS) was used as the surrounding medium. Using an HR-ICP-MS instrument (Thermo Scientific, Bremen, Germany), a total metal ion content (cobalt, chromium, molybdenum and nickel) of 20.5 mg/L was determined, which was adjusted to the desired target concentration of 200 μg/L using PBS. The target concentration of 200 μg/L was based on a study in which the metal ion level of a joint puncture of patients was analyzed prior to revision surgery, which showed average concentrations in the range of 200–250 μg/L in the joint fluid [22].

2.2. Elimination of Endotoxin

To avoid the influence of adherent endotoxins on the results, the generated particles and ions had to be free of endotoxins. To remove endotoxins, metal particles were cleaned by an ethanol washing process, and the metal ion solution was heat shocked. The elimination of endotoxins was proven by the Limulus Amebocyte Lysate (LAL) test (Lonza, Cologne, Germany).

2.3. Animals and Intraarticular Injection

Forty female BALB/c mice (Charles River Wiga Company, Sulzbach, Germany), weighing 23.5 ± 1.9 g, were used to establish an animal model in which biological reactions could be evaluated. These mice were randomly assigned to three groups: the PBS control group (n = 10), MP group (n = 10) and MI group (n = 10). To exclude the possibility that the process of the intra-articular injection itself produces an inflammation reaction in subsequent experiments, the remaining untreated mice (n = 10) were regarded as the negative control (NC) group. All experimental steps involving animals were performed according to the rules and regulations of the Animal Protection Laboratory Animal Regulations (2013) and European Directive 2010/63/EU Act, which is in accordance with the National Animal Protection Law (protocol number 55.2-1-54-2532-82.12, Government of Bavaria, Germany).

Prior to the intra-articular injection, all solutions were sonicated for at least 60 min to prevent possible agglomeration and precipitation. Under sterile conditions, 50 μL PBS-suspension, 50 μL of a 0.1 vol% CoCrMo particle suspension and 50 μL of 200 μg/L CoCrMo ions were injected into the left knees of the mice. After a 7-day incubation period, animals were euthanized by an intracardial pentobarbital injection (Merial GmbH, Hallbergmoos, Germany).

2.4. Immunohistochemistry

Both knees (left treated and right untreated) of all groups were removed and fixed in 4% paraformaldehyde for 24 h. All knee joints were decalcified using Osteosoft solution (Merck KGaA, Darmstadt, Germany) for four days at room temperature. Decalcification of the knees was followed by dehydration in a Spin Tissue Processor-120 (Especialidades Médicas Myr, S.L., Tarragona, Spain) in an ascending alcohol series (7%, 96%, 100%, and xylene) and, finally, a transfer to two successive paraffin baths at 60 °C. After being embedded in paraffin, solidified blocks were cut into 2 μm thick sections. Subsequently, the prepared tissue sections were immunochemically analyzed using six different monoclonal mouse antibodies: TNF-a (1: 200 dilution), IL -1b (1: 150 dilution), IL -6 (1: 200 dilution), CD68 (1: 100 dilution), CD45 (1: 400 dilution) and CD3 (1: 100 dilution) (Biorbyt Ltd., Cambridge, United Kingdom). Factors from the spleen (TNF-a), lung (IL -1β and IL-6) and tonsil (CD68, CD45 and CD3), which always react positively with the corresponding antibodies, served as positive controls in our study (data not shown).

As shown in many studies, the cytokines TNF-α, IL-1β and IL-6 are important inflammatory triggers in peri-implant tissues and are even involved in subsequent prosthetic loosening [23,24]. Therefore, in this study, TNF-α, IL-1β and IL-6 were used as inflammatory markers. According to the pro-inflammatory cytokine expression (TNF-α, IL-6 and IL-1β) in immunohistochemistry, we assessed the degree of inflammatory response in each group.

CD45 can be expressed in various immunocompetent cells, including dendritic cells, T lymphocytes, B lymphocytes, macrophages, etc. [25,26]. CD45 was used as a general marker of immunocompetent cells in the synovial layer in this study. The CD3 antigen binds to the membranes of all T lymphocytes and virtually no other cell types, which makes it a useful immunohistochemical marker of T lymphocytes in tissue sections [27]. The CD68 antigen is mainly expressed by monocytes and macrophages. When monocytes migrate into local tissues, they differentiate into macrophages [28,29]. Therefore, we used CD68 as a marker of macrophages in this study. To analyze the systemic biological reactions caused by metal particles and metal ions in our murine model, the immunohistochemical markers mentioned above were analyzed in the right knees of all groups.

Images of the stained samples were collected under a light microscope at 200× magnification (Carl Zeiss Micro Imaging GmbH, Oberkochen, Germany). Every image included the most synovial tissue (region of interest) of each sample. If cells or tissues were stained from light yellow to brown, positive immunostaining was recorded. Two observers independently performed manual counts of the obtained images with the assistance of Image J software (National Institutes of Health, Bethesda, MD, USA) (Ver.1.43, available at rsbweb.nih.gov/ij). Preceding manual counting, images were cropped, scaled to μm and separated by a color channel, and artifacts were removed. An area tool was used to select the synovial region and calculate the area. The Image J cell counter tool recorded mouse clicks on cells that were labeled with colored dots. The results were saved to a spreadsheet and screen shots were used to record the session. Positive cells from the synovial membrane were counted by each observer. If the results were inconsistent, specific samples were collected and counted by both observers for a third time.

2.5. Statistics

The data obtained from the immunohistochemical evaluation were evaluated using IBM SPSS®Statistics 22 (IBM Deutschland GmbH, Ehningen, Germany). Statistical analyses were carried out with the non-parametric Kruskal–Wallis test for independent samples. P-values below 0.05 were considered statistically significant and were adjusted by Bonferroni correction. A graphical presentation of the results was produced using box plots.

3. Results

3.1. Characterization of Metal Particles and Ions

For the generated metal particles, the mean size was in the nanometer range (ECD: 61.25 ± 18.47 nm). Meanwhile, their aspect ratio was 1.69 ± 0.66 and roundness was 0.64 ± 0.16 (Table 1). The particle shape was predominantly round and oval; in addition, a small proportion of acicular particles were formed (shape: round, 44%; oval, 49%; needle, 7%). This size and morphology of the particles were consistent with clinically found metal particles after a MoM total hip replacement [30,31].

Table 1. Morphological parameters of particles. The CoCr29Mo6 particle shape is predominantly oval and round with a small proportion of acicular particles (shape: round, 44%; oval, 49%; needle, 7%). Equivalent circular diameter (ECD).

Material	ECD	Aspect Ratio	Roundness
CoCr29Mo6 alloy	61.25 ± 18.47nm	1.69 ± 0.66	0.64 ± 0.16

The CoCr29Mo6 particle shape was predominantly oval and round with a small proportion of acicular particles (shape: round, 44%; oval, 49%; needle, 7%).

Using HR-ICP-MS, a total metal ion content (cobalt, chromium, nickel and molybdenum) of 20.5 mg/L was determined, which was adjusted to the desired target concentration of 200 µg/L using PBS (Table 2) [32]. The target concentration of ions was based on the concentration measured in the synovial fluid of patients with endoprosthesis during revision surgery [22].

Table 2. Total ion concentrations.

Content in	Co	Cr	Mo	Ni
stock solution	12.0 ± 2.4 mg/L	3.9 ± 0.6 mg/L	0.9 ± 0.1 mg/L	1.3 ± 0.6 mg/L
experimental solution	120 ± 24 µg/L	39 ± 5.7 µg/L	8.8 ± 1.1 µg/L	12.8 ± 6.0 µg/L

Total ion concentrations of the CoCrMo stock solution as well as in the experimental solution (200 µg/L in total) according to Jonitz-Heincke et al. [32].

3.2. *Results of the Left Knee Joints*

3.2.1. Expression of TNF-α, IL-1β and IL-6

With regard to the staining results of TNF-α, IL-1β and IL-6, the MP group showed a significantly increased number of positive cells compared with the PBS group ($p < 0.05$). However, the MI group did not express significantly more TNF-α ($p = 0.056$), IL-1β ($p = 0.420$) and IL-6 ($p = 0.124$) than the PBS group. Therefore, according to these results, the MP group had stronger inflammatory reactions in the synovial layers of the left knees (Figure 1).

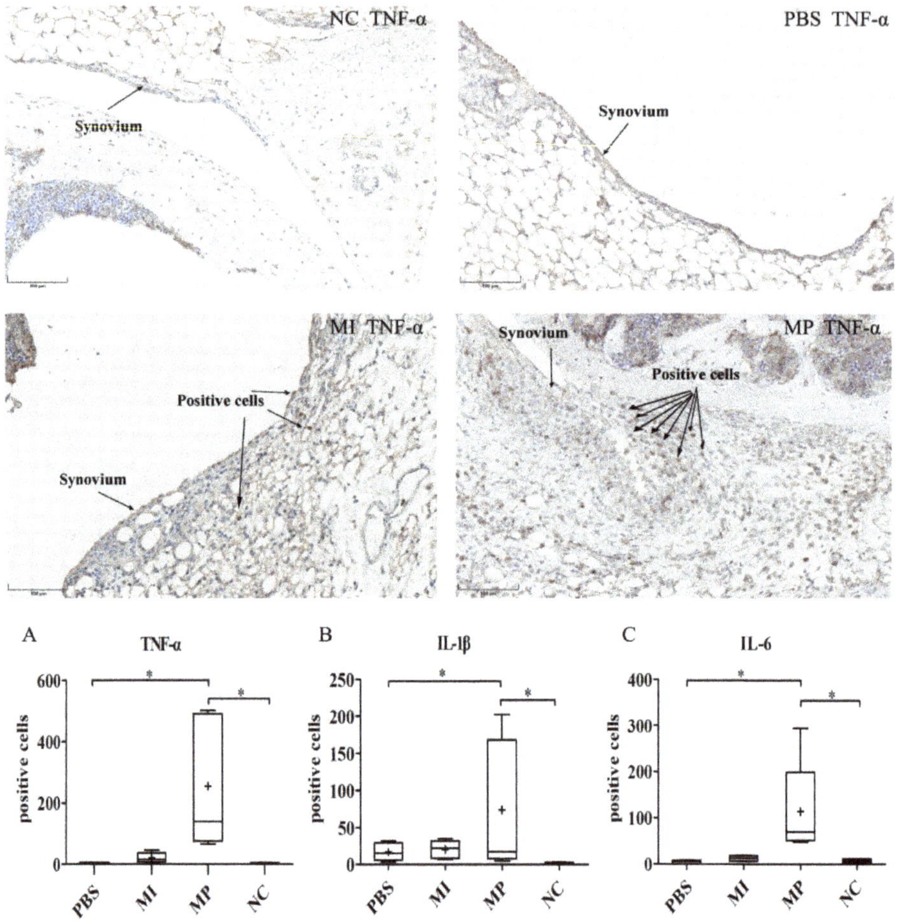

Figure 1. Expression of inflammatory markers (IL-1β, IL-6 and TNF-α) in the synovial layer of left murine knees. The staining results of TNF-α in the negative control (NC), phosphate-buffered saline (PBS), metal ion (MI) and metal particles (MP) group. Both the strongest expression and a thickened synovial layer are found in the MP group. (**A**) The expression of TNF-α in the synovial layer of left murine knees. (**B**) The expression of IL-1β in the synovial layer of left murine knees. (**C**) The expression of IL-6 in the synovial layer of left murine knees. The staining results of IL-1β, TNF-α and IL-6 are consistent. The MP group had the strongest inflammatory reactions. (Magnification: 200x; * $p < 0.05$).

3.2.2. Expression of CD68, CD3 and CD45

The MP group showed a significantly higher number of CD45-positive cells compared with the PBS group ($p < 0.05$). Likewise, the MI group had a significantly different number of CD45-positive cells compared with the PBS group ($p < 0.05$). Interestingly, in terms of CD45, there was no significant difference between the MP group and the MI group (Figure 2A).

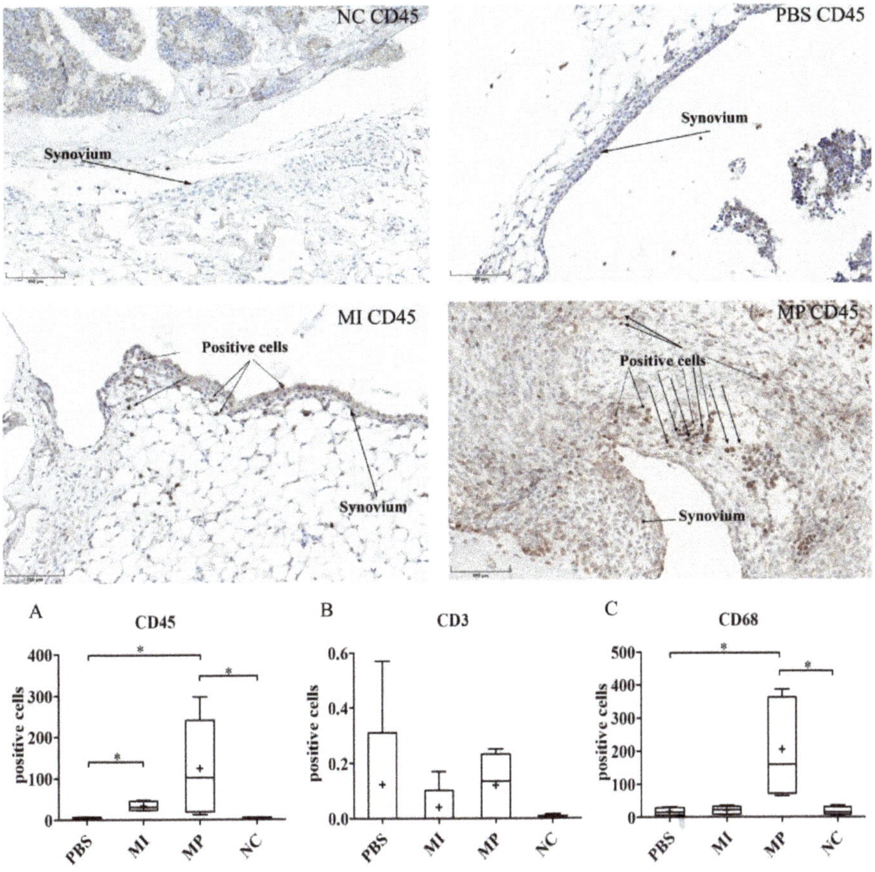

Figure 2. Expression of CD45, CD3 and CD68 in the synovial layer of left murine knees. The staining results of CD45 in the NC, PBS, MI and MP group. Numerous CD45 positive cells were found in the MP and MI group. MP and MI groups had significantly increased positive cells compared with the PBS group ($p < 0.05$). (**A**) The expression of CD45 in the synovial layer of left murine knees. (**B**) The expression of CD3 in the synovial layer of left murine knees. No significant difference was found in all groups. (**C**) The expression of CD68 in the synovial layer of left murine knees. (Magnification: 200x; * $p < 0.05$).

With respect to the expression of CD3 positive cells in our study, there was no statistically significant difference among the three groups ($p = 0.45$) (Figure 2B).

The MP group showed a significantly increased number of CD68 positive cells compared with the PBS group ($p < 0.05$), and the MI group ($p < 0.05$) in the synovial layer of the left knees. There was no significant difference between the MI group and the PBS group ($p = 1.0$). Considering that macrophages play an important role in inflammatory reactions, these results indicate that the MP group had the strongest inflammatory reactions, which is consistent with the TNF-α, IL-1β and IL-6 results (Figure 2C).

3.3. *Results of the Right Knee Joints*

3.3.1. Expression of TNF-α, IL-1β and IL-6

For TNF-α and IL-1β, the number of positive cells in the MP group was only significantly increased compared with the PBS group ($p < 0.05$). There was no significant difference among all groups in the expression of IL-6 in the synovial layer of the right murine knee joints, ($p = 0.13$) (Figure 3A–C).

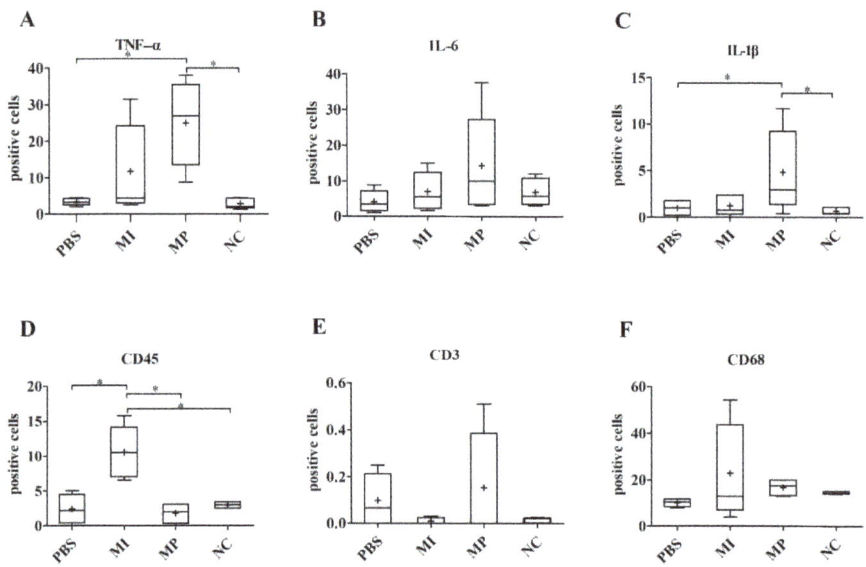

Figure 3. Expression of all biological markers in the synovial layer of right murine knees. The aim of examining the right knees was to evaluate systemic reactions caused by the dissemination of metal particles and metal ions. (**A**) The expression of TNF-α in the synovial layer of right murine knees. (**B**) The expression of IL-6 in the synovial layer of right murine knees. (**C**) The expression of IL-1β in the synovial layer of right murine knees. (**D**) The expression of CD45 in the synovial layer of right murine knees. The MI group exhibited a significant difference compared with the PBS groups ($p < 0.05$). (**E**) The expression of CD3 in the synovial layer of right murine knees. (**F**) The expression of CD68 in the synovial layer of right murine knees. (* $p < 0.05$).

3.3.2. Expression of CD68, CD3 and CD45

The expression of CD45-positive cells in the MI group was significantly increased compared with all other groups (PBS: $p < 0.05$, MP: $p < 0.05$). No statistically significant difference was found in the direct comparison of the PBS and MP groups ($p = 1.0$). Meanwhile, there was no significant difference among all groups in the expression of CD68 and CD3 in the synovial layer of the right knees (CD68: $p = 0.55$, CD3: $p = 0.91$) (Figure 3E–G).

4. Discussion

Our initial hypothesis for this study could not be proved. No increased T-lymphocyte markers were found in either the MP or MI group. However, in terms of proinflammatory reactions (IL-1β, IL-6 and TNF-α), the MP group had stronger reactions than the MI group. In addition, in terms of systemic reactions, the MP group and the MI group both had significantly increased biological reactions in the synovial membrane of right-sided knee joints compared to the PBS (control) group.

Histopathological studies can decisively contribute to the determination of the main cell populations underlying the biological mechanisms of aseptic inflammation and osteolysis [33]. In addition, the key protein molecules (such as IL-1β, IL-6 and TNF-α) involved in aseptic loosening can also be precisely detected by immunohistochemistry [34,35]. Therefore, periprosthetic tissues harvested during revision surgery, especially the bone–implant interface membrane and pseudo-synovial tissues, are frequently examined via histological techniques in clinical work [14,36]. However, periprosthetic tissues from revision surgeries mainly reflect end-stage failure and may not adequately reveal the primary biological reactions caused by metal particles and ions released immediately after the initial surgery. Nevertheless, the primary biological reactions caused by metal products in vivo may be crucial to elucidate the evolution of the pathophysiological events that lead to prosthetic osteolysis. In light of the limitations mentioned above, an established murine model was used in this study, which mimics the initial biological reactions in the synovial-like tissues around prostheses caused by metal particles and ions [35,37–39]. In contrast to various in vitro cell culture studies that focus on one cell type [32,40], the murine model can not only reflect complex cellular and tissue interactions, but can also mimic the dynamic process of joints. Additionally, compared with other models, such as hamsters' skinfold-chamber models [41] and "air pouch" models [42], results from this model, as the suspensions were injected intraarticularly, are easier to translate into a clinical scenario because the generated wear debris primarily accumulates in the joint that has been replaced. In terms of the characteristics of metal materials, the CoCrMo particles and ions used in our study were consistent with those found in some clinical studies.

CD45 can be expressed on many types of cells in the immune system, such as lymphocytes, natural killer cells, granulocytes, dendritic cells and monocytes/macrophages [26,43]. Therefore, CD45 was chosen as a general marker of immunocompetent cells in the synovial layer. For the left (treated) mice knees, the MP group and MI group both showed significantly increased CD45-positive cells compared with the PBS group, assuming that some immunocompetent cells were recruited to the synovial layer as a result of the metal particles and metal ions. However, it is difficult to distinguish specific immunocompetent cells and biological reactions solely from marking CD45. CD3 and CD68 were used as specific markers for T lymphocytes and macrophages [27], while IL-1β, IL-6 and TNF-α were used as common inflammatory markers [16].

For the groups of left knees that received an intra-articular injection, the MP group's inflammatory reactions (IL-1β, TNF-α, IL-6 and CD68) increased considerably more than those of the other groups. Considering the physical properties of CoCrMo particles, released nanometer-sized particles probably cause physical harm, especially when the knee joint is in motion [21]. This may be one reason why CoCrMo particles had very intensive inflammatory reactions in this in vivo study. Additionally, via various released cellular mediators, damaged cells can lead to the activation and recruitment of immunocompetent cells, especially macrophages. Subsequently, recruited macrophages phagocytize some small CoCrMo particles, while foreign body multinucleated giant cells surround very large particles [44]. Inside the cells, the particles are exposed to oxidative attacks, and consequently, high levels of ions might be released during the chemical corrosion processes [45]. Therefore, CoCrMo particles not only have a destructive effect because of their physical properties—the corrosion process and the presence of metal ions can also cause some biological reactions in the surrounding tissues [46]. Furthermore, because of the presence of phagocytized debris, cellular necrosis can occur and many mediators of inflammation are released by macrophages, which can recruit more macrophages, amplifying the inflammatory cascade [20]. Numerous macrophages (CD68 positive cells) were found in the MP group in this study. Although metal ions can also initiate oxidation-reduction reactions, the metal ions were rapidly quenched, while metal particles offered a reservoir of redox-reactive metals for continuous Reactive oxygen species (ROS) generation, which could further explain why metal particles resulted in the strongest local inflammatory reactions in this study.

However, in contrast with the inflammatory reaction results in our murine model, some in vitro studies have shown different results. Chamaon et al. indicated that the treatment of a human monocytic cell line with ionic cobalt led to a decrease in metabolic activity [Water Soluble Tetrazolium-1(WST-1) assay], while CoCrMo particles had no effect [47]. They explained that the rather large abrasive particles (from 200 nm to several micrometers) used in their study might have resulted in the relatively low impact of CoCrMo particles [47]. Because particle size is considered a critical parameter that influences the biological reactions to wear particles, the majority of MoM-produced particles are usually from 40 to 60 nm [2,48]. One easily neglected factor should also be considered: the in vitro cell cultivation process was static and could not reflect physical injury from metal particles in vivo. Moreover, Caicedo et al. showed that similar increases of IL-1β, TNF-α and IL-6 were evoked by cobalt, molybdenum ions, and Co-Cr-Mo alloy particles in human monocytes/macrophages [49]. Meanwhile, they found that ions and cobalt alloy particles induced inflammasome activation in vitro in a dose-dependent manner [50]. In our murine model, metal ions might be unavoidably systemically disseminated via lymphatics and blood vessels and then circulated throughout the host body, probably leading to a more general distribution than the metal particles. The gradually decreased metal ion level could attenuate local inflammatory stimulation effects, which might be one important reason why the level of inflammatory mediators (IL-1β, IL-6 and TNF-α) and the number of macrophages (CD68+ cells) did not significantly increase in the MI group in our study.

According to the literature, a cell-mediated (type-IV delayed hypersensitivity) response, which is mainly characterized by the activation of T lymphocytes, can be triggered by metal particles and ions during the process of aseptic loosening [19,51]. At least theoretically, metal ions are able to activate type-IV hypersensitivity by forming haptens with host proteins [16,52]. Compared with toxic reactions, type-IV hypersensitivity reactions do not have a simple dose–response relationship, with higher doses being more potent than lower doses, because even small amounts of antigens can cause strong reactions [17]. In our study, we attempted to count T lymphocytes (CD3-positive cells) to verify the existence of a type-IV hypersensitivity response. However, unexpectedly, regarding T lymphocytes, there were no significant differences among the groups. A delayed hypersensitivity response typically occurred between 24 and 48 h after exposure to an antigen [53]. One possibility was that a delayed-type hypersensitivity response did not play an important role in aseptic implant loosening, and that T lymphocytes were scarce in the primary process. Many researchers support this opinion, because there is a relatively low revision rate related to "allergic" loosening [20]. Approximately 10% of the population is hypersensitive to the materials found in jewelry and joint replacements [54]. If metallic debris and ions can induce a serious hypersensitivity response in the periprosthetic tissues, then more revision surgeries might have been performed as a result of "allergic osteolysis and aseptic loosening"; however, this is not actually the case [20,54].

Some studies have indicated that small nanoparticles and high levels of metal ions from implants can be detected in remote organs, including the liver, lung, spleen and kidney [55]. Whether the systemic dissemination of wear particles and metal ions can cause side effects is a matter of debate [20,23]. As such, in our study, the synovial membrane of each right-sided knee joint after the intra-articular particle or ion injection (left knee) was also analyzed for the presence of biological markers (IL-1β, IL-6 TNF-α, CD45, CD68 and CD3). For IL-1β and TNF-α, only the right knees of the MP group had a stronger response than the PBS group, which means that a systemic response caused by metal particles occurred. By contrast, IL-6 was not significantly increased in the right knees of the MP group. One possible explanation for this result is the presence of a temporal factor, because IL-6 secretion in the periprosthetic membrane is preceded by the expression of TNF-α and IL-1β [56]. With regard to CD45, the results of the right knees in the MI group were consistent with those of the left knees, indicating that some immunocompetent cells (CD45-positive cells) were activated by metal ions, and that metal ions also led to systemic reactions. In terms of CD3 and CD68, there were no significant differences in the right knees of all groups, which means that no macrophages or T lymphocytes were recruited in the right knees of all groups.

There are some limitations to this study. Although the murine model can reflect inflammatory reactions caused by implant materials, it cannot imitate the osteolysis process around implants.

Additionally, the single-injection murine model used in this study might not completely reflect the chronic production of wear debris in patients with joint replacements. Some researchers have used osmotic pumps in a murine model to achieve a continuous infusion of degeneration products [57]. This model is more like the clinical scenario, but irritation from osmotic pumps might be an interference factor. In a further study, we will aim to establish one continuous infusion model in combination with existing techniques. To investigate the systemic reactions caused by particles and ions, we used untreated knees. In the future, the impact of metal particles and ions on other organs, such as the liver, spleen, heart and kidney, will also be evaluated, which would further reflect the overall reactions caused by metal debris. Additionally, only one concentration of metal particles and ions was used, referring to the national animal laws. In a future study, different metal particle and ion concentrations, and the impact of different concentrations on aseptic inflammation, will be investigated.

5. Conclusions

The results of this study clearly demonstrate that the primary process which occurs after the release of metal particle and ions, especially CoCrMo particles, can lead to an intensive proinflammatory response in vivo. Metal ions can also cause the recruitment of immunocompetent cells but, in view of local inflammatory reactions, macrophages and inflammatory mediators were scarce in vivo. During the process by which metal particles and ions were released in present study, T lymphocytes were not recruited in our murine model. Systemic reactions by metal particles and ions were found according to the observation of untreated right knees.

Author Contributions: Data curation, X.C. and S.C.D.; formal analysis, X.C. and S.C.D.; funding acquisition, J.P.K., R.B., S.U. and A.C.P.; investigation, S.H.; methodology, S.H. and A.C.P.; project administration, S.U. and A.C.P.; resources, J.P.K., S.U. and A.C.P.; supervision, A.B.-M. and S.U.; validation, S.C.D.; writing—original draft, X.C.; writing—review & editing, S.U. and A.C.P. All authors have read and agreed to the published version of the manuscript.

Funding: This research was funded by German Research Foundation, grant number UT 119/3-1, BA 3347/12-1, KR 3755/5-1. And Xiangyun Cheng's doctoral work is funded by the China Scholarship Council (CSC) under the State Scholarship Fund (Grant No.201708140085).

Acknowledgments: The authors would like to thank the German Research Foundation (DFG) and China Scholarship Council (CSC) for funding of this project. We also gratefully thank Dr. Paul Johan Høl for his assistance in ICP-MS techniques.

Conflicts of Interest: The authors declare no conflict of interest.

References

1. Eliaz, N. Corrosion of Metallic Biomaterials: A Review. *Materials* **2019**, *12*, 407. [CrossRef]
2. Catelas, I.; Bobyn, J.D.; Medley, J.B.; Krygier, J.J.; Zukor, D.J.; Huk, O.L. Size, shape, and composition of wear particles from metal-metal hip simulator testing: Effects of alloy and number of loading cycles. *J. Biomed. Mater. Res. A* **2003**, *67*, 312–327. [CrossRef]
3. Merola, M.; Affatato, S. Materials for Hip Prostheses: A Review of Wear and Loading Considerations. *Materials* **2019**, *12*, 495. [CrossRef]
4. Grieco, P.W.; Pascal, S.; Newman, J.M.; Shah, N.V.; Stroud, S.G.; Sheth, N.P.; Maheshwari, A.V. New alternate bearing surfaces in total hip arthroplasty: A review of the current literature. *J. Clin. Orthop. Trauma* **2018**, *9*, 7–16. [CrossRef]
5. Xu, J.; Yang, J.; Nyga, A.; Ehteramyan, M.; Moraga, A.; Wu, Y.; Zeng, L.; Knight, M.M.; Shelton, J.C. Cobalt (II) ions and nanoparticles induce macrophage retention by ROS-mediated down-regulation of RhoA expression. *Acta Biomater.* **2018**, *72*, 434–446. [CrossRef]
6. Sidaginamale, R.P.; Joyce, T.J.; Bowsher, J.G.; Lord, J.K.; Avery, P.J.; Natu, S.; Nargol, A.V.; Langton, D.J. The clinical implications of metal debris release from the taper junctions and bearing surfaces of metal-on-metal hip arthroplasty: Joint fluid and blood metal ion concentrations. *Bone Jt. J.* **2016**, *98*, 925–933. [CrossRef]

7. Kretzer, J.P.; Mueller, U.; Streit, M.R.; Kiefer, H.; Sonntag, R.; Streicher, R.M.; Reinders, J. Ion release in ceramic bearings for total hip replacement: Results from an in vitro and an in vivo study. *Int. Orthop.* **2018**, *42*, 65–70. [CrossRef]
8. Utzschneider, S.; Paulus, A.; Datz, J.C.; Schroeder, C.; Sievers, B.; Wegener, B.; Jansson, V. Influence of design and bearing material on polyethylene wear particle generation in total knee replacement. *Acta Biomater.* **2009**, *5*, 2495–2502. [CrossRef]
9. Pourzal, R.; Catelas, I.; Theissmann, R.; Kaddick, C.; Fischer, A. Characterization of Wear Particles Generated from CoCrMo Alloy under Sliding Wear Conditions. *Wear* **2011**, *271*, 1658–1666. [CrossRef]
10. Sansone, V.; Pagani, D.; Melato, M. The effects on bone cells of metal ions released from orthopaedic implants. A review. *Clin. Cases Miner. Bone Metab.* **2013**, *10*, 34–40. [CrossRef]
11. Yu, F.Y.; Xie, C.Q.; Sun, J.T.; Peng, W.; Huang, X.W. Overexpressed miR-145 inhibits osteoclastogenesis in RANKL-induced bone marrow-derived macrophages and ovariectomized mice by regulation of Smad3. *Life Sci.* **2018**, *202*, 11–20. [CrossRef]
12. Gallo, J.; Vaculova, J.; Goodman, S.B.; Konttinen, Y.T.; Thyssen, J.P. Contributions of human tissue analysis to understanding the mechanisms of loosening and osteolysis in total hip replacement. *Acta Biomater.* **2014**, *10*, 2354–2366. [CrossRef]
13. Pajarinen, J.; Jamsen, E.; Konttinen, Y.T.; Goodman, S.B. Innate immune reactions in septic and aseptic osteolysis around hip implants. *J. Long Term Eff. Med. Implants* **2014**, *24*, 283–296. [CrossRef]
14. Campbell, P.; Ebramzadeh, E.; Nelson, S.; Takamura, K.; De Smet, K.; Amstutz, H.C. Histological features of pseudotumor-like tissues from metal-on-metal hips. *Clin. Orthop. Relat. Res.* **2010**, *468*, 2321–2327. [CrossRef]
15. Mao, X.; Tay, G.H.; Godbolt, D.B.; Crawford, R.W. Pseudotumor in a well-fixed metal-on-polyethylene uncemented hip arthroplasty. *J. Arthroplast.* **2012**, *27*, 493.e13. [CrossRef]
16. Goodman, S.B. Wear particles, periprosthetic osteolysis and the immune system. *Biomaterials* **2007**, *28*, 5044–5048. [CrossRef]
17. Granchi, D.; Savarino, L.M.; Ciapetti, G.; Baldini, N. Biological effects of metal degradation in hip arthroplasties. *Crit. Rev. Toxicol.* **2018**, *48*, 170–193. [CrossRef]
18. Granchi, D.; Savarino, L.; Ciapetti, G.; Cenni, E.; Rotini, R.; Mieti, M.; Baldini, N.; Giunti, A. Immunological changes in patients with primary osteoarthritis of the hip after total joint replacement. *J. Bone Jt. Surg. Br.* **2003**, *85*, 758–764. [CrossRef]
19. Konttinen, Y.T.; Pajarinen, J. Adverse reactions to metal-on-metal implants. *Nat. Rev. Rheumatol.* **2013**, *9*, 5–6. [CrossRef]
20. Gallo, J.; Goodman, S.B.; Konttinen, Y.T.; Wimmer, M.A.; Holinka, M. Osteolysis around total knee arthroplasty: A review of pathogenetic mechanisms. *Acta Biomater.* **2013**, *9*, 8046–8058. [CrossRef]
21. Abu-Amer, Y.; Darwech, I.; Clohisy, J.C. Aseptic loosening of total joint replacements: Mechanisms underlying osteolysis and potential therapies. *Arthritis. Res. Ther.* **2007**, *9* (Suppl. 1), S6. [CrossRef]
22. De Smet, K.; De Haan, R.; Calistri, A.; Campbell, P.A.; Ebramzadeh, E.; Pattyn, C.; Gill, H.S. Metal ion measurement as a diagnostic tool to identify problems with metal-on-metal hip resurfacing. *J. Bone Jt. Surg. Am.* **2008**, *90* (Suppl. 4), 202–208. [CrossRef]
23. Posada, O.M.; Tate, R.J.; Grant, M.H. Toxicity of cobalt-chromium nanoparticles released from a resurfacing hip implant and cobalt ions on primary human lymphocytes in vitro. *J. Appl. Toxicol.* **2015**, *35*, 614–622. [CrossRef]
24. Lin, T.H.; Tamaki, Y.; Pajarinen, J.; Waters, H.A.; Woo, D.K.; Yao, Z.; Goodman, S.B. Chronic inflammation in biomaterial-induced periprosthetic osteolysis: NF-kappaB as a therapeutic target. *Acta Biomater.* **2014**, *10*, 1–10. [CrossRef]
25. Tchilian, E.Z.; Beverley, P.C. CD45 in memory and disease. *Arch. Immunol. Ther. Exp.* **2002**, *50*, 85–93.
26. Hermiston, M.L.; Xu, Z.; Weiss, A. CD45: A critical regulator of signaling thresholds in immune cells. *Annu. Rev. Immunol.* **2003**, *21*, 107–137. [CrossRef]
27. Martin-Romero, C.; Santos-Alvarez, J.; Goberna, R.; Sanchez-Margalet, V. Human leptin enhances activation and proliferation of human circulating T lymphocytes. *Cell. Immunol.* **2000**, *199*, 15–24. [CrossRef]
28. Xu, W.; Holzhuter, G.; Sorg, H.; Wolter, D.; Lenz, S.; Gerber, T.; Vollmar, B. Early matrix change of a nanostructured bone grafting substitute in the rat. *J. Biomed. Mater. Res. B Appl. Biomater.* **2009**, *91*, 692–699. [CrossRef]

29. Ramprasad, M.P.; Terpstra, V.; Kondratenko, N.; Quehenberger, O.; Steinberg, D. Cell surface expression of mouse macrosialin and human CD68 and their role as macrophage receptors for oxidized low density lipoprotein. *Proc. Natl. Acad. Sci. USA* **1996**, *93*, 14833–14838. [CrossRef]
30. Reinders, J.; Sonntag, R.; Vot, L.; Gibney, C.; Nowack, M.; Kretzer, J.P. Wear testing of moderate activities of daily living using in vivo measured knee joint loading. *PLoS ONE* **2015**, *10*, e0123155. [CrossRef]
31. Schroder, C.; Reinders, J.; Zietz, C.; Utzschneider, S.; Bader, R.; Kretzer, J.P. Characterization of polyethylene wear particle: The impact of methodology. *Acta Biomater.* **2013**, *9*, 9485–9491. [CrossRef]
32. Jonitz-Heincke, A.; Tillmann, J.; Klinder, A.; Krueger, S.; Kretzer, J.P.; Hol, P.J.; Paulus, A.C.; Bader, R. The Impact of Metal Ion Exposure on the Cellular Behavior of Human Osteoblasts and PBMCs: In Vitro Analyses of Osteolytic Processes. *Materials* **2017**, *10*, 734. [CrossRef]
33. Holt, G.; Murnaghan, C.; Reilly, J.; Meek, R.M. The biology of aseptic osteolysis. *Clin. Orthop. Relat. Res.* **2007**, *460*, 240–252. [CrossRef]
34. Al Saffar, N.; Revell, P.A. Interleukin-1 production by activated macrophages surrounding loosened orthopaedic implants: A potential role in osteolysis. *Br. J. Rheumatol.* **1994**, *33*, 309–316. [CrossRef]
35. Utzschneider, S.; Becker, F.; Grupp, T.M.; Sievers, B.; Paulus, A.; Gottschalk, O.; Jansson, V. Inflammatory response against different carbon fiber-reinforced PEEK wear particles compared with UHMWPE in vivo. *Acta Biomater.* **2010**, *6*, 4296–4304. [CrossRef]
36. Smith, A.J.; Dieppe, P.; Vernon, K.; Porter, M.; Blom, A.W. Failure rates of stemmed metal-on-metal hip replacements: Analysis of data from the National Joint Registry of England and Wales. *Lancet* **2012**, *379*, 1199–1204. [CrossRef]
37. Lorber, V.; Paulus, A.C.; Buschmann, A.; Schmitt, B.; Grupp, T.M.; Jansson, V.; Utzschneider, S. Elevated cytokine expression of different PEEK wear particles compared to UHMWPE in vivo. *J. Mater. Sci. Mater. Med.* **2014**, *25*, 141–149. [CrossRef]
38. Paulus, A.C.; Frenzel, J.; Ficklscherer, A.; Rossbach, B.P.; Melcher, C.; Jansson, V.; Utzschneider, S. Polyethylene wear particles induce TLR 2 upregulation in the synovial layer of mice. *J. Mater. Sci. Mater. Med.* **2014**, *25*, 507–513. [CrossRef]
39. Utzschneider, S.; Lorber, V.; Dedic, M.; Paulus, A.C.; Schroder, C.; Gottschalk, O.; Schmitt-Sody, M.; Jansson, V. Biological activity and migration of wear particles in the knee joint: An in vivo comparison of six different polyethylene materials. *J. Mater. Sci. Mater. Med.* **2014**, *25*, 1599–1612. [CrossRef]
40. Klinder, A.; Seyfarth, A.; Hansmann, D.; Bader, R.; Jonitz-Heincke, A. Inflammatory Response of Human Peripheral Blood Mononuclear Cells and Osteoblasts Incubated With Metallic and Ceramic Submicron Particles. *Front. Immunol.* **2018**, *9*, 831. [CrossRef]
41. Kraft, C.N.; Diedrich, O.; Burian, B.; Schmitt, O.; Wimmer, M.A. Microvascular response of striated muscle to metal debris. A comparative in vivo study with titanium and stainless steel. *J. Bone Jt. Surg. Br.* **2003**, *85*, 133–141. [CrossRef]
42. Anabtawi, M.; Beck, P.; Lemons, J. Biocompatibility testing of simulated total joint arthoplasty articulation debris. *J. Biomed. Mater. Res. B Appl. Biomater.* **2008**, *84*, 478–485. [CrossRef]
43. Tchilian, E.Z.; Beverley, P.C. Altered CD45 expression and disease. *Trends Immunol.* **2006**, *27*, 146–153. [CrossRef]
44. Bijukumar, D.R.; Segu, A.; Souza, J.C.; Li, X.; Barba, M.; Mercuri, L.G.; Jacobs, J.J.; Mathew, M.T. Systemic and local toxicity of metal debris released from hip prostheses: A review of experimental approaches. *Nanomedicine* **2018**, *14*, 951–963. [CrossRef]
45. Delgado-Ruiz, R.; Romanos, G. Potential Causes of Titanium Particle and Ion Release in Implant Dentistry: A Systematic Review. *Int. J. Mol. Sci.* **2018**, *19*, 3585. [CrossRef]
46. Haeri, M.; Wllert, T.; Langford, G.M.; Gilbert, J.L. Electrochemical control of cell death by reduction-induced intrinsic apoptosis and oxidation-induced necrosis on CoCrMo alloy in vitro. *Biomaterials* **2012**, *33*, 6295–6304. [CrossRef]
47. Chamaon, K.; Schonfeld, P.; Awiszus, F.; Bertrand, J.; Lohmann, C.H. Ionic cobalt but not metal particles induces ROS generation in immune cells in vitro. *J. Biomed. Mater. Res. B Appl. Biomater.* **2019**, *107*, 1246–1253. [CrossRef]
48. Hallab, N.J.; Jacobs, J.J. Biologic effects of implant debris. *Bull. NYU Hosp. Jt. Dis.* **2009**, *67*, 182–188.
49. Caicedo, M.S.; Pennekamp, P.H.; McAllister, K.; Jacobs, J.J.; Hallab, N.J. Soluble ions more than particulate cobalt-alloy implant debris induce monocyte costimulatory molecule expression and release of

proinflammatory cytokines critical to metal-induced lymphocyte reactivity. *J. Biomed. Mater. Res. A* **2010**, *93*, 1312–1321. [CrossRef]

50. Caicedo, M.S.; Desai, R.; McAllister, K.; Reddy, A.; Jacobs, J.J.; Hallab, N.J. Soluble and particulate Co-Cr-Mo alloy implant metals activate the inflammasome danger signaling pathway in human macrophages: A novel mechanism for implant debris reactivity. *J. Orthop. Res.* **2009**, *27*, 847–854. [CrossRef]
51. Pandit, H.; Vlychou, M.; Whitwell, D.; Crook, D.; Luqmani, R.; Ostlere, S.; Murray, D.W.; Athanasou, N.A. Necrotic granulomatous pseudotumours in bilateral resurfacing hip arthoplasties: Evidence for a type IV immune response. *Virchows Arch. Int. J. Pathol.* **2008**, *453*, 529–534. [CrossRef] [PubMed]
52. Dapunt, U.; Giese, T.; Prior, B.; Gaida, M.M.; Hansch, G.M. Infectious versus non-infectious loosening of implants: Activation of T lymphocytes differentiates between the two entities. *Int. Orthop.* **2014**, *38*, 1291–1296. [CrossRef] [PubMed]
53. Warrington, R.; Watson, W.; Kim, H.L.; Antonetti, F.R. An introduction to immunology and immunopathology. *Allergy Asthma Clin. Immunol.* **2011**, *7* (Suppl. 1), S1. [CrossRef] [PubMed]
54. Burkandt, A.; Katzer, A.; Thaler, K.; Von Baehr, V.; Friedrich, R.E.; Ruther, W.; Amling, M.; Zustin, J. Proliferation of the synovial lining cell layer in suggested metal hypersensitivity. *In Vivo* **2011**, *25*, 679–686.
55. Urban, R.M.; Tomlinson, M.J.; Hall, D.J.; Jacobs, J.J. Accumulation in liver and spleen of metal particles generated at nonbearing surfaces in hip arthroplasty. *J. Arthroplast.* **2004**, *19* (Suppl. 3), 94–101. [CrossRef]
56. Dinarello, C.A. Blocking IL-1 in systemic inflammation. *J. Exp. Med.* **2005**, *201*, 1355–1359. [CrossRef]
57. Ma, T.; Huang, Z.; Ren, P.G.; McCally, R.; Lindsey, D.; Smith, R.L.; Goodman, S.B. An in vivo murine model of continuous intramedullary infusion of polyethylene particles. *Biomaterials* **2008**, *29*, 3738–3742. [CrossRef]

materials

MDPI

Article

Long-Term Results of a Second-Generation, Small-Diameter, Metal-On-Metal Bearing in Primary Total Hip Arthroplasty at 14-Year Follow-Up

Tobias Reiner [1,*], Matthias C. Klotz [1], Kirsten Seelmann [1], Fabian Hertzsch [1], Moritz M. Innmann [1], Marcus R. Streit [1], Timo A. Nees [1], Babak Moradi [1], Christian Merle [1], Jan Philippe Kretzer [2] and Tobias Gotterbarm [1,3]

1 Center for Orthopedics, Trauma Surgery and Spinal Cord Injury, Heidelberg University Hospital, Schlierbacher Landstraße 200a, 69118 Heidelberg, Germany; office@drklotz.org (M.C.K.); ks1105@gmx.de (K.S.); fabianhertzsch@yahoo.de (F.H.); moritz.innmann@med.uni-heidelberg.de (M.M.I.); marcus.streit@med.uni-heidelberg.de (M.R.S.); timo.nees@med.uni-heidelberg.de (T.A.N.); babak.moradi@med.uni-heidelberg.de (B.M.); christian.merle@med.uni-heidelberg.de (C.M.); tobias.gotterbarm@kepleruniklinikum.at (T.G.)

2 Laboratory of Biomechanics and Implant Research, Center for Orthopedics, Trauma Surgery and Spinal Cord Injury, Heidelberg University Hospital, 69118 Heidelberg, Germany; philippe.kretzer@med.uni-heidelberg.de

3 Department of Orthopedics, Kepler University Hospital, Krankenhausstr. 7a, 4020 Linz, Austria

* Correspondence: tobias.reiner@med.uni-heidelberg.de; Tel.: +49-6221-5634803; Fax: +49-6221-5626360

Received: 18 November 2019; Accepted: 20 January 2020; Published: 24 January 2020

check for updates

Abstract: (1) Background: The objective of the present study was to review the clinical and radiological results of a small-head, MoM bearing in primary THA and to determine blood metal ion levels at long-term follow-up. (2) Methods: We retrospectively evaluated the clinical and radiological results of 284 small-diameter, MoM 28-mm Metasul THA at a mean follow-up of 14.5 years, and measured blood metal ion concentrations in 174 of these patients. (3) Results: After 14 years, survival free for revision due to any reason was 94%. Proximal femoral osteolysis was seen in 23% of hips, and MRI demonstrated ARMD in 27 of the 66 investigated hips (41%). Mean cobalt, chromium, and titanium ion concentrations were 0.82 μg/L (range 0.22–4.45), 1.51 μg/L (0.04–22.69), and 2.68 μg/L (0.26–19.56) in patients with unilateral THA, and 2.59 μg/L (0.43–24.75), 2.50 μg/L (0.26–16.75), and 3.76 μg/L (0.67–19.77), respectively in patients with bilateral THA. Twenty-nine percent of patients showed cobalt or chromium ion levels > 2 μg/L. (4) Conclusions: Despite good clinical long-term results, increased blood metal ion levels (cobalt or chromium > 2 μg/L) were found in approximately one-third of asymptomatic patients, and proximal femoral osteolysis and ARMD were frequently seen in this cohort. Blood metal ion analysis appears helpful in the long-term follow-up of these patients in order to identify individuals at risk. In accordance with contemporary consensus statements, symptomatic patients with elevated metal ion levels and/or progressive osteolysis should be considered for additional CT or MARS MRI to determine the extent of soft tissue affection prior to revision surgery. Further studies are necessary to investigate the clinical relevance of ARMD in asymptomatic patients with small-head, MoM THA.

Keywords: Metasul; 28 mm small head; metal-on-metal THA; cobalt; chromium; titanium; blood metal ions

1. Introduction

Second-generation, small-head, metal-on-metal (MoM) total hip replacements were reintroduced in 1988 by Weber [1], and initiated the rise of metal-on-metal hip arthroplasties at the beginning of this century. Metal-on-metal bearings were commonly implanted in younger patients hoping to overcome the polyethylene-wear-related complications of periprosthetic osteolysis and aseptic implant loosening. In 2008, metal-on-metal articulations were used in approximately 35% of all hip replacements in the United States [2]. High early failure rates, especially in large-diameter, metal-on-metal total hip arthroplasties (THA), and the growing incidence of adverse local tissue reactions related to metal wear, led to a swift decrease in the use of those implants in the subsequent years [3–6]. Accumulating metal ions in the joint cavity, which are generated by corrosive degradation of metal wear products, are able to influence both bone metabolism and the immune system through different pathways, contributing to the pathogenesis of periprosthetic osteolysis and the formation of adverse local soft tissue reactions, also referred to as ARMD (adverse reaction to metal debris). Although MoM bearings are rarely used nowadays, the systematic follow-up of these patients will continue to be of clinical importance due to the large number of metal-on-metal articulations that were implanted in past decades, especially in younger patients [7]. Risk stratification algorithms for the management of patients with MoM bearings have been provided by different regulatory authorities [8–10], and published guidelines suggest that small-diameter (< 36 mm) MoM implants are at low risk of developing ARMD. In contrast to large-diameter MoM articulations, a systematic long-term follow-up comparable to conventional THA with routine follow-up intervals of 3 to 5 years in the long term is considered sufficient, and blood metal ion analysis is not recommended in the follow-up routine of patients with small-diameter, MoM articulations [10]. Although some authors have recently raised concerns about the late onset of ARMD associated with increased metal wear of small-diameter, MoM implants [11–13], the results of metal ion analyses in the long-term follow-up of these patients are not clear.

The objective of the present study was (i) to evaluate the clinical and radiological results of small-head, MoM THA at long-term follow-up, (ii) to determine blood metal ion concentrations in a large cohort of patients at a minimum follow-up of 10 years, and (iii) to investigate potential risk factors associated with elevated blood metal ion levels in these patients.

2. Materials and Methods

2.1. Study Design and Patients

In this cross-sectional study, we retrospectively evaluated a consecutive series of 262 patients (284 hips) following cementless THA with a 28-mm Metasul metal-on-metal articulation. The study was approved by the ethics committee of the Heidelberg school of medicine (No. S-365/2013), and informed written consent was obtained prior to inclusion of each patient. Surgery was performed consecutively between April 1995 and November 2001 at Heidelberg University Hospital using either a modified Watson-Jones or a transgluteal lateral approach. The indication for the use of a MoM bearing at that time was young patient age and a high expected physical activity level. The mean age of patients at time of surgery was 52 years (range 21 to 74 years). At a mean follow-up of 14.5 years, 44 patients (17%, 33 male, 11 female) had died and 14 (5%) were lost to follow-up, leaving 193 patients (211 hips) who were available for review (Figure 1). Up to the latest follow-up, fourteen hips (5%) had undergone revision surgery. Of the remaining cohort, 174 patients (189 hips) agreed to participate in blood metal ion analysis, which was performed at a mean follow-up duration of 14.5 years (range 10.3 to 18.8 years) after surgery. In order to eliminate other sources of cobalt or chromium ion release, eighteen patients (19 hips) were excluded due to additional metal implants such as total knee replacements [14], and seventeen patients (17 hips) had to be excluded because of femoral components made of cobalt-chromium-alloys. Of the remaining cohort, 113 patients with unilateral THA and 26 patients with bilateral THA were available for further statistical analysis.

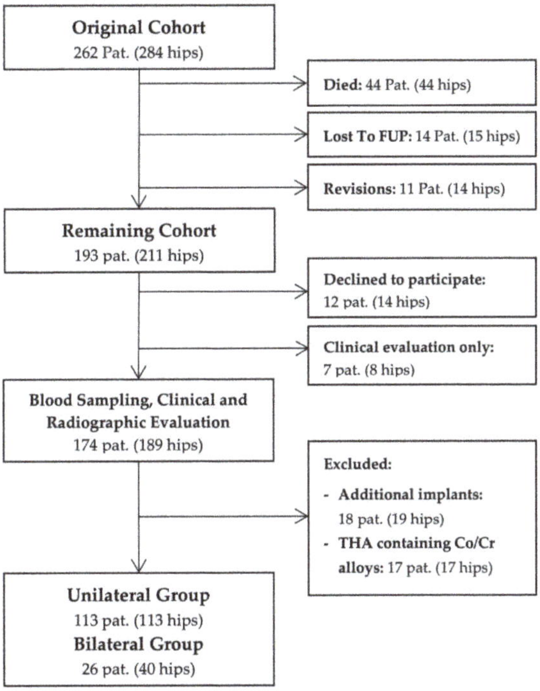

Figure 1. Flowchart summarizing clinical follow-up and patient selection for metal ion analysis.

A 28-mm Metasul (Zimmer, Winterthur, Switzerland) MoM articulation was used in all hips. The acetabular component of this implant consists of a forged, high-carbide (0.2–0.25%) cobalt-chromium alloy liner, which is embedded in a polyethylene insert. It was used in combination with a cementless, press-fit titanium acetabular shell; 95 hips received an Allofit acetabular cup (Zimmer, Winterthur, Switzerland) and 58 received a Fitmore acetabular component (Zimmer, Winterthur, Switzerland). An uncemented straight-tapered titanium stem with a standard 12/14 mm Euro taper was used in all hips for femoral reconstruction; 128 hips received a CLS Spotorno stem (Zimmer, Winterthur, Switzerland) and 25 received a G2 stem (Depuy Orthopaedics, Warsaw, Poland).

2.2. Clinical and Radiographic Follow-up

Clinical examination was performed using the Harris Hip Score. Standard pelvis anteroposterior and lateral radiographs of the hip were evaluated with regard to radiolucent lines and osteolysis. We defined periprosthetic osteolysis as a lucent zone absent of trabecular bone, which was not visible on the immediate postoperative radiograph [15]. Radiolucencies and osteolysis were evaluated according to the zones established by Gruen et al. [16] and the classification system of DeLee and Charnley [17]. Cup inclination angles were determined using the TraumaCad software (TraumaCad®, Voyant Health, Columbia, SC, USA), taking the inter-teardrop line as a fixed landmark [18]. In addition, cross-sectional imaging with metal artifact reduction sequence magnetic resonance imaging (MARS MRI) was available in 53 patients (66 hips) of the study cohort, which were retrospectively evaluated regarding ARMD formation. The indication to perform MARS MRI in these patients was blood cobalt or chromium ion level > 1 μg/L. A total of 107 patients in the study cohort fulfilled these inclusion criteria and were invited for MRI as part of a previously published study [13].

2.3. Metal Ion Analysis

Blood samples were taken using a blood collection system specific for trace metal ion analysis (Sarstedt, Nuembrecht, Germany; Refs. 58.1162.600 and 01.1604.400). The first 5ml of blood were discarded and blood samples were stored at −20 °C. Whole blood metal ion analysis was performed at the Geochemical Laboratories at Heidelberg University using high-resolution, inductively-coupled, plasma-mass spectrometry (HR-ICP-MS, Element 2, Thermo Fisher Scientific, Bremen, Germany). ICP-MS is currently considered one of the preferred techniques for blood metal ion measurement [10]. All samples were analyzed at the same time in order to minimize calibration errors arising from the spectrometer. Metal ion analysis was repeated three times in every sample and mean values were calculated. Detection limits of 0.005 µg/L for cobalt, 0.02 µg/L for chromium, and 0.06 µg/L for titanium were established for this method [19]. Additionally, the glomerular filtration rate (GFR) was calculated using the CKD-EPI formula based on the serum creatinine values of each patient.

2.4. Statistical Methods

Statistical analysis was performed using the software SPSS® for Windows® (version 22.0; SPSS IBM Corp., Chicago, IL, USA) and Graphpad Prism® (version 6.0, Graphpad Software, San Diego, CA, USA). Data were evaluated descriptively as arithmetic mean, standard deviation, median, minimum, and maximum. Demographic data and mean metal ion levels were compared between the bilateral and the unilateral group using the student's t-test. For comparison of categorical variables between the two groups, the chi-square test was used. Kaplan-Meier survivorship analysis was performed with revision for any reason as the endpoint. In the unilateral group, correlation analysis was performed using Spearman correlation coefficient and multivariate linear regression analysis in order to investigate the correlation between blood metal ion concentration and potential risk factors associated with elevated cobalt ion levels, which were defined as gender, cup inclination angle, body mass index, and follow-up length. Additionally, the relationship between periprosthetic osteolysis and blood metal ion concentrations of cobalt, chromium, and titanium was assessed using logistic regression analysis. Correlation was defined as poor (0.00 to 0.20), fair (0.21 to 0.40), moderate (0.41 to 0.60), good (0.61 to 0.80), or excellent (0.81 to 1.00). All tests were two-sided and a p-value < 0.05 was considered significant.

3. Results

3.1. Survival Analysis

The cumulative survival rate at 10 years, using revision for any reason as the endpoint, was 96% (95% confidence interval (CI); 92–98%; 235 hips at risk) and 94% (95% CI; 90–96%; 112 hips at risk) at a mean follow-up of 14 years (Figure 2). Of the 14 hips requiring revision surgery, four (1.4%) were revised for adverse reaction to metal debris (ARMD) and four (1.4%) were revised for aseptic loosening of either the femoral (n = 2) or acetabular component (n = 2). Another four hips (1.4%) were revised for infection, and two were revised for late periprosthetic fracture (0.7%). The mean time to revision surgery for ARMD was 10.5 years (range 7 to 15 years), and the mean time to revision surgery for aseptic loosening was 6.2 years (range 3.5 to 11 years).

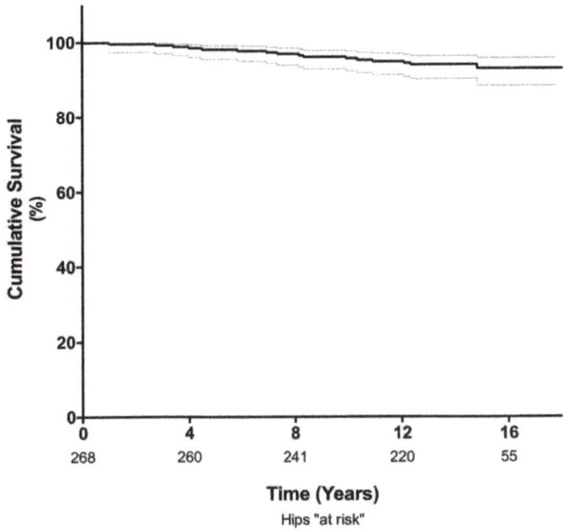

Figure 2. Kaplan-Meier analysis showing the survival free of revision for any cause was 96% (95% CI 92–98%) at 10 years and 94% (95% CI; 90–96%) at a mean follow-up of 14 years.

3.2. Clinical and Radiographic Evaluation

The mean Harris Hip Score of the cohort was 90 points (range 40 to 100) at the time of follow-up. The mean inclination angle of the acetabular component was 42 degrees (range 29–50 degrees). No femoral component showed radiographic signs of loosening. Radiographs demonstrated femoral osteolysis in 23% of the hips and radiolucent lines > 2 mm in 13% of hips. Osteolysis and radiolucent lines were predominantly located in the proximal Gruen zones. Their distribution is illustrated in Figure 3. Periacetabular osteolysis was rarely seen, with an overall frequency of 2%. MARS-MRI demonstrated pseudotumor formation in 27 of the 66 investigated patients (41%). ARMD were generally small and predominantly cystic in nature. More detailed results of this investigation were previously published in another study of this research group [13].

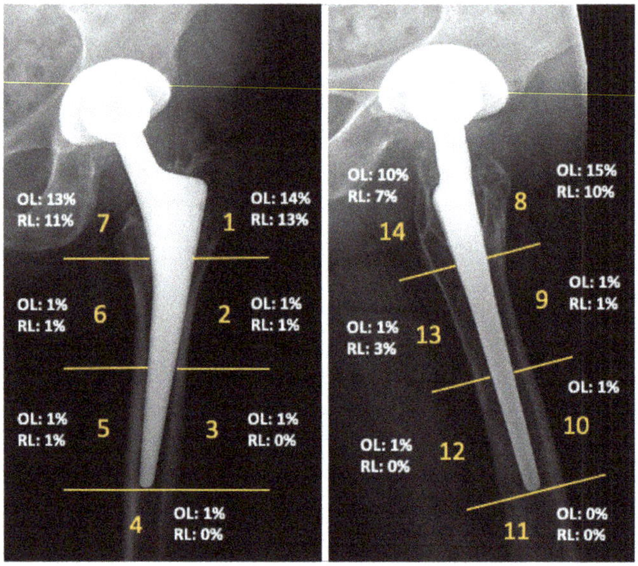

Figure 3. Results of the radiographic evaluation showing the distribution of radiolucent lines (RL) and osteolysis (OL), as seen on anteroposterior and lateral radiographs according to Gruen zones at a mean follow-up of 14 years.

3.3. Metal Ion Analysis

A total of 139 patients were eligible for blood metal ion analysis, with the Metasul bearing being the only known source for cobalt or chromium ion release (Figure 1). The demographic data of the study cohort are summarized in Table 1. The results of blood metal ion analysis are shown in Table 2 and Figure 4. Patients with bilateral THA showed higher mean cobalt and chromium levels; however, this difference was not statistically significant. Forty-one patients (29%) had either cobalt or chromium ion levels > 2 µg/L, and 23 (17%) showed cobalt or chromium ion levels > 3 µg/L. Ninety-four patients (68%) demonstrated titanium ion levels > 2 µg/L and 26 (19%) had titanium ion levels > 4 µg/L. Four patients showed radiological evidence of femoral neck impingement without disassociation of the acetabular liner as a possible source for increased metal wear, which was visible as a little notch at the femoral neck on the lateral radiograph. Three of the four patients were asymptomatic with a mean HHS of 98 points. Mean cobalt, chromium, and titanium ion levels were 3.23 µg/L, 2.84 µg/L, and 8.69 µg/L, respectively. All other patients with increased metal ion levels showed no evidence of mechanical failure or component loosening on plain radiographs. No patient in the study cohort showed severe chronic kidney disease (GFR < 30 ml/min). Univariate analysis revealed moderate correlation between cobalt and chromium ion concentrations ($\varrho = 0.465$, $p < 0.001$), and fair correlation between chromium and titanium ion levels ($\varrho = 0.228$, $p = 0.015$) and between body mass index and cobalt ion levels ($\varrho = -0.224$, $p = 0.017$). However, in multivariate analysis, none of the tested variables was proven as a risk factor for elevated metal ion levels. Logistic regression analysis showed no association between the presence of periprosthetic osteolysis and blood metal ion levels of cobalt (odds ratio, 0.94; 95% CI, 0.50–1.77; $p = 0.941$), chromium (OR, 1.01; 95% CI, 0.85–1.21; $p = 0.905$), or titanium (OR, 0.88; 95% CI, 0.66–1.16; $p = 0.362$).

Table 1. Demographic data of the unilateral and bilateral group for metal ion analysis.

	Unilateral Group (n = 113) Mean (Range)	Bilateral Group (n = 26) Mean (Range)	*p*-Value
Age at follow-up (years)	67 (34–86)	64 (48–79)	0.117
Female gender (%)	35	50	0.167
Body mass index (kg/m^2)	26 (17–40)	27 (18–39)	0.887
Time of follow-up (years)	14.3 (10.2–18.8)	14.4 (11.9–17.7)	0.960
GFR (ml/min)	72 (31–116)	81 (54–106)	0.021*
Cup Inclination (°)	43 (25–62)	45 (27–62)	0.064
Harris Hip Score	91 (40–100)	90 (46–100)	0.765

* indicating statistically significant differences between the two groups

Table 2. Results of blood metal ion analysis.

		Unilateral Group (n = 113) µg/L	Bilateral Group (n = 26) µg/L	*p*-Value
Cobalt	Mean (SD)	0.82 (0.78)	2.59 (4.81)	0.082
	Median	0.55	1.30	
	Range	0.22–4.45	0.43–24.75	
Chromium	Mean (SD)	1.51 (2.47)	2.50 (3.22)	0.092
	Median	0.85	1.38	
	Range	0.04–22.69	0.26–16.75	
Titanium	Mean (SD)	2.68 (2.50)	3.76 (3.70)	0.079
	Median	2.01	2.71	
	Range	0.26–19.56	0.67–19.77	

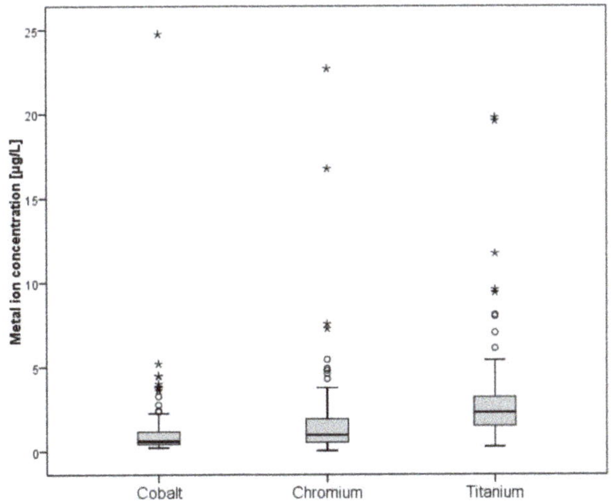

Figure 4. Box-and-whisker plots showing whole blood ion concentrations of cobalt, chromium, and titanium. The box marks the range between first and third quartile, with the band inside the box indicating the median and whiskers indicating minimum and maximum data respectively.

4. Discussion

Small-diameter, metal-on-metal implants are supposed to be at low risk of developing ARMD, and a systematic follow-up comparable to conventional THA is considered to be sufficient due to the good clinical mid- and long-term results reported in the literature [10]. Current guidelines recommend additional imaging using CT-scan or MARS-MRI to rule out potential ARMD in patients with blood cobalt ion levels > 2 μg/L [10]. However, little is known about the metal ion exposure in patients with small-head, MoM THA at long-term follow-up. The aim of this study was to report clinical and radiological results and to investigate blood metal ion levels in a large cohort of patients with small-diameter, MoM THA at long-term follow-up. The results of this study show that despite good clinical results, radiological findings of femoral osteolysis and ARMD were frequently seen in this cohort of patients with well-functioning small-head, metal-on-metal THA, and 29% of patients demonstrated elevated cobalt or chromium ion levels, i.e., > 2 μg/L, at long term follow-up.

To our knowledge, the present study represents the largest cohort of patients following small-head, MoM THA investigated with blood metal ion analysis at long term follow-up. Metal ion levels may vary significantly depending on the medium (e.g., whole blood, serum, or erythrocytes) and the technique (AAS vs. ICP-MS) used for analysis, which limits comparison among published studies [20]. Migaud et al. investigated whole blood metal ion concentrations in 26 patients following small-diameter, Metasul, metal-on-metal THA at a mean follow-up of 12 years. They reported median cobalt and chromium levels of 0.95 μg/L (range 0.4–4.8 μg/L) and 1.2 μg/L (range 0.1–5.6 μg/L), respectively [21]. Comparable results were reported by Ayoub et al., with mean cobalt ion levels of 1.85 μg/L (range 0.35–13.6 μg/L) and chromium ion levels of 1.32 μg/L (range 0.1–7.9 μg/L) at a mean follow-up of 15.9 years [22]. Our results of metal ion analysis at a mean follow-up period of 14 years are consistent with these findings, with mean metal ion levels being within the range of < 2 μg/L. However, approximately one-third of patients in our cohort demonstrated metal ion levels above 2 μg/L, and therefore, should undergo further imaging with ultrasound, CT-scan, and/or MARS-MRI in order to rule out ARMD, according to current guidelines [10]. In the study of Ayoub et al., only three patients demonstrated cobalt ion levels > 3 μg/L, and no ARMD was seen in this group of 42 female patients using ultrasound assessment. We presume that the higher proportion of patients with elevated cobalt and chromium ion concentrations seen in our study might be attributed to the larger patient cohort. The prevalence of ARMD in asymptomatic patients with small-diameter, MoM THA at long-term follow-up still is not clear, and larger cohort studies using CT or MRI should be performed to address that question. In accordance with our findings, a study by Hwang et al. investigated the prevalence of ARMD in patients following 28-mm Metasul MoM THA using computed tomography, and found ARMD to be present in 20% of the hips at a mean follow-up of 15 years [11].

For hip resurfacing and large-head, metal-on-metal THA, different risk factors for implant failure and elevated metal ion levels could be identified, such as high cup inclination angles or female sex [23,24]. Sidaginamale et al. [25] found a correlation between elevated ion levels and abnormal wear patterns in retrievals of resurfacing components. Langton et al. [24] analyzed 278 asymptomatic patients with hip resurfacing devices, and found elevated cobalt ion concentrations and female sex to be associated with early implant failure secondary to ARMD. Hart et al. [26] showed that increased blood metal ion concentrations were associated with implant failure in patients after hip resurfacing and large-diameter, metal-on-metal THA. In accordance with the findings of Lass et al. [27], we could not identify any risk factors associated with elevated blood metal ion levels in this cohort of patients with small-head, metal-on-metal implants. Impingement between the femoral neck and the Metasul liner is a known phenomenon, which can lead to increased metal wear or disassociation of the acetabular liner [28,29]. Four patients in our cohort showed radiological signs of impingement, with a visible notch at the femoral neck of the titanium stem on the lateral radiographs; titanium ion levels were increased in these patients. Therefore, titanium ion analysis can be beneficial to detect excessive wear due to impingement, in particular because it can be difficult to diagnose acetabular impingement on plain radiographs in some cases.

We found an acceptable clinical outcome for this bearing type according to the NICE recommendations [30], with a cumulative rate of implant survival of 96% with revision for any reason as the end point at 10 years. The mean patient age of 52 years at time of surgery was relatively young in this cohort. This was mainly attributed to the fact that the indication for THA in combination with a small-head, metal-on-metal bearing at that time was advanced osteoarthritis in young patients with a high activity level, which can be considered a potential selection bias when comparing our data to other reports on implant survival. Comparable long-term results for the Metasul bearing have been reported by Lass et al. [27] (survival rate of 87% at 18.8 years) and Hwang et al. [31] (survival rate of 97.8% at 18.4 years for acetabular cup revision for any reason). However, there is concern about the high rate of proximal femoral osteolysis and ARMD, as well as the high prevalence of elevated metal ion concentrations found in this cohort. We abandoned the use of metal-on-metal articulations in favor of alternative bearings such as ceramic on highly cross-linked polyethylene, as the local and systemic long-term effects associated with metal debris and metal ion release are still not fully understood [15].

There are some limitations to our study. Five percent of patients were lost to follow-up and a further 10% declined to participate in blood metal ion analysis. Also, no CT-scans were carried out in order to avoid additional radiation exposure. As a consequence, the rate of osteolysis could have been underestimated, especially around the acetabular components. Furthermore, MRI was only performed in 66 of the 189 hips (35%) that were available for clinical and radiological assessment, which could have resulted in a selection bias regarding the prevalence of ARMD. The fact that only 66 of the 107 advised patients with elevated metal ion levels agreed to participate in MRI assessment was mainly attributed to long travel distances and/or the absence of symptoms [13]. Further studies with larger patient cohorts using CT or MRI should be performed to investigate the prevalence of ARMD in asymptomatic patients with small-diameter, MoM THA at long-term follow-up. In addition, due to the cross-sectional study design, metal ion analysis was performed at a single time point, with a mean follow-up of 14.4 years after surgery; no sequential analysis was performed for each patient. However, longitudinal studies showed that blood metal ion levels in patients with well-functioning small-head, metal-on-metal bearings did not tend to increase over time [32–34].

5. Conclusions

The present study demonstrates good clinical results for cementless, 28 mm, MoM, THA at long-term follow-up, with a cumulative survival rate of 94% after 14 years. However, increased blood metal ion levels (cobalt or chromium > 2 μg/L) were found in approximately one-third of asymptomatic patients, and proximal femoral osteolysis and ARMD were frequently seen in this cohort. Blood metal ion analysis appears helpful in the long-term follow-up of these patients in order to identify individuals at risk. In accordance with contemporary consensus statements [10], symptomatic patients with elevated metal ion levels and/or progressive osteolysis should be considered for additional CT or MARS MRI to determine the extent of soft tissue affection prior to revision surgery. Further studies are necessary to investigate the clinical relevance of ARMD in asymptomatic patients with small-head, MoM THA.

Author Contributions: Conceptualization, T.R. and M.C.K.; Methodology, T.R. and M.C.K.; Formal Analysis, T.R.; Investigation, T.R., M.C.K., K.S., F.H.; Resources, T.R., M.C.K., J.P.K., T.G., C.M.; Writing—Original Draft Preparation, T.R.; Writing—Review and Editing, T.R., T.G., M.M.I., M.R.S., T.A.N., B.M., J.P.K., C.M.; Visualization, T.R.; Supervision, T.R., M.C.K., T.G.; Project Administration, T.R. and T.G.; Funding Acquisition, T.R. and T.G. All authors have read and agreed to the published version of the manuscript.

Funding: This work was funded by the non-commercial research fund of Stiftung Endoprothetik (Hamburg, Germany), grant number 57,000 €, and by the Baden-Württemberg Ministry of Science, Research and the Arts and by Ruprecht-Karls-Universität Heidelberg.

Acknowledgments: We like to thank Thomas Bruckner PhD, for his assistance with the statistical analysis and Stefan Rheinberger for his help with blood metal ion analysis.

Conflicts of Interest: The authors declare no conflict of interest. The funders had no role in the design of the study; in the collection, analyses, or interpretation of data; in the writing of the manuscript, or in the decision to publish the results.

References

1. Weber, B.G. Experience with the Metasul total hip bearing system. *Clin. Orthop. Relat. Res.* **1996**, *329*, S69–S77. [CrossRef] [PubMed]

2. Bozic, K.J.; Kurtz, S.; Lau, E.; Ong, K.; Chiu, V.; Vail, T.P.; Rubash, H.E.; Berry, D.J. The epidemiology of bearing surface usage in total hip arthroplasty in the United States. *J. Bone Joint Surg. Am.* **2009**, *91*, 1614–1620. [CrossRef] [PubMed]
3. Hart, A.J.; Sabah, S.; Henckel, J.; Lewis, A.; Cobb, J.; Sampson, B.; Mitchell, A.; Skinner, J.A. The painful metal-on-metal hip resurfacing. *J. Bone Joint Surg. Br.* **2009**, *91*, 738–744. [CrossRef] [PubMed]
4. Pandit, H.; Glyn-Jones, S.; McLardy-Smith, P.; Gundle, R.; Whitwell, D.; Gibbons, C.L.; Ostlere, S.; Athanasou, N.; Gill, H.S.; Murray, D.W. Pseudotumours associated with metal-on-metal hip resurfacings. *J. Bone Joint Surg. Br.* **2008**, *90*, 847–851. [CrossRef]
5. Smith, A.J.; Dieppe, P.; Vernon, K.; Porter, M.; Blom, A.W.; National Joint Registry of, England and Wales. Failure rates of stemmed metal-on-metal hip replacements: Analysis of data from the National Joint Registry of England and Wales. *Lancet* **2012**, *379*, 1199–1204. [CrossRef]
6. Campbell, P.; Ebramzadeh, E.; Nelson, S.; Takamura, K.; De Smet, K.; Amstutz, H.C. Histological features of pseudotumor-like tissues from metal-on-metal hips. *Clin. Orthop. Relat. Res.* **2010**, *468*, 2321–2327. [CrossRef]
7. Lombardi, A.V., Jr.; Barrack, R.L.; Berend, K.R.; Cuckler, J.M.; Jacobs, J.J.; Mont, M.A.; Schmalzried, T.P. The Hip Society: Algorithmic approach to diagnosis and management of metal-on-metal arthroplasty. *J. Bone Joint Surg. Br.* **2012**, *94*, 14–18. [CrossRef]
8. Medical and Healthcare Products Regulatory Agency (MHRA). *Medical Device Alert: All Metal-on-metal (MoM) Hip Replacements: Updated Advice for Follow-up of Patients*; MDA/2017/018; Medicines and Healthcare products Regulatory Agency: London, UK, 2017. Available online: http://www.mhra.gov.uk/ (accessed on 21 January 2020).
9. Medical Devices. Metal-on-metal Hip Implants. Information for Orthopaedic Surgeons. Available online: http://www.fda.gov/MedicalDevices/ProductsandMedicalProcedures/ImplantsandProsthetics/MetalonMetalHipImplants/ucm241667.htm (accessed on 21 January 2020).
10. Hannemann, F.; Hartmann, A.; Schmitt, J.; Lutzner, J.; Seidler, A.; Campbell, P.; Delaunay, C.P.; Drexler, H.; Ettema, H.B.; Garcia-Cimbrelo, E.; et al. European multidisciplinary consensus statement on the use and monitoring of metal-on-metal bearings for total hip replacement and hip resurfacing. *Orthop. Traumatol. Surg. Res.* **2013**, *99*, 263–271. [CrossRef]
11. Hwang, K.T.; Kim, Y.H.; Kim, Y.S.; Ryu, J.A. Prevalence of a soft-tissue lesion after small head metal-on-metal total hip replacement: 13- to 19-year follow-up study. *Bone Joint J.* **2014**, *96*, 1594–1599. [CrossRef]
12. Lombardi, A.V., Jr.; Berend, K.R.; Adams, J.B.; Satterwhite, K.L. Adverse Reactions to Metal on Metal Are Not Exclusive to Large Heads in Total Hip Arthroplasty. *Clin. Orthop. Relat. Res.* **2016**, *474*, 432–440. [CrossRef]
13. Reiner, T.; Do, T.D.; Klotz, M.C.; Hertzsch, F.; Seelmann, K.; Gaida, M.M.; Weber, M.A.; Gotterbarm, T. MRI Findings in Patients After Small-Head Metal-on-Metal Total Hip Arthroplasty with a Minimum Follow-up of 10 Years. *J. Bone Joint Surg. Am.* **2017**, *99*, 1540–1546. [CrossRef] [PubMed]
14. Lons, A.; Putman, S.; Pasquier, G.; Migaud, H.; Drumez, E.; Girard, J. Metallic ion release after knee prosthesis implantation: A prospective study. *Int. Orthop.* **2017**, *41*, 2503–2508. [CrossRef] [PubMed]
15. Innmann, M.M.; Gotterbarm, T.; Kretzer, J.P.; Merle, C.; Ewerbeck, V.; Weiss, S.; Aldinger, P.R.; Streit, M.R. Minimum ten-year results of a 28-mm metal-on-metal bearing in cementless total hip arthroplasty in patients fifty years of age and younger. *Int. Orthop.* **2014**, *38*, 929–934. [CrossRef] [PubMed]
16. Gruen, T.A.; McNeice, G.M.; Amstutz, H.C. "Modes of failure" of cemented stem-type femoral components: A radiographic analysis of loosening. *Clin. Orthop. Relat. Res.* **1979**, *141*, 17–27. [CrossRef]
17. DeLee, J.G.; Charnley, J. Radiological demarcation of cemented sockets in total hip replacement. *Clin. Orthop. Relat. Res.* **1976**, *121*, 20–32. [CrossRef]
18. Streit, M.R.; Schroder, K.; Korber, M.; Merle, C.; Gotterbarm, T.; Ewerbeck, V.; Aldinger, P.R. High survival in young patients using a second generation uncemented total hip replacement. *Int. Orthop.* **2012**, *36*, 1129–1136. [CrossRef] [PubMed]
19. Krachler M, H.C.; Kretzer, J.P. Validation of ultratrace analysis of Co, Cr, Ni and Mo in whole blood, serum and urine using ICP-SMS. *J. Anal. Spectrom.* **2009**, *24*, 605–610. [CrossRef]
20. Engh, C.A.; MacDonald, S.J.; Sritulanondha, S.; Korczak, A.; Naudie, D.; Engh, C. Metal ion levels after metal-on-metal total hip arthroplasty: A five-year, prospective randomized trial. *J. Bone Joint Surg. Am.* **2014**, *96*, 448–455. [CrossRef]

21. Migaud, H.; Putman, S.; Krantz, N.; Vasseur, L.; Girard, J. Cementless metal-on-metal versus ceramic-on-polyethylene hip arthroplasty in patients less than fifty years of age: a comparative study with twelve to fourteen-year follow-up. *J. Bone Joint Surg. Am.* **2011**, *93*, 137–142. [CrossRef]
22. Ayoub, B.; Putman, S.; Cholewinski, P.; Paris, A.; Migaud, H.; Girard, J. Incidence of Adverse Reactions to Metal Debris From 28-mm Metal-on-Metal Total Hip Arthroplasties With Minimum 10 Years of Follow-Up: Clinical, Laboratory, and Ultrasound Assessment of 44 Cases. *J. Arthroplasty.* **2017**, *32*, 1206–1210. [CrossRef]
23. De Haan, R.; Pattyn, C.; Gill, H.S.; Murray, D.W.; Campbell, P.A.; De Smet, K. Correlation between inclination of the acetabular component and metal ion levels in metal-on-metal hip resurfacing replacement. *J. Bone Joint Surg. Br.* **2008**, *90*, 1291–1297. [CrossRef] [PubMed]
24. Langton, D.J.; Joyce, T.J.; Jameson, S.S.; Lord, J.; Van Orsouw, M.; Holland, J.P.; Nargol, A.V.; De Smet, K.A. Adverse reaction to metal debris following hip resurfacing: The influence of component type, orientation and volumetric wear. *J. Bone Joint Surg. Br.* **2011**, *93*, 164–171. [CrossRef] [PubMed]
25. Sidaginamale, R.P.; Joyce, T.J.; Lord, J.K.; Jefferson, R.; Blain, P.G.; Nargol, A.V.; Langton, D.J. Blood metal ion testing is an effectivescreening tool to identify poorly performing metal-on-metal bearingsurfaces. *J. Bone Joint. Res.* **2013**, *2*, 84–95. [CrossRef] [PubMed]
26. Hart, A.J.; Sabah, S.A.; Sampson, B.; Skinner, J.A.; Powell, J.J.; Palla, L.; Pajamaki, K.J.; Puolakka, T.; Reito, A.; Eskelinen, A. Surveillance of Patients with Metal-on-Metal Hip Resurfacing and Total Hip Prostheses: A Prospective Cohort Study to Investigate the Relationship Between Blood Metal Ion Levels and Implant Failure. *J. Bone Joint Surg. Am.* **2014**, *96*, 1091–1099. [CrossRef] [PubMed]
27. Lass, R.; Grubl, A.; Kolb, A.; Domayer, S.; Csuk, C.; Kubista, B.; Giurea, A.; Windhager, R. Primary cementless total hip arthroplasty with second-generation metal-on-metal bearings: a concise follow-up, at a minimum of seventeen years, of a previous report. *J. Bone Joint Surg. Am.* **2014**, *96*, e37. [CrossRef] [PubMed]
28. Dastane, M.; Wan, Z.; Deshmane, P.; Long, W.T.; Dorr, L.D. Primary hip arthroplasty with 28-mm Metasul articulation. *J. Arthroplasty.* **2011**, *26*, 662–664. [CrossRef]
29. Malik, A.; Dorr, L.D.; Long, W.T. Impingement as a mechanism of dissociation of a metasul metal-on-metal liner. *J. Arthroplasty.* **2009**, *24*, 323.e13–3323.e16. [CrossRef]
30. National Institute for Health and Clinical Excellence (NICE). *Guidance on the Selection of Prostheses for Primary Total Hip Replacement*; National Institute for Health and Clinical Excellence (NICE): London, UK, 2000. Available online: https://www.nice.org.uk/guidance/ta304 (accessed on 21 January 2020).
31. Hwang, K.T.; Kim, Y.H.; Kim, Y.S.; Choi, I.Y. Is second generation metal-on-metal primary total hip arthroplasty with a 28 mm head a worthy option?: A 12- to 18-year follow-up study. *J. Arthroplasty.* **2013**, *28*, 1828–1833. [CrossRef]
32. Brodner, W.; Bitzan, P.; Meisinger, V.; Kaider, A.; Gottsauner-Wolf, F.; Kotz, R. Serum cobalt levels after metal-on-metal total hip arthroplasty. *J. Bone Joint Surg. Am.* **2003**, *85*, 2168–2173. [CrossRef]
33. Lazennec, J.Y.; Boyer, P.; Poupon, J.; Rousseau, M.A.; Roy, C.; Ravaud, P.; Catonne, Y. Outcome and serum ion determination up to 11 years after implantation of a cemented metal-on-metal hip prosthesis. *Acta Orthop.* **2009**, *80*, 168–173. [CrossRef]
34. Savarino, L.; Padovani, G.; Ferretti, M.; Greco, M.; Cenni, E.; Perrone, G.; Greco, F.; Baldini, N.; Giunti, A. Serum ion levels after ceramic-on-ceramic and metal-on-metal total hip arthroplasty: 8-year minimum follow-up. *J. Orthop. Res.* **2008**, *26*, 1569–1576. [CrossRef] [PubMed]

materials

MDPI

Article

Titanium Acetabular Component Deformation under Cyclic Loading

Nicholas A. Beckmann [1,2,*], Rudi G. Bitsch [3], Theresa Bormann [4], Steffen Braun [4] and Sebastian Jaeger [4]

1 Clinic for Orthopedics and Trauma Surgery, Heidelberg University Hospital, Heidelberg University, 69118 Heidelberg, Germany
2 Department of Orthopaedic Surgery and Traumatology, Inselspital, Bern University Hospital, 3010 Bern, Switzerland
3 National Joint Center, ATOS Clinics, 69115 Heidelberg, Germany; rudi.bitsch@atos.de
4 Laboratory of Biomechanics and Implant Research, Clinic for Orthopedics and Trauma Surgery, Heidelberg University Hospital, Heidelberg University, 69118 Heidelberg, Germany; Therese.Bormann@med.uni-heidelberg.de (T.B.); Steffen.Braun@med.uni-heidelberg.de (S.B.); Sebastian.Jaeger@med.uni-heidelberg.de (S.J.)
* Correspondence: nicholasandreas.beckmann@insel.ch or nicholas.beckmann@med.uni-heidelberg.de

Received: 19 November 2019; Accepted: 17 December 2019; Published: 20 December 2019

check for updates

Abstract: Acetabular cup deformation may affect liner/cup congruency, clearance and/or osseointegration. It is unclear, whether deformation of the acetabular components occurs during load and to what extent. To evaluate this, revision multi-hole cups were implanted into six cadaver hemipelvises in two scenarios: without acetabular defect (ND); with a large acetabular defect (LD) that was treated with an augment. In the LD scenario, the cup and augment were attached to the bone and each other with screws. Subsequently, the implanted hemipelvises were loaded under a physiologic partial-weight-bearing modality. The deformation of the acetabular components was determined using a best-fit algorithm. The statistical evaluation involved repeated-measures ANOVA. The mean elastic distension of the ND cup was 292.9 μm (SD 12.2 μm); in the LD scenario, 43.7 μm (SD 11.2 μm); the mean maximal augment distension was 79.6 μm (SD 21.6 μm). A significant difference between the maximal distension of the cups in both scenarios was noted ($F(1, 10) = 11.404$; $p = 0.007$). No significant difference was noted between the compression of the ND and LD cups, nor between LD cups and LD augments. The LD cup displayed significantly lower elastic distension than the ND cup, most likely due to increased stiffness from the affixed augment and screw fixation.

Keywords: total hip arthroplasty; implant deformation; acetabulum

---

1. Introduction

Total hip arthroplasty (THA) is considered to be one of the most successful operations performed in orthopedic surgery and the treatment of choice for end-stage osteoarthritis of the hip [1]. Consequently, the frequency of hip joint arthroplasty continues to increase worldwide. In 2007 Kurtz et al. estimated that the demand for primary total hip arthroplasty in the USA will increase by 174% to 572,000 by 2030 [2]. Concurrently they projected that the demand for revision THA would increase between 2005 and 2030 from 40,800 to 96,700 procedures representing an increase of 137% [2]. Generally, the longevity of revision THA is less than that of primary THA [3]. Lie et al. in 2004 analyzed 4762 revisions reported to the Norwegian Arthroplasty Register with a mean follow up of 3.2 years and found a 26% risk of failure after 10 years for cases without prior infection [3].

Cementless acetabular components have achieved widespread acceptance in THA as a result of their improved and reliable long-term results [4]. Primary stability is achieved through press-fit fixation that requires 1–3 mm under-reaming of the acetabular cavity and forceful impaction [5]. The forces utilized are substantial and generally result in some degree of deformation of the metal cup [6,7], particularly when the cup is thin-walled and of large diameter [8]. Meding et al. have shown that implant deformation resulting from the implantation process is non-uniform and results in diametrical pinching close to the implant rim, which is ascribed to impact pressure against the ischial and ileal columns of the acetabulum [8]. Prior studies have focused on assessing the pattern and degree of deformation resulting from the implantation process [7,9–14]. A few studies have assessed deformation after press-fit implantation and load application [15,16], however, these studies did not investigate cup deformation during load on the cup, as would be the case once the patient moves his/her hip.

The goal of our study was to evaluate and compare the elastic deformation of in-vitro titanium press-fit cups in two revision THA scenarios during subjection to cyclic gait loading for varying time periods. One sample group consisted of cadaver bone implanted with only a revision Gription cup (Pinnacle Multihole with Gription coating, Depuy/Synthes) (GC), whereas the second group was implanted with a revision Gription cup (GCS) and Gription augment (GAS) construct (Depuy/Synthes) both fixed to the pelvis with screws. We considered the diametrical change of the component rim during cyclic loading to relate to and reflect elastic deformation of the component that could potentially influence stable seating of the implant and the bone/implant apposition. This could contribute to micromotion on the implant-bone interface and between the cup and modular liner, respectively, which in turn might affect the osseointegration of the implant or influence the backside wear of the liner.

2. Materials and Methods

This study was approved by the local ethics committee (Ethikkommission Medizinische Fakultät Heidelberg, S309/2011).

Six fresh frozen cadaver hemipelvises were thawed at room temperature, dissected free of soft tissue. BMD measurements were done on attached femoral neck fragments prior to their removal from the hemipelvises. BMD was evaluated using Dual X-ray Absorptiometry (DXA) (QDR-2000 DXA densitometer; Hologic Inc, Waltham, MA, USA) and AP radiographs were also obtained on all specimens for pre-operative planning, and to exclude relevant pathology.

The average donor age was 78 years (range 51 to 96 years); the average donor mass was 66 kg (range 49 to 86 kg); and the average donor height was 171 cm (range 147.3 to 177.8 cm) resulting in a mean BMI of 23 kg/m^2 (range 19.0 to 28.9 kg/m^2). The mean bone mineral density, BMD, was 0.8 g/cm^2 (range 0.641 to 0.924 g/cm^2). Cadaver pelvises with a BMD below 0.6 g/cm^2 were excluded.

We compared two component scenarios that are frequently used in revision surgery.

2.1. Scenario 1: Revision Cup without Substantial Bony Defect (ND)

In the first scenario, a revision Gription cup (GC) (Pinnacle® Multihole with Gription® coating, DePuy Synthes Companies, Warsaw, IN, USA) was implanted press-fit in six hemipelvises without bone defects according to manufacturer recommendations as described in previous studies [17,18]. The acetabula were reamed in a concentric fashion in 2 mm increments removing all cartilage. The last reamer used was 1 mm smaller than the corresponding cup size, allowing a press-fit insertion of the cup. The cups were implanted at 45° of inclination and 15° of anteversion. Cup sizes of 50, 52, 54, 56, and 58 mm (58 mm cup was implanted in two cases) were used to correspond to the respective acetabular cavity. No additional screws were used as the press-fit allows instant stability of the cup in the bone.

2.2. Scenario 2: Revision Cup with Large Bony Defect Treated Using an Augment (LD)

In the second scenario, the same six hemipelvises received a Paprosky 2b defect of 10 mm depth, which was created in a standardized manner at the posterolateral aspect of the acetabulum. 30% of the circumference of the rim was involved and the edge of the defect bordered on the anterior inferior iliac spine. The defects were subsequently covered with a 10 mm Gription augment (GAS), which was fixed to the host bone with two 5.5 mm × 30 mm screws, the so-called "augment-first technique". A Gription cup (GCS, Gription Cup with Screws) was then fixed to the host bone using three 6.5 mm × 30 mm screws and to the augment with one 6.5 mm × 15 mm screw. Cup diameters referred to 50, 52, 54, 56, 58 and 58 mm. No cement was used. Marathon® cross-linked polyethylene liners (DePuy Synthes Companies, Warsaw, IN, USA) corresponding to the respective cup and a 28 mm diameter metal head were used in all cases. THA's were performed by a highly trained and experienced surgeon (RGB). Post-operative radiographs were obtained and confirmed the positioning of the implants and exclude fractures in all cases.

The implanted hemi-pelvises were fixated in a container using polyurethane foam (RenCast FC 53 A/B, Goessel + Pfaff GmbH, Karlskron/Brautlach, Germany) and integrated into a custom-made multi-axial testing machine (TD, testing device). This customized multi-axial testing machine enabled us to apply the changing loads and force vectors generated during a normal gait cycle as described in prior studies [17,18].

Loads applied were taken from the Bergmann et al. OrthoLoad data set [19]. Bergmann et al. [20] divided the gait cycle into phases and for each phase load components in the x, y and z axes were given according to a defined coordinate system. Using the load data (F_x, F_y, F_z) and their respective angles, the orientation of the resultant force vectors during normal gait could be replicated. Using this data in our testing machine, the magnitude and direction of the force vectors were controlled by the MTS regulator (MTS headquarters, 14000 Technology Drive, Eden Prairie, MN, USA).

Utilizing our TD, our specimens were subjected to our adjusted loads in a cyclic physiologic manner that mimicked the normal gait cycle with the difference that we limited the applied load to 30% of that experienced in the normal gait cycle. We chose 30% of the normal load as an estimate of the partial weight-bearing allowed in patients during the immediate postoperative phase of revision hip arthroplasty. The loads we applied varied from a maximum of 69.93% to a minimum of 8.71% of body weight compared to the respective values at 233.1% and 29.02%, which have been determined with full weight bearing in the normal gait cycle [20,21]. One thousand load cycles were carried out at 1 Hz.

Marker points (size 0.8 mm, GOM Gmbh, Braunschweig, Germany) trackable by an optical measuring system were placed around the circumference of the cup rim. They were also placed along the rim of the augment that constituted 30% of the entire cup circumference (see Figure 1).

Figure 1. Photograph of acetabular cup, augment and liner. Note the attached optical marker points along the rim of the cup and augment.

The measurements were taken at 0, 50, 100, 200, 400, 600, 800, and 1000 cycles using a frame rate of 15Hz. For each set of load cycles, we used the optical readings from the cup rim and augment marker points, respectively, to calculate best-fit circles by utilizing the Best Fit Algorithm that then enabled us to measure the maximal and minimal circle diameters at each set of load cycles (see Figure 2) [22,23].

We considered elastic deformation to be the change in diameter of these "best-fit" circles from the circle diameter measured initially after implantation without load, which was calculated from the positional changes of the rim markers. Mean maximum and minimum diameter changes (mean maximum and minimum deformation) were calculated for both GC (measurement 1), and for GCS and GAS (measurement 2) construct groups for each of 0–1000 cycle sets. The results for the three groups were then compared.

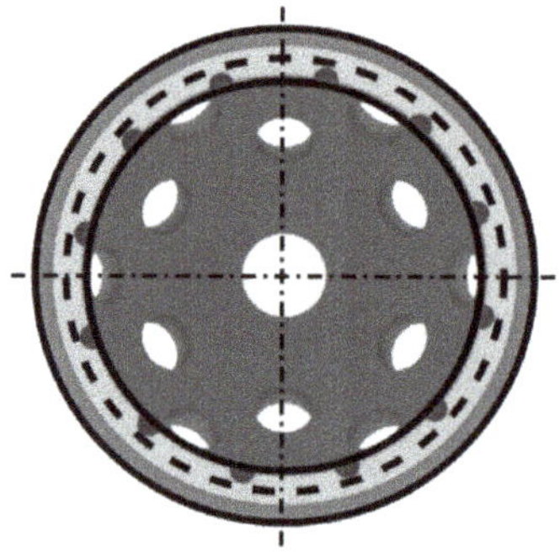

Figure 2. Representation of best-fit circle, denoted by the dotted lines around the rim of the acetabular cup, which was calculated by the relative motion of the optical markers.

3. Statistics

The data were evaluated descriptively using the arithmetic mean, standard deviation, minimum and maximum. A repeated-measures analysis of variance (ANOVA) was performed to test for significant differences for the parameter of deformation (primary cup vs. cup with screws and cup with screws vs. augment). Prior to data analysis, the normal distribution of the data was evaluated using a Shapiro–Wilk test, which was chosen over other statistical tests of normal distribution since a prior study has shown that this test provides more power given a known significance than other tests of normal distribution [24]. Subsequently, the homogeneity of variance was verified using the Levene test, which is a prerequisite for ANOVA. The results allowed for the use of the ANOVA test. The Greenhouse–Geisser adjustment was used to correct for violations of sphericity. The data were analyzed using SPSS 25 (IBM, Armonk, New York, NY, USA).

4. Results

The results are displayed below in Table 1 and Figure 3.

Table 1. Mean elastic deformation (μm) for scenario 1 (GC) and scenario 2 (GCS and GAS) during all cycle sets.

Deformation (μm)	Scenario 1 without Defect	Scenario 2 with Paprosky 2b Defect	
	Cup (GC)	Cup (GCS)	Augment (GAS)
Mean Compression ± SD	−182.4 ± 15.0	−79.9 ± 9.2	−73.2 ± 4.3
Mean Distension ± SD	292.9 ± 12.2	43.7 ± 11.2	79.6 ± 21.6
Overall Deformation	475.3	123.6	152.8

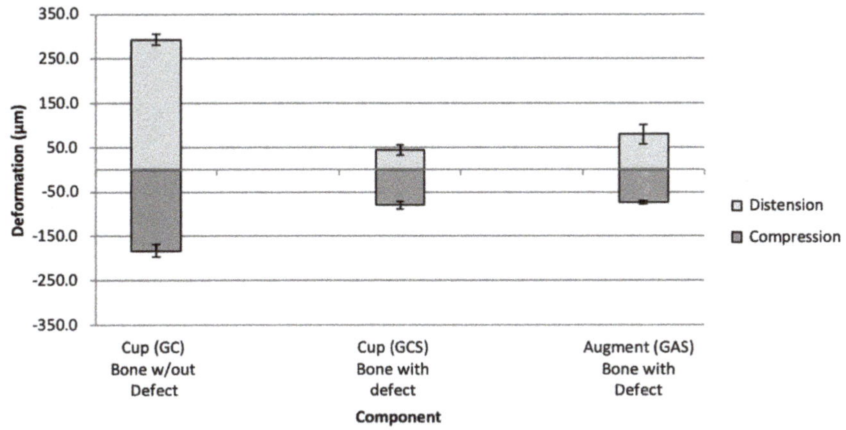

Figure 3. Bar graph of mean elastic deformation (distension and distension) of GC (no bony defect; scenario 1) and GCS and GAS (with bony defect; scenario 2).

Comparison of the compression and distension deformation between primary and revision cup:

The cyclical loading showed no statistically significant effect under the standardized loading conditions with regards to the compression deformation, $F(3.027, 30.266) = 0.484$, $p = 0.698$. There was no statistically significant difference for the compression cup deformation between the primary and revision groups, $F(1, 10) = 1.740$, $p = 0.217$, yet a tendency for higher elastic deformation of the ND cup (GC) can be found.

The duration of load (i.e., the later load cycles) had no significant influence on the distension deformation, $F(1.718, 17.176) = 0.368$, $p = 0.666$. The repeated measures ANOVA with Greenhouse-Geisser correction determined that the difference in distension deformation of the cups in scenario 1 (No Defect) and scenario 2 (Large Defect) was statistically significant, $F(1, 10) = 11.404$, $p = 0.007$.

Comparison of the compression and distension deformation between the revision cup and augment:

Under the standardized loading conditions the cyclical loading showed no statistically significant effect with regards to the compression deformation, $F(1.889, 18.892) = 1.048$, $p = 0.367$. There was no statistically significant difference for the compression deformation between the revision cup and augment, $F(1, 10) = 0.015$, $p = 0.904$.

The duration of load (the later load cycles) had no significant influence on the distension deformation, $F(1.515, 15.152) = 1.104$, $p = 0.340$. The repeated measures ANOVA with Greenhouse-Geisser correction determined that there was no statistically significant difference between LD cup (GCS) and augment (GAS) with regards to the distension deformation of each, $F(1, 10) = 0.389$, $p = 0.547$.

The results are displayed in Table 1 and Figure 3.

5. Discussion

A large body of existing research has played an important role in the current success of both primary and revision THA, and resulted in a steadily increasing number of procedures and a decreasing age of patients [25,26]. Loosening and dislocation have been identified as the major causes of implant failure [26–28] and have been the focus of most research. Micro-motion at the bone/implant interface with subsequent particle production, tissue reaction, and osteolysis has been well documented [17,29–32]. Primary stability of the implant is recognized as critically important for ultimate surgical success [33–35]. Early osseointegration of all porous metal implants requires only minimal relative motion between implant and host bone. Micro-motion or bone/implant gap size of up to 50 μm has been shown to result in successful osseointegration, but above 150 μm there is attachment by fibrous tissue [33,36–38].

In contrast to studies of micromotion and other contributing factors leading to implant failure, implant deformation has received relatively little research attention. Existing studies have focused almost exclusively on deformation changes that occur during the process of implantation, and have disregarded deformation occurring afterward. The deformation of press-fit acetabular cups into 1–3 mm under-reamed sockets has been documented in several prior studies [8,37]. Prior researchers noted that the acetabular bone is most dense at the anterosuperior and posteroinferior margins (ileal and ischial columns) and the pubic area, constituting 3-point support [28]. The ileal and ischial columns are the most unyielding foci of the acetabular rim during the implantation process [12,28], and post-implantation provide the most bone/implant contact, foci of greatest load transfer and ultimately the most support for the implant [6,28]. Studies have shown that contact at the pelvic rim/implant area constitutes 25–50% of the total host bone/ implant apposition [9,36]. The host bone implant apposition decreases from rim to pole [39]. It has been found that even under ideal circumstances, total bone/implant apposition is never achieved and is unnecessary for successful osseointegration [9]. Cadaver studies of previously well-functioning components have shown that bone/implant contact areas have varied widely in extent [36]. The pelvic rim is also the area of greatest implant support during weight-bearing/gait loading [28].

During forceful press-fit implantation (previously measured at 400 N with porous titanium acetabular cups in-vivo) [8,37], the ileal and ischial columns exert a pinching effect on the cup component [7,9,12,14,35,37]. This causes the cup to assume a hemi-elliptical rather than a hemispherical shape that can lead to incongruity, diminished apposition and increased gap areas at the bone/implant interface [9,11,35]. These gaps, if excessive, can be associated with a number of negative consequences, including improper depth and angle of implant seating with subsequent potential dislocation, increased micro-motion at the bone/implant interface [35], facilitation of particle accumulation and increased tissue fluid, impaired liner insertion secondary to distortion of the cup locking mechanism and diminished clearance that adversely affects joint lubrication and liner wear [8,9,11,35,37,40]. Other factors that have been shown to influence the degree of cup deformation are the reaming process prior to implantation and the geometry of the acetabular cavity, characteristics of the implanted cup, bone density/hardness, the force applied during implantation and the seating of the cup [8,9,37,41,42]. Prior studies have shown that manual reaming results in a cavity that is usually slightly larger than the last reaming instrument and is hemi-elliptical in shape as a result of the varying degrees of stiffness throughout the host acetabular cavity [7,9,14,42]. Lin et al. found large errors in hand-reamed cavities [34]. Therefore, careful reaming of the acetabular cavity and accurate cup seating have been identified as significant modifiable factors in reducing the degree of implant deformation [9] that occurs during the implantation process, and we hypothesize that they may also influence the degree of elastic deformation that is the focus of our study. Characteristics of the implant that have been found to influence the potential to deform are the type of metal used, and diameter and thickness of the cup. Meding et al. found that titanium cups deform more than those of CoCr, and increasing the diameter and thinness of the wall are associated with increased potential to deform [8,35].

The focus and methodology of our current study differ in several respects from all prior studies of deformation. Our study does not evaluate deformation occurring from the process of implantation, but looks at elastic deformation occurring as a result of cyclic loading applied several days post-implantation. We utilized cadaver bone rather than the synthetic bone to permit the most realistic testing scenario, and measurements were made several days after implantation to minimize the elliptical distortion of the implantation process since it has been demonstrated that for several days post-implantation there is a visco-elastic relaxation of the pelvic bone and implant that reduces the deformation of titanium components [7]. We focused on the rim since it has been shown to be the site of the most bone/implant apposition and the region of maximal loading with gait [6,28]. Since our methodology utilized the derivation of a "best-fit circle" based on an assumption of symmetry of the acetabular implant, our results are not direct measurements of implant deformation or of gap size and can only be used for comparison of our two study groups.

Our results showed that all our study specimens displayed some elastic deformation during cyclic loading, but there was a distinct difference between the two groups. We found statistically significantly more elastic deformation in Scenario 1 (revision cup only, lack of bony defect) than in Scenario 2 (revision cup plus augment and screws implanted into a Paprosky 2b defect). Values for the revision cup with screws and values for the augment with screws were not significantly different from each other, indicating that cup plus augment and screws tend to function as a unit.

For all specimens, maximal elastic deformation (compression and distension) was reached after approximately 50 cycles of loading with no significant additional deformation noted with increased cycles.

We hypothesize that the addition of screws and screws plus augment effectively increases the rigidity of the construct, which is consistent with findings of prior studies on deformation that occurred secondary to implantation, which showed that rigid CoCr cups deformed less than titanium cups [8]. Additionally, the compression of the cup and augment in scenario 2 proved more substantial than the distension.

Our results are clinically significant in several ways. The use of ancillary screws decreased the degree of elastic deformation of the implant rim, allowing better bone/implant apposition, reduced gap areas, and potentially improved osseointegration. The presence of an augment did not negatively impact the degree of elastic deformation, and the cup/augment construct effectively functioned as a unit. Our findings also indicate that cyclic loading that mimics normal gait is associated with increased elastic deformation and validates prior findings [9,35]. We propose that the elastic deformation we identified, and the deformation of implantation may both be potential influencing factors in the occurrence of backside wear of modular liners. Furthermore, acetabular component deformation during load may affect clearance at the articulating interface and may play a role in the rare, albeit relevant liner dislocation. One case series reported on 23 liner dissociations after Pinnacle implantation [43], which may be a result of overly elastic cups.

Cup diameter and wall thickness have repeatedly been shown to affect cup deformation. Large (jumbo) porous cups are an alternative option to cup/augment combination in revision THA associated with deficient bone stock and acetabular defects, and their large diameter and thin walls predispose to increased deformation, although ancillary screw fixation may limit this deformation. Oversized porous cups have been reported to have an increased risk of dislocation of multifactorial cause [6,8,44]. Deformation of the cups has not been considered a contributing factor but may be worthy of further consideration.

6. Limitations

Our study has several limitations. The cohorts were small, and only two revision constructs were assessed, the first without substantial osseous defect, and the second using a substantial (Paprosky 2b) bony defect.

The optical markers covered only half the implant rim, since the remaining rim was concealed from the optical cameras by the moving prosthesis neck—we assumed that the cup rim was symmetrical. Our methodology utilized the optical marker data to derive a "best-fit circle" and the change of circle diameter was considered to reflect elastic deformation. Therefore, our results are not a direct measure of deformation or bone/implant gap regions, but serve only as an indirect indicator, and can only be used as a means of comparison between our two study groups. This is the first study to utilize this methodology and evaluate deformation during loading, and there are therefore no currently existing comparable data for verification.

We utilized cadaver bone since the polyurethane models were validated only for deformation occurring during implantation [8] and we considered cadaver bone to be more physiologic. However, it lacks some of the viscoelastic properties of live bone, and may not reflect the clinical scenario [12,36].

There is tentative evidence that hardness of bone, related to BMD, necessitates increased compressive force for implantation that may affect results. Our cadaver sample was from older patients with a mean BMD of 0.8g/cm^2 (range 0.641 to 0.924 g/cm^2) and values beyond this range may have different results.

In our experimental set-up, only Paprosky 2b defects were examined, and defects of other grades may have different results.

Multi-hole cups have been shown to present more deformation than single-hole cups during implantation [16], which allows one to assume that a similar effect may be seen under dynamic loading after implantation. For the purpose of comparability, a multi-hole cup was used in both scenarios, although this type of cup would tend not to be used clinically in a scenario without a bony defect.

Variations in surgical technique cannot be excluded although the same highly experienced surgeon (RGB) performed all implantations with supplied manufacturers tools.

In conclusion, our in-vitro study utilizing revision constructs in cadaver bone is the first to compare the elastic deformation occurring at the rim of two different implant constructs during cyclic loading that replicates the limited loading of normal gait as experienced in the early postoperative period. Our results show that the use of adjunctive screws significantly decreases the degree of elastic deformation under these conditions, and the inclusion of an augment does not adversely impact the degree of elastic deformation.

Author Contributions: Conceptualization, N.A.B., R.G.B. and S.J.; Data curation, N.A.B., R.G.B., S.B. and S.J.; Formal analysis, N.A.B., T.B. and S.J.; Funding acquisition, S.J.; Investigation, N.A.B., S.B. and S.J.; Methodology, N.A.B., R.G.B. and S.J.; Project administration, N.A.B. and S.J.; Supervision, S.J.; Validation, N.A.B. and S.J.; Writing—original draft, N.A.B.; Writing—review & editing, N.A.B., R.G.B., T.B., S.B. and S.J. All authors have read and agreed to the published version of the manuscript.

Funding: Funding for this project was provided by the University of Heidelberg, and the Laboratory of Biomechanics and Implant Research.

Conflicts of Interest: The authors declare no pertinent conflict of interest. N.A.B. reports research grants from Johnson & Johnson DePuy Synthes that are not related to the current study. R.G.B. and S.J. report grants from B Braun Aesculap, Johnson & Johnson Depuy Synthes, Heraeus Medical, Waldemar Link, and Zimmer Biomet that are not related to the current study. In addition, S.J. reports grants from Peter Brehm GmbH, which is also unrelated to the current project. Both T.B. and S.B. report no conflict of interest.

References

1. Learmonth, I.D.; Young, C.; Rorabeck, C. The operation of the century: Total hip replacement. *Lancet* **2007**, *370*, 1508–1519. [CrossRef]
2. Kurtz, S.; Ong, K.; Lau, E.; Mowat, F.; Halpern, M. Projections of primary and revision hip and knee arthroplasty in the United States from 2005 to 2030. *J. Bone Joint Surg. Am.* **2007**, *89*, 780–785. [CrossRef]
3. Lie, S.A.; Havelin, L.I.; Furnes, O.N.; Engesaeter, L.B.; Vollset, S.E. Failure rates for 4762 revision total hip arthroplasties in the Norwegian Arthroplasty Register. *J. Bone Joint Surg. Br.* **2004**, *86*, 504–509. [CrossRef]
4. Ng, F.Y.; Zhu, Y.; Chiu, K.Y. Cementless acetabular component inserted without screws—The effect of immediate weight-bearing. *Int. Orthop.* **2007**, *31*, 293–296. [CrossRef]
5. Kroeber, M.; Ries, M.D.; Suzuki, Y.; Renowitzky, G.; Ashford, F.; Lotz, J. Impact biomechanics and pelvic deformation during insertion of press-fit acetabular cups. *J. Arthroplast.* **2002**, *17*, 349–354. [CrossRef]
6. Bone, M.C.; Dold, P.; Flohr, M.; Preuss, R.; Joyce, T.J.; Deehan, D.; Holland, J. A novel method for measuring acetabular cup deformation in cadavers. *Proc. Inst. Mech. Eng. Part H* **2013**, *227*, 1341–1344. [CrossRef]
7. Springer, B.D.; Habet, N.A.; Griffin, W.L.; Nanson, C.J.; Davies, M.A. Deformation of 1-piece metal acetabular components. *J. Arthroplast.* **2012**, *27*, 48–54. [CrossRef]
8. Meding, J.B.; Small, S.R.; Jones, M.E.; Berend, M.E.; Ritter, M.A. Acetabular cup design influences deformational response in total hip arthroplasty. *Clin. Orthop. Relat. Res.* **2013**, *471*, 403–409. [CrossRef]
9. Ong, K.L.; Lehman, J.; Notz, W.I.; Santner, T.J.; Bartel, D.L. Acetabular cup geometry and bone-implant interference have more influence on initial periprosthetic joint space than joint loading and surgical cup insertion. *J. Biomech. Eng.* **2006**, *128*, 169–175. [CrossRef]

10. Bone, M.C.; Dold, P.; Flohr, M.; Preuss, R.; Joyce, T.J.; Aspden, R.M.; Holland, J.; Deehan, D. The influence of the strength of bone on the deformation of acetabular shells: A laboratory experiment in cadavers. *Bone Joint J.* **2015**, *97-B*, 473–477. [CrossRef]
11. Panjeton, G.D.; Kim, S.E.; Chang, K.; Palm, L.S.; Ifju, P.G. Deformation of the Zurich cementless acetabular cup caused by implantation in a canine cadaver model. *Vet. Surg.* **2017**, *46*, 441–447. [CrossRef] [PubMed]
12. Markel, D.; Day, J.; Siskey, R.; Liepins, I.; Kurtz, S.; Ong, K. Deformation of metal-backed acetabular components and the impact of liner thickness in a cadaveric model. *Int. Orthop.* **2011**, *35*, 1131–1137. [CrossRef] [PubMed]
13. Liu, F.; Chen, Z.; Gu, Y.; Wang, Q.; Cui, W.; Fan, W. Deformation of the Durom acetabular component and its impact on tribology in a cadaveric model—A simulator study. *PLoS ONE* **2012**, *7*, e45786. [CrossRef] [PubMed]
14. Fritsche, A.; Bialek, K.; Mittelmeier, W.; Simnacher, M.; Fethke, K.; Wree, A.; Bader, R. Experimental investigations of the insertion and deformation behavior of press-fit and threaded acetabular cups for total hip replacement. *J. Orthop. Sci.* **2008**, *13*, 240–247. [CrossRef]
15. Ong, K.L.; Rundell, S.; Liepins, I.; Laurent, R.; Markel, D.; Kurtz, S.M. Biomechanical modeling of acetabular component polyethylene stresses, fracture risk, and wear rate following press-fit implantation. *J. Orthop. Res.* **2009**, *27*, 1467–1472. [CrossRef]
16. Messer-Hannemann, P.; Campbell, G.M.; Morlock, M.M. Deformation of acetabular press-fit cups: Influence of design and surgical factors. *Clin. Biomech. (Bristol, Avon)* **2019**, *69*, 96–103. [CrossRef]
17. Beckmann, N.A.; Bitsch, R.G.; Janoszka, M.B.; Klotz, M.C.; Bruckner, T.; Jaeger, S. Treatment of High-Grade Acetabular Defects: Do Porous Titanium Cups Provide Better Stability Than Traditional Titanium Cups When Combined With an Augment? *J. Arthroplast.* **2018**, *33*, 1838–1843. [CrossRef]
18. Beckmann, N.A.; Jaeger, S.; Janoszka, M.B.; Klotz, M.C.; Bruckner, T.; Bitsch, R.G. Comparison of the Primary Stability of a Porous Coated Acetabular Revision Cup With a Standard Cup. *J. Arthroplast.* **2018**, *33*, 580–585. [CrossRef]
19. OrthoLoad. Available online: http://www.OrthoLoad.com (accessed on 15 January 2016).
20. Bergmann, G.; Graichen, F.; Rohlmann, A. Hip joint loading during walking and running, measured in two patients. *J. Biomech.* **1993**, *26*, 969–990. [CrossRef]
21. Bergmann, G.; Graichen, F.; Rohlmann, A.; Bender, A.; Heinlein, B.; Duda, G.N.; Heller, M.O.; Morlock, M.M. Realistic loads for testing hip implants. *Biomed. Mater. Eng.* **2010**, *20*, 65–75. [CrossRef]
22. Weißmann, V.; Boss, C.; Schulze, C.; Hansmann, H.; Bader, R. Experimental Characterization of the Primary Stability of Acetabular Press-Fit Cups with Open-Porous Load-Bearing Structures on the Surface Layer. *Metals* **2018**, *8*, 839. [CrossRef]
23. Mueller, U.; Lee, C.; Heisel, C.; Thomsen, M.; Bitsch, R.G.; Kretzer, J.P. Failure of Polyethylene Inlays in Cementless Total Hip Arthroplasty: A Retrieval Analysis. *BioMed. Res. Int.* **2016**, *2016*, 5496396. [CrossRef] [PubMed]
24. Razali, N.M.; Yap, B. Power Comparisons of Shapiro-Wilk, Kolmogorov-Smirnov, Lilliefors and Anderson-Darling Tests. *J. Stat. Model. Anal.* **2011**, *2*, 21–33.
25. Ulrich, S.D.; Seyler, T.M.; Bennett, D.; Delanois, R.E.; Saleh, K.J.; Thongtrangan, I.; Kuskowski, M.; Cheng, E.Y.; Sharkey, P.F.; Parvizi, J.; et al. Total hip arthroplasties: What are the reasons for revision? *Int. Orthop.* **2008**, *32*, 597–604. [CrossRef] [PubMed]
26. Bozic, K.J.; Kurtz, S.M.; Lau, E.; Ong, K.; Vail, T.P.; Berry, D.J. The epidemiology of revision total hip arthroplasty in the United States. *J. Bone Joint Surg. Am.* **2009**, *91*, 128–133. [CrossRef] [PubMed]
27. Widmer, K.H.; Zurfluh, B.; Morscher, E.W. Load transfer and fixation mode of press-fit acetabular sockets. *J. Arthroplast.* **2002**, *17*, 926–935. [CrossRef] [PubMed]
28. Schmalzried, T.P.; Jasty, M.; Harris, W.H. Periprosthetic bone loss in total hip arthroplasty. Polyethylene wear debris and the concept of the effective joint space. *J. Bone Joint Surg. Am.* **1992**, *74*, 849–863. [CrossRef]
29. Beckmann, N.A.; Bitsch, R.G.; Gondan, M.; Schonhoff, M.; Jaeger, S. Comparison of the stability of three fixation techniques between porous metal acetabular components and augments. *Bone Joint Res.* **2018**, *7*, 282–288. [CrossRef]
30. Wirtz, D.C.; Niethard, F.U. Etiology, diagnosis and therapy of aseptic hip prosthesis loosening—A status assessment. *Z. Orthop. Ihre Grenzgeb.* **1997**, *135*, 270–280. [CrossRef]

31. Aspenberg, P.; Herbertsson, P. Periprosthetic bone resorption. Particles versus movement. *J. Bone Joint Surg. Br.* **1996**, *78*, 641–646. [CrossRef]
32. Pilliar, R.M. Porous-surfaced metallic implants for orthopedic applications. *J. Biomed. Mater. Res.* **1987**, *21* (Suppl. A1), 1–33. [PubMed]
33. Wetzel, R.; Simnacher, M.; Scheller, G. Initial stability of press-fit acetabular cups—An in-vitro study. *Biomed. Technol. (Berl)* **2005**, *50*, 400–403. [CrossRef] [PubMed]
34. Lin, Z.M.; Meakins, S.; Morlock, M.M.; Parsons, P.; Hardaker, C.; Flett, M.; Isaac, G. Deformation of press-fitted metallic resurfacing cups. Part 1: Experimental simulation. *Proc. Inst. Mech. Eng. Part H* **2006**, *220*, 299–309.
35. Squire, M.; Griffin, W.L.; Mason, J.B.; Peindl, R.D.; Odum, S. Acetabular component deformation with press-fit fixation. *J. Arthroplast.* **2006**, *21* (6 Suppl. 2), 72–77. [CrossRef] [PubMed]
36. Schwartz, J.T., Jr.; Engh, C.A.; Forte, M.R.; Kukita, Y.; Grandia, S.K. Evaluation of initial surface apposition in porous-coated acetabular components. *Clin. Orthop. Relat. Res.* **1993**, *293*, 174–187. [CrossRef]
37. Engh, C.A.; O'Connor, D.; Jasty, M.; McGovern, T.F.; Bobyn, J.D.; Harris, W.H. Quantification of implant micromotion, strain shielding, and bone resorption with porous-coated anatomic medullary locking femoral prostheses. *Clin. Orthop. Relat. Res.* **1992**, *285*, 13–29. [CrossRef]
38. Pilliar, R.M.; Lee, J.M.; Maniatopoulos, C. Observations on the effect of movement on bone ingrowth into porous-surfaced implants. *Clin. Orthop. Relat. Res.* **1986**, *208*, 108–113. [CrossRef]
39. MacKenzie, J.R.; Callaghan, J.J.; Pedersen, D.R.; Brown, T.D. Areas of contact and extent of gaps with implantation of oversized acetabular components in total hip arthroplasty. *Clin. Orthop. Relat. Res.* **1994**, *298*, 127–136. [CrossRef]
40. Lachiewicz, P.F.; Watters, T.S. The jumbo acetabular component for acetabular revision: Curtain Calls and Caveats. *Bone Joint J.* **2016**, *98* (1 Suppl. A), 64–67. [CrossRef]
41. Spears, I.R.; Morlock, M.M.; Pfleiderer, M.; Schneider, E.; Hille, E. The influence of friction and interference on the seating of a hemispherical press-fit cup: A finite element investigation. *J. Biomech.* **1999**, *32*, 1183–1189. [CrossRef]
42. Yew, A.; Jin, Z.M.; Donn, A.; Morlock, M.M.; Isaac, G. Deformation of press-fitted metallic resurfacing cups. Part 2: Finite element simulation. *Proc. Inst. Mech. Eng. Part H* **2006**, *220*, 311–319. [CrossRef] [PubMed]
43. Yun, A.; Koli, E.N.; Moreland, J.; Iorio, R.; Tilzey, J.F.; Mesko, J.W.; Lee, G.C.; Froimson, M. Polyethylene Liner Dissociation Is a Complication of the DePuy Pinnacle Cup: A Report of 23 Cases. *Clin. Orthop. Relat. Res.* **2016**, *474*, 441–446. [CrossRef] [PubMed]
44. Dold, P.; Bone, M.C.; Flohr, M.; Preuss, R.; Joyce, T.J.; Deehan, D.; Holland, J. Validation of an optical system to measure acetabular shell deformation in cadavers. *Proc. Inst. Mech. Eng. Part H* **2014**, *228*, 781–786. [CrossRef] [PubMed]

materials

MDPI

Article

Calcium Phosphate Bone Graft Substitutes with High Mechanical Load Capacity and High Degree of Interconnecting Porosity

Georg Hettich [1,*], Ronja A. Schierjott [1,2], Matthias Epple [3], Uwe Gbureck [4], Sascha Heinemann [5], Hadi Mozaffari-Jovein [6] and Thomas M. Grupp [1,2]

1 Aesculap AG, Research & Development, Am Aesculap-Platz, 78532 Tuttlingen, Germany; ronja_alissa.schierjott@aesculap.de (R.A.S.); thomas.grupp@aesculap.de (T.M.G.)
2 Department of Orthopaedics, Physical Medicine and Rehabilitation, Department of the Ludwig-Maximilians-Universität München Marchioninistrasse 15, 81377 Munich, Germany
3 Inorganic Chemistry and Center for Nanointegration Duisburg-Essen (CeNIDE), University of Duisburg-Essen, Universitaetsstr. 5-7, 45117 Essen, Germany; matthias.epple@uni-due.de
4 Department for Functional Materials in Medicine and Dentistry, University of Wuerzburg, Pleicherwall 2, 97070 Wuerzburg, Germany; uwe.gbureck@fmz.uni-wuerzburg.de
5 INNOTERE biomaterial, Meissner Str. 191, 01445 Radebeul, Germany; sascha.heinemann@innotere.de
6 Institute of Materials Science and Engineering Tuttlingen (IWAT), Furtwangen University, Kronenstraße 16, 78532 Tuttlingen, Germany; Hadi.MozaffariJovein@hs-furtwangen.de
* Correspondence: Georg.hettich@aesculap.de; Tel.: +49-7461-9531708

Received: 23 September 2019; Accepted: 18 October 2019; Published: 23 October 2019

check for updates

Abstract: Bone graft substitutes in orthopedic applications have to fulfill various demanding requirements. Most calcium phosphate (CaP) bone graft substitutes are highly porous to achieve bone regeneration, but typically lack mechanical stability. This study presents a novel approach, in which a scaffold structure with appropriate properties for bone regeneration emerges from the space between specifically shaped granules. The granule types were tetrapods (TEPO) and pyramids (PYRA), which were compared to porous CaP granules (CALC) and morselized bone chips (BC). Bulk materials of the granules were mechanically loaded with a peak pressure of 4 MP; i.e., comparable to the load occurring behind an acetabular cup. Mechanical loading reduced the volume of CALC and BC considerably (89% and 85%, respectively), indicating a collapse of the macroporous structure. Volumes of TEPO and PYRA remained almost constant (94% and 98%, respectively). After loading, the porosity was highest for BC (46%), lowest for CALC (25%) and comparable for TEPO and PYRA (37%). The pore spaces of TEPO and PYRA were highly interconnected in a way that a virtual object with a diameter of 150 μm could access 34% of the TEPO volume and 36% of the PYRA volume. This study shows that a bulk of dense CaP granules in form of tetrapods and pyramids can create a scaffold structure with load capacities suitable for the regeneration of an acetabular bone defect.

Keywords: calcium phosphate; granules; biomechanics; bone graft substitutes

1. Introduction

Materials for bone defect treatment in orthopedic applications have to fulfill various demanding requirements. In most cases the materials are subjected to a considerable mechanical load. Therefore materials with high mechanical load capacities are required, such as metallic biomaterials, including titanium or tantalum in the form of porous implants [1–3] or high strength acrylic bone cement polymethylmethacrylate (PMMA) [4,5]. Since these materials are not degradable in vivo, only a replacement of the anatomical function is achieved rather than a regeneration of the bone or the joint. Furthermore, due to a stress-shielding effect related to the high stiffness of the aforementioned materials, additional bone in the periphery of these materials may be lost [6,7]. When the implant has to be revised, the size of the bone defect further increases and the fixation of the revision implant subsequently becomes more difficult [8,9]. Decreasing the extent of the bone defect is crucial to obtaining better implant survival rates in revision joint replacement [10,11].

Common techniques to achieve the regeneration of bone defects involve the transplantation of autografts or allografts, as well as the implantation of synthetic bone graft substitutes predominantly based on calcium phosphate (CaP) [12,13]. CaP bone graft substitutes include grafts obtained by the (hydro)thermal processing of native bone [14] and fully synthetic CaP, such as hydroxyapatite (HA) or tricalcium phosphates (TCP) [15]. The bone graft substitutes are usually applied in form of putties [16], granules [17], blocks [18], or self-setting calcium phosphate cements [19]. The material design is predominantly focused on fast tissue ingrowth and bone regeneration rather than on mechanical aspects, since the application site is either mechanically unloaded or additional mechanical support is provided; for example, by plate osteosynthesis techniques [20,21]. Hence, such materials are usually highly porous with an intrinsic microporosity and macropores for a fast ingrowth of bone forming osteoblasts and material resorbing osteoclasts. A minimum macropore size of approximately 100 μm is required [22–25], whereby a structure with macropores and interconnections larger than 300 μm is recommended for enhanced new bone formation and the invasion of capillaries [15,26]. The beneficial effect of such porous scaffold structures on tissue ingrowth and bone regeneration has been demonstrated in numerous in vitro and in vivo studies [27–32]. However, these highly porous bone graft substitutes are unsuitable for a mechanically-loaded orthopedic application, due to the well-known detrimental effect of porosity on mechanical integrity [33].

Previous approaches to increase the mechanical capacity of CaP were mainly focused on the development of composites with an additional ductile and/or degradable phase [34]. This was demonstrated by the fabrication of TCP–iron composites by either mechanical alloying [35] or plasma-spark sintering [36], as well as by a combination with polymers, such as polylactic-co-glycolic acid or polycaprolactone [37]. Although these approaches were successful to a certain extent at achieving improved mechanical properties, they were associated with other complications, such as the formation of acidic (polymers) or particulate (iron oxide) degradation products with potential inflammation reactions at the implantation sites.

Therefore, the objectives of this study were to develop a novel approach of a "pure" CaP bone graft substitute with load capacities and to compare this approach with commonly used materials for bone defect filling in orthopedic applications. Dense granules in specific shapes were used as bulk material in a way that (a) granules have contact to one another in order to provide mechanical load capacity of the bulk material, and (b) the space between the granules is retained in order to provide an appropriate scaffold structure for bone regeneration. It was hypothesized that with this novel approach, the contradiction between porosity and mechanical integrity of CaP scaffolds could be solved such that sufficient mechanical load capacity and scaffold structure (porosity, pore sizes, and interconnectivity) are achieved. Two different granule designs, i.e., tetrapods and pyramids, were chosen and fabricated either by powder injection molding techniques (tetrapods) or by additive manufacturing techniques (pyramids). Tetrapods were made of β-tricalcium phosphate (β-TCP) and pyramids of a biphasic material consisted of HA and β-TCP. The intra-granular space, i.e., the space within the granule material itself, was minimized by an applied sintering regime, leading to mostly dense ceramics with only marginal microporosity. Different granule designs, fabrication techniques, and CaP formulations were chosen to consider different variations of the approach.

Commercially available, porous CaP granules and morselized bone chips were used as reference materials. These reference materials represented two commonly used orthopedic treatment options for bone defect regeneration. The commercial CaP granules are used as filling material for revision hip surgery; e.g., for the reconstruction of the acetabulum and to fill the shaft [38]. Morselized bone chips are used within the impaction bone grafting technique in revision joint replacements [39–42].

These four materials were analyzed by physico-chemical (scanning electron microscopy and X-ray diffraction), mechanical (impactions and cyclic compressions), and micro-computer tomography (micro-CT) techniques. The obtained data were evaluated with a special focus on the mechanical stability (i.e., changes in bulk height) and the scaffold structure after mechanical loading (i.e., bulk porosity, pore size and interconnectivity).

2. Materials and Methods

2.1. Material Synthesis and Granule Fabrication

Four different bulk materials were tested in this study (Table 1). Shapes of tetrapods (TEPO) and pyramids (PYRA) were inspired by previous studies on specifically-shaped granular bone graft substitutes [43,44]. Screening tests revealed that the TEPO and PYRA shapes provide an optimal compromise between mechanical stability and bulk porosity. More hollow shapes (e.g., hollow tetrahedrons) provided higher porosity but not enough mechanical stability, whereas more spherical shapes (e.g., cylinders) provided higher mechanical stability but not enough bulk porosity.

Table 1. Overview of the materials tested.

	TEPO	PYRA	CALC	BC
Shape	tetrapods	pyramids	granules	bone chips
Manufacturing	powder injection molding	extrusion-based, cement reaction	cement reaction	bone nibbler
Size	4–5 mm	3–4 mm	3–4 mm	5–8 mm
Material	ß-TCP	68% HA; 32% ß-TCP	nano-HA; Monetite	HA

Tetrapods with an edge length of 3.3 mm were fabricated by the Fraunhofer Institute for Manufacturing Technology and Advanced Materials (IFAM, Bremen, Germany) using powder injection molding (BoyXS, Dr. BOY GmbH and Co. KG, Neustadt-Fernthal, Germany). Powder injection molding allowed the usage of multiple tetrapod cavities in one mold, increasing the economic efficiency of the manufacturing process. A commercially available β-TCP powder (Plasma Biotal, Derbyshire, UK) and a polyethylene-based binder were used as feedstock. TCP is an inherently soluble CaP with a higher solubility than HA. The β-phase is thermodynamically stable up to 1125 °C, while above this temperature the α-phase occurs [45]. Debinding of the green parts was performed by ethanol washing and sintering in air. The sintered specimens had a relative density of approximately 95% of the theoretical density of 3.07 g/cm^3. The overall crystallinity was about 93%, leaving a small amorphous share.

Pyramids with an edge length of 3.5 mm were fabricated by INNOTERE (INNOTERE biomaterial, Radebeul, Germany) using extrusion-based 3D printing at room temperature, followed by carrying out the cement reaction and washing. The pyramids were manufactured using a specific CaP paste and a custom-made printing protocol [46]. The density resulted from the material density of the cement paste and was about 2 g/cm^3. The shape was determined by the printing process on a flat surface, where the contour tapered from the flat bottom to top without undercuts. This allowed high throughput using a multichannel printer and high density. The green parts were sintered afterwards such that a biphasic material consisting of HA and β-TCP resulted.

Commercially available, porous CaP granules (CALC; Calcibon; Zimmer Biomet, Warsaw, IN, USA) with a size between 3 and 4 mm consisted of nanocrystalline HA and $CaHPO_4$ (monetite) were used as the first reference material. Morselized, cancellous bone chips (BC) with an edge length between 5 and 8 mm were prepared from three fresh frozen human femoral heads and were used as the second reference material. The usage of human bone was approved by the Ethics Committee of the University of Heidelberg (S-170/2016). According to the recommended clinical preparation procedure, the femoral heads were first sawed into slices and the cortical bone was removed before the spongy bone was manually morselized using a bone nibbler [41,47].

2.2. Material Characterization Techniques

Scanning electron microscopy (SEM) of CaP granules prior to mechanical loading was performed with an ESEM Quanta 400 instrument (ThermorFisher Scientific, Waltham, MA, USA) with gold/palladium-sputtered samples. SEM of bone chips was performed with a Zeiss Evo 50 instrument with gold sputtered samples. X-ray powder diffraction (XRD) was performed on a Bruker D8 Advance diffractometer (Bruker, Billerica, MA, USA) with Cu Kα radiation (1.54184 Å) in Bragg–Brentano geometry on thoroughly ground sample on a silicon sample holder. The software TOPAS 4.2 (Bruker, Billerica, MA, USA) was used for quantitative XRD analysis by Rietveld refinement analysis.

2.3. Mechanical Analysis

The mechanical analysis was designed to approximate the mechanical load that would occur when the material is applied to reconstruct an acetabular bone defect during hip joint replacement. Two different mechanical loads were applied (impaction and compression procedure) and three different timepoints (t_0, t_1, and t_2) were considered (Figure 1a). The impaction procedure was designed to simulate the load occurring during the impaction bone grafting technique. The compression procedure was designed to simulate the load behind an acetabular cup occurring during daily activities after patient remobilization. Timepoint t_0 represents the situation where the material is filled in the defect without any load, timepoint t_1 represents the situation after the impaction procedure, and timepoint t_2 represents the situation after impaction and compression procedure.

The materials were filled in a cylindrical container with a diameter of 23 mm and filling heights of 14 mm for granules and 30 mm for bone chips. These heights were chosen such that a height of approximately 13 mm resulted after the impaction and compression procedures. For the impaction procedure, the containers were placed on a rubber pad, a metallic punch was placed on the materials (22 mm diameter, 75 mm height), and ten impactions with an orthopedic hammer (30 cm, 410 g) were applied [48–50]. For the compression procedure, a spherical punch with a diameter of 22 mm was used to apply a load with a peak force of 1.52 kN (corresponding to a peak pressure of 4 MPa) in as sinusoidal waveform with a frequency of 3 Hz. The chosen peak pressure of 4 MPa corresponds to the estimated contact pressure of acetabular cartilage and the loads around acetabular cups [51–55]. Screening tests revealed a settling behavior of the materials before 100,000 load cycles, such that a number of cycles was limited to 100,000. The compression procedure was conducted using a resonance pulsator (Newline 10 kN, PneuSys software, SincoTec, Clausthal-Zellerfeld, Germany).

Six filled containers of each bulk material were tested. The height of each bulk material was measured at each timepoint. The heights were normalized to that at t_1 at 100% in order to compare the decrease in height between the materials from timepoint t_1 to t_2, which might be an indicator for primary stability of the implant.

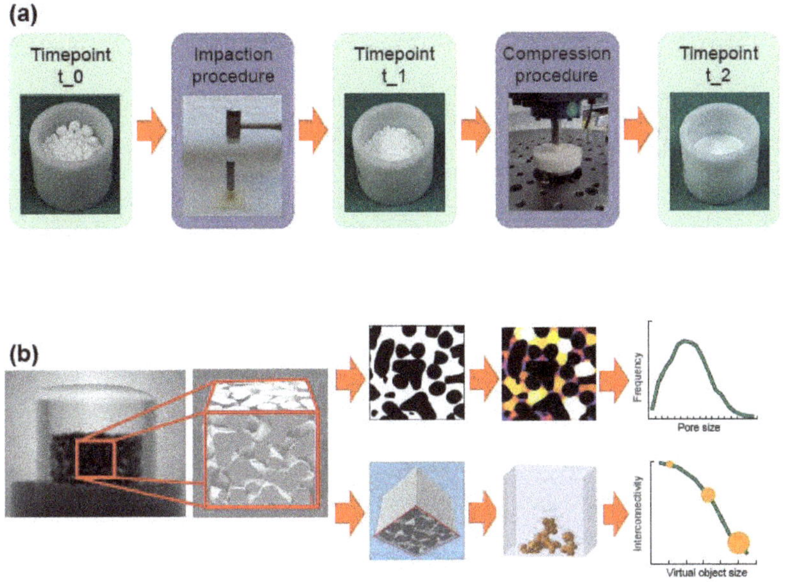

Figure 1. Workflow of the mechanical and morphological analyses. (**a**) The material, loosely filled in a container (timepoint t_0), was loaded according to the impaction bone grafting technique (impaction procedure). After the impaction procedure (timepoint t_1), the material was compressed by applying the estimated loads behind an acetabular cup (compression procedure) resulting in a realistic load scenario (timepoint t_2). (**b**) A cube with a 10 mm edge length in the center of the container was analyzed with a micro-CT scanner. In a first analysis, dense material (shown in black) was separated from the pore space (shown in white/colored) to obtain the porosity, the average pore size, and the pore size distribution. In a second analysis, dense material was also separated from pore space but the cube was additionally "closed" on five sides. By simulating the penetration of a virtual object with various diameters into the porous system from the remaining "open" side, the interconnectivity in terms of accessible volume (shown in orange) with respect to the virtual object size could be obtained.

2.4. Morphological Analysis

Morphological analysis (Figure 1b) was performed using a micro-CT system (phoenix v|tome|x, General Electric, Boston, MA, USA). In total, eight scans per material were performed. For each bulk material, a scan of one representative container at t_0 and t_1 and scans of all six containers at t_2 were conducted. The containers were fixed in the micro-CT scanner by a custom-made device so that the containers rotated exactly around their vertical axes. Images were taken at 120 kV, 140 mA, and 0.25° rotation steps. The focus was adjusted to cover a 15 mm × 15 mm × 15 mm cube in the center of the container. 1000 images with 1000 × 1000 pixels were captured, resulting in a final voxel size of 15 μm.

Image processing was performed with the software ImageJ (Version 1.51k, National Institutes of Health, Bethesda, MD, USA). Segmentation of dense material (i.e., CaP and trabecular bone) from the pore system (i.e., inter-granular pores, intra-granular pores and trabecular spacing) was performed with manually-defined global density thresholds. Inaccuracies at the image boundaries were eliminated by cutting the image stack into 667 images with 667 × 667 pixels in the center of the container, resulting in a "region of interest" (ROI) of 10 mm × 10 mm × 10 mm. A median filter with the radius of three pixels was applied to reduce image noise. Porosity, pore size, and pore size distribution were analyzed with the BoneJ functions "Volume Fraction," "Thickness," and "Histogram" in ImageJ [56,57]. Porosity represents the volume fraction of pore volume with respect to the total volume in percentage. Pore size represents the diameter of the greatest sphere that fits within the pore space. Pore size distribution represents the frequency of voxels dedicated to specific pore sizes. This is displayed as a histogram in which the dedicated voxels were summarized in 100 μm intervals and plotted relative to all pore voxels in percentages. Interconnectivity in terms of accessible volume was defined as the volume accessible for a virtual object which penetrates the scaffold from one side, a method called "directional shrink-wrap" [58,59]. It was calculated using the function "ROI Shrink-Wrap" in the CTAn software (Version 1.18, Bruker, Kontich, Belgium). For different object diameters, the interconnectivity was defined as the ratio of accessible volume with respect to the total volume in percent. The virtual object diameter was chosen as 10 voxels, 30 voxels, and 50 voxels, representing diameters of 150 μm, 450 μm, and 750 μm. Five surfaces of the ROI were closed such that only one surface was accessible for the virtual object. The accessible surface was chosen to be at the opposite side of the force application, since this side would be directed towards the native bone, and hence represents the side for potential bone ingrowth (Figure 1b).

2.5. Statistical Analysis

All data results are presented as means ± standard deviations. Heights are shown at all timepoints, whereas porosity, pore size, pore size distribution, and interconnectivity are only shown at timepoint t_2. Statistical differences between the materials were analyzed by a Mann–Whitney U test with a level of significance of $p < 0.05$ using the function ranksum in Matlab (The Mathworks, Natick, MA, USA).

3. Results

3.1. Material Characterization Results

Figure 2 shows SEM images of the three granule types and the bone chips. TEPO showed a slide edge, as expected for the powder injection fabrication (Figure 2a). In contrast, PYRA showed signs of the extrusion-based molding process where the pasty starting material was applied. At higher magnification, both objects showed primary crystals after sintering with a diameter of a few μm. In contrast, CALC showed an irregular porous structure wherein the rectangular/cubic pores were obviously created by a leaching process. At higher magnification, primary crystallites in the sub-μm range indicated a recrystallization in the presence of water. BC showed the irregular shape and the open-porous structure of trabecular bone.

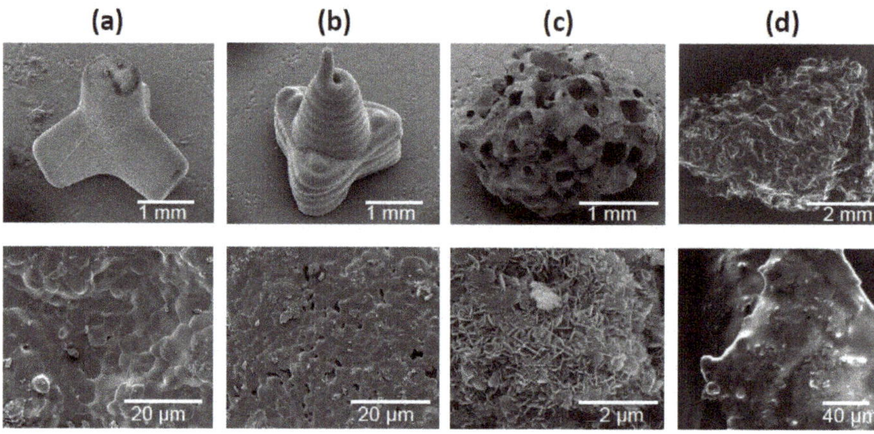

Figure 2. Scanning electron micrographs of granules and bone chips used in this study. (**a**) Tetrapods, (**b**) pyramids, (**c**) Calcibon granules, and (**d**) morselized bone chips.

The compositions of the objects prepared were assessed by X-ray powder diffraction after grinding (Figure 3). TEPO consisted of phase-pure β-TCP. In contrast, PYRA contained both β-TCP and hydroxyapatite in a weight ratio of about 68:32. CALC consisted of approximately 92 wt% hydroxyapatite and 8 wt% monetite. Bone chips consist mainly of collagen and nanoscopic calcium phosphate particles [12]. In the powder diffractogram of bone chips, only calcium phosphate particles are visible (data not shown) [60].

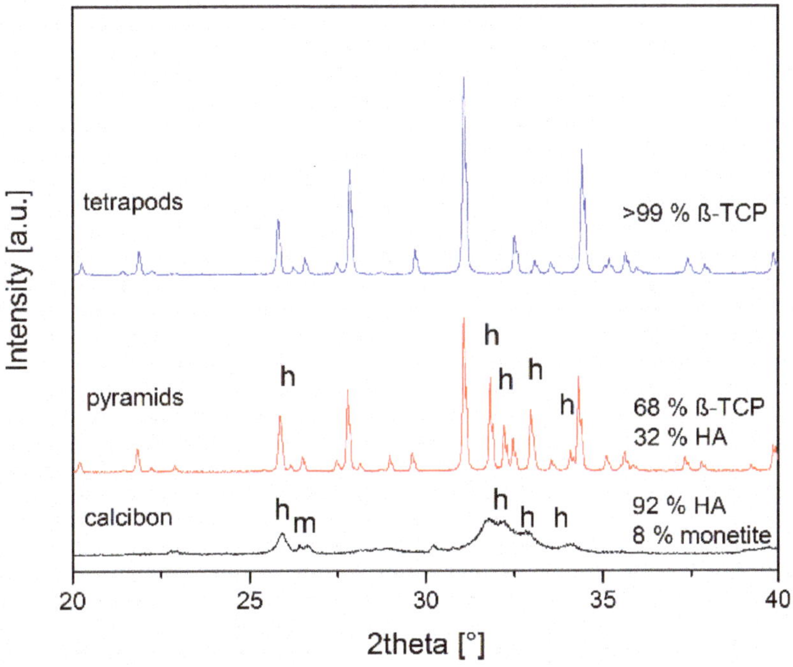

Figure 3. X-ray diffraction patterns and quantitative phase composition according to Rietveld refinement analysis of the granules. All peaks for tetrapods correspond to ß-tricalcium phosphate, the other peaks result from hydroxyapatite (h) and monetite (m).

3.2. Mechanical Results

Heights of the four bulk materials within the containers were used to characterize mechanical stability after mechanical loading (Figure 4). The height of each sample was normalized to the height of t_1 with 100%. The averaged normalized height of TEPO at t_0 was 113% ± 8% and decreased to 94% ± 1% at t_2; that of PYRA was 104% ± 1% and decreased to 98 ± 1%; that of CALC at t_0 was 147% ± 4% and decreased to 89% ± 1%; and that of BC was 138% ± 8% and decreased to 85% ± 3%. Statistically significant differences at timepoint t_2 were observed between tetrapods and bone chips ($p < 0.05$), between pyramids and Calcibon ($p < 0.01$), and between pyramids and bone chips ($p < 0.01$).

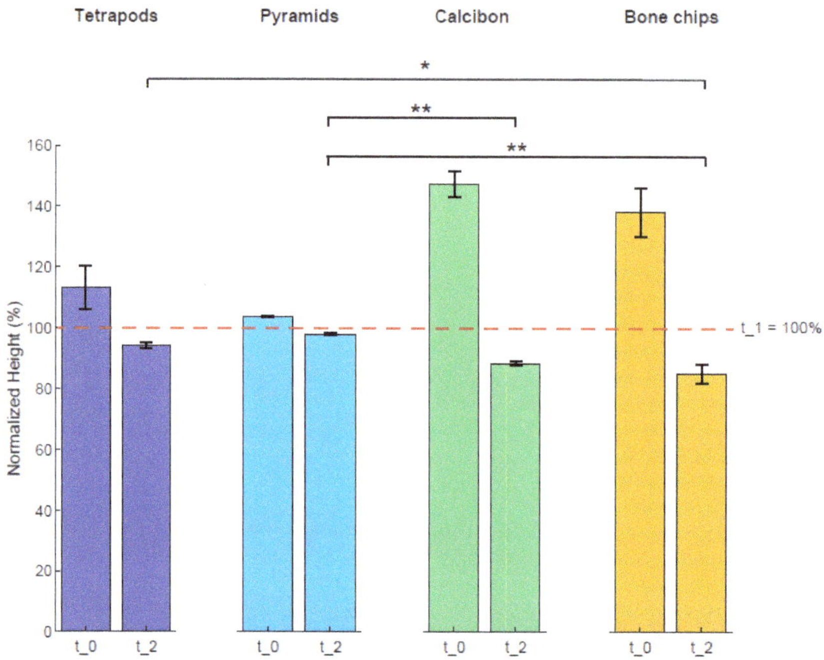

Figure 4. Heights of the four bulk materials before and after mechanical loading. The height of each sample was normalized to that of timepoint t_1 at 100%. Shown are mean and standard deviation values at t_0 and t_2. Statistically significant differences between the materials are shown for timepoint t_2 with $* = p < 0.05$ and $** = p < 0.01$.

3.3. Morphological Results

Color-coded micro-CT images show the pore system of one typical example of each bulk material at each timepoint (Figure 5). Dense material is shown in black (i.e., CaP or trabecular bone) and pores are color-coded according to their size and the given color bar. TEPO and PYRA only showed inter-granular pores (i.e., pores between the granules) and no intra-granular pores (i.e., pores within the granules themselves). TEPO and PYRA remained mainly stable during the impaction and compression procedures such that only few fragments were clogging the pore space. Therefore, inter-granular pore space only slightly decreased due to mechanical loading. In contrast, CALC showed inter-granular pores and intra-granular pores at t_0. After the impaction procedure (t_1), inter-granular pores diminished due to granule crushing and only intra-granular pores remained. After the compression procedure (t_2), a complete granule disruption led to an almost dense material without any pores. BC showed inter-granular pores and intra-granular pores at t_0 and intra-granular pores at t_1. Since the trabecular bone structure mainly withstood the mechanical loading, BC was the only material which showed intra-granular pores after the mechanical loadings.

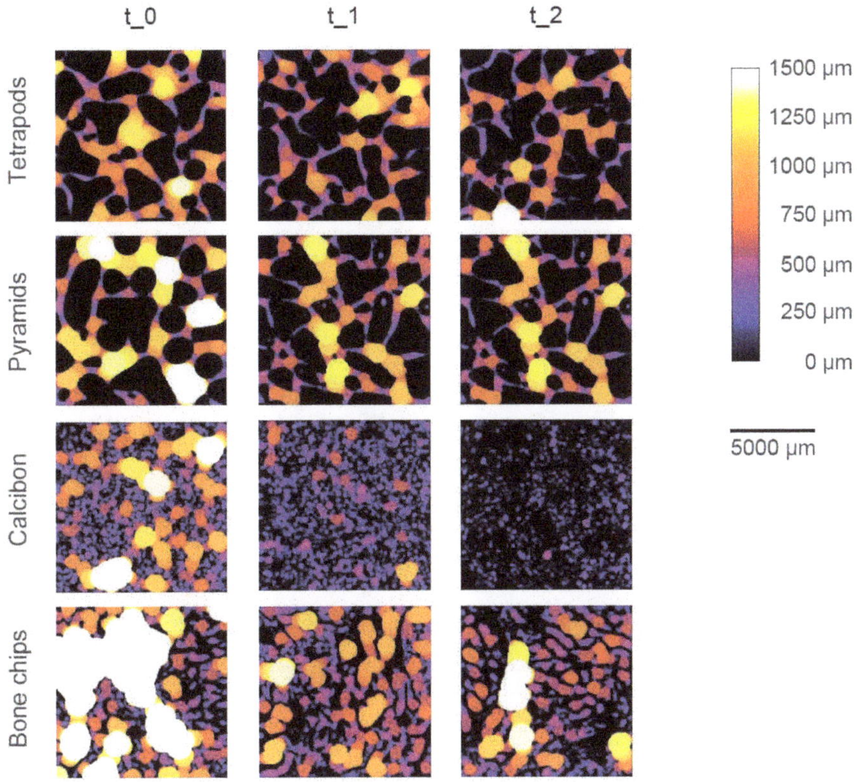

Figure 5. Graphical representation of the pore space of one typical specimen of each bulk material at each timepoint. The size of the pore space is color-coded according to the color bar. Dense material is shown in black. With increasing pore size, the brightness increases, until pore size above 1500 μm is shown in white.

These qualitative findings were confirmed by quantitative analyses at timepoint t_2 (Figure 6). The mean porosity at t_2 was highest for BC (46% ± 10%), lowest for CALC (25% ± 2%), and similar for TEPO and PYRA (37% ± 1% and 37% ± 2%, respectively) (Figure 6a). The mean pore size was highest for TEPO (581 ± 15 μm) and PYRA (640 ± 30 μm), whereas the mean pore size was lower for BC (360 ± 140 μm) and CALC (141 ± 4 μm) (Figure 6b). The pore size distributions are shown as histograms summarized in 100 μm intervals (Figure 6c). The pore space of CALC could be attributed to pore sizes below 500 μm and the pore space of BC to pore sizes below 800 μm. For both materials, most pores were around 200 μm. In contrast, the pores of TEPO and PYRA were more uniformly distributed and could be attributed to pore sizes between 400 μm and 1000 μm.

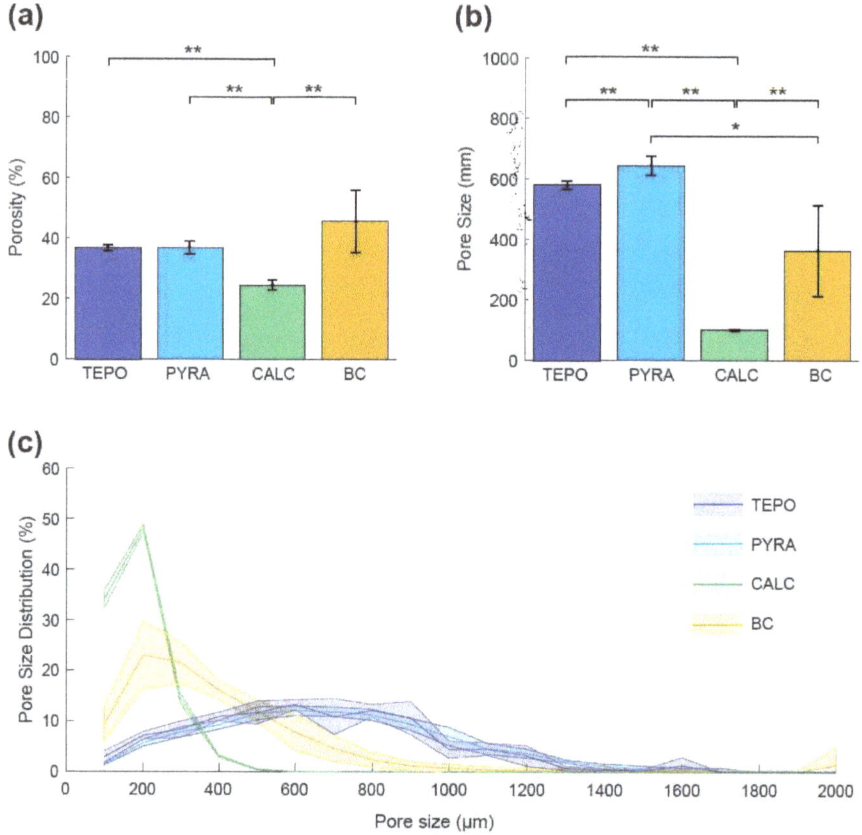

Figure 6. Porosity, pore size, and pore size distribution of tetrapods (TEPO), pyramids (PYRA), Calcibon granules (CALC), and bone chips (BC) at timepoint t_2. (**a**) Porosities and (**b**) pore sizes as mean and standard deviation values; statistically significant differences are indicated as * = $p < 0.05$ and ** = $p < 0.01$. (**c**) Pore size distribution histogram summarized in 100 μm steps as mean values (solid lines) and standard deviations (shaded areas).

The interconnectivity in terms of accessible volume was assessed using the "directional shrink-wrap" transport pathway analysis (see Methods and Figure 7). The accessible volumes for the virtual object diameters of 150 μm, 450 μm, and 750 μm revealed large differences between the four materials (accessible volume is shown as brown structure within the ROI). A virtual object with a diameter of 150 μm could access 34% of the overall volume of TEPO, 36% of PYRA, 0.1% of CALC, and 35% of BC. The values of TEPO and PYRA are close to their porosity value of 37%, indicating an open-porous system with interconnections larger than 150 μm in both materials. When the virtual object diameter was increased to a diameter of 450 μm, a considerable amount of the TEPO and PYRA volumes were still accessible (18% and 23%, respectively), whereas the accessible volume of BC decreased to 4% and that of CALC completely diminished. With a further increase of virtual object diameter to 750 μm, the accessible volumes of TEPO, PYRA, and BC were reduced to 2%, 4%, and 1%, respectively.

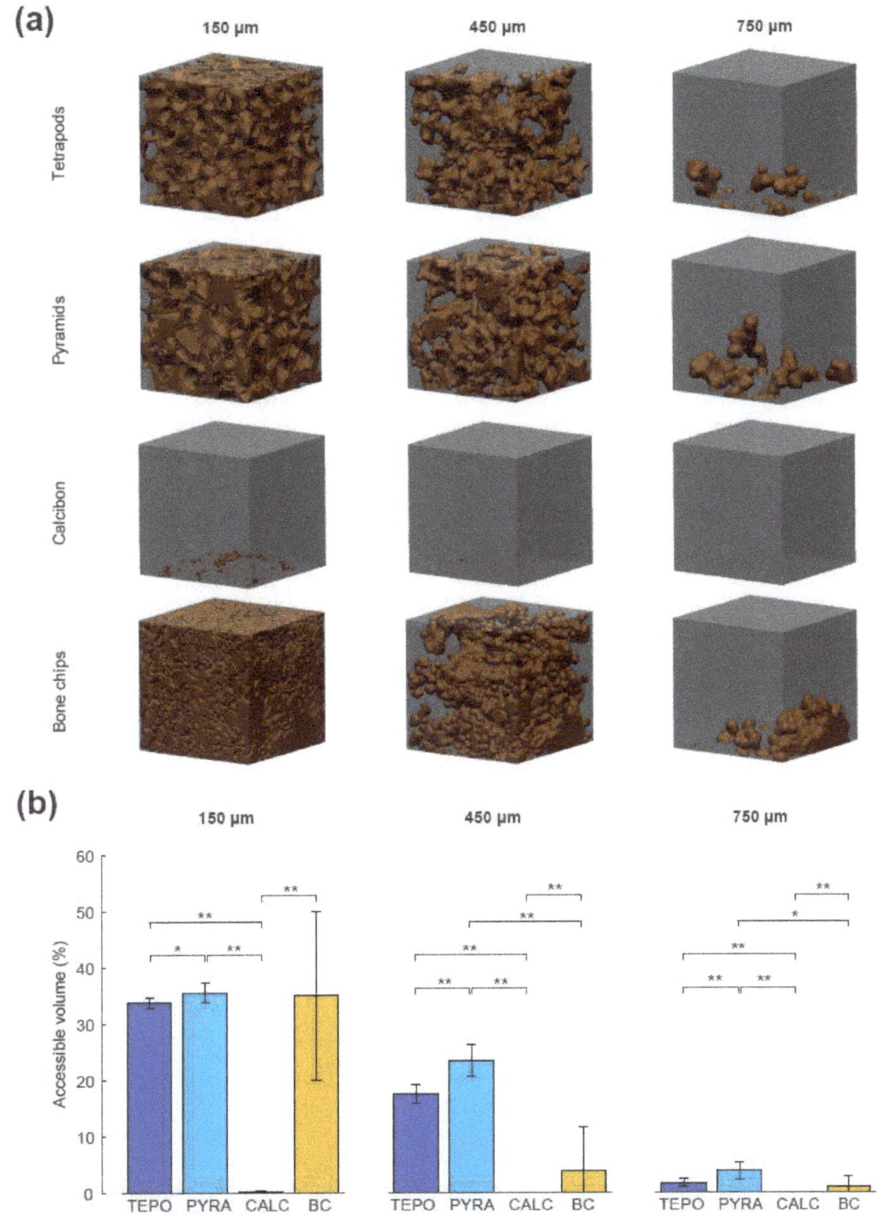

Figure 7. Interconnectivity in terms of accessible volume of the four materials at timepoint t_2 for the virtual object diameters of 150 µm, 450 µm, and 750 µm. (**a**) The accessible volumes of the four materials and the virtual object diameters are graphically shown as brown structures within the regions of interest (ROIs). (**b**) The accessible volumes of the four materials and the virtual object diameters as the ratio between accessible volume, with respect to overall volume in percentage; statistically significant differences are indicated as $* = p < 0.05$ and $** = p < 0.01$.

4. Discussion and Conclusions

The objectives of this study were to develop a novel approach for CaP bone graft substitutes with sufficient load capacity for orthopedic applications and to compare this approach with commonly used bone grafts and a bone graft substitute. The comparison was based on physicochemical (scanning electron microscopy and X-ray diffraction), mechanical (impaction and cyclic compression), and morphological (micro-computer tomography) methods. The physicochemical characteristics were analyzed as surface topology and phase composition; mechanical characteristics as decreases in height at three timepoints; and morphological characteristics as porosity, pore size, pore size distribution, and interconnectivity. Pore size and interconnectivity are independent measures of scaffold structure [58], whereby pore size defines the diameter of the greatest sphere that fits within the pore space, and interconnectivity defines the volume accessible for a virtual object with a certain diameter.

Most of the commonly used porous CaP bone graft substitutes are inherently brittle, which leads to insufficient mechanical load capacities for orthopedic applications. Much effort has been made to improve mechanical load capacities using different reinforcement strategies [15,61,62]. The strength deficit of synthetic CaP scaffolds compared to human bone is mainly attributed the composite nature of bone and the hierarchical bone architecture from nano-scale to macro-scale, which is very difficult to replicate in a synthetic material [63–65]. Another important drawback of CaP scaffolds is that the progress of an initiated crack will not be hindered by the deformation of the material ahead of the crack (as would be the case in a ductile material), causing a complete disruption of the scaffold [15].

In the approach presented here, a bulk of tetrapods or pyramids created a network of interconnected macropores by their inter-granular space, whereby the absent microporosity (intra-granular space) enhanced the overall mechanical load capacity. The approach bypasses the above-described problem of the inherent brittleness of bioceramic scaffolds in the way that a crack of one granule would not lead to a catastrophic failure of the whole structure. It was shown that this scaffold structure is able to withstand mechanical load in an impaction and compression model, suggesting its viability in application, as an acetabular bone-defect-filling material. The load capacity could be seen by the almost constant height of tetrapods and pyramids from timepoint t_1 (100%) to timepoint t_2 (94% and 98% respectively; Figure 4). At the same time, the network of interconnected macropores with pore sizes and interconnections larger than 450 μm remained intact, demonstrating the potential for vascularization and bone remodeling, as proposed by several authors [15,26,31]. This was not the case for commercially available, porous CaP granules, which collapsed under mechanical load, leading to diminished porosity, pore size, and interconnectivity. Although bone chips showed a higher total porosity compared to tetrapods or pyramids after dynamic loading (46% versus 37%), they had smaller mean pore sizes and interconnectivity values for virtual object diameters above 450 μm. Even more important was the decrease in height from t_1 (100%) to t_2 (85%) of BC, which might be a contributing factor to the reduced survival rate of bone chips when applied in large or loaded acetabular bone defects [47,66]. Changes in the bulk heights of tetrapods and pyramids after mechanical loading could be mainly attributed to changes in granule morphology (i.e., a few distal parts of the granules were broken, so the granules moved closer together) and not on the granule material itself. In contrast, changes in bulk heights of porous CaP and bone chips could be attributed to changes in morphology and material (i.e., also the intra-granular pore space decrease under mechanical loading).

Previous studies which already addressed the use of specifically-shaped CaP granules for bone defect filling were mainly focused on scaffold structures rather than on mechanical aspects. These granules were either not intended for load-bearing applications [44] or failed at a compressive load of approximately 3–4 N for a single granule [43]. Studies considering mechanical aspects also compared stacks of macroporous β-TCP granules with different porosities [22,67]. It was shown that a stack of the high porosity granules resembles trabecular bone in its porous microarchitecture [67] and that granules with 4 mm edge length could be used for impaction grafting technique [22]. However, in both studies a single load of 3–5 MPa led to a large compression displacement. A CaP bone graft substitute of dense granules which combines a mechanical load capacity up to 4 MPa with appropriate scaffold structure for bone ingrowth (porosity of 37%; accessible volumes for 150 μm for objects of 34% and 36%, and for a 450 μm objects, 18% and 23%) is novel to the best of our knowledge.

This study was focused on mechanical and morphological aspects and did not include biological considerations. Even though the sintered materials consist of established β-TCP and HA, whose biocompatibility was proven in numerous in vitro, in vivo and clinical studies [65,68], the next step should include a basic cell-culture investigation (e.g., using MTT-assay). Furthermore, the approach should be tested in a more advanced in vitro test setup with a realistic bone model and in combination with the intended prosthesis and direction of load. A clinical application of the approach definitely requires in vivo studies in large animal models with sufficient bone defect size and mechanical loading. Such studies should not only consider the primary stability in vivo, but must also address the change of mechanical stability due to bone tissue ingrowth and subsequent granule degradation.

Author Contributions: Conceptualization, G.H., R.A.S., M.E., U.G., S.H., H.M.-J., and T.M.G.; investigation, G.H., R.A.S., M.E., U.G., S.H., H.M.-J., and T.M.G.; methodology, G.H., R.A.S., M.E., U.G., S.H., H.M.-J., and T.M.G.; visualization G.H.; supervision T.M.G.; writing—original draft preparation, G.H. and R.A.S.; writing—review and editing, M.E., U.G., S.H., H.M.-J., and T.M.G.

Funding: This research received no external funding.

Acknowledgments: The authors would like to thank Geraldine Porini, Walter Keller and Dennis Pede for their help in conducting the micro-CT analysis. We thank Oleg Prymak for the X-ray diffraction analysis.

Conflicts of Interest: Three of the authors (G.H., R.A.S., T.M.G.) are employees of Aesculap AG Tuttlingen, a manufacturer of orthopedic implants. One author (S.H.) is employee of INNOTERE GmbH Radebeul, a manufacturer of medical devices for bone void filling. Two of the authors (M.E., U.G.) are scientific advisors in Aesculap R&D projects.

References

1. Lietaert, K.; Wauthle, R.; Schrooten, J. Porous Metals in Orthopedics. In *Biomaterials in Clinical Practice: Advances in ClinIcal Research and Medical Devices*; Zivic, F., Affatato, S., Trajanovic, M., Schnabelrauch, M., Grujovic, N., Choy, K.L., Eds.; Springer: Cham, Switzerland, 2018; pp. 281–301. ISBN 978-3-319-68024-8.
2. Levine, B.R.; Fabi, D.W. Porous metals in orthopedic applications-A review. Poröse Metalle in orthopädischen Anwendungen-Eine Übersicht. Mat.-wiss. u. *Werkstofftech* **2010**, *41*, 1001–1010. [CrossRef]
3. Geetha, M.; Singh, A.K.; Asokamani, R.; Gogia, A.K. Ti based biomaterials, the ultimate choice for orthopaedic implants—A review. *Prog. Mater. Sci.* **2009**, *54*, 397–425. [CrossRef]
4. Saleh, K.J.; El Othmani, M.M.; Tzeng, T.H.; Mihalko, W.M.; Chambers, M.C.; Grupp, T.M. Acrylic bone cement in total joint arthroplasty: A review. *J. Orthop. Res.* **2016**, *34*, 737–744. [CrossRef] [PubMed]
5. Charnley, J. Anchorage of the femoral head prosthesis to the shaft of the femur. *J. Bone Jt. Surg. Br.* **1960**, *42*, 28–30. [CrossRef]
6. Noyama, Y.; Miura, T.; Ishimoto, T.; Itaya, T.; Niinomi, M.; Nakano, T. Bone Loss and Reduced Bone Quality of the Human Femur after Total Hip Arthroplasty under Stress-Shielding Effects by Titanium-Based Implant. *Mater. Trans.* **2012**, *53*, 565–570. [CrossRef]
7. Panegrossi, G.; Ceretti, M.; Papalia, M.; Casella, F.; Favetti, F.; Falez, F. Bone loss management in total knee revision surgery. *Int. Orthop.* **2014**, *38*, 419–427. [CrossRef]
8. Springer, B.D.; Fehring, T.K.; Griffin, W.L.; Odum, S.M.; Masonis, J.L. Why revision total hip arthroplasty fails. *Clin. Orthop. Relat. Res.* **2009**, *467*, 166–173. [CrossRef]
9. Sporer, S.M.; Paprosky, W.G.; O'Rourke, M.R. Managing bone loss in acetabular revision. *Instr. Course Lect.* **2006**, *55*, 287–297. [CrossRef]
10. Koob, S.; Scheidt, S.; Randau, T.M.; Gathen, M.; Wimmer, M.D.; Wirtz, D.C.; Gravius, S. Biologisches Downsizing: Azetabulare Knochendefekte in der Huftrevisionsendoprothetik. *Orthopade* **2017**, *46*, 158–167. [CrossRef] [PubMed]
11. Van Egmond, N.; de Kam, D.C.J.; Gardeniers, J.W.M.; Schreurs, B.W. Revisions of extensive acetabular defects with impaction grafting and a cement cup. *Clin. Orthop. Relat. Res.* **2011**, *469*, 562–573. [CrossRef]
12. Dorozhkin, S.V.; Epple, M. Biological and Medical Significance of Calcium Phosphates. *Angew. Chem. Int. Ed.* **2002**, *41*, 3130–3146. [CrossRef]

13. Laurencin, C.; Khan, Y.; El-Amin, S.F. Bone graft substitutes. *Expert Rev. Med. Devices* **2006**, *3*, 49–57. [CrossRef] [PubMed]
14. Ratnayake, J.T.B.; Gould, M.L.; Shavandi, A.; Mucalo, M.; Dias, G.J. Development and characterization of a xenograft material from New Zealand sourced bovine cancellous bone. *J. Biomed. Mater. Res. B Appl. Biomater.* **2017**, *105*, 1054–1062. [CrossRef]
15. Dorozhkin, S. Calcium Orthophosphate-Based Bioceramics. *Materials* **2013**, *6*, 3840–3942. [CrossRef] [PubMed]
16. Bohner, M. Design of ceramic-based cements and putties for bone graft substitution. *eCM* **2010**, *20*, 1–12. [CrossRef] [PubMed]
17. Maté Sánchez de Val, J.E.; Calvo-Guirado, J.L.; Gómez-Moreno, G.; Pérez-Albacete Martínez, C.; Mazón, P.; Aza, P.N.D. Influence of hydroxyapatite granule size, porosity, and crystallinity on tissue reaction in vivo. Part A: Synthesis, characterization of the materials, and SEM analysis. *Clin. Oral Implant. Res.* **2016**, *27*, 1331–1338. [CrossRef] [PubMed]
18. Tanaka, T.; Kumagae, Y.; Saito, M.; Chazono, M.; Komaki, H.; Kikuchi, T.; Kitasato, S.; Marumo, K. Bone formation and resorption in patients after implantation of beta-tricalcium phosphate blocks with 60% and 75% porosity in opening-wedge high tibial osteotomy. *J. Biomed. Mater. Res. B Appl. Biomater.* **2008**, *86*, 453–459. [CrossRef]
19. Dorozhkin, S. Calcium Orthophosphate Cements and Concretes. *Materials* **2009**, *2*, 221–291. [CrossRef]
20. Huber, F.-X.; Belyaev, O.; Hillmeier, J.; Kock, H.-J.; Huber, C.; Meeder, P.-J.; Berger, I. First histological observations on the incorporation of a novel nanocrystalline hydroxyapatite paste OSTIM in human cancellous bone. *BMC Musculoskelet. Disord.* **2006**, *7*, 50. [CrossRef]
21. Uhthoff, H.K.; Poitras, P.; Backman, D.S. Internal plate fixation of fractures: Short history and recent developments. *J. Orthop. Sci.* **2006**, *11*, 118–126. [CrossRef]
22. Hsu, Y.H.; Turner, I.G.; Miles, A.W. Fabrication and mechanical testing of porous calcium phosphate bioceramic granules. *J. Mater. Sci. Mater. Med.* **2007**, *18*, 1931–1937. [CrossRef] [PubMed]
23. Bohner, M.; Baumgart, F. Theoretical model to determine the effects of geometrical factors on the resorption of calcium phosphate bone substitutes. *Biomaterials* **2004**, *25*, 3569–3582. [CrossRef] [PubMed]
24. Bohner, M.; Baroud, G.; Bernstein, A.; Döbelin, N.; Galea, L.; Hesse, B.; Heuberger, R.; Meille, S.; Michel, P.; von Rechenberg, B.; et al. Characterization and distribution of mechanically competent mineralized tissue in micropores of β-tricalcium phosphate bone substitutes. *Mater. Today* **2017**, *20*, 106–115. [CrossRef]
25. Hulbert, S.F.; Young, F.A.; Mathews, R.S.; Klawitter, J.J.; Talbert, C.D.; Stelling, F.H. Potential of ceramic materials as permanently implantable skeletal prostheses. *J. Biomed. Mater. Res.* **1970**, *4*, 433–456. [CrossRef]
26. Karageorgiou, V.; Kaplan, D. Porosity of 3D biomaterial scaffolds and osteogenesis. *Biomaterials* **2005**, *26*, 5474–5491. [CrossRef]
27. Von Doernberg, M.-C.; von Rechenberg, B.; Bohner, M.; Grünenfelder, S.; van Lenthe, G.H.; Müller, R.; Gasser, B.; Mathys, R.; Baroud, G.; Auer, J. In vivo behavior of calcium phosphate scaffolds with four different pore sizes. *Biomaterials* **2006**, *27*, 5186–5198. [CrossRef]
28. Flautre, B.; Descamps, M.; Delecourt, C.; Blary, M.C.; Hardouin, P. Porous HA ceramic for bone replacement: Role of the pores and interconnections-experimental study in the rabbit. *J. Mater. Sci. Mater. Med.* **2001**, *12*, 679–682. [CrossRef]
29. Sweedy, A.; Bohner, M.; Baroud, G. Multimodal analysis of in vivo resorbable CaP bone substitutes by combining histology, SEM, and microcomputed tomography data. *J. Biomed. Mater. Res. B Appl. Biomater.* **2018**, *106*, 1567–1577. [CrossRef]
30. Diez-Escudero, A.; Espanol, M.; Beats, S.; Ginebra, M.-P. In vitro degradation of calcium phosphates: Effect of multiscale porosity, textural properties and composition. *Acta Biomater.* **2017**, *60*, 81–92. [CrossRef]
31. Feng, B.; Jinkang, Z.; Zhen, W.; Jianxi, L.; Jiang, C.; Jian, L.; Guolin, M.; Xin, D. The effect of pore size on tissue ingrowth and neovascularization in porous bioceramics of controlled architecture in vivo. *Biomed. Mater.* **2011**, *6*, 15007. [CrossRef]
32. Kasten, P.; Beyen, I.; Niemeyer, P.; Luginbühl, R.; Bohner, M.; Richter, W. Porosity and pore size of beta-tricalcium phosphate scaffold can influence protein production and osteogenic differentiation of human mesenchymal stem cells: An in vitro and in vivo study. *Acta Biomater.* **2008**, *4*, 1904–1915. [CrossRef] [PubMed]

33. Zhang, J.; Liu, W.; Schnitzler, V.; Tancret, F.; Bouler, J.-M. Calcium phosphate cements for bone substitution: Chemistry, handling and mechanical properties. *Acta Biomater.* **2014**, *10*, 1035–1049. [CrossRef] [PubMed]
34. Lu, J.; Yu, H.; Chen, C. Biological properties of calcium phosphate biomaterials for bone repair: A review. *Rsc Adv.* **2018**, *8*, 2015–2033. [CrossRef]
35. Swain, S.K.; Gotman, I.; Unger, R.; Gutmanas, E.Y. Bioresorbable β-TCP-FeAg nanocomposites for load bearing bone implants: High pressure processing, properties and cell compatibility. Mater. *Sci. Eng. C Mater. Biol. Appl.* **2017**, *78*, 88–95. [CrossRef] [PubMed]
36. Montufar, E.B.; Horynová, M.; Casas-Luna, M.; Diaz-de-la-Torre, S.; Celko, L.; Klakurková, L.; Spotz, Z.; Diéguez-Trejo, G.; Fohlerová, Z.; Dvorak, K.; et al. Spark Plasma Sintering of Load-Bearing Iron–Carbon Nanotube-Tricalcium Phosphate CerMets for Orthopaedic Applications. *JOM* **2016**, *68*, 1134–1142. [CrossRef]
37. Pilia, M.; Guda, T.; Appleford, M. Development of composite scaffolds for load-bearing segmental bone defects. *BioMed Res. Int.* **2013**, *2013*, 458253. [CrossRef]
38. Biomet Deutschland GmbH. Bone Substitute Materials. Available online: http://www.biomet.co.uk/userfiles/files/Biomaterials/Bone%20substitute%20materials.pdf (accessed on 29 August 2019).
39. Slooff, T.J.; Huiskes, R.; van Horn, J.; Lemmens, A.J. Bone grafting in total hip replacement for acetabular protrusion. *Acta Orthop. Scand.* **1984**, *55*, 593–596. [CrossRef]
40. Slooff, T.J.; Buma, P.; Schreurs, B.M. Acetabular and Femoral Reconstruction with Impacted Graft and Cement. *Clin. Orthop. Relat. Res.* **1996**, *324*, 108–115. [CrossRef]
41. Toms, A.D.; Barker, R.L.; Jones, R.S.; Kuiper, J.H. Impaction bone-grafting in revision joint replacement surgery. *J. Bone Jt. Surg. Am.* **2004**, *86*, 2050–2060. [CrossRef]
42. Blom, A.W.; Wylde, V.; Livesey, C.; Whitehouse, M.R.; Eastaugh-Waring, S.; Bannister, G.C.; Learmonth, I.D. Impaction bone grafting of the acetabulum at hip revision using a mix of bone chips and a biphasic porous ceramic bone graft substitute. *Acta Orthop.* **2009**, *80*, 150–154. [CrossRef]
43. Choi, S.; Liu, I.-l.; Yamamoto, K.; Igawa, K.; Mochizuki, M.; Sakai, T.; Echigo, R.; Honnami, M.; Suzuki, S.; Chung, U.-i.; et al. Development and evaluation of tetrapod-shaped granular artificial bones. *Acta Biomater.* **2012**, *8*, 2340–2347. [CrossRef] [PubMed]
44. Clarke, S.A.; Hoskins, N.L.; Jordan, G.R.; Henderson, S.A.; Marsh, D.R. In vitro testing of Advanced JAX Bone Void Filler System: Species differences in the response of bone marrow stromal cells to beta tri-calcium phosphate and carboxymethylcellulose gel. *J. Mater. Sci. Mater. Med.* **2007**, *18*, 2283–2290. [CrossRef] [PubMed]
45. Perera, F.H.; Martínez-Vázquez, F.J.; Miranda, P.; Ortiz, A.L.; Pajares, A. Clarifying the effect of sintering conditions on the microstructure and mechanical properties of β-tricalcium phosphate. *Ceram. Int.* **2010**, *36*, 1929–1935. [CrossRef]
46. Lode, A.; Meissner, K.; Luo, Y.; Sonntag, F.; Glorius, S.; Nies, B.; Vater, C.; Despang, F.; Hanke, T.; Gelinsky, M. Fabrication of porous scaffolds by three-dimensional plotting of a pasty calcium phosphate bone cement under mild conditions. *J. Tissue Eng. Regen. Med.* **2014**, *8*, 682–693. [CrossRef] [PubMed]
47. Jones, S.A. Impaction Grafting Made Easy. *J. Arthroplast.* **2017**, *32*, S54–S58. [CrossRef] [PubMed]
48. Cornu, O.; Bavadekar, A.; Godts, B.; van Tomme, J.; Delloye, C.; Banse, X. Impaction bone grafting with freeze-dried irradiated bone. Part II. Changes in stiffness and compactness of morselized grafts: Experiments in cadavers. *Acta Orthop. Scand.* **2003**, *74*, 553–558. [CrossRef]
49. Bavadekar, A.; Cornu, O.; Godts, B.; Delloye, C.; van Tomme, J.; Banse, X. Stiffness and compactness of morselized grafts during impaction: An in vitro study with human femoral heads. *Acta Orthop. Scand.* **2001**, *72*, 470–476. [CrossRef]
50. Brewster, N.T.; Gillespie, W.J.; Howie, C.R.; Madabhushi, S.P.G.; Usmani, A.S.; Fairbairn, D.R. Mechanical considerations in impaction bone grafting. *J. Bone Jt. Surg.* **1999**, *81*, 118–124. [CrossRef]
51. Day, W.H.; Swanson, S.A.; Freeman, M.A. Contact pressures in the loaded human cadaver hip. *J. Bone Jt. Surg. Br.* **1975**, *57*, 302–313. [CrossRef]
52. Greenwald, A.S.; Haynes, D.W. Weight-bearing areas in the human hip joint. *J. Bone Jt. Surg. Br.* **1972**, *54*, 157–163. [CrossRef]
53. Hodge, W.A.; Fijan, R.S.; Carlson, K.L.; Burgess, R.G.; Harris, W.H.; Mann, R.W. Contact pressures in the human hip joint measured in vivo. *Proc. Natl. Acad. Sci. USA* **1986**, *83*, 2879–2883. [CrossRef] [PubMed]

54. Wu, H.H.; Wang, D.; Ma, A.B.; Gu, D.Y. Hip joint geometry effects on cartilage contact stresses during a gait cycle. In Proceedings of the 36th Annual International Conference of the IEEE Engineering in Medicine and Biology Society, Orlando, FL, USA, 16–20 August 2016; IEEE: Piscataway, NJ, USA, 2016. ISBN 9781457702198.
55. Widmer, K.-H.; Zurfluh, B.; Morscher, E.W. Load transfer and fixation mode of press-fit acetabular sockets. *J. Arthroplast.* **2002**, *17*, 926–935. [CrossRef] [PubMed]
56. Doube, M.; Klosowski, M.M.; Arganda-Carreras, I.; Cordelieres, F.P.; Dougherty, R.P.; Jackson, J.S.; Schmid, B.; Hutchinson, J.R.; Shefelbine, S.J. BoneJ: Free and extensible bone image analysis in Image. *J. Bone* **2010**, *47*, 1076–1079. [CrossRef] [PubMed]
57. Bouxsein, M.L.; Boyd, S.K.; Christiansen, B.A.; Guldberg, R.E.; Jepsen, K.J.; Muller, R. Guidelines for assessment of bone microstructure in rodents using micro-computed tomography. *J. Bone Miner. Res.* **2010**, *25*, 1468–1486. [CrossRef]
58. Ashworth, J.C.; Mehr, M.; Buxton, P.G.; Best, S.M.; Cameron, R.E. Cell Invasion in Collagen Scaffold Architectures Characterized by Percolation Theory. *Adv. Healthc. Mater.* **2015**, *4*, 1317–1321. [CrossRef]
59. Ashworth, J.C.; Mehr, M.; Buxton, P.G.; Best, S.M.; Cameron, R.E. Parameterizing the Transport Pathways for Cell Invasion in Complex Scaffold Architectures. *Tissue Eng. Part C Methods* **2016**, *22*, 409–417. [CrossRef]
60. Peters, F.; Schwarz, K.; Epple, M. The structure of bone studied with synchrotron X-ray diffraction, X-ray absorption spectroscopy and thermal analysis. *Thermochim. Acta* **2000**, *361*, 131–138. [CrossRef]
61. Geffers, M.; Groll, J.; Gbureck, U. Reinforcement Strategies for Load-Bearing Calcium Phosphate Biocements. *Materials* **2015**, *8*, 2700–2717. [CrossRef]
62. Roohani-Esfahani, S.I.; Dunstan, C.R.; Li, J.J.; Lu, Z.; Davies, B.; Pearce, S.; Field, J.; Williams, R.; Zreiqat, H. Unique microstructural design of ceramic scaffolds for bone regeneration under load. *Acta Biomater.* **2013**, *9*, 7014–7024. [CrossRef]
63. Ritchie, R.O. The conflicts between strength and toughness. *Nat. Mater.* **2011**, *10*, 817–822. [CrossRef]
64. Munch, E.; Launey, M.E.; Alsem, D.H.; Saiz, E.; Tomsia, A.P.; Ritchie, R.O. Tough, bio-inspired hybrid materials. *Science* **2008**, *322*, 1516–1520. [CrossRef] [PubMed]
65. Habraken, W.; Habibovic, P.; Epple, M.; Bohner, M. Calcium phosphates in biomedical applications: Materials for the future? *Mater. Today* **2016**, *19*, 69–87. [CrossRef]
66. Van Haaren, E.H.; Heyligers, I.C.; Alexander, F.G.M.; Wuisman, P.I.J.M. High rate of failure of impaction grafting in large acetabular defects. *J. Bone Jt. Surg. Br.* **2007**, *89*, 296–300. [CrossRef] [PubMed]
67. Chappard, D.; Terranova, L.; Mallet, R.; Mercier, P. 3D Porous Architecture of Stacks of beta-TCP Granules Compared with That of Trabecular Bone: A microCT, Vector Analysis, and Compression Study. *Front. Endocrinol.* **2015**, *6*, 161. [CrossRef] [PubMed]
68. Epple, M. Review of potential health risks associated with nanoscopic calcium phosphate. *Acta Biomater.* **2018**, *77*, 1–14. [CrossRef]

materials

Article

Biomechanical Study of a Novel, Expandable, Non-Metallic and Radiolucent CF/PEEK Vertebral Body Replacement (VBR)

Daniel Adler [1,†], Michael Akbar [1,*,†], Anna Spicher [2], Stephanie-Alice Goerke [3] and Werner Schmoelz [2]

1 Spine Center, Department of Orthopaedic Surgery, Trauma Surgery and Division of Spinal Cord Injury, Ruprecht-Karls-University Heidelberg, Schlierbacher Landstraße 200A, 69118 Heidelberg, Germany
2 Department of Trauma Surgery, Medical University of Innsbruck, Innrain 52, 6020 Innsbruck, Austria
3 Department of Anatomy, Medical University of Innsbruck, Innrain 52, 6020 Innsbruck, Austria
* Correspondence: michael.akbar@med.uni-heidelberg.de; Tel.: +49-6221-56-26305; +49-6221-56-27630
† Adler and Akbar contributed equally to the article and share first authorship.

Received: 7 August 2019; Accepted: 23 August 2019; Published: 26 August 2019

Abstract: Vertebral body replacement is well-established to stabilize vertebral injuries due to trauma or cancer. Spinal implants are mainly manufactured by metallic alloys; which leads to artifacts in radiological diagnostics; as well as in radiotherapy. The purpose of this study was to evaluate the biomechanical data of a novel carbon fiber reinforced polyetheretherketone (CF/PEEK) vertebral body replacement (VBR). Six thoracolumbar specimens were tested in a six degrees of freedom spine tester. In all tested specimens CF/PEEK pedicle screws were used. Two different rods (CF/PEEK versus titanium) with/without cross connectors and two different VBRs (CF/PEEK prototype versus titanium) were tested. In lateral bending and flexion/extension; range of motion (ROM) was significantly reduced in all instrumented states. In axial rotation; the CF/PEEK combination (rods and VBR) resulted in the highest ROM; whereas titanium rods with titanium VBR resulted in the lowest ROM. Two cross connectors reduced ROM in axial rotation for all instrumentations independently of VBR or rod material. All instrumented states in all planes of motion showed a significantly reduced ROM. No significant differences were detected between the VBR materials in all planes of motion. Less rigid CF/PEEK rods in combination with the CF/PEEK VBR without cross connectors showed the smallest reduction in ROM. Independently of VBR and rod material; two cross connectors significantly reduced ROM in axial rotation. Compared to titanium rods; the use of CF/PEEK rods results in higher ROM. The stiffness of rod material has more influence on the ROM than the stiffness of VBR material.

Keywords: Vertebral body replacement (VBR); non metallic; radiolucent; CF/PEEK; biomechanics; tumor; vertebral fracture; spine

1. Introduction

Vertebral body replacement (VBR) combined with dorsal instrumentation is the preferred treatment option to achieve decompression and restore stability of the spinal column either in patients with a traumatic fracture (>AO Spine A3) [1–6] or major pathological instability due to infection [7–10], spinal metastases [11–13] or primary malignancy [7,8,11–13]. Metastatic spinal lesions occur in 5%–10% of all cancer patients while primary spinal tumors are a rare entity [12]. Most spinal metastases are located anteriorly in the vertebral body and in the peridural space, while a dorsal location is rare [4,6,10–12,14]. The incidence of clinically apparent metastatic disease in the spine has increased since advances in tumor treatment generally have improved the life expectancy of cancer patients [10,15]. Local tumor progression can lead to vertebral body collapse with decompensation of the sagittal profile, causing pain and neurological deficits [10,15,16]. Surgical treatment options are mainly palliative and adapted to the patients' general condition, aiming for minimal peri- and postoperative morbidity [10,15]. However, more radical treatment options with removal of spinal lesions showed better results in patients with good prognosis [17–20]. A loss of correction [4,7,8] or implant failures due to loosening at the implant–bone interface with recurrent instability [1,9,10,13,16] are well described after isolated posterior instrumentation. 360° instrumentations provide the highest stiffness in biomechanical studies, where a rotationally unstable corpectomy defect model is used [4,6,7,10]. Postoperative local radiotherapy and chemotherapies may complicate the postoperative course by prolonged instability with wound infections (threefold higher risk), hardware failure, or progressive implant loosening [21,22]. In tumor patients, general condition and life expectancy have to be taken into account. Patients with prolonged life expectancy (>12 months) are recommended to be treated with a 360° instrumentation [15,17–20]. To provide sufficient rotational stability and load-bearing capacity, the vast majority of common VBRs are manufactured from different metallic alloys, preferably titanium. But imaging artifacts due to the presence of metal components in CT or MRI imaging adversely complicates postoperative radiological diagnostics in local tumor recurrence or local radiation therapy. High strength, non-metallic dorsal and ventral implants made from carbon fiber reinforced polyetheretherketone (CF/PEEK) have recently become available to avoid these artifacts [10]. Precise knowledge of biomechanical characteristics (VBR combined with dorsal instrumentation) is essential for safe clinical application and long-term spinal stability [10,11].

Various biomechanical [1,4–6,11,12] and clinical studies [2,3,23,24] have addressed spinal stability after implantation of a metallic VBR, while biomechanical studies evaluating non-metallic VBR are rare. The objective of the present study was to quantitatively analyze the biomechanical data of a new, non-metallic CF/PEEK VBR in combination with a dorsal CF/PEEK screw and rod system, with and without cross connectors. The implant combinations with the novel material were compared to a common titanium VBR in combination with titanium rods. Construct stiffness and flexibility (range of motion (ROM)) were tested in a six degrees of freedom spine simulator. To our best knowledge, this is the first study to evaluate the biomechanical characteristics of an in-situ, expandable, non-metallic, radiolucent VBR in combination with non-metallic, radiolucent posterior pedicle screw instrumentation.

2. Materials and Methods

Six (n = 6) fresh frozen human thoracolumbar spine units (Th11–L3) were tested with a mean age at death of 59.2 years (ranging from 56 to 65 years). The bodies were donated by people who had given their written informed consent prior to death to use their bodies for scientific and educational purposes. For standardization and homogeneous study conditions, all tested vertebral bodies were analyzed via pre-interventional CT scan (General Electrics, Lightspeed VCT 16, qCT including EFP calibration, GE Medical Systems, USA). Trabecular bone mineral density (BMD) was measured between 66.8 mg/ccm and 100.4 mg/ccm (mean BMD: 82.0 mg/ccm, SD: ± 27.6). Spinal specimens with deformities, previous spinal surgery, structural disorders or post traumatic disorders were excluded. The vacuum sealed, frozen specimens (minus 20 °C) were thawed overnight at 6 °C before all soft tissues were removed, preserving supporting spinal ligaments and joint capsules.

For posterior instrumentation, pedicle entry points were identified according to the anatomical landmarks and controlled using bi-planar fluoroscopy. Nonmetallic, radiolucent 6.5 × 45 mm CF/PEEK pedicle screws (VADER®, icotec, Altstätten, Switzerland) were inserted in all specimens at the Th11/12 level, as well as at the L2/3 level for all investigated cases (Figure 1). Pedicle screws were combined with two different rod types in the various cases tested: A nonmetallic and radiolucent 6 mm CF/PEEK rod (icotec: VADER®, Altstatten, Switzerland) and a 6.0 mm standard titanium rod system (icotec ag, Altstätten, Switzerland). Titanium cross connectors (icotec ag, Altstätten, Switzerland) completed the setup.

Figure 1. Non-metallic X-ray-translucent carbon fiber reinforced polyetheretherketone (CF/PEEK) pedicle screws (icotec: VADER®, Altstätten, Switzerland).

The cranial (TH11) and the caudal vertebra (L3) were embedded in polymethylmethacrylate (PMMA: Heraeus Kulzer GmbH, Technovit 3040, Wehrheim, Germany) with sufficient clearance for pedicle instrumentation and with the midline of all five vertebrae aligned horizontally. Flanges in the upper and lower PMMA blocks provided a rigid fixation to the spine simulator. To measure intersegmental motion of the treated (Th12–L2) and adjacent segments (Th11/12 and L2/3), an ultrasound-based 3D motion analysis system (Winbiomechanics, Zebris, Isny, Germany, resolution 0.1°) was fixed to the ventral side of the vertebrae (Figure 2). According to international standards, all tests were carried out at room temperature and specimens were kept moist with isotonic saline solution for the study period [25–28]. With respect to the recommendations for testing of spinal implants [25–28], biomechanical testing was performed in a six degrees of freedom spine simulator as described by Knop et al. [6] and Schmoelz et al. [29], equipped with a six-component load cell (Schunk FT Delta SI 660-60, Lauffen/Neckar, Germany) with feedback control and a connection to a stepper motor for load application (Figure 2). For all tests, pure moments of ±7.5 Nm were applied in the three main motion planes: Flexion/extension, lateral bending (left/right) and axial rotation (left/right).

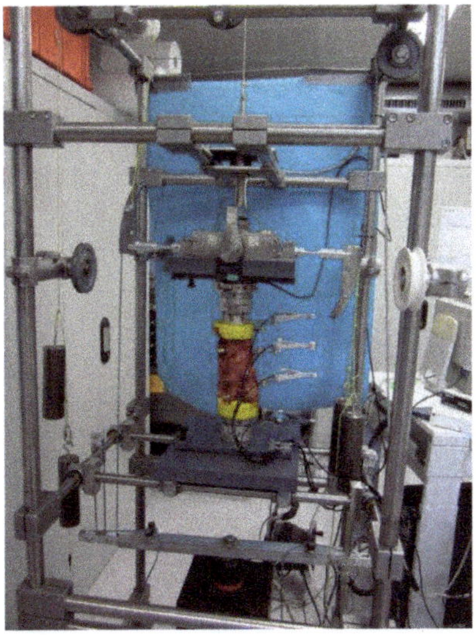

Figure 2. Spine tester (Schunk FT Delta SI 660-60, Lauffen/Neckar, Germany) with six degrees of freedom; three in translation (green) and three in rotation (orange). Specimen embedded in PMMA.

At first, the intact specimens (Th11-L3) were loaded with pure moments of 7.5 Nm to record a baseline. Afterwards a corpectomy of the target vertebra L1 with removal of the cranial and caudal discs was conducted with standard surgical tools, according to clinical routine. For the reconstruction of the anterior spinal column, either an expandable, nonmetallic radiolucent CF/PEEK VBR (prototype, icotec Altstätten, Switzerland) (Figure 3) or an expandable titanium VBR (X-Core, Nuvasive, Bremen, Germany) was implanted. Both VBRs consist of an in situ expandable centerpiece, to which modular endplates can be attached. While the CF/PEEK VBR prototype is nonmetallic and radiolucent with macrostructured endplates to prevent dislocation, the X-Core VBR is manufactured from titanium alloy and its endplates are equipped with spikes to prevent dislocation. For both implants, height can be adjusted continuously within a clinically relevant range by a gear wheel drive unit. The desired height can be locked in position using a locking screw. The VBR endplate´s size was determined via templates and the VBR was placed in typical technique.

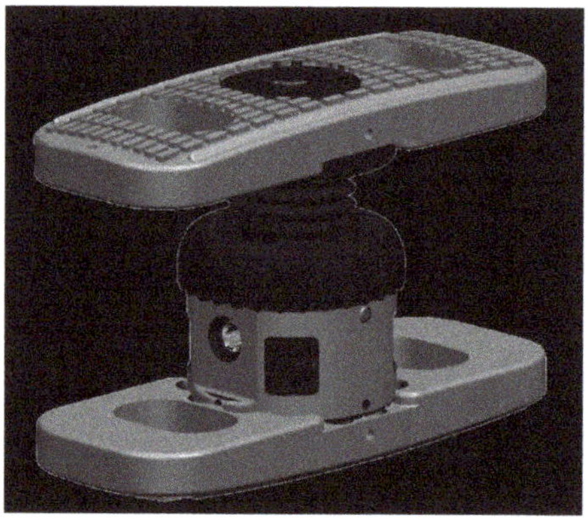

Figure 3. Prototype of the non-metallic, X-ray-translucent CF/PEEK expandable VBR. icotec, Altstätten, Switzerland.

All surgical procedures were performed by the first author (experienced senior spine surgeon). Two plane (anterior, posterior and lateral view) native radiographs were taken to control and document correct positioning of pedicle screws and VBRs (Figure 4a,b). Final fixation of posterior rods to the pedicle screws was carried out in a standardized fashion with an axial preload of the spine simulator of 100 N.

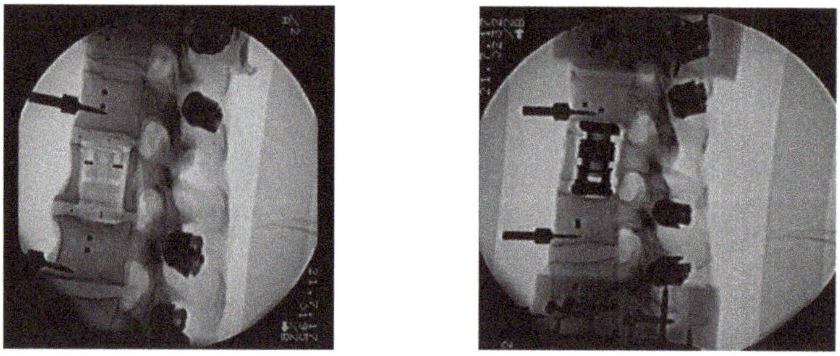

(a) (b)

Figure 4. Lateral view native radiographs documenting correct positioning of the (**a**) prototype non-metallic, X-ray-translucent CF/PEEK expandable vertebral body replacement (VBR) and the expandable titanium VBR (**b**).

3. Study Protocol

For each specimen, the following states were tested with alternatives for the VBR (CF/PEEK or titanium), the posterior rod material (CF/PEEK or titanium), and the application of cross connectors.

(1) Flexibility test, Native (**native**): *Corpectomy and instrumentation with VBR and pedicle fixation with CF/PEEK rod.*
(2) Flexibility test—CF/PEEK *VBR* instrumented with *CF/PEEK rod* (**CF_CF**): *Change of posterior rod fixation to titanium rod.*
(3) Flexibility test—CF/PEEK *VBR* Instrumented with *titanium rod,* (**CF_Ti**): *Change of VBR to titanium.*
(4) Flexibility test—titanium *VBR* instrumented with *titanium rod* (**Ti_Ti**): *Addition of two cross connectors to the posterior rods dissecting the ligamentum supraspinale/interspinale.*
(5) Flexibility test—titanium *VBR* instrumented with *titanium rod* and 2 cross connectors (**Ti_Ti_cc**): *Change to CF/PEEK VBR.*
(6) Flexibility test—*CF/PEEK VBR* instrumented with *titanium rod* and 2 cross connectors (**CF_Ti_cc**): *Change rods to CF/PEEK.*
(7) Flexibility test—*CF/PEEK VBR* Instrumented with *CF/PEEK rod* and 2 cross connectors (**CF_CF_cc**).

Statistical analysis of the ROM was performed using the SPSS software (Microsoft Windows release 24, SPSS Inc., Chicago, IL, USA). Data and results of the ROM were evaluated for the three motion directions, normalized and compared to the motion of the native segment. A general linear model (GLM) with repeated measures was used for statistical comparison. p-values were calculated with adjustment for multiple corrections.

4. Results

Results of the ROM are displayed in absolute metrics (Table 1) and as boxplot normalized to non-instrumented native condition (Figure 5) in the three planes of motion for the index segments (Th12–L2).

Table 1. Results of the range of motion (ROM) test.

	Lateral Bending Th12-L2		Flexion/Extension Th12-L2		Axial Rotation Th12-L2	
	Mean	*SD*	*Mean*	*SD*	*Mean*	*SD*
Native	11.17	3.26	10.32	2,40	4.83	0.95
CF_CF	0.27	0.18	0.33	0.18	3.95	0.90
CF_CF_cc	0.30	0.21	0.30	0.20	2.74	0.46
CF_Ti	0.27	0.20	0.30	0.12	2.88	0.40
CF_Ti_cc	0.24	0.18	0.29	0.17	1.95	0.33
Ti_Ti	0.25	0.21	0.31	0.17	2.55	0.53
Ti_Ti_cc	0.24	0.21	0.35	0.22	1.84	0.33

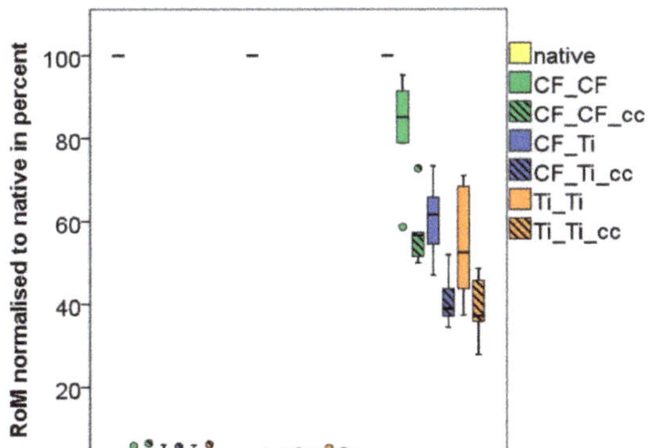

Figure 5. Boxplot showing the median and quartiles of the index segment for all tested states in the three motion directions normalized in percentage of the native state.

4.1. Index Segments (Th12-L2)

All instrumented states in all planes of motion showed a significantly ($p < 0.05$) reduced ROM compared to the native state.

In lateral bending, the mean ROM in the native state was 11.2° (SD 3.26). The various instrumentations reduced ROM to 0.24°–0.30° (SD 0.18–0.21). No significant differences (<0.1°, $p = 1.0$) in ROM were measured for the different VBR material (CF/PEEK versus titanium), posterior rod material (CF/PEEK versus titanium) or additional cross connectors.

In flexion/extension, the mean ROM in the native state was 10.3° (SD 2.4). Varying instrumentations reduced ROM to 0.29°–0.35° (SD 0.12–0.31). No significant differences (<0.1°, $p = 1.0$) in ROM were measured for the different VBR material (CF/PEEK versus titanium), posterior rod material (CF/PEEK versus titanium), or additional cross connectors.

In axial rotation the mean ROM in native state was 4.83° (SD 0.95). The various instrumentations reduced ROM to 1.84°–3.95° (SD 0.33–0.9). Less rigid CF/PEEK rods combined with CF/PEEK VBR without the use of cross connectors showed the smallest ROM reduction to 3.95° (SD 0.9) in axial rotation. Titanium rods with titanium VBR and the use of two cross connectors reduced ROM to 1.84° (SD 0.33). Independent of type of VBR or rod material (CF/PEEK or titanium), two additional cross connectors reduced significantly, ($p < 0.05$) the ROM in axial rotation. Posterior rod material (CF/PEEK or titanium) had greater effects to the ROM than VBR material (CF/PEEK or titanium) in axial rotation.

4.2. Caudal Adjacent Segment (L2/3)

Equally to the index level, all instrumented states showed in all planes of motion a significantly ($p < 0.05$) reduced ROM compared to native state.

In lateral bending the mean ROM in native state was 7.9° (SD 2.64). The various instrumentations reduced ROM to 1.38°–1.76° (SD 0.66–1.02). No significant differences (<0.4°, $p > 0.78$) in ROM were measured for the different VBR materials (CF/PEEK versus titanium), posterior rod materials (CF/PEEK versus titanium) or additional cross connectors. Similar to the index segment, mean ROM of the L2/3 segment in lateral bending was slightly higher (0.2°) with the use of two cross connectors.

In flexion/extension the mean ROM in native state was 7.5° (SD 2.76). The various instrumentations reduced ROM to 1.13°–1.50° (SD 0.58–0.95). No significant differences in ROM were measured for the different VBR material (CF/PEEK versus titanium), posterior rod material (CF/PEEK versus titanium) or additional cross connectors. Similar to lateral bending mean ROM was slightly (0.2°) higher with the use of two cross connectors.

In axial rotation the mean ROM in native state was 3.88° (SD 2.33). The various instrumentations reduced ROM to 1.27°–1.96° (SD 0.48–0.85). Less rigid CF/PEEK rods combined with CF/PEEK VBR without the use of cross connectors again showed smallest ROM reduction to 1.96° (SD 0.85) in axial rotation. Titanium rods with titanium VBR and the use of two cross connectors reduced ROM to 1.27° (SD 0.48). Independently to VBR or rod material (CF/PEEK or titanium) two cross connectors significantly reduced ROM in axial rotation ($p < 0.05$). Posterior rod material (CF/PEEK or titanium) had greater effects to the ROM than VBR material (CF/PEEK or titanium) in axial rotation.

4.3. Cranial Adjacent Segment (Th11/12)

Equally to the index and caudal adjacent level, all instrumented states showed, in all planes of motion, a significantly ($p < 0.05$) reduced ROM compared to native state.

In lateral bending the mean ROM in native state was 4.49° (SD 3.40). The various instrumentations reduced ROM to 2.61°–2.96° (SD 1.41–1.63). No significant differences in ROM were measured for the different VBR material (CF/PEEK versus titanium), posterior rod material (CF/PEEK versus titanium) or additional cross connectors. Similar to the index and caudal adjacent level mean ROM of Th11/12 segment was slightly (0.3°) higher with the use of two cross connectors in lateral bending.

In flexion/extension the mean ROM in native state was 6.59° (SD 2.80). The various instrumentations reduced ROM to 1.87°–2.28° (SD 1.34–1.90). No significant differences in ROM were measured for the different VBR material (CF/PEEK versus titanium), posterior rod material (CF/PEEK versus titanium) or additional cross connectors. Similar to lateral bending the mean ROM was slightly (0.3°) higher with the use of two cross connectors in lateral bending.

In axial rotation the mean ROM in native state was 4.27° (SD 1.93). The various instrumentations reduced ROM to 2.70°–3.25° (SD 1.27–1.47). Less rigid CF/PEEK rods combined with CF/PEEK VBR without the use of cross connectors again showed smallest ROM reduction to 3.25° (SD 1.45) in axial rotation. Titanium rods with titanium VBR and the use of two cross connectors reduced ROM to 2.70° (SD 1.42). Independently to VBR or rod material (CF/PEEK or titanium) two cross connectors reduced ($p < 0.05$) ROM in axial rotation slightly. Posterior rod material (CF/PEEK or titanium) had greater effects on the ROM than VBR material (CF/PEEK or titanium) in axial rotation.

5. Discussion

This is the first report of a biomechanical testing series utilizing a novel, nonmetallic, radiolucent and expandable CF/PEEK VBR for 360° instrumentation in combination with an established screw and rod system (rod material CF/PEEK and titanium). For this purpose a human cadaveric corpectomy defect model was tested in a spine tester with three-dimensional motion measurement of each segment. The implants presented in this study were designed to reconstruct spinal stability after corpectomy in traumatic or malignant vertebral fractures. Reconstruction was varied by anterior VBR material (CF/PEEK versus titanium) and dorsal instrumentation's material (CF/PEEK versus titanium) with or without additional cross connectors.

Our results indicate the less rigid CF/PEEK rods combined with CF/PEEK VBR without the use of cross connectors reduces the ROM in axial rotation by only 18% ($\gamma = 3.95°$, SD 0.9) compared to an intact specimen ($\gamma = 4.83°$, SD 0.95). In all other modes (flexion/extension and lateral bending) the CF/PEEK rods combined with CF/PEEK VBR provided comparable reduction in ROM when compared to titanium rods in combination with a titanium VBR.

Titanium rods in combination with a titanium VBR and the use of two cross connectors demonstrated significantly more stiffness in axial rotation with a decrease of ROM of 62% ($\gamma = 1.84°$, SD 0.33). Independently to VBR or rod material (CF/PEEK or titanium) two cross connectors significantly reduced ($p < 0.05$) the ROM in axial rotation. Posterior rod material (CF/PEEK or titanium) had greater effects on the ROM than VBR material (CF/PEEK or titanium) in axial rotation (Table 1, Figure 5). In lateral bending and flexion/extension varying instrumentations significantly ($p < 0.05$) reduced mean ROM by 97% and 96% compared to the intact specimen, respectively. There were no significant differences between different VBR and posterior rod system (CF/PEEK versus titanium).

Different studies [11,12,29–31] evaluated the biomechanical behavior of VBRs in combination with screw and rod systems. The length of the posterior instrumentation was shown to be the major determinant for the constructs' stability/stiffness [30,31]. Longer posterior instrumentation (two adjacent levels, cranial and caudal to the VBR) provided significant higher stiffness compared to bi-segmental instrumentation even with an additional antero-lateral plate. Therefore, only posterior instrumentations with two adjacent levels above and below the VBR were tested in this study. Further studies [4–8,13] compared the stability of different in situ expandable VBRs and non-expandable VBRs in combination with/without posterior screw and rod systems and/or anterior instrumentation with a locked angular stable plate. Isolated anterior instrumentation (VBR and an antero-lateral plate) revealed a significantly lower stiffness compared to intact specimens. A significant increase of stiffness in all motion planes was detected after additional posterior screw and rod instrumentation [4–8,13]. Isolated anterior instrumentation (VBR combined with an additional anterior polyaxial or angular stable plating) cannot be recommended for stabilization of vertebral corpectomy defects [4–8,13]. In clinical routine, isolated anterior spinal instrumentation in metastatic disease to the spine is inappropriate. Consistent with other authors [4,6–8,12,23] we recommend either posterior instrumentation with/without decompression (patients with limited life expectancy or bad general condition) or a 360° instrumentation in patients with adequate general condition and life expectancy.

Carbon fiber reinforced polymer intervertebral implants were already described by Brantigan and Steffee in 1991 with excellent biomechanical results and fusion rates [32,33]. Schulte et al. [10] described a vertebral body replacement with a bioglass-polyurethane spacer fixed with a ventral plate of carbon-fiber reinforced polyethererherketone (CF-PEEK). Biomechanical testing detected a significant reduction of ROM in all three motion planes. In the course of the study, one patient died 18 months post operation. After biopsy and biomechanical testing of the explanted spinal segments the ROM in all motion planes demonstrated values comparable to the previous biomechanical testing with cadavers. Early signs of osteointegration at the bone-endplate interface in combination with mechanical interlocking by bony heterotypic ossifications resulted in an even improved stability. In comparison to titanium implants this osseous integration might provide additional stability in CF/PEEK implants.

Various non-expandable and expandable VBRs were tested [8,34,35] on primary stiffness. In 360° instrumentations the material of VBR was shown to have a minor effect in the treated segment. Anterior PMMA constructs [35], as well as titanium mesh cages [34], in combination with multilevel posterior instrumentation provided higher stiffness than intact specimens. Consistent to these results in the present study, no significant differences between the VBR materials (CF/PEEK versus titanium) were detected.

360° instrumentations are proven to be biomechanically superior to isolated posterior instrumentations regarding stability and stiffness in treated spinal segments [8,31,34,36]. But it has to be kept in mind that an additional ventral stabilization increases surgical risk factors like an enlarged surgical approach, higher blood loss, increased risk of infection, and prolonged operation time [11]. Therefore, these procedures should be performed after carefully individualized decision making depending on the patient's general condition, and only in experienced spine centers [11,16].

Common limitations of biomechanical in vitro testing also apply to the present study. Due to the lack of influence on biomechanical characteristics of vital spinal muscles, in vitro models are reduced to bony and ligamentary structures [11,34,35]. Tissue healing and consolidation of the bone as in vivo factors cannot be displayed and analyzed. Comparing the results with other studies is difficult, as variable testing conditions, sequences, and specimen characteristics (level, BMD, age, and species) can be found in the literature [11]. Another limitation is the relatively small sample size which was used to investigate the various reconstruction options. However, common inter-individual variables found in clinical routine can be excluded in the controlled laboratory environment. If the biomechanical effect of an intervention is not provable in controlled laboratory standards with a limited size of samples it is assumed rather unlikely to have clinical impact [37]. Conclusions concerning intermediate and long term stability of spinal reconstructions cannot be drawn from the present study as no cyclic loading was performed. Primary stability of the implants was determined with assessment of ROM using pure moments in a six degrees of freedom spine tester. The use of pure moments has well described limitations depending on the set-up features. Nevertheless, implant testing with pure moments is a worldwide [25–27,38] accepted method to compare various types of spinal instrumentations.

6. Conclusions

Compared to the native state, all instrumented states showed a significantly ($p < 0.05$) reduced ROM in all planes of motion. No significant differences were detected between the VBR materials (CF/PEEK versus titanium) in all planes of motion. Less rigid CF/PEEK rods in combination with the CF/PEEK VBR without cross connectors showed the smallest reduction in ROM. Independently of VBR and rod material (CF/PEEK versus titanium), two cross connectors significantly reduced ROM in axial rotation and are therefore highly recommended. Compared to titanium rods, the use of CF/PEEK rods results in higher ROM. The stiffness of rod material has more influence on the ROM than the stiffness of VBR material.

Author Contributions: D.A., surgical specimens preparation, methodology, writing—original draft preparation, visualization, review & editing; M.A., writing—review and editing, supervision, project administration; A.S. and S.-A.G., anatomical specimens preparation, formal analysis, Data curation; W.S., Conceptualization, data curation, formal analysis, resources, validation, supervision, project administration, investigation, writing—review and editing.

Funding: The laboratory costs of the study were supported by institutional funds of icotec, Altstätten, Switzerland. The funding sources had no involvement in the study design; in the collection, analysis and interpretation of data; or in the decision to submit the paper for publication.

Acknowledgments: Implants used in the present study were provided free of charge by icotec, Altstätten, Switzerland. We thank Ralf Klabunde, employed by icotec ag, who contributed his profound expertise. The authors would like to express their gratitude to individuals who donated their bodies and tissues for the advancement of education and research. We acknowledge financial support by Deutsche Forschungsgemeinschaft within the funding programme Open Access Publishing, by the Baden-Württemberg Ministery of Science, Research and the Arts and by Ruprecht-Karls-Universität Heidelberg.

Conflicts of Interest: None of the remaining authors has a conflict of interest and none received payment of any sort for their contributions to this work.

References

1. James, K.S.; Wenger, K.H.; Schlegel, J.D.; Dunn, H.K. Biomechanical evaluation of the stability of thoracolumbar burst fractures. *Spine* **1994**, *19*, 1731–1740. [CrossRef] [PubMed]
2. Gradl, G. Combined stabilization of thoracolumbar spine fractures. *Eur. J. Trauma* **2006**, *32*, 249–252. [CrossRef]
3. Schnake, K.J.; Stavridis, S.I.; Kandziora, F. Five-year clinical and radiological results of combined anteroposterior stabilization of thoracolumbar fractures. *J. Neurosurg. Spine* **2014**, *20*, 497–504. [CrossRef] [PubMed]
4. Ulmar, B.; Erhart, S.; Unger, S.; Weise, K.; Schmoelz, W. Biomechanical analysis of a new expandable vertebral body replacement combined with a new polyaxial antero-lateral plate and/or pedicle screws and rods. *Eur. Spine J.* **2012**, *21*, 546–553. [CrossRef] [PubMed]
5. Rohlmann, A.; Zander, T.; Fehrmann, M.; Klockner, C.; Bergmann, G. Influence of implants for vertebral body replacement on the mechanical behavior of the lumbar spine. *Orthopade* **2002**, *31*, 503–507. [CrossRef] [PubMed]
6. Knop, C.; Lange, U.; Bastian, L.; Blauth, M. Three-dimensional motion analysis with Synex. Comparative biomechanical test series with a new vertebral body replacement for the thoracolumbar spine. *Eur. Spine J.* **2000**, *9*, 472–485. [CrossRef] [PubMed]
7. Khodadadyan-Klostermann, C.; Schaefer, J.; Schleicher, P.; Pflugmacher, R.; Eindorf, T.; Haas, N.P.; Kandziora, F. Expandable cages: biomechanical comparison of different cages for ventral spondylodesis in the thoracolumbar spine. *Chirurg* **2004**, *75*, 694–701. [PubMed]
8. Pflugmacher, R.; Schleicher, P.; Schaefe, J.; Scholz, M.; Ludwig, K.; Khodadadyan-Klostermann, C.; Haas, N.P.; Kandziora, F. Biomechanical comparison of expandable cages for the vertebral body replacement in the thoracolumbar spine. *Spine* **2004**, *29*, 1413–1419. [CrossRef]
9. Wilke, H.J.; Wenger, K.; Claes, L. Testing criteria for spinal implants: recommendations for the standardization of in vitro stability testing of spinal implants. *Eur. Spine J.* **1998**, *7*, 148–154. [CrossRef]
10. Schulte, M.; Schultheiss, M.; Hartwig, E.; Wilke, H.J.; Wolf, S.; Sokiranski, R.; Fleiter, T.; Kinzl, L.; Claes, L. Vertebral body replacement with a bioglass-polyurethane composite in spine metastases—Clinical, radiological and biomechanical results. *Eur. Spine J.* **2000**, *9*, 437–444. [CrossRef]
11. Disch, A.C.; Luzatti, A.; Melcher, I.; Schaser, K.D.; Feraboli, F.; Schmoelz, W. Three-dimensional stiffness in a thoracolumbar en-bloc spondylectomy model: A biomechanical in vitro study. *Clin. Biomech.* **2007**, *22*, 957–964. [CrossRef] [PubMed]
12. Disch, A.C.; Schaser, K.D.; Melcher, I.; Luzzati, A.; Feraboli, F.; Schmoelz, W. En bloc spondylectomy reconstructions in a biomechanics in-vitro study. *Eur. Spine J.* **2008**, *17*, 715–725. [CrossRef] [PubMed]
13. Disch, A.C.; Knop, C.; Schaser, K.D.; Blauth, M.; Schmoelz, W. Angular stable anterior plating following thoracolumbar corpectomy reveals superior segmental stability compared to conventional polyaxial plate fixation. *Spine* **2008**, *33*, 1429–1437. [CrossRef] [PubMed]

14. Harrington, K.D. Orthopedic surgical management of skeletal complications of malignancy. *Cancer* **1997**, *80*, 1614–1627. [CrossRef]
15. Adler, D.; Kriegsmann, M.; Sinn, P.; Schneeweiss, A.; Almansour, H.; Lehner, B.; Akbar, M. Metastatic breast cancer in the spine: Molecular predictors for choosing adequate treatment strategies. *Orthopade* **2018**, *47*, 594–603. [CrossRef]
16. Adler, D.; Almansour, H.; Akbar, M. What is actually adult spinal deformity? Development, classification, and indications for surgical treatment. *Orthopade* **2018**, *47*, 276–287. [CrossRef] [PubMed]
17. Boriani, S.; Biagini, R.; De Lure, F.; Bertoni, F.; Malaguti, M.C.; Di Fiore, M.; Zanoni, A. En bloc resections of bone tumors of the thoracolumbar spine. A preliminary report on 29 patients. *Spine* **1996**, *21*, 1927–1931. [CrossRef]
18. Sundaresan, N.; Steinberger, A.A.; Moore, F.; Sachdev, V.P.; Krol, G.; Hough, L.; Kelliher, K. Indications and results of combined anterior-posterior approaches for spine tumor surgery. *J. Neurosurg.* **1996**, *85*, 438–446. [CrossRef]
19. Tomita, K.; Kawahara, N.; Baba, H.; Tsuchiya, H.; Nagata, S.; Toribatake, Y. Total en bloc spondylectomy for solitary spinal metastases. *Int. Orthop.* **1994**, *18*, 291–298. [CrossRef]
20. Tokuhashi, Y.; Matsuzaki, H.; Oda, H.; Oshima, M.; Ryu, J. A revised scoring system for preoperative evaluation of metastatic spine tumor prognosis. *Spine* **2005**, *30*, 2186–2191. [CrossRef]
21. Bouchard, J.A.; Koka, A.; Bensusan, J.S.; Stevenson, S.; Emery, S.E. Effects of irradiation on posterior spinal fusions. A rabbit model. *Spine* **1994**, *19*, 1836–1841. [CrossRef]
22. Ghogawala, Z.; Mansfield, F.L.; Borges, L.F. Spinal radiation before surgical decompression adversely affects outcomes of surgery for symptomatic metastatic spinal cord compression. *Spine* **2001**, *26*, 818–824. [CrossRef]
23. Thongtrangan, I.; Balabhadra, R.S.; Le, H.; Park, J.; Kim, D.H. Vertebral body replacement with an expandable cage for reconstruction after spinal tumor resection. *Neurosurg. Focus* **2003**, *15*, 1–6. [CrossRef]
24. Brandão, R.A.C.S.; Martins, W.C.D.S.; Arantes Jr, A.A.; Gusmão, S.N.S.; Perrin, G.; Barrey, C. Titanium versus polyetheretherketone implants for vertebral body replacement in the treatment of 77 thoracolumbar spinal fractures. *Surg. Neurol. Int.* **2017**, *8*, 191. [CrossRef]
25. Panjabi, M.M.; Krag, M.; Summers, D.; Videman, T. Biomechanical time-tolerance of fresh cadaveric human spine specimens. *J. Orthop. Res.* **1985**, *3*, 292–300. [CrossRef]
26. Wilke, H.J.; Jungkunz, B.; Wenger, K.; Claes, L.E. Spinal segment range of motion as a function of in vitro test conditions: effects of exposure period, accumulated cycles, angular deformation rate, and moisture condition. *Anat. Rec.* **1998**, *251*, 15–19. [CrossRef]
27. Panjabi, M.M. Biomechanical evaluation of spinal fixation devices: I. A conceptual framework. *Spine* **1988**, *13*, 1129–1134. [CrossRef]
28. Wippermann, B.W.; Schratt, H.E.; Steeg, S.; Tscherne, H. Complications of spongiosa harvesting of the ilial crest. A retrospective analysis of 1191 cases. *Chirurg* **1997**, *68*, 1286–1291. [CrossRef]
29. Schmoelz, W.; Schaser, K.D.; Knop, C.; Blauth, M.; Disch, A.C. Extent of corpectomy determines primary stability following isolated anterior reconstruction in a thoracolumbar fracture model. *Clin. Biomech.* **2010**, *25*, 16–20. [CrossRef]
30. Eichholz, K.M.; Hitchon, P.W.; From, A.; Rubenbauer, P.; Nakamura, S.; Lim, T.H.; Torner, J. Biomechanical testing of anterior and posterior thoracolumbar instrumentation in the cadaveric spine. Invited submission from the joint section meeting on disorders of the spine and peripheral nerves. *J. Neurosurg. Spine* **2004**, *1*, 116–121. [CrossRef]
31. Vahldiek, M.J.; Panjabi, M.M. Stability potential of spinal instrumentations in tumor vertebral body replacement surgery. *Spine* **1998**, *23*, 543–550. [CrossRef]
32. Brantigen, J.W.; Steffee, A.D.; Geiger, J.M. A carbon Fiber implant to aid interbody lumbar fusion. Mechanical testing. *Spine* **1991**, *16*, 277–282. [CrossRef]
33. Brantigen, J.W.; Steffee, A.D. A carbon Fiber implant to aid interbody lumbar fusion. Two-year clinical results in the first 26 patients. *Spine* **1993**, *18*, 2106–2107. [CrossRef]
34. Oda, I.; Cunningham, B.W.; Abumi, K.; Kaneda, K.; McAfee, P.C. The stability of reconstruction methods after thoracolumbar total spondylectomy. An in vitro investigation. *Spine* **1999**, *24*, 1634–1638. [CrossRef]
35. Shannon, F.J.; DiResta, G.R.; Ottaviano, D.; Castro, A.; Healey, J.H.; Boland, P.J. Biomechanical analysis of anterior poly-methylmethacrylate reconstruction following total spondylectomy for metastatic disease. *Spine* **2004**, *29*, 2012–2096. [CrossRef]

36. Knoller, S.M.; Meyer, G.; Eckhardt, C.; Lill, C.A.; Schneider, E.; Linke, B. Range of motion in reconstruction situations following corpectomy in the lumbar spine: a question of bone mineral density? *Spine* **2005**, *30*, E229–E235. [CrossRef]
37. Lange, T.; Schulte, T.L.; Gosheger, G.; Boevingloh, A.S.; Mayr, R.; Schmoelz, W. Effects of multilevel posterior ligament dissection after spinal instrumentation on adjacent segment biomechanics as a potential risk factor for proximal junctional kyphosis: a biomechanical study. *BMC Musculoskelet. Disord.* **2018**, *19*, 57. [CrossRef]
38. Gedet, P.; Thistlethwaite, P.A.; Ferguson, S.J. Minimizing errors during in vitro testing of multisegmental spine specimens: considerations for component selection and kinematic measurement. *J. Biomech.* **2007**, *40*, 1881–1885. [CrossRef]

materials

MDPI

Article

Impact of Electrocautery on Fatigue Life of Spinal Fusion Constructs—An In Vitro Biomechanical Study

Haidara Almansour [1], Robert Sonntag [2], Wojciech Pepke [1], Thomas Bruckner [3], Jan Philippe Kretzer [2] and Michael Akbar [1,*]

1 Clinic for Orthopedics and Trauma Surgery, Heidelberg University Hospital, 69118 Heidelberg, Germany
2 Laboratory of Biomechanics and Implant Research, Clinic for Orthopedics and Trauma Surgery, Heidelberg University Hospital, 69118 Heidelberg, Germany
3 Institute of Medical Biometry and Informatics, University of Heidelberg, 69118 Heidelberg, Germany
* Correspondence: michael.akbar@med.uni-heidelberg.de; Tel.: +49-6221-56-26305 or +49-6221-56-27630; Fax: +49-6221-56-27307

Received: 5 July 2019; Accepted: 31 July 2019; Published: 3 August 2019

check for updates

Abstract: Instrumentation failure in the context of spine surgery is attributed to cyclic loading leading to formation of fatigue cracks, which later propagate and result in rod fracture. A biomechanical analysis of the potential impact of electrocautery on the fatigue life of spinal implants has not been previously performed. The aim of this study was to assess the fatigue life of titanium (Ti) and cobalt-chrome (CoCr) rod-screw constructs after being treated with electrocautery. Twelve spinal constructs with CoCr and Ti rods were examined. Specimens were divided into four groups by rod material (Ti and CoCr) and application of monopolar electrocautery on the rods' surface (control-group and electrocautery-group). Electrocautery was applied on each rod at three locations, then constructs were cyclically tested. Outcome measures were load-to-failure, total number of cycles-to-failure, and location of rod failure. Ti-rods treated with electrocautery demonstrated a significantly decreased fatigue life compared to non-treated Ti-rods. Intergroup comparison of cycles-to-failure revealed a significant mean decrease of almost 9×10^5 cycles ($p = 0.03$). No CoCr-rods failed in this experiment. Electrocautery application on the surface of Ti-rods significantly reduces their fatigue life. Surgeons should exercise caution when using electrocautery in the vicinity of Ti-rods to mitigate the risk of rod failure.

Keywords: electrocautery; titanium alloy; cobalt-chrome alloy; fatigue behavior; biomechanical study

---

1. Introduction

Electrocautery and spinal implants are principal tools in the spine surgeon's armamentarium. These tools are utilized in complex surgeries to treat a multitude of spinal conditions and to ameliorate patients' disabilities [1,2].

Technological advancements in biomechanics and biomaterials have revolutionized the sphere of spinal instrumentation, enabling surgeons to widen the indications for corrective surgery, and broadened the array of surgical techniques. In this context, spinal fusion was first introduced by Hibbs and Peltier to stabilize a spine affected by tuberculosis [3,4]. Albee used the tibia as graft material for spine stabilization [5]. King [6] and Lange [7] attempted internal fixation [1,4]. Harrington [8] introduced an innovative rod distraction system to correct coronal deformity [4], and Cotrel and Dubousset metamorphosed the profession with their segmental instrumentation for simultaneous correction of coronal and sagittal planes, thereby facilitating a three-dimensional (3D) approach to treat a 3D spine deformity [9]. These surgeries demand sophisticated technical aptitude and hence require meticulous preoperative planning. Consequently, one of the major preoperative decisions is choosing the correct deformity-specific, pathology-appropriate, spinal fusion construct [1].

Spinal implants must fulfil many criteria to be considered safe and efficient. Qualities such as biocompatibility, osseointegration, high strength, low Young's modulus, high corrosion, and wear resistance are pivotal to implants' long-term performance. Because no current material can fulfil all the clinical and biomedical requirements, material scientists have endeavored to enhance these qualities utilizing many surface modification techniques [10]. These processes include mechanical treatment, sol-gel application, thermal spraying, chemical and electrochemical treatment, micro-arc oxidation, laser surface modification, friction stir processing, and ion application [11,12].

Heating techniques and plasmochemical techniques occupy an important role in the sphere of biomedical engineering and materials science. Heat treatment is utilized to increase the fatigue strength of an alloy and to orchestrate an optimal balance of the material's ductility, machinability, and stability [10]. Plasma implementation is non-toxic because it is free of solvents [13]. Implementing techniques such as plasma assisted microwave chemical vapor deposition, plasma etching, plasma nitriding, and deposition of a-DLC (amorphous diamond-like carbon) layers inoculated with nitrogen and silicon have been shown to have a significant impact on the microstructure and surface characteristics of alloys [13].

Accurate and appropriate surface modification techniques diversify and optimize the clinical use of alloys in surgical fields. However, spine surgeries are overburdened with post-operative complications [2], including implant-related failure and rod fracture (RF) [1]. The overall complication rates of surgical corrections of spine-deformity are reported to be 30% within two-years after surgery [14–18]. An important source of complications remains the intrinsic limitation of the robustness of spinal implants [15,19–24]. In a prospective multicenter assessment of risk factors for early RF following adult spinal deformity (ASD) surgery with one-year follow-up, Smith et al. [16] identified that RF occurred in as many as 9% of ASD patients, and up to 22% in patients who underwent pedicle subtraction osteotomy (PSO), a powerful technique for the correction of sagittal spino-pelvic malalignment. In another retrospective study, Smith et al. [23] found that most RF occurred within one year of index surgery.

In the literature to date, several risk factors for RF have been analyzed. They could be patient-related—e.g., higher Body Mass Index (BMI), patient age, history of spine surgery, development of pseudarthrosis, and greater baseline sagittal spinopelvic malalignment (Sagittal Vertical Axis (SVA), Pelvic Tilt (PT), and Pelvic Incidence-Lumbar Lordosis (PI-LL) mismatch), and a greater need for sagittal correction—or implant-related (implant material, or in situ rod contouring in the context of PSO) [16,21,23,24]. Notably, the arthroplasty literature provided a clue to an underlying confounder camouflaged behind the subtleties of RF: the thermal damage of electrocautery to the microstructure of implants. Huber et al. [25] were the first to document a case of shaft breakage of a hip endoprosthesis secondary to contact with an electrocautery device. Subsequently, Konrads et al. [26] reported on four similar cases which occurred after revision surgery. These authors argued that the thermal microstructural disruption ensued from an intra-operative electric arc seen post-contact of the electrosurgical electrode with the implant; this could have eventually led to breakage of the implant at the contact point. In addition, Yuan et al. [27] examined 1859 explanted hip implants and concluded that "iatrogenic arc melting" due to electrocautery was responsible for the color changes and corrosion on the surface of the studied implants. Furthermore, via fatigue testing, metallurgic analysis, and electron microscopy, Sonntag et al. [28] delineated a microstructural change in the Ti6A14V base material with high-frequency electrosurgery on hip endoprosthesis, resulting in a decreased load-to-failure when the electrocautery tip was applied to high-load areas of the stem during revision surgery.

Clinically speaking, it is difficult to avoid contact between the electrocautery tip and the implant during spine surgery. Particularly in cases of revision surgery, these electrical arcs could be deemed common.

To the best of our knowledge, this is the first biomechanical study that addresses this topic in spine surgery. In concordance with the American Standard for Testing and Materials, our objective was to assess the impact of electrocautery contact on the fatigue life of titanium (Ti) and cobalt-chrome (CoCr) spinal implants.

2. Materials and Methods

2.1. Spinal Fusion Constructs and Vertebrectomy Models

Twelve unused specimens of CoCr and Ti implants were included (Expedium Spine, DePuy, Raynham, MA, USA). All rods and pedicle screw constructs had the same diameter (5 mm) and were cut to a length of 100 mm. Monoaxial pedicle screws (Ti6-Al4-V, 5.5 mm × 7 mm × 30 mm) and locking screws (Ti6-Al4-V, 7 mm) were used to rigidly couple the rods on the vertebrectomy model according to manufacturer-specific instructions using original instruments. Models were composed of dual Polypoxymethylene (POM) with dimensions as described by the American Standard for Testing and Materials (ASTM) F1717 [29]. This standard provides a uniform approach to conduct fatigue testing on spinal implants. The two blocks comprising each vertebrectomy model simulate two adjacent vertebrae fixed by posterior-instrumentation. The standard distance between the two blocks was 56 mm.

Specimens were divided into four groups by rod material and application of monopolar-electrocautery device on the rods' surface (Table 1).

Table 1. Specimen grouping matrix.

Group	Description	Samples
Ti-CG	Titanium rods control group (without electrocautery application)	$n = 3$
Ti-EG	Titanium electrocautery group	$n = 3$
CoCr-CG	Cobalt-Chrome control group (without electrocautery application)	$n = 3$
CoCr-EG	Cobalt-chrome electrocautery group	$n = 3$

Ti = titanium; CoCr = cobalt-chrome; CG = control group; EG = electrocautery group.

2.2. Electrocautery Contact

Electrocautery application (VIO-300, Erbe Elektromedizin, Tübingen, Germany) was performed by an experienced spine surgeon. Analogous settings used in the operation room of our institution were utilized (mode: swift-coag, 90 Watt, effect: 3, duration: ~4 s). High-frequency monopolar-electrocautery was applied on the surface of each rod at three defined positions (Figure 1). Under these conditions electric light arc was visible in all cases and discoloration of the rod surface was observed.

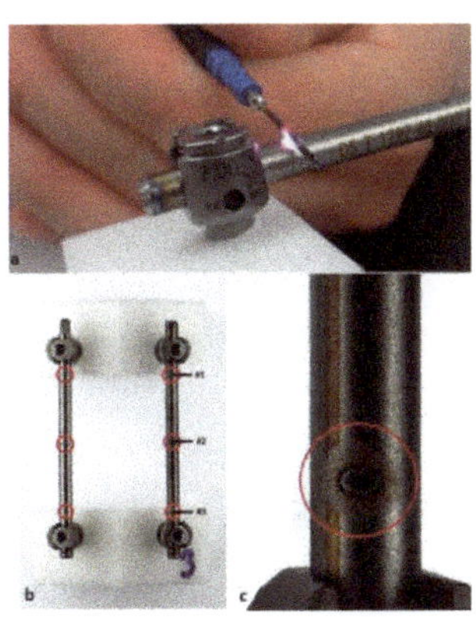

Figure 1. Electrocautery application. (**a**) Electric light arc, (**b**) sites, and (**c**) magnified view of the surface impact post electrocautery application. The first site was approximately 5 mm from the cephalic pedicle screw (#1). (#2 was at the center and #3 was analogous to #1 on the other side).

2.3. *Dynamic Mechanical Testing*

Models were mounted in a servo-hydraulic uniaxial testing apparatus (HCE, Bosch-Rexroth, Lohr a. Main, Germany) under dry conditions such that the rods were aligned in the direction of vertical force application. The upper and lower vertebrae blocks were free to rotate in order to account for any bending of the rods during dynamic testing. A multistep fatigue test was performed at a frequency of 12 *Hz* (Figure 2).

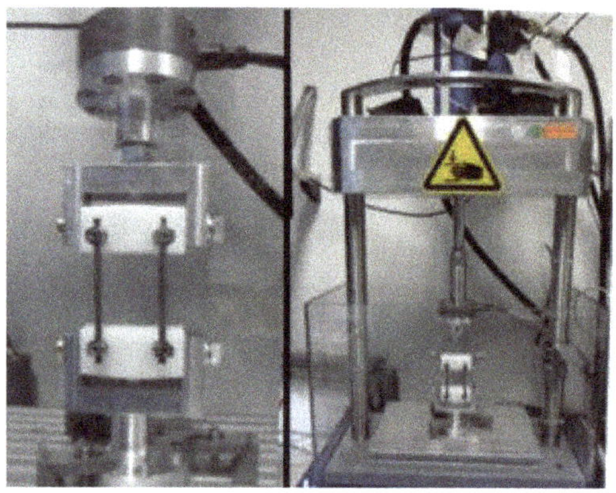

Figure 2. Frontal view of the fatigue testing setup.

Tests started at an initial maximum sinusoidal load of 100 *N* (minimal load: 10 *N*). After one million cycles, the maximum load was increased by 50 *N*, and the load ratio *R* was kept constant at 0.1, until construct failure occurred. Based on the total number of steps and cycles completed, the estimated load-to-failure was calculated using this formula:

$$F_D = F_{-1} + 50N \cdot \frac{\mathrm{n}}{1,000,000} \tag{1}$$

where F_{-1} is the maximum force of the previous step before fracture, and n is the number of cycles in the step where the specimen failed.

2.4. *Outcome Measures*

Primary outcomes estimated load-to-failure, total number of cycles to failure, and location of rod failure.

2.5. *Statistical Analysis*

One-way analysis of variance (ANOVA) with post-hoc-test correction of least statistical difference was used to compare continuous values of the outcome measures. All data are presented as mean values ± standard deviation and 95% confidence interval. Statistical tests were computed using statistical package (SPSS-v.24, IBM, Armonk, NY, USA). The threshold of statistical significance was set at 0.05.

3. Results

Results of the biomechanical test are summarized in Tables 2 and 3.

Table 2. Summary of the dynamic biomechanical test results.

Spinal Construct	Electrocautery Application	Completed Load Level	Min/Max Load at Failure (N)	No. Cycles to Failure	Failure Site	Fatigue Strength (N)
Ti-CG1	No	#4	30/300	**4,473,034**	Unilateral rod-screw junction	273.6
Ti-CG2	No	#4	30/300	**4,388,472**	Bilateral rod-screw junction	269.4
Ti-CG3	No	#3	25/250	**3,995,938**	Unilateral rod-screw junction	249.8
Ti-EG1	Yes	#3	25/250	**3,093,921**	Bilateral peripheral rod fractures	204.7
Ti-EG2	Yes	#3	25/250	**3,328,583**	Unilateral peripheral rod fracture *	216.4
Ti-EG3	Yes	#3	25/250	**3,770,073**	Bilateral central rod fractures	238.5
CoCr-CG1	No	#6	40/400	**6,351,621**	Bilateral pedicle screws **	367.6
CoCr-CG2	No	#7	40/400	**7,000,000**	Bilateral pedicle screws **	400
CoCr-CG3	No	#7	45/450	**7,183,433**	Bilateral pedicle screws **	409.2
CoCr-EG1	Yes	#7	45/450	**7,079,071**	Bilateral pedicle screws **	403.9
CoCr-EG2	Yes	#7	45/450	**7,118,378**	Bilateral pedicle screws **	405.9189
CoCr-EG3	Yes	#6	40/400	**6,112,167**	Bilateral pedicle screws **	355.60835

CG = Control Group; EG = Electrocautery Group; N = Newton; Ti = Titanium; CoCr = Cobalt-Chrome. * Ti-EG2 fracture site occurred at the level of the locking screws, not at the site of electrocautery application. ** all pedicle screws were made of titanium. No CoCr rods failed in this biomechanical investigation.

Table 3. Summary of load-to-failure and total number of cycles to failure for all tested groups.

	Load to Failure (N) *	95% CI	*p*	No. Cycles to Failure *	95% CI	*p*
Ti-CG	264.3 ± 12.7	[232.7–295.9]	**$p = 0.02$**	$4.3 \times 10^6 \pm 25 \times 10^3$	[3.6×10^6–4.9×10^6]	**$p = 0.03$**
Ti-EG	219.8 ± 17.2	[177.2–262.5]		$3.4 \times 10^6 \pm 34 \times 10^3$	[2.5×10^6–4.2×10^6]	
CoCr-CG	392.2 ± 21.8	[338.1–446.5]	$p > 0.05$	$6.8 \times 10^6 \pm 43 \times 10^3$	[5.7×10^6–7.9×10^6]	$p > 0.05$
CoCr-EG	388.5 ± 28.5	[317.7–459.3]		$6.8 \times 10^6 \pm 57 \times 10^3$	[5.3×10^6–8.2×10^6]	

* Mean ± standard deviation, CI = Confidence Interval, N = Newton, *p*: **Bold** denotes statistical significance, Ti = Titanium, CoCr = Cobalt-Chrome, CG = Control Group, EG = Electrocautery Group.

3.1. *Titanium Control Group (Ti-CG)*

All specimens exhibited fatigue fracture of the rod at the rod-screw junction (Figure 3).

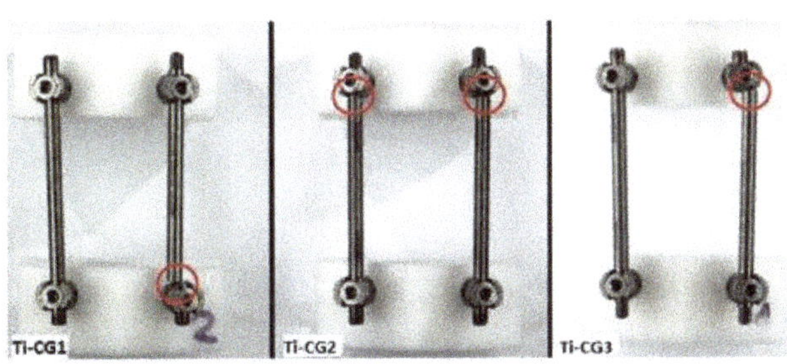

Figure 3. Sites of rod fracture of titanium control group (Ti-CG) post biomechanical testing and fatigue fractures of the rods at the rod-screw junction.

3.2. *Titanium Electrocautery Group (Ti-EG)*

In two specimens (Ti-EG1 and Ti-EG3), failure sites corresponded to the sites of electrocautery application (Figure 4).

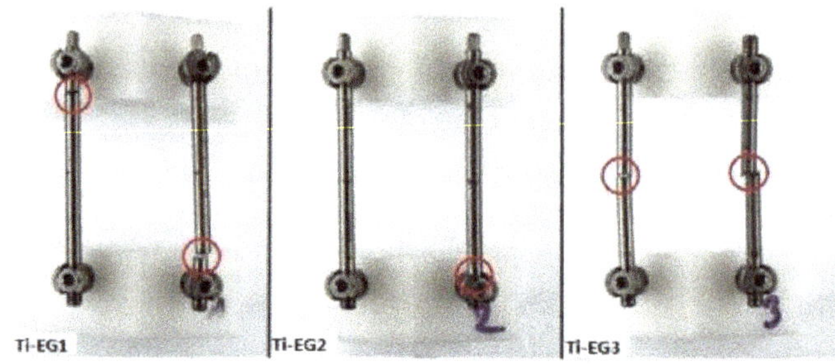

Figure 4. Sites of rod fracture of titanium electrocautery group (Ti-EG) post biomechanical testing.

Ti-EG revealed a significantly lower load-to-failure than Ti-CG (17% decrease, $p = 0.023$). Intergroup comparison of cycles to failure revealed a significant mean decrease of almost 9×10^5 cycles ($p = 0.03$). (Figure 5).

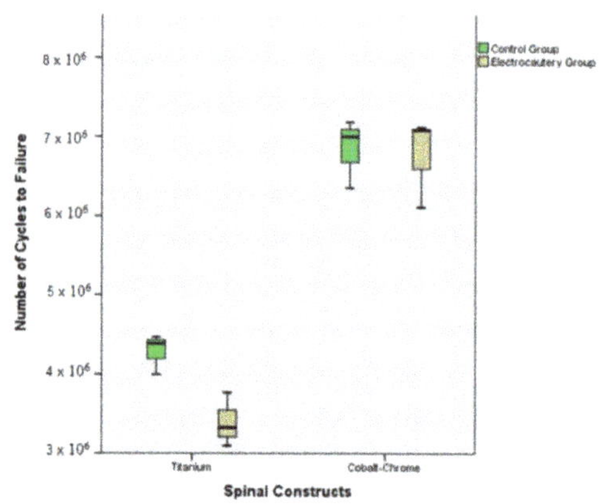

Figure 5. Boxplot representing the mean and 25% and 75% interquartile range of the total number of cycles to failure among the four tested groups.

Fatigue failure was confirmed by microscopic analysis of the fractured surface (Figure 6).

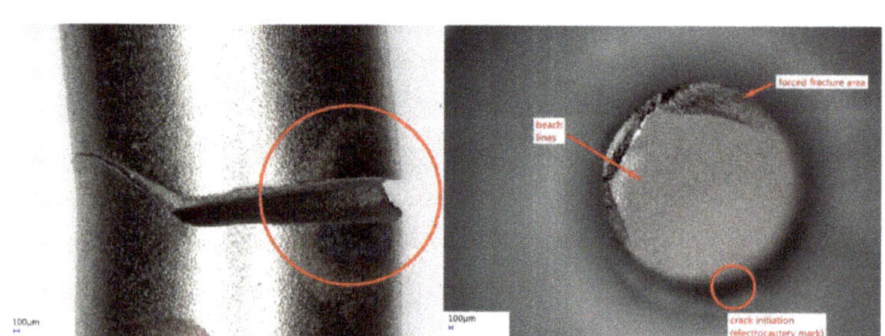

(a) (b)

Figure 6. Microscopic analysis of fractured titanium rod after electrocautery (VHX-5000, Keyence, Osaka, Japan). (**a**) Post-fracture situation at the electrocautery mark. (**b**) Fracture surface with typical signs of fatigue fracture (beach lines and forced fracture area). The circle shows the location of crack initiation at the electrocautery mark.

Notably, Ti-EG2 failed at the rod-screw junction similar to all of the Ti-CG rods.

3.3. CoCr Control Group (CoCr-CG) and Electrocautery Group (CoCr-EG)

In all tested constructs, the Ti pedicle screw failed. None of the CoCr-CG or CoCr-EG rods had fractured, exhibiting an increased load-to-failure in comparison to their Ti counterparts.

Intergroup differences were insignificant across all outcome measures. This implies that, in these tested constructs, the pedicle screw proved to be the weakest component under cyclic loading. In all specimens of these two groups, fatigue fracture transpired at the first thread of the titanium pedicle screw (Figures 7 and 8).

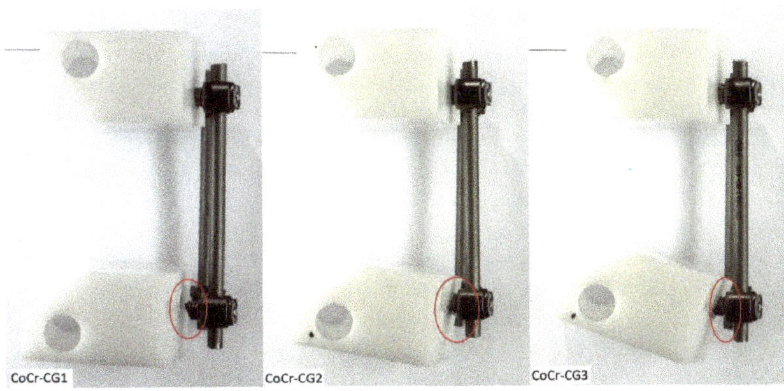

Figure 7. Sites of the titanium screw fracture of CoCr control group (CoCr-CG) post biomechanical testing.

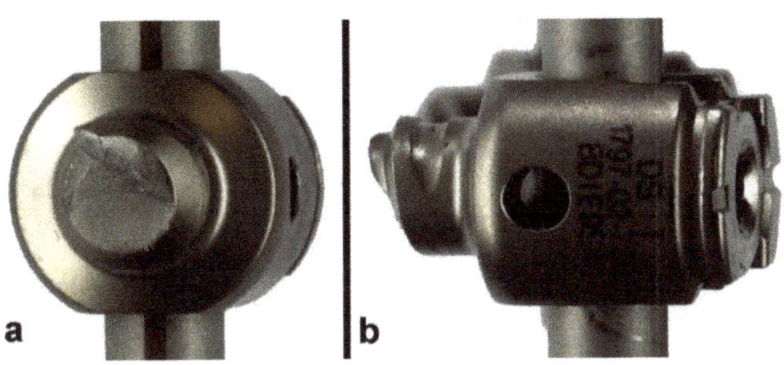

Figure 8. Magnified frontal (**a**) and lateral (**b**) views of the titanium screw fracture of one of CoCr electrocautery group constructs (CoCr-EG 3).

4. Discussion

The human body is an intricate biomechanical composite that exerts cyclic loading on spinal instrumentation, which could lead to their failure [21,30]. The predominant use of electrocautery in operating-rooms mandates spine surgeons and implant engineers to contemplate the potential negative impact of electrocautery on the long-term integrity of spinal fusion constructs. Thus, addressing differences between the mechanical properties of these materials and studying the thermal effect of electrocautery are essential to predict the in vivo behavior of implants.

The micro-architecture of Ti and CoCr materials dictates their biomechanical performance and fatigue life [21,22,31]. Instrumentation failure is thought to be caused by recurring loading which generates oscillating stresses that may be well below the yield stress of the material. Unfortunately, this process is unnoticeable, either by the patient during daily activities or by the surgeon throughout follow-up visits, until final construct failure develops [21,32,33].

In vivo loading on spinal implants has been previously measured and reported to be 250 N in the prone position, 400 N whilst walking, and up to 700 N during exercise [34–37]. Notably, Ti-EG constructs failed at a maximum load of 250 N. This underlines the potential detrimental impact of electrocautery on the safety of Ti constructs, even within daily physiological loading.

4.1. Thermal Damage and Notch Sensitivity: A Threat to Ti Biomechanical Integrity

Previous investigations have emphasized that the load-to-failure is contingent upon a material's resistance to surface alterations. Fatigue fractures almost always nucleate at the exact location of notches or "discontinuity of geometrical structure" on the material's surface [21,24,31,38,39]. This is particularly important when the site of the notch corresponds to areas where maximal loading is applied. Very small cracks relative to the dimensions of the micro-structure have been shown to cause failure faster than a long crack [40,41]. Therefore, the radius of the notch is, as a geometrical function, important [42]. Jang et al. demonstrated that thermal damage of laser etching on hip implants led to their early fracture [43]. Huber et al. [25] and Konrads et al. [26] reported a decrease in the fatigue life of Ti hip implants due to the thermal damage of electrocautery. In this context, Sonntag et al. [28] further demonstrated that this thermal damage caused a significant alteration of Ti microstructure. Ti6Al4V alloy has a bimodal microstructure consisting of a globular α-matrix phase, enveloped by the so-called Widmannstätten structure (α + β) [32,44–46]. Thermal energy affects the internal kinetics of theses phases and triggers a cascade of diffusion processes, as well as transforming the dimensions of the α and β phases. Alteration of those dimensions correlates with their load-to-failure and susceptibility to internal crack initiation. It could even be regarded as an internal notch [33,47–49].

In the light of these findings, the thermal energy applied on the tested Ti rods and the marks seen on their surface, which may be due to a corrosive attack [27] could set the stage for premature rod failure.

4.2. Pathogenesis of Ti Rod Fractures at the Rod-Screw Junction

Among all Ti-CG and one of the Ti-EG specimens, a crack nucleation site was visually detected in the proximity of the locking screw. Tightening of the screw leaves surface marks on the rods and could account for this failure (Figure 9). The presented results echo previously published findings: Lindsey et al. and Ngyuen et al. [21,22] detected an increased trend of CoCr rod failure at the same location of our tested constructs. Yamanaka et al. [40] described the same phenomenon on Ti rods. Dick et al. [31] suggested that surgeons should avoid severe tightening of the screws as this would lead to notch formation on the surface of Ti rods. Hence, a modification of the operating manual might be indicated to avoid severe tightening of the screw. This, however, requires specific testing and is a secondary outcome of this study.

Figure 9. Magnified lateral view of Ti-CG post biomechanical testing: fatigue fracture of the rod at the rod-screw junction, illustrating surface marks of the tightened screws on the rod.

4.3. CoCr Versus Ti Rods

CoCr exhibited a higher fatigue life than Ti, which is in line with previous investigations [23,50,51]. However, we are not aware of previous studies addressing spinal CoCr susceptibility to thermal modification of electrocautery. Uniquely, our tested constructs were not affected by this thermal modification as the pedicle screws failed. This could be explained by their superior rigidity, corrosion, and wear resistance [52]. This biomechanical superiority is also the reason for their implementation in demanding surgical techniques such as PSO. Paradoxically, in the context of PSO, CoCr constructs were associated with higher rates of RF. This could be due to the notch effect of in situ bending inherent within the PSOs, and the unusually high loading they need to withstand [16]. However, the superior performance of CoCr comes at the expense of greater stiffness at both ends of the construct, resulting in increased incidence of adjacent segment disease [53–55]. This could be attributed to the fact that Ti-6Al-4V alloy has a much lower Young's modulus (~112 GPa) when compared to Co-Cr alloys (~210 GPa) [56,57].

One of the strengths of our study could be its clinical implications. Patient expectations of surgical outcomes have reached unparalleled heights [58]. However, implant-related complications have significant drawbacks on post-operative Health-Related Quality of Life (HRQoL) and were associated with slower rates of improvement [1,23,59]. Moreover, the negative economic impact of instrumentation failure and post-operative complications on the health-care system in the context of spine surgery is high [60,61].

This study provides a biomechanical strongpoint by testing constructs under high sinusoidal loading levels ($n = 8$, each 10^6 cycles) in which rod diameter was uniform, which makes the results comparable. Previous biomechanical investigations have tested constructs with different rod diameters and were tested under lower load levels [21,22]. Higher loading levels would give more confidence to the conclusion of long-term fatigue endurance. Intuitively, the longer a spinal implant persists intact in the body, the more cyclic loading it has to withstand. Testing constructs under short- and long-term mechanical loading enabled the conclusion that CoCr rods, even under long-term cyclic loading (8×10^6 cycles), proved unaffected by thermal damage. Furthermore, constructs were examined according to a standard for spinal instrumentation testing [29]. The application of the ASTM F1717 standard enabled the delineation of electrocautery's impact on the mechanical properties of materials and efficiently eliminated potential confounding factors, such as implant design or donor variability in the context of cadaver studies.

On the other hand, intrinsic to in vitro biomechanical investigations, this study was conducted within a scaffold of limitations and simplifying premises that are important to discuss. First, severity of the electrocautery attack cannot be fully controlled during manual application which could lead to heterogeneity of thermal damage on the tested rods. This could explain why the Ti-EG specimens fractured at different locations. Second, the use of only one cautery mode precludes conclusions on other electrocautery modes. Third, in vivo clinical translation of the findings is impeded by the dry conditions of these experiments which were conducted at room temperature; fatigue life could differ when implants are at body temperature and surrounded by body fluids [39]. Fourth, the application of the ASTM F1717 sinusoidal testing environment represents a "worst case scenario"; constructs alone bore the burden of the axial loading which does not fully represent the in vivo state where the load is also shared by the spine and neighboring soft tissue. However, physiological stresses on implants could be as high as 700 N which permits us to draw our conclusions, as Ti rods failed in the range of 250 N. Fifth, tested constructs underwent only axial loading, resulting in a bending moment which mimics one modality of human movement, flexion-extension, and does not take into account lateral bending and axial rotation. Finally, we did not perform a power analysis prior to conducting this study to determine the minimum sample size required for the CI of 95%. Also, the wide CI is attributed to the small sample size. However, this limitation means that the impact of electrocautery could have been underestimated in our results. Subsequently, a larger sample size would enable calculation of a statistically significant hazard ratio to quantify the negative impact of thermal damage on the fatigue life of Ti rods. Future biomechanical studies with a larger sample size are merited to analyze different settings of coagulation and cutting modes, and to assess loadings exerted by bending and axial rotation of vertebrae. Randomized controlled studies are required to determine the absolute effect of electrocautery on spinal implants. Nevertheless, our findings disentangle a potential culprit underlying post-operative Ti rod fracture.

5. Conclusions

This biomechanical study showed that the impact of electrocautery on Ti rods could have significant clinical and biomechanical repercussions for patient safety and satisfaction, surgical training, and implant design. Ti rods exhibited decreased fatigue life and failed at the site of electrocautery application. Spine surgeons should exercise caution in the vicinity of spinal implants, especially during revisions. Similarly, it is also reasonable for manufacturers to devise strategies against thermal damage and explore methods to increase the long-term structural integrity of Ti rods.

Author Contributions: H.A: Investigation, formal analysis, writing the original manuscript, review and editing. R.S. and W.P.: Conceptualization, methodology and resource acquisition, review and editing. T.B.: Statistical analysis support. J.-P.K. and M.A.: Project administration, funding acquisition, resources and supervision. All authors critically reviewed the manuscript.

Funding: This research was funded by Deutsche Arthrose-Hilfe. Grant number P399-A326 and the APC was partially funded by Deutsche Forschungsgemeinschaft within the funding program Open Access Publishing, by the Baden-Württemberg Ministry of Science, Research and the Arts and by Ruprecht-Karls-Universität Heidelberg.

Conflicts of Interest: The authors declare no conflicts of interest. The sponsors had no role in the design, execution, interpretation, or writing of the study.

References

1. Soroceanu, A.; Diebo, B.G.; Burton, D.; Smith, J.S.; Deviren, V.; Shaffrey, C.; Kim, H.J.; Mundis, G.; Ames, C.; Errico, T. Radiographical and implant-related complications in adult spinal deformity surgery: Incidence, patient risk factors, and impact on health-related quality of life. *Spine* **2015**, *40*, 1414–1421. [CrossRef]
2. Sciubba, D.M.; Yurter, A.; Smith, J.S.; Kelly, M.P.; Scheer, J.K.; Goodwin, C.R.; Lafage, V.; Hart, R.A.; Bess, S.; Kebaish, K.; et al. A Comprehensive Review of Complication Rates after Surgery for Adult Deformity: A Reference for Informed Consent. *Spine Deform.* **2015**, *3*, 575–594. [CrossRef]
3. Hibbs, R.A.; Peltier, L.F. A report of fifty-nine cases of scoliosis treated by the fusion operation. *Clin. Orthop. Relat. Res.* **1988**, *229*, 4–19.
4. Heary, R.F.; Madhavan, K. The History of Spinal Deformity. *Neurosurgery* **2008**, *63*, A5–A15. [CrossRef]
5. Albee, F.H. The classic: Transplantation of a portion of the tibia into the spine for Pott's disease: A preliminary report. *Clin. Orthop. Relat. Res.* **2007**, *460*, 14–16. [CrossRef]
6. King, D. Internal fixation for lumbosacral fusion. *JBJS* **1948**, *30*, 560–578. [CrossRef]
7. Lange, F. Support for the spondylitic spine by means of buried steel bars, attached to the vertebrae. *JBJS* **1910**, *2*, 344–361.
8. Harrington, P.R. Treatment of scoliosis: Correction and internal fixation by spine instrumentation. *JBJS* **1962**, *44*, 591–634. [CrossRef]
9. Cotrel, Y.; Dubousset, J. A new technic for segmental spinal osteosynthesis using the posterior approach. *Revue de Chirurgie Orthopédique et Réparatrice de L'appareil Moteur* **1984**, *70*, 489–494. [CrossRef]
10. Kaur, M.; Singh, K. Review on titanium and titanium based alloys as biomaterials for orthopaedic applications. *Mater. Sci. Eng. C* **2019**, *102*, 844–862. [CrossRef]
11. Liu, X.; Chu, P.K.; Ding, C. Surface modification of titanium, titanium alloys, and related materials for biomedical applications. *Mater. Sci. Eng. R Rep.* **2004**, *47*, 49–121. [CrossRef]
12. Liu, W.; Liu, S.; Wang, L. Surface Modification of Biomedical Titanium Alloy: Micromorphology, Microstructure Evolution and Biomedical Applications. *Coatings* **2019**, *9*, 249. [CrossRef]
13. Kyzioł, K.; Kaczmarek, Ł.; Brzezinka, G.; Kyzioł, A. Structure, characterization and cytotoxicity study on plasma surface modified Ti–6Al–4V and γ-TiAl alloys. *Chem. Eng. J.* **2014**, *240*, 516–526. [CrossRef]

14. Smith, J.S.; Sansur, C.A.; Donaldson, W.F., 3rd; Perra, J.H.; Mudiyam, R.; Choma, T.J.; Zeller, R.D.; Knapp, D.R., Jr.; Noordeen, H.H.; Berven, S.H.; et al. Short-term morbidity and mortality associated with correction of thoracolumbar fixed sagittal plane deformity: A report from the Scoliosis Research Society Morbidity and Mortality Committee. *Spine* **2011**, *36*, 958–964. [CrossRef]
15. Scheer, J.K.; Tang, J.A.; Deviren, V.; Acosta, F.; Buckley, J.M.; Pekmezci, M.; McClellan, R.T.; Ames, C.P. Biomechanical analysis of cervicothoracic junction osteotomy in cadaveric model of ankylosing spondylitis: Effect of rod material and diameter. *J. Neurosurg. Spine* **2011**, *14*, 330–335. [CrossRef]
16. Smith, J.S.; Klineberg, E.; Lafage, V.; Shaffrey, C.I.; Schwab, F.; Lafage, R.; Hostin, R.; Mundis, G.M.; Errico, T.J.; Kim, H.J.; et al. Prospective multicenter assessment of perioperative and minimum 2-year postoperative complication rates associated with adult spinal deformity surgery. *J. Neurosurg. Spine* **2016**, *25*, 1–14. [CrossRef]
17. Bridwell, K.H.; Baldus, C.; Berven, S.; Edwards, C., 2nd; Glassman, S.; Hamill, C.; Horton, W.; Lenke, L.G.; Ondra, S.; Schwab, F.; et al. Changes in radiographic and clinical outcomes with primary treatment adult spinal deformity surgeries from two years to three- to five-years follow-up. *Spine* **2010**, *35*, 1849–1854. [CrossRef]
18. Cho, S.K.; Bridwell, K.H.; Lenke, L.G.; Yi, J.S.; Pahys, J.M.; Zebala, L.P.; Kang, M.M.; Cho, W.; Baldus, C.R. Major complications in revision adult deformity surgery: Risk factors and clinical outcomes with 2- to 7-year follow-up. *Spine* **2012**, *37*, 489–500. [CrossRef]
19. Bagchi, K.; Mohaideen, A.; Thomson, J.D.; Foley, L.C. Hardware complications in scoliosis surgery. *Pediatr. Radiol.* **2002**, *32*, 465–475. [CrossRef]
20. Bago, J.; Ramirez, M.; Pellise, F.; Villanueva, C. Survivorship analysis of Cotrel-Dubousset instrumentation in idiopathic scoliosis. *Eur. Spine J.* **2003**, *12*, 435–439. [CrossRef]
21. Lindsey, C.; Deviren, V.; Xu, Z.; Yeh, R.F.; Puttlitz, C.M. The effects of rod contouring on spinal construct fatigue strength. *Spine* **2006**, *31*, 1680–1687. [CrossRef]
22. Nguyen, T.Q.; Buckley, J.M.; Ames, C.; Deviren, V. The fatigue life of contoured cobalt chrome posterior spinal fusion rods. *Proc. Inst. Mech. Eng. Part H J. Eng. Med.* **2011**, *225*, 194–198. [CrossRef]
23. Smith, J.S.; Shaffrey, C.I.; Ames, C.P.; Demakakos, J.; Fu, K.M.; Keshavarzi, S.; Li, C.M.; Deviren, V.; Schwab, F.J.; Lafage, V.; et al. Assessment of symptomatic rod fracture after posterior instrumented fusion for adult spinal deformity. *Neurosurgery* **2012**, *71*, 862–867. [CrossRef]
24. Tang, J.A.; Leasure, J.M.; Smith, J.S.; Buckley, J.M.; Kondrashov, D.; Ames, C.P. Effect of severity of rod contour on posterior rod failure in the setting of lumbar pedicle subtraction osteotomy (PSO): A biomechanical study. *Neurosurgery* **2013**, *72*, 276–282. [CrossRef]
25. Huber, G.; Weik, T.; Morlock, M.M. Schädigung eines Hüftendoprothesenschafts durch Einsatz eines Hochfrequenzmessers. *Der Orthopäde* **2009**, *38*, 622–625. [CrossRef]
26. Konrads, C.; Wente, M.N.; Plitz, W.; Rudert, M.; Hoberg, M. Implantatschädigung durch Einsatz eines Hochfrequenzmessers. *Der Orthopäde* **2014**, *43*, 1106–1111. [CrossRef]
27. Yuan, N.; Park, S.H.; Luck, J.V.; Campbell, P.A. Revisiting the concept of inflammatory cell-induced corrosion. *J. Biomed. Mater. Res. Part B Appl. Biomater.* **2017**, *106B*, 1148–1155. [CrossRef]
28. Sonntag, R.G.J.; Pulvermacher, S.; Mueller, U.; Eckert, J.; Braun, S.; Reichkendler, M.; Kretzer, J.P. Electrocautery damage can reduce implant fatigue strength: Cases and in vitro investigation. *J. Bone Jt. Surg.* **2019**, *101*, 868–878. [CrossRef]
29. ASTM-International. Standard Test Methods for Spinal Implant Constructs in a Vertebrectomy Model. Available online: https://www.astm.org/Standards/F1717.htm (accessed on 20 May 2019).
30. Ashman, R. Mechanical testing of spinal implants. *Semin. Spine Surg.* **1993**, *5*, 73–80.
31. Dick, J.; Bourgeault, C. Notch sensitivity of titanium alloy, commercially pure titanium, and stainless steel spinal implants. *Spine* **2001**, *26*, 1668–1672. [CrossRef]
32. Wang, S.; Wei, M. Tensile properties of gas tungsten arc weldments in commercially pure titanium, Ti–6Al–4V and Ti–15V–3Al–3Sn–3Cr alloys at different strain rates. *Sci. Technol. Weld. Join.* **2004**, *9*, 415–422. [CrossRef]
33. Lütjering, G. Influence of processing on microstructure and mechanical properties of ($\alpha + \beta$) titanium alloys. *Mater. Sci. Eng. A* **1998**, *243*, 32–45. [CrossRef]
34. Rohlmann, A.; Graichen, F.; Bender, A.; Kayser, R.; Bergmann, G. Loads on a telemeterized vertebral body replacement measured in three patients within the first postoperative month. *Clin. Biomech.* **2008**, *23*, 147–158. [CrossRef]

35. Rohlmann, A.; Bergmann, G.; Graichen, F. Loads on an internal spinal fixation device during walking. *J. Biomech.* **1997**, *30*, 41–47. [CrossRef]
36. Rohlmann, A.; Claes, L.; Bergmann, G.; Graichen, F.; Neef, P.; Wilke, H.-J. Comparison of intradiscal pressures and spinal fixator loads for different body positions and exercises. *Ergonomics* **2001**, *44*, 781–794. [CrossRef]
37. D'Lima, D.D.; Fregly, B.J.; Colwell, C.W. Implantable sensor technology: Measuring bone and joint biomechanics of daily life in vivo. *Arthritis Res. Ther.* **2013**, *15*, 203. [CrossRef]
38. Tang, J.A. Comparison of a novel pedicle subtraction osteotomy model using the traditional American Society of Testing and Materials standard for spinal biomechanics fatigue testing [RETRACTED]. *J. Neurosurg. Spine* **2012**. [CrossRef]
39. Chen, P.Q.; Lin, S.J.; Wu, S.S.; So, H. Mechanical performance of the new posterior spinal implant: Effect of materials, connecting plate, and pedicle screw design. *Spine* **2003**, *28*, 881–886. [CrossRef]
40. Yamanaka, K.; Mori, M.; Yamazaki, K.; Kumagai, R.; Doita, M.; Chiba, A. Analysis of the Fracture Mechanism of Ti-6Al-4V Alloy Rods That Failed Clinically After Spinal Instrumentation Surgery. *Spine* **2015**, *40*, E767–E773. [CrossRef]
41. Ravichandran, K. Three-dimensional crack-shape effects during the growth of small surface fatigue cracks in a titanium-base alloy. *Fatigue Fract. Eng. Mater. Struct.* **1997**, *20*, 1423–1442. [CrossRef]
42. Cook, R.D.; Young, W.C. *Advanced Mechanics of Materials*; Prentice Hall: Upper Saddle River, NJ, USA, 1999.
43. Jang, B.; Kanawati, A.; Brazil, D.; Bruce, W. Laser etching causing fatigue fracture at the neck–shoulder junction of an uncemented femoral stem: A case report. *J. Orthop.* **2013**, *10*, 95–98. [CrossRef]
44. Malinov, S.; Sha, W. Application of artificial neural networks for modelling correlations in titanium alloys. *Mater. Sci. Eng. A* **2004**, *365*, 202–211. [CrossRef]
45. Balasubramanian, M.; Jayabalan, V.; Balasubramanian, V. Effect of microstructure on impact toughness of pulsed current GTA welded α–β titanium alloy. *Mater. Lett.* **2008**, *62*, 1102–1106. [CrossRef]
46. Pederson, R.; Babushkin, O.; Skystedt, F.; Warren, R. Use of high temperature X-ray diffractometry to study phase transitions and thermal expansion properties in Ti-6Al-4V. *Mater. Sci. Technol.* **2003**, *19*, 1533–1538. [CrossRef]
47. Oguma, H.; Nakamura, T. The effect of microstructure on very high cycle fatigue properties in Ti–6Al–4V. *Scr. Mater.* **2010**, *63*, 32–34. [CrossRef]
48. Everaerts, J.; Verlinden, B.; Wevers, M. The influence of the alpha grain size on internal fatigue crack initiation in drawn Ti-6Al-4V wires. *Procedia Struct. Integr.* **2016**, *2*, 1055–1062. [CrossRef]
49. Gil, F.; Rodriguez, D.; Planell, J. Grain growth kinetics of pure titanium. *Scr. Metall. Mater.* **1995**, 33. [CrossRef]
50. Shinohara, K.; Takigawa, T.; Tanaka, M.; Sugimoto, Y.; Arataki, S.; Yamane, K.; Watanabe, N.; Ozaki, T.; Sarai, T. Implant Failure of Titanium Versus Cobalt-Chromium Growing Rods in Early-onset Scoliosis. *Spine* **2016**, *41*, 502–507. [CrossRef]
51. Doulgeris, J.J.; Aghayev, K.; Gonzalez-Blohm, S.A.; Del Valle, M.; Waddell, J.; Lee, W.E.; Vrionis, F.D. Comparative analysis of posterior fusion constructs as treatments for middle and posterior column injuries: An in vitro biomechanical investigation. *Clin. Biomech.* **2013**, *28*, 483–489. [CrossRef]
52. Wojnar, L.; Dąbrowski, J.R.; Oksiuta, Z. Porosity structure and mechanical properties of vitalium-type alloy for implants. *Mater. Charact.* **2001**, *46*, 221–225. [CrossRef]
53. Serhan, H.; Mhatre, D.; Newton, P.; Giorgio, P.; Sturm, P. Would CoCr rods provide better correctional forces than stainless steel or titanium for rigid scoliosis curves? *J. Spinal Disord. Tech.* **2013**, *26*, E70–E74. [CrossRef]
54. Lamerain, M.; Bachy, M.; Delpont, M.; Kabbaj, R.; Mary, P.; Vialle, R. CoCr rods provide better frontal correction of adolescent idiopathic scoliosis treated by all-pedicle screw fixation. *Eur. Spine J. Off. Publ. Eur. Spine Soc. Eur. Spinal Deform. Soc. Eur. Sect. Cerv. Spine Res. Soc.* **2014**, *23*, 1190–1196. [CrossRef]
55. Han, S.; Hyun, S.J.; Kim, K.J.; Jahng, T.A.; Kim, H.J. Comparative Study Between Cobalt Chrome and Titanium Alloy Rods for Multilevel Spinal Fusion: Proximal Junctional Kyphosis More Frequently Occurred in Patients Having Cobalt Chrome Rods. *World Neurosurg.* **2017**, *103*, 404–409. [CrossRef]
56. Geetha, M.; Singh, A.K.; Asokamani, R.; Gogia, A.K. Ti based biomaterials, the ultimate choice for orthopaedic implants–A review. *Prog. Mater. Sci.* **2009**, *54*, 397–425. [CrossRef]
57. Niinomi, M.; Liu, Y.; Nakai, M.; Liu, H.; Li, H. Biomedical titanium alloys with Young's moduli close to that of cortical bone. *Regen. Biomater.* **2016**, *3*, 173–185. [CrossRef]

58. Saban, K.L.; Penckofer, S.M. Patient expectations of quality of life following lumbar spinal surgery. *J. Neurosci. Nurs.* **2007**, *39*, 180–189. [CrossRef]
59. Cho, K.-J.; Suk, S.-I.; Park, S.-R.; Kim, J.-H.; Kim, S.-S.; Choi, W.-K.; Lee, K.-Y.; Lee, S.-R. Complications in Posterior Fusion and Instrumentation for Degenerative Lumbar Scoliosis. *Spine* **2007**, *32*, 2232–2237. [CrossRef]
60. McCarthy, I.; Hostin, R.; O'Brien, M.; Saigal, R.; Ames, C.P. Health Economic Analysis of Adult Deformity Surgery. *Neurosurg. Clin. N. Am.* **2013**, *24*, 293–304. [CrossRef]
61. Yeramaneni, S.; Robinson, C.; Hostin, R. Impact of spine surgery complications on costs associated with management of adult spinal deformity. *Curr. Rev. Musculoskelet. Med.* **2016**, *9*, 327–332. [CrossRef]

MDPI
St. Alban-Anlage 66
4052 Basel
Switzerland
Tel. +41 61 683 77 34
Fax +41 61 302 89 18
www.mdpi.com

Materials Editorial Office
E-mail: materials@mdpi.com
www.mdpi.com/journal/materials

www.ingramcontent.com/pod-product-compliance
Lightning Source LLC
Chambersburg PA
CBHW061602220926
51850CB00005B/397

* 9 7 8 3 0 3 6 5 1 2 8 1 5 *